Lecture Notes in Computer Science   9168

Commenced Publication in 1973
Founding and Former Series Editors:
Gerhard Goos, Juris Hartmanis, and Jan van Leeuwen

More information about this series at http://www.springer.com/series/7407

Igor Potapov (Ed.)

Developments
in Language Theory

19th International Conference, DLT 2015
Liverpool, UK, July 27–30, 2015
Proceedings

 Springer

Editor
Igor Potapov
University of Liverpool
Liverpool
UK

ISSN 0302-9743   ISSN 1611-3349 (electronic)
Lecture Notes in Computer Science
ISBN 978-3-319-21499-3   ISBN 978-3-319-21500-6   (eBook)
DOI 10.1007/978-3-319-21500-6

Library of Congress Control Number: 2015943441

LNCS Sublibrary: SL1 – Theoretical Computer Science and General Issues

Springer Cham Heidelberg New York Dordrecht London

Printed on acid-free paper

Springer International Publishing AG Switzerland is part of Springer Science+Business Media
(www.springer.com)

Preface

The 19th International Conference on Developments in Language Theory (DLT 2015) was organized by the University of Liverpool, UK, during July 27–30, 2015.

The DLT conference series is one of the major international conference series in language theory and related areas. The Developments in Language Theory (DLT) conference was established by G. Rozenberg and A. Salomaa in 1993. Since then, the DLT conferences were held on every odd year: Magdeburg, Germany (1995), Thessaloniki, Greece (1997), Aachen, Germany (1999), and Vienna, Austria (2001). Since 2001, a DLT conference takes place in Europe on every odd year and outside Europe on every even year. The locations of DLT conferences since 2002 were: Kyoto, Japan (2002), Szeged, Hungary (2003), Auckland, New Zealand (2004), Palermo, Italy (2005), Santa Barbara, California, USA (2006), Turku, Finland (2007), Kyoto, Japan (2008), Stuttgart, Germany (2009), London, Ontario, Canada (2010), Milan, Italy (2011), Taipei, Taiwan (2012), Marne-la-Vallée, France (2013), Ekaterinburg, Russia (2014).

The series of International Conferences on Developments in Language Theory provides a forum for presenting current developments in formal languages and automata. Its scope is very general and includes, among others, the following topics and areas: combinatorial and algebraic properties of words and languages; grammars, acceptors and transducers for strings, trees, graphs, arrays; algebraic theories for automata and languages; codes; efficient text algorithms; symbolic dynamics; decision problems; relationships to complexity theory and logic; picture description and analysis; polyominoes and bidimensional patterns; cryptography; concurrency; cellular automata; bio-inspired computing; quantum computing.

This volume of *Lecture Notes in Computer Science* contains the papers that were presented at DLT 2015. There were 54 qualified submissions. Each submission was reviewed by at least three Program Committee members. The committee decided to accept 31 papers. The volume also includes the abstracts and extended abstracts of five invited speakers:

- Mikolaj Bojanczyk: "Recognisable Languages over Monads"
- Patrick Dehornoy: "Garside and Quadratic Normalisation: A Survey"
- Vesa Halava: "Proofs of Undecidability"
- Markus Lohrey: "Grammar-Based Tree Compression"
- Wolfgang Thomas "Finite Automata and Transitive Closure Logic"

This year the Steering Committee of the DLT agreed to introduce a Best Paper Award to encourage and reward high-quality research in language theory and related areas. The Program Committee members after careful consideration of all accepted papers decided to give the Best Paper Award to Jorge Almeida, Jana Bartonova, Ondrej Klima, and Michal Kunc for their paper "On Decidability of Intermediate Levels of Concatenation Hierarchies."

We warmly thank all the invited speakers and all the authors of the submitted papers. We would also like to thank all the members of the Program Committee and the external referees (listed in the proceedings) for their hard work in evaluating the papers. We also thank all members of the Organizing Committee at the University of Liverpool. We wish to express our sincere appreciation to the conference sponsors: NeST Software Lab at the University of Liverpool and the European Association for Theoretical Computer Science and the editors of the *Lecture Notes in Computer Science* series and Springer, in particular Alfred Hofmann, for their help in publishing this volume. The reviewing process was organized using the EasyChair conference system created by Andrei Voronkov. We would like to acknowledge that this system helped greatly to improve the efficiency of the committee work.

July 2015   Igor Potapov

Organization

Program Committee

Srecko Brlek	Université du Québec à Montréal, Canada
Manfred Droste	University of Leipzig, Germany
Rusins Freivalds	University of Riga, Latvia
Mika Hirvensalo	Turku University, Finland
Markus Holzer	University of Giessen, Germany
Juraj Hromkovic	ETH Zürich, Switzerland
Artur Jez	University of Wroclaw, Poland
Natasha Jonoska	University of South Florida, USA
Juhani Karhumaki	Turku University, Finland
Gregory Kucherov	CNRS/LIGM, France
Pierre McKenzie	University of Montreal, Canada
Jean-Éric Pin	CNRS/LIAFA, France
Igor Potapov, Chair	University of Liverpool, UK
Daniel Reidenbach	Loughborough University, UK
Marinella Sciortino	University of Palermo, Italy
Rick Thomas	University of Leicester, UK
Mikhail Volkov	Ural Federal University, Russia
Hsu-Chun Yen	National Taiwan University, Taiwan

Steering Committee

Marie-Pierre Béal	Université Paris-Est-Marne-la-Vallée, France
Cristian S. Calude	University of Auckland, New Zealand
Volker Diekert	University of Stuttgart, Germany
Juraj Hromkovic	ETH Zürich, Switzerland
Oscar H. Ibarra	UCA, Santa Barbara, USA
Masami Ito	Kyoto Sangyo University, Japan
Natasha Jonoska	University of South Florida, USA
Juhani Karhumaki (Chair)	Turku University, Finland
Martin Kutrib	University of Giessen, Germany
Michel Rigo	University of Liege, Belgium
Antonio Restivo	University of Palermo, Italy
Grzegorz Rozenberg	Leiden Institute of Advanced Computer Science, The Netherlands
Arto Salomaa	Turku University, Finland
Kai Salomaa	Queen's University, Canada

Wojciech Rytter   Warsaw University, Poland
Mikhail Volkov   Ural Federal University, Russia
Takashi Yokomori   Waseda University, Japan

Organizing Committee

Igor Potapov   University of Liverpool, UK
Leszek Gasieniec   University of Liverpool, UK
Russell Martin   University of Liverpool, UK
Reino Niskanen   University of Liverpool, UK
Dave Shield   University of Liverpool, UK

Additional Reviewers

Anashin, Vladimir   Kutrib, Martin
Anselmo, Marcella   Labelle, Gilbert
Bala, Sebastian   Lo Bosco, Giosue'
Blondin Massé, Alexandre   Malcher, Andreas
Blondin, Michael   Maletti, Andreas
Boeckenhauer, Hans-Joachim   Manea, Florin
Bucci, Michelangelo   Mccolm, Gregory
Cadilhac, Michaël   Meckel, Katja
Carton, Olivier   Newman, Alantha
Chang, Yi-Jun   Nickson, Thomas
Chen, Ho-Lin   Niskanen, Reino
Choffrut, Christian   Okhotin, Alexander
Colcombet, Thomas   Otop, Jan
Day, Joel   Otto, Friedrich
Farruggia, Andrea   Panella, Federica
Fici, Gabriele   Paperman, Charles
Freydenberger, Dominik D.   Paun, Andrei
Frid, Anna   Penelle, Vincent
Gawrychowski, Pawel   Place, Thomas
Gogacz, Tomasz   Prianychnykova, Olena
Halava, Vesa   Provençal, Xavier
Han, Yo-Sub   Quaas, Karin
Harju, Tero   Restivo, Antonio
Jurdzinski, Tomasz   Saarela, Aleksi
Kanazawa, Makoto   Saari, Kalle
Kari, Jarkko   Salomaa, Kai
Konstantinidis, Stavros   Schewe, Sven
Kosolobov, Dmitry   Sebastien, Labbe
Kozen, Dexter   Seki, Shinnosuke
Kuske, Dietrich   Steiner, Wolfgang

Truthe, Bianca
Törmä, Ilkka
Walen, Tomasz
Wang, Bow-Yaw

Yassawi, Reem
Yu, Tian-Li
Zeitoun, Marc

Abstracts of Invited Talks

Recognisable Languages Over Monads

Mikołaj Bojańczyk

University of Warsaw

This paper[1] proposes monads as a framework for algebraic language theory. Examples of monads include words and trees, finite and infinite. Each monad comes with a standard notion of an algebra, called an Eilenberg-Moore algebra, which generalises algebras studied in language theory like semigroups or ω-semigroups. On the abstract level of monads one can prove theorems like the Myhill-Nerode theorem, the Eilenberg theorem; one can also define profinite objects.

Mikołaj Bojańczyk—Supported by the Polish NCN grant 2014-13/B/ST6/03595.

[1] A full version of this paper is available at http://arxiv.org/abs/1502.04898 . The full version includes many examples of monads, proofs, stronger versions of theorems from this extended abstract, and entirely new theorems.

Grammar-Based Tree Compression

Markus Lohrey

Universität Siegen
`lohrey@eti.uni-siegen.de`

This paper gives a survey on recent progress in grammar-based compression for trees. Also algorithms that directly work on grammar-compressed trees will be surveyed.

This research is supported by the DFG-project LO 748/10-1.

Garside and Quadratic Normalisation: A Survey

Patrick Dehornoy

Laboratoire de Mathématiques Nicolas Oresme, CNRS UMR 6139, Université de Caen, 14032 Caen cedex, France, and Institut Universitaire de France
patrick.dehornoy@unicaen.fr,
www.math.unicaen.fr/~dehornoy

Starting from the seminal example of the greedy normal norm in braid monoids, we analyze the mechanism of the normal form in a Garside monoid and explain how it extends to the more general framework of Garside families. Extending the viewpoint even more, we then consider general quadratic normalisation procedures and characterise Garside normalisation among them.

Keywords: normal form; normalisation; regular language; fellow traveller property; greedy decomposition; Garside family; quadratic rewriting system; braid monoids; Artin–Tits monoids; plactic monoids

Proofs of Undecidability
Extended Abstract

Vesa Halava

Department of Mathematics and Statistics, University of Turku, Finland
Department of Computer Science, University of Liverpool, UK
vesa.halava@utu.fi

In the theory of computation and computability the existence of undecidable problems is a fundamental property. Already in the famous paper by A. Turing in 1936 where he defined the machines we now know as Turing machines, Turing showed that there exist problems which are unsolvable. Indeed, undecidability was around even before Turing machines as the celebrated result of K. Gödel (1931), known as the Incompleteness Theorem, can be considered as a computational undecidability result.

In this talk we focus on the proofs of undecidability and study techniques of reductions of the proofs. Two types of reductions are studied:

(i) reductions (or codings) between two computational machinery, and
(ii) reductions mapping an undecidable problem to a particular subclass of instances of the same problem.

We will pass through notable problems such as the halting problem of the Turing machines, the word problem and the termination problem of the semi-Thue systems, the Post's correspondence problem and the assertion problem of the Post's Normal systems. Also, we will notice that the Incompleteness Theorem by Gödel is around the corner.

As examples for the reductions studied here we give the following:

For type (i), we will study the original undecidability proof by E. Post (1946) for the correspondence problem using the Normal systems. We compare Post's technique to the standard textbook reduction from the Turing machines (or the semi-Thue systems).

For type (ii), we consider a method for proving undecidability of the word problem for the 3-rule semi-Thue systems (Matiyasevich and Sénizergues 2005). We will compare their reduction to the reduction by Ceĭtin (1958) from the word problem of Thue system into word problem of 7-rule Thue system. Moreover, we study the difference of these techniques in the case of the termination problem of the semi-Thue systems.

Finite Automata and Transitive Closure Logic

Wolfgang Thomas

RWTH Aachen University
Lehrstuhl Informatik 7, Aachen, 52056
thomas@informatik.rwth-aachen.de

The most prominent connection between automata theory and logic is the expressive equivalence between finite automata over words and monadic second-order logic MSO. In this lecture, we focus on a more restricted, not as popular, but also fundamental logic, "monadic transitive closure logic" MTC, which over words has the same expressive power as MSO. In MTC one can proceed from a formula $F(z, z')$ to $F^*(x, y)$ which expresses that a sequence exists from x to y such that each step is in accordance with F. Thus, MTC is a very natural framework to express reachability properties. We survey old and recent results which clarify the expressive power and possible uses of MTC. We start with the equivalence between MTC and MSO over word models. Then we discuss MTC over trees, grids, and graphs, addressing the relation to tree walking automata and analogous automata over grids and graphs. We conclude with some open questions.

Contents

Recognisable Languages over Monads

Mikołaj Bojańczyk$^{(\boxtimes)}$

University of Warsaw, Warsaw, Poland
bojan@mimuw.edu.pl

Abstract. This paper proposes monads as a framework for algebraic language theory. Examples of monads include words and trees, finite and infinite. Each monad comes with a standard notion of an algebra, called an Eilenberg-Moore algebra, which generalises algebras studied in language theory like semigroups or ω-semigroups. On the abstract level of monads one can prove theorems like the Myhill-Nerode theorem, the Eilenberg theorem; one can also define profinite objects.

The principle behind algebraic language theory for various kinds of objects, such as words or trees, is to use a "compositional" function from the objects into a finite set. To talk about compositionality, one needs objects with some kind of substitution. It so happens that category theory has an abstract concept of objects with substitution, namely a monad. The goal of this paper is to propose monads as a unifying framework for discussing existing algebras and designing new algebras. To introduce monads and their algebras, we begin with two examples, which use a monad style to present algebras for finite and infinite words.

Example 1. Consider the following non-standard definition of a semigroup. Define a +-algebra **A** to be a set A called its *universe*, together with a *multiplication operation* $\mathrm{mul}_\mathbf{A} : A^+ \to A$, which is the identity on single letters, and which is associative in the sense that the following diagram commutes.

$$
\begin{array}{ccc}
(A^+)^+ & \xrightarrow{\ \mu_A\ } & A^+ \\
{\scriptstyle (\mathrm{mul}_\mathbf{A})^+} \downarrow & & \downarrow {\scriptstyle \mathrm{mul}_\mathbf{A}} \\
A^+ & \xrightarrow[\ \mathrm{mul}_\mathbf{A}\ ]{} & A
\end{array} \ ,
$$

In the diagram, $(\mathrm{mul}_\mathbf{A})^+$ is the function that applies $\mathrm{mul}_\mathbf{A}$ to each label of a word, and μ_A is the function which flattens a word of words into a word, e.g.

$$
(abc)(aa)(acaa) \qquad \mapsto \qquad abcaaacaa.
$$

M. Bojańczyk—Author supported by the Polish NCN grant 2014-13/B/ST6/03595. A full version of this paper is available at http://arxiv.org/abs/1502.04898. The full version includes many examples of monads, proofs, stronger versions of theorems from this extended abstract, and entirely new theorems.

I. Potapov (Ed.): DLT 2015, LNCS 9168, pp. 1–13, 2015.
DOI: 10.1007/978-3-319-21500-6_1

Restricting the multiplication operation in a +-algebra to words of length two (the semigroup binary operation) is easily seen to be a one-to-one correspondence between +-algebras and semigroups. □

The second example will be running example in the paper.

Running Example 1. Let us define an algebra for infinite words in the spirit of the previous example. Define A^∞ to be the finite and ω-words over A, i.e. $A^+ \cup A^\omega$. Define an ∞-algebra **A** to be a set A, called its *universe*, together with a *multiplication operation* $\mathrm{mul}_\mathbf{A} : A^\infty \to A$, which is the identity on single letters, and which is associative in the sense that the following diagram commutes.

$$
\begin{array}{ccc}
(A^\infty)^\infty & \xrightarrow{\ \mu_A\ } & A^\infty \\
{\scriptstyle (\mathrm{mul}_\mathbf{A})^\infty}\big\downarrow & & \big\downarrow{\scriptstyle \mathrm{mul}_\mathbf{A}} \\
A^\infty & \xrightarrow[\ \mathrm{mul}_\mathbf{A}\ ]{} & A
\end{array}
$$

In the diagram, $(\mathrm{mul}_\mathbf{A})^\infty$ is the function that applies $\mathrm{mul}_\mathbf{A}$ to the label of every position in a word from $(A^\infty)^\infty$, and μ_A is defined in analogy to mul_{A^+}, with the following proviso: if the argument of μ_A contains an infinite word on some position, then all subsequent positions are ignored, e.g.

$$(abc)(aa)(a^\omega)(abca)(ab^\omega) \qquad \mapsto \qquad abca^\omega$$

An ∞-algebra is essentially the same thing as an ω-semigroup, see [PP04], with the difference that ω-semigroups have separate sorts for finite and infinite words. There is also a close connection with Wilke semigroups [Wil91], which will be described as the running example develops. □

The similarities in the examples suggest that the should be an abstract notion of algebra, which would cover the examples and possibly other settings, e.g. trees. A closer look at the examples reveals that concepts of algebraic language theory such as "algebra", "morphism", "language", "recognisable language" can be defined only in terms of the following four basic concepts (written below in the notation appropriate to +-algebras):

1. how a set A is transformed into a set A^+;
2. how a function $f : A \to B$ is lifted to a function $f^+ : A^+ \to B^+$;
3. a flattening operation from $(A^+)^+ \to A^+$;
4. how to represent an element of A as an element of A^+.

These four concepts are what constitutes a monad, a fundamental concept in category theory and, recently, programming languages like Haskell. However, unlike for Haskell, in this paper the key role is played by Eilenberg-Moore algebras.

The point of this paper is that, based on a monad one can also define things like: "syntactic algebra", "pseudovariety", "MSO logic", "profinite object", and even prove some theorems about them. Furthermore, monads as an abstraction cover practically every setting where algebraic language theory has been applied so far, including labelled scattered orderings [BR12], labelled countable total orders [CCP11], ranked trees [Ste92], unranked trees [BW08], preclones [ÉW03]. These applications are discussed at length in the full version. The full version also shows how new algebraic settings can be easily produced using monads, as illustrated on a monad describing words with a distinguished position, where standard theorems and definitions come for free by virtue of being a monad. A related paper is [Ési10], which gives an abstract language theory for Lawvere theories, and proves that Lawvere theories admit syntactic algebras and a pseudovariety theorem. Lawvere theories can be viewed as the special case of finitary monads, e.g. finite words are Lawvere theories, but infinite words are not.

This paper shows that several results of formal language theory can be stated and proved on the abstract level of monads, including: the Myhill-Nerode theorem on syntactic algebras, the Eilenberg pseudovariety theorem, or the Reiterman theorem on profinite identities defining pseudovarieties. Another example is decidability of MSO, although here monads only take care of the symbol-pushing part, leaving out the combinatorial part that is specific to individual monads, like applying the Ramsey theorem in the case of infinite words. When proving such generalisations of classical theorems, one is naturally forced to have a closer look at notions such as "derivative of a language", or "finite algebra", which are used in the assumptions of the theorems.

Much effort is also devoted to profinite constructions. It is shown that every monad has a corresponding profinite monad, which, like any monad, has its own notion of recognisability, which does not reduce to recognisability in the original monad. For example, the monad for finite words has a corresponding monad of profinite words, and recognisable languages of profinite words turn out to be a generalisation of languages of infinite words definable in the logic MSO+U.

Thanks. I would like to thank Bartek Klin (who told me what a monad is), Szymon Toruńczyk and Marek Zawadowski for discussions on the subject.

1 Monads and Their Algebras

This paper uses only the most rudimentary notions of category theory: the definitions of a category (objects and composable morphisms between them), and of a functor (something that maps objects to objects and morphisms to morphisms in a way that is consistent with composition). All examples in this paper use the category of sets, where objects are sets and morphisms are functions; or possibly the category of sorted sets, where objects are sorted sets for some fixed set of sort names, and morphisms are sort-preserving functions.

A monad over a category is defined to be a functor T from the category to itself, and for every object X in the category, two morphisms

$$\eta_X : X \to \mathsf{T}X \qquad \text{and} \qquad \mu_X : \mathsf{TT}X \to \mathsf{T}X,$$

which are called the unit and multiplication operations. The monad must satisfy the axioms given in Figure 1.

We already saw two monads in Example 1 and in the running example.

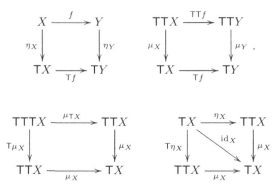

Fig. 1. The axioms of a monad are that these four diagrams commute for every object X in the category and every morphism $f : X \to Y$. The upper diagrams say that the unit and multiplication are natural. The lower left diagram says that multiplication is associative, and the lower right says that the unit is consistent with multiplication.

For this paper, the most important thing about monads is that they have a natural corresponding algebras. An *Eilenberg-Moore algebra for a monad* T, or simply T-algebra, is a pair **A** consisting of a *universe* A, which is an object in the underlining category, together with a multiplication morphism

$$\mathrm{mul}_{\mathbf{A}} : \mathsf{T}A \to A,$$

such that the $\mathrm{mul}_{\mathbf{A}} \circ \eta_A$ is the identity, and which is associative in the sense that the following diagram commutes.

$$
\begin{array}{ccc}
\mathsf{TT}A & \xrightarrow{\mu_A} & \mathsf{T}A \\
{\scriptstyle \mathsf{T}\mathrm{mul}_{\mathbf{A}}}\downarrow & & \downarrow{\scriptstyle \mathrm{mul}_{\mathbf{A}}} \\
\mathsf{T}A & \xrightarrow[\mathrm{mul}_{\mathbf{A}}]{} & A
\end{array}
$$

Observe that this associativity is similar to the lower left axiom in Figure 1. In fact, the lower left axiom in Figure 1 and the upper half of the lower right axiom say that $\mathsf{T}X$ equipped with the operation μ_X forms a T-algebra, called the free T-algebra over X.

We use the convention that an algebra is denoted by a boldface letter, while its universe is written without boldface. A T-*morphism* between two T-algebras **A** and **B** is a function h between their universes which respects their multiplication operations in the sense that the following diagram commutes.

$$
\begin{array}{ccc}
\mathsf{T}A & \xrightarrow{\mathsf{T}h} & \mathsf{T}B \\
{\scriptstyle \mathrm{mul}_{\mathbf{A}}}\downarrow & & \downarrow{\scriptstyle \mathrm{mul}_{\mathbf{B}}} \\
A & \xrightarrow[h]{} & B
\end{array}
$$

This completes the definition of monads and their algebras.

Languages and colorings. To develop the basic definitions of recognisable languages over a monad, we require the following parameters, which we call the *setting*: the underlying category, the monad, a notion of finite alphabet, and a notion of finite T-algebra. So far, we do not place any restrictions on the notions of finiteness, e.g. when considering sets with infinitely many sorts, reasonable settings will often have finite algebras whose universe is not finite in the same sense as a finite algebra. Actually, for some monads, it is not clear what a finite algebra should be, e.g. this is the case for infinite trees, and this paper sheds little new light on the question. Fix a setting, with the monad being called T, for the following definitions.

A *coloring* of a T-algebra is defined to be a morphism from its universe to some object in the underlying category. A coloring is said to be *recognised* by a T-morphism if the coloring factors through the morhpism. A coloring is called T-*recognisable* if it is recognised by some T-morphism with a finite target, according to the notion of finite T-algebra given in the setting.

When the underlying category is, possibly sorted, sets we can talk about languages, as defined below. Consider a finite alphabet, according to the notion of finite alphabet given in the setting. In all of the examples of this paper, a finite alphabet will be a possibly sorted set with finitely many elements. In particular, if there are infinitely many sorts, then a finite alphabet will use only finitely many. A T-language over a finite alphabet Σ is defined to be any subset $L \subseteq T\Sigma$. Notions of recognisability are inherited from colorings, using the characteristic function of a language.

The Myhill-Nerode Theorem. We present a monad generalisation of the Myhill-Nerode theorem. That is, we give a sufficient condition for colorings, and therefore also languages, to have a syntactic (i.e. minimal) morphism. The generalisation only works in the setting of sorted sets, and therefore also in the setting of normal sets. Fix the setting of sorted sets, for some choice of, possibly infinitely many, sort names.

Define a (possibly sorted) set A to be *finitary* if for every $w \in TA$, there is some finite $A_w \subseteq A$ such that $w \in TA_w$. A monad is called finitary if every set is finitary, e.g. this is the case for the monad of finite words.

Theorem 1.1. *[Syntactic Morphism Theorem] Consider a monad T in the setting of sorted sets. Let f be a coloring of a T-algebra $\mathbf{A}$, which is recognised by a T-morphism h into some T-algebra with finitary universe. There exists a surjective T-morphism into a T-algebra*

$$\mathsf{synt} f : \mathbf{A} \to \mathbf{A}_f,$$

called the syntactic morphism *of f, which recognises f and which factors through every surjective T-morphism recognising f. Furthermore, $\mathsf{synt} f$ is unique up to isomorphisms on $\mathbf{A}_f$.*

If $\mathbf{A}$ itself has finitary universe, then f is recognised by the identity T-morphism on $\mathbf{A}$. Therefore, if the monad is finitary, then every T-language has a syntactic morphism. This covers monads for objects such as finite words, ultimately periodic words, or various kinds of finite trees. In monads describing truly infinite objects, e.g. the monad for ∞-words used in the running example, a syntactic morphism might not exist.

Running Example 2. Consider the following ∞-language

$$L = \{a^{n_1}ba^{n_2}b\cdots : \text{the sequence } n_i \text{ is unbounded, i.e. } \limsup n_i = \infty.\}$$

One can show that this language does not have a syntactic morphism, not even if the target algebra is allowed to have infinite universe. The idea is that for every n, there is a recognising morphism which identifies all words in $\{a^1, a^2, \ldots, a^n\}$, but there is no morphism which identifies all words in a^+. $\square$

The Eilenberg Theorem. Another result which can be stated and proved on the level of monads is Eilenberg's pseudovariety theorem. This theorem says that, in the case of semigroups, language pseudovarieties and algebra pseudovarieties, which will be defined below, are in bijective correspondence. The theorem implies that if $\mathbb{L}$ is a language pseudovariety, then the membership problem $L \in \mathbb{L}$ can be decided only by looking at the syntactic semigroup of L, and one need not look at the accepting set, nor at the information about which letters are mapped to which elements of the semigroup.

Surely Eilenberg must have known that the pseudovariety works for monads in general and not just for monoids and semigroups, since he invented both the pseudovariety theorem and algebras in abstract monads. However, such a generalisation is not in his book [Eil74]. The generalisation subsumes pseudovariety theorems for: finite words in both monoid and semigroup variants [Eil74], ∞-words [Wil91], scattered linear orderings [BR12], finite trees [Ste92].

Fix a setting where the category is sets with finitely many sorts. Assume that the notion of finite T-algebra is simply that the universe is finite, and a finite alphabet is one with finitely many elements. A T-algebra with a finite universe is finitary, and therefore every T-recognisable language has a syntactic algebra by the Syntactic Morphism Theorem. Define a *derivative*[1] of a T-recognisable T-language $L \subseteq T\Sigma$ to be any other subset of $T\Sigma$ that is recognised by the syntactic algebra of L. Define a T-*language pseudovariety* to be a class of T-recognisable T-languages which is closed under Boolean combinations, derivatives, and pre-images under T-morphisms. Since complementation is a form of derivative, Boolean combinations could be replaced by unions in the definition of a T-language pseudovariety. As usual for pseudovarieties, a T-language is formally treated as its characteristic function, which means that a language comes with a description of its input alphabet. Define a T-*algebra pseudovariety* to

---

[1] This notion of derivative is nonstandard. For example, in the case of finite words, the more accepted notion is that a derivative of L is any language of the form $w^{-1}Lv^{-1}$. Because of this nonstandard definition, Theorem 1.2 does not directly generalise the classical Pseudovariety Theorem by Eilenberg. This is discussed in the full version, which contains a proper generalisation of the classical Pseudovariety Theorem.

be a class of finite T-algebras which is closed under finite products, images of surjective T-morphisms, and subalgebras.

For a class $\mathbb{L}$ of recognisable T-languages, define Alg $\mathbb{L}$ to be the class of finite T-algebras that only recognise T-languages from $\mathbb{L}$. For a class $\mathbb{A}$ of finite T-algebras, define Lan $\mathbb{A}$ to be the T-languages recognised by T-algebras from $\mathbb{A}$. The Pseudovariety theorem says that these mappings are mutual inverses.

Theorem 1.2. *The mappings* Lan *and* Alg *are mutual inverses, when restricted to pseudovarieties.*

Running Example 3. Call an ∞-language *definite* if there is some $n \in \mathbb{N}$ such that membership in the language depends only on the first n letters. Examples of definite languages include: "words that begin with a", or "words of length at least two". Call an ∞-algebra **A** *definite* if there is some $n \in \mathbb{N}$ such that

$$\mathrm{mul}_\mathbf{A}(x_1 \cdots x_n x_{n+1}) = \mathrm{mul}_\mathbf{A}(x_1 \cdots x_n)$$

holds for every $x_1, \ldots, x_{n+1}$ in the universe of the algebra. It is not difficult to show that a recognisable ∞-language is definite if and only if its syntactic algebra is definite, definite ∞-languages form a language pseudovariety, definite ∞-algebras form an algebra pseudovariety, and the two are in correspondence as in the Pseudovariety Theorem. $\square$

2   Deciding Monadic Second-Order Logic

An important part of language theory is the equivalence of recognisability and definability in monadic second-order logic MSO. Examples where this equivalence holds include finite words and trees, and more interestingly from a combinatorial perspective, infinite words and trees. There are common parts in all of the proofs, and parts that are specific to each domain. We show that the common parts can be stated and proved on the abstract level of monads.

Representing an algebra. To give an algorithm for MSO satisfiability that uses algebras, one needs a representation of finite algebras so that they can be manipulated by algorithms. We propose such a representation; this is the main part of this section. Fix a setting for the rest of this section, with the category being sets, possibly sorted, but with finitely many sorts.

In most interesting cases, the monad T produces infinite sets, even on finite arguments. Therefore, the finiteness of the universe of a T-algebra **A** does not, on its own, imply that the algebra has a finite representation, because one needs to also represent its multiplication operation, whose type is $TA \to A$. To represent algebras, we will use an assumption, to made more precise below, which roughly says that if an algebra has finite universe A, then the the multiplication is determined by its values on a small finite subset of TA. For instance, in the monad of finite words from Example 1, one chooses from A^+ only the length two words, because a +-algebra, i.e. a semigroup, is uniquely determined by its binary multiplication. We now describe these notions in more detail.

Define a *subfunctor* of T to be a mapping T_0 which takes each set X to a subset $T_0 X \subseteq TX$. A subfunctor on its own is not a monad, however it can be used to generate a monad as follows. For a set X, define $T_0^* X$ to be the least set which contains the units of X, and which also contains any multiplication (flattening) of an element in $T_0 T_0^* X$. It is not difficult to show that T_0^* is a submonad of T, i.e. a subfunctor with a monad structure as inherited from T. A subfunctor T_0 is said to *span* a T-algebra **A** if for every subset X of the universe, $\mathrm{mul_A}$ has the same image over $T_0^* X$ and over TX, i.e.

$$\mathrm{mul_A} T_0^* X = \mathrm{mul_A} TX.$$

A subfunctor is *complete* if it spans every T-algebra, and *finitely complete* if it spans every finite T-algebra; the latter depends on the notion of finite T-algebra.

Consider a subfunctor T_0 that is finitely complete for a monad T. For a finite T-algebra **A**, define its T_0-*reduct* to be the pair consisting of the universe A of **A**, and the restriction of the multiplication operation from **A** to the subfunctor:

$$\mathrm{mul_A}|_{T_0 A} : T_0 A \to A$$

It is easy to show, using associativity, that if T_0 spans **A**, then **A** is uniquely determined by its T_0-reduct. In particular, if T_0 is complete, then every operation $T_0 A \to A$ extends to at most one T-algebra with universe A. Note the "at most one", e.g. not every binary operation extends to a semigroup operation, for this associativity is needed. The same holds for finite completeness and finite algebras. The point of using T_0-reducts is that sometimes T_0 can be chosen so that it preserves finiteness, and therefore finite T-algebras can be represented in a finite way as functions $T_0 A \to A$.

Running Example 4. As in [Wil91], one can use the Ramsey theorem to show that a finite ∞-algebra is uniquely determined by the values of its multiplication operation on arguments of the form xy and x^ω. Stated differently,

$$\mathsf{W} X \quad \overset{\text{def}}{=} \quad \{xy, x^\omega : x, y \in X\} \subseteq X^\infty.$$

is finitely complete subfunctor of the ∞-functor. The submonad W^* maps an alphabet X to the finite and ultimately periodic words over X. A W-reduct of a finite ∞-algebra is essentially the same thing as a Wilke semigroup, modulo the difference that Wilke semigroups are two-sorted. In [Wil91], Wilke shows axioms which describe when a W-algebra extends to an ∞-algebra. □

A subfunctor T_0 is called *effective* if it satisfies the following two conditions.

1. If X is a finite set then $T_0 X$ is finite and can be computed. This means that a finite T-algebra with universe A be represented as a function $T_0 A \to A$.
2. For every finite set Σ and every $w \in T_0 \Sigma$, one can compute a T-morphism into a finite T-algebra that recognises $\{w\}$, with the T-algebra represented as in item 1, and the T-morphism represented by its values on units of Σ.

The second condition is maybe less natural, it will be used in deciding MSO.

Running Example 5. We claim that the functor W in the running example is effective. For the first condition, the set WX is isomorphic to the disjoint union $X^2 + X$, and can therefore clearly be computed. For the second condition, one needs to show that for every ultimately periodic ∞-word w, there is a finite ∞-algebra that recognises the singleton $\{w\}$, and its W-reduct can be computed. The universe of this algebra consists of suffixes of w, finite infixes modulo repetitions, and an error element. $\square$

Monadic second-order logic. To establish the connection between MSO and recognisability on the level of monads, we use a definition of MSO which does not talk about "positions" or "sets of positions" of a structure, but which is defined in purely language theoretic terms. In the abstract version, predicates, such as the successor predicate, are modelled by languages. For a set $\mathcal{L}$ of T-languages, define MSO$_\mathsf{T}(\mathcal{L})$ to be the smallest class of T-languages which contains $\mathcal{L}$, is closed under Booolean operations, images and inverse images of T-morphisms.

Running Example 6. The standard notion of MSO for ∞-words, as studied by Büchi, is equivalent to MSO$_\mathsf{T}(\mathcal{L})$ where $\mathcal{L}$ contains only two recognisable ∞-languages over alphabet $\{0, 1\}$, namely the language of words which contain only zeros, and the language of words where every one is followed only by ones. $\square$

One can show that if $\mathcal{L}$ contains only T-recognisable T-languages, then so does MSO$_\mathsf{T}(\mathcal{L})$. This uses the assumptions on the setting being sorted sets, and finite algebras being ones with finite universes. A non-example is the category of nominal sets with orbit-finite sets, where powerset does not preserve orbit-finiteness, and also MSO contains non-recognisable languages, see [Boj13].

A language in MSO$_\mathsf{T}(\mathcal{L})$ can be represented as a tree where leaves are languages from $\mathcal{L}$, binary nodes are labelled by union or intersection, complementation is represented by unary nodes, and for every T-morphism $h : \mathsf{T}\Sigma \to \mathsf{T}\Gamma$ there are two kinds of unary nodes, one for image and the other for inverse image. Therefore, if $\mathcal{L}$ is finite (or has some fixed enumeration) then it makes sense to consider the following decision problem, called MSO$_\mathsf{T}(\mathcal{L})$ *satisfiability*: an instance is a tree representing a language from MSO$_\mathsf{T}(\mathcal{L})$, and the question is whether the corresponding language is nonempty. We provide below a sufficient criterion for the decidability the problem.

Theorem 2.1. *Consider a setting with finitely sorted sets, and let $\mathcal{L}$ be all recognisable languages. If there is a subfunctor that is effective and finitely complete, then* MSO$_\mathsf{T}(\mathcal{L})$ *satisfiability is decidable.*

As mentioned at the beginning of the section, the theorem takes care of the symbol pushing part in deciding satisfiability of MSO, and leaves only the combinatorial part. The combinatorial part is finding an effective and finitely complete subfunctor, this is typically done using some kind of Ramsey theorem.

Running Example 7. Applying Theorem 2.1 to the observations made so far in the running example, we conclude that MSO satisfiability over ∞-words is decidable. The proof obtained this way follows the same lines as Büchi's original proof. $\square$

3   Profinite Monads

Profinite constructions are an important tool in the study of recognisable languages. Example applications include: lattices of word languages correspond to implications on profinite words [GGP08], pseudovarieties in universal algebra correspond to profinite identities [Rei82], recognisable word languages are the only ones which give a uniformly continuous multiplication in the profinite extension [GGP10]. In this section we show that these results can be stated and proved on the abstract level of monads, thus covering cases like profinite trees or profinite ∞-words. Furthermore, we show that profinite objects, e.g. profinite words, form a monad as well, which has its own notion of recognisable language, a notion that seems interesting and worthy of further study.

The Stone dual. A short definition of the profinite object uses Stone duality. Define an *ultrafilter* in a Boolean algebra to be a subset of its universe which is closed under ∧, and which contains every element of the Boolean algebra or its complement, but not both. The set of ultrafilters is called the *Stone dual* of the Boolean algebra. The Stone dual also comes with a topological structure, but since we do not use it here, we do not describe it.

From the monad point of view, we are interested in the special case of the Boolean algebra of recognisable languages in a T-algebra. Let $\mathbf{A}$ be a T-algebra, not necessarily finite. Define the Stone dual of $\mathbf{A}$, denoted by $\mathsf{Stone}\mathbf{A}$, to be the Stone dual of the Boolean algebra of those subsets of $\mathbf{A}$ that are recognised by T-morphisms from $\mathbf{A}$ into finite T-algebras. This definition generalises the well-known space of profinite words, i.e. if the monad is the monad of finite words. Note how the definition depends on the notion of finite algebra.

Running Example 8. For an alphabet Σ and $w \in \Sigma^+$, define $w^{\#}$ to be the set of languages $L \subseteq \Sigma^\infty$ which are recognised by some finite ∞-algebra and which contain $w^{n!}$ for all but finitely many n. This set is clearly closed under intersection, and a standard pumping argument shows that for every recognisable language L, it contains L or $\Sigma^+ - L$. Therefore $w^{\#}$ is an ultrafilter, i.e. a profinite ∞-word. A common notation in profinite words would be to use ω, but this would conflict with the infinite power in the context of ∞-words. ☐

Defining pseudovarieties by identities. Our first result on Stone duals is a monad generalisation of the Reiterman Theorem [Rei82]. The original Reiterman Theorem, which uses terminology of universal algebra, says that pseudovarieties can be characterised by profinite identities. In the full version, we present a notion of profinite identity, and prove the following theorem.

Theorem 3.1. *Let $\mathbb{L}$ be class of recognisable T-languages. Then $\mathbb{L}$ is a pseudovariety if and only if it is defined by a set of profinite identities.*

Running Example 9. Recall the pseudovariety of definite ∞-languages. This pseudovariety is defined by a single profinite identity, namely

$$x^\omega = x^{\#}.$$

One proves this the same way as in the classical result on definite languages of finite words, which are characterised by the identity $x^\# y = x^\#$. The latter identity is implied by the former, because $x^\omega y = x^\omega$ is true in all ∞-words. $\square$

The full version contains other results on Stone duals, in particular monad generalisations of results from [GGP08] and [GGP10].

A profinite monad. We now explain how to convert a monad T into another monad, called $\overline{\mathsf{T}}$, that describes profinite objects over T. The functor $\overline{\mathsf{T}}$ maps a set Σ to the Stone dual of the T-algebra $\mathsf{T}\Sigma$. For a function $f : \Sigma \to \Gamma$, the mapping $\overline{\mathsf{T}}f$ takes an ultrafilter $U \in \overline{\mathsf{T}}\Sigma$ to the set

$$V = \{L \subseteq \mathsf{T}\Gamma : (\mathsf{T}f)^{-1}(L) \in U\}$$

which is easily seen to be an ultrafilter, and therefore an element of $\overline{\mathsf{T}}\Gamma$. This definition is a special case of the functor in the classical theorem on duality of Boolean algebras and Stone spaces; in particular $\overline{\mathsf{T}}$ is a functor. It remains to provide the monad structure, namely the multiplication and unit. These will be given in Theorem 3.2, using the following notion of profinite completion. Define the *profinite completion* of a T-morphism $h : \mathsf{T}\Sigma \to \mathbf{A}$ into a finite T-algebra to be the mapping

$$\bar{h} : \overline{\mathsf{T}}\Sigma \to A$$

defined as follows: an ultrafilter U is mapped to the unique element a in the universe of $\mathbf{A}$ such that the ultrafilter contains the language $h^{-1}(a)$.

Theorem 3.2. *For every set Σ, there are unique operations*

$$\bar{\eta}_\Sigma : \Sigma \to \overline{\mathsf{T}}\Sigma \qquad \bar{\mu}_\Sigma : \overline{\mathsf{T}}\,\overline{\mathsf{T}}\Sigma \to \overline{\mathsf{T}}\Sigma$$

such that for every finite T-algebra $\mathbf{A}$ and every T-morphism $h : \mathsf{T}\Sigma \to \mathbf{A}$, the following diagrams commute

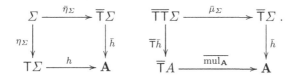

Furthermore, equipped with the above operations, $\overline{\mathsf{T}}$ is a monad.

What is the benefit of seeing $\overline{\mathsf{T}}$ as a monad? Since $\overline{\mathsf{T}}$ is itself a monad, it can have its own notion of finite algebra and recognisable language. We illustrate how these notions can be interesting, using the special case of profinite words. Consider the monad $\Sigma \mapsto \Sigma^+$ for finite words. Let us denote by $\Sigma \mapsto \Sigma^{\bar{+}}$ the profinite version of this monad. An element of $\Sigma^{\bar{+}}$ is a simply profinite word over the alphabet Σ. One way of creating a finite $\bar{+}$-algebra is to take a finite $+$-algebra, i.e. a finite semigroup, and extend its multiplication operation to profinite words using profinite completion. More interesting $\bar{+}$-algebras are not obtained this way, here is one example.

Example 2. We say that a profinite word $w \in \{0,1\}^{\bar{+}}$ has *exactly n ones* if, when seen as an ultrafilter, it contains the recognisable language of words with at least n ones. If a profinite word has exactly n ones for some n, then we say that it has *a bounded number* of ones. For example, $1^{\#}$ does not have a bounded number of ones. A certain amount of calculations shows that the set of profinite words in $\{0,1\}^{\bar{+}}$ which have a bounded number of ones is recognised by a $\bar{+}$-morphism into a finite $\bar{+}$-algebra. The finite $\bar{+}$-algebra has three elements in its universe, standing for: "only zeros", "a bounded number of ones", and "not a bounded number of ones".

Let us define MSO+inf by applying the abstract notion of MSO defined in Section 2, with the predicates being the following languages of profinite words over alphabet $\{0,1\}$: "a bounded number of ones", "only zeros", "every one is followed by only zeros". Under a different terminology, this class of languages was considered in [Tor12]. Since the operators of MSO preserve recognisability, it follows that MSO+inf contains only $\bar{+}$-recognisable languages. It is not clear if MSO+inf contains all $\bar{+}$-recognisable languages, but Corollary 2 from [Tor12] and a new undecidability result from [MB15] imply that MSO+inf has undecidable satisfiability. $\square$

References

[Boj13] Bojanczyk, M.: Nominal monoids. Theory Comput. Syst. **53**(2), 194–222 (2013)

[BR12] Bedon, N., Rispal, C.: Schützenberger and Eilenberg theorems for words on linear orderings. J. Comput. Syst. Sci. **78**(2), 517–536 (2012)

[BW08] Bojanczyk, M., Walukiewicz, I.: Forest algebras. In: Logic and Automata: History and Perspectives [in Honor of Wolfgang Thomas], pp. 107–132 (2008)

[CCP11] Carton, O., Colcombet, T., Puppis, G.: Regular languages of words over countable linear orderings. In: Aceto, L., Henzinger, M., Sgall, J. (eds.) ICALP 2011, Part II. LNCS, vol. 6756, pp. 125–136. Springer, Heidelberg (2011)

[Eil74] Eilenberg, S.: Automata, languages, and machines, vol. A (1974)

[Ési10] Ésik, Z.: Axiomatizing the equational theory of regular tree languages. The Journal of Logic and Algebraic Programming **79**(2), 189–213 (2010)

[ÉW03] Ésik, Z., Weil, P.: On logically defined recognizable tree languages. In: Pandya, P.K., Radhakrishnan, J. (eds.) FSTTCS 2003. LNCS, vol. 2914, pp. 195–207. Springer, Heidelberg (2003)

[GGP08] Gehrke, M., Grigorieff, S., Pin, J.É.: Duality and equational theory of regular languages. In: Aceto, L., Damgård, I., Goldberg, L.A., Halldórsson, M.M., Ingólfsdóttir, A., Walukiewicz, I. (eds.) ICALP 2008, Part II. LNCS, vol. 5126, pp. 246–257. Springer, Heidelberg (2008)

[GGP10] Gehrke, M., Grigorieff, S., Pin, J.É.: A topological approach to recognition. In: Abramsky, S., Gavoille, C., Kirchner, C., Meyer auf der Heide, F., Spirakis, P.G. (eds.) ICALP 2010. LNCS, vol. 6199, pp. 151–162. Springer, Heidelberg (2010)

[MB15] Toruńczyk, S., Bojańczyk, M., Parys, P.: The MSO+U theory of $(\mathbb{N}, <)$ is undecidable (2015). CoRR, arXiv:1502.04578

[PP04] Perrin, D., Pin, J.-É.: Infinite Words: Automata, Semigroups, Logic and Games. Elsevier (2004)

[Rei82] Reiterman, J.: The Birkhoff theorem for finite algebras. Algebra Universalis **14**(1), 1–10 (1982)

[Ste92] Steinby, M.: A theory of tree language varieties. In: Tree Automata and Languages, pp. 57–82 (1992)

[Tor12] Toruńczyk, S.: Languages of profinite words and the limitedness problem. In: Czumaj, A., Mehlhorn, K., Pitts, A., Wattenhofer, R. (eds.) ICALP 2012, Part II. LNCS, vol. 7392, pp. 377–389. Springer, Heidelberg (2012)

[Wil91] Wilke, T.: An Eilenberg theorem for infinity-languages. In: Leach Albert, J., Monien, B., Rodríguez-Artalejo, M. (eds.) ICALP 1991. LNCS, vol. 510, pp. 588–599. Springer, Heidelberg (1991)

Garside and Quadratic Normalisation: A Survey

Patrick Dehornoy[1,2]([⊠])

[1] Laboratoire de Mathématiques Nicolas Oresme, CNRS UMR 6139,
Université de Caen, 14032 Caen cedex, France
patrick.dehornoy@unicaen.fr
http://www.math.unicaen.fr/~dehornoy
[2] Institut Universitaire de France, Paris, France

Abstract. Starting from the seminal example of the greedy normal norm in braid monoids, we analyze the mechanism of the normal form in a Garside monoid and explain how it extends to the more general framework of Garside families. Extending the viewpoint even more, we then consider general quadratic normalisation procedures and characterise Garside normalisation among them.

Keywords: Normal form · Normalisation · Regular language · Fellow traveller property · Greedy decomposition · Garside family · Quadratic rewriting system · Braid monoids · Artin–Tits monoids · Plactic monoids

This text is an essentially self-contained survey of a general approach of normalisation in monoids developed in recent years by the author in collaboration with several co-authors and building on the seminal example of the greedy normal form of braids independently introduced by S. Adjan [1] and by M. El-Rifai and H. Morton [22]. The main references are the book [17], written with F. Digne, E. Godelle, D. Krammer, and J. Michel, the recent preprint [19], written with Y. Guiraud, and, for algorithmic aspects, the article [16], written with V. Gebhardt.

If M is a monoid (or a semigroup), and S is a generating subfamily of M, then, by definition, every element of M is the evaluation of some S-word. A *normal form* for M with respect to S is a map that assigns to each element of M a distinguished representative S-word, hence a (set theoretic) section for the evaluation map from S^* to M. The interest of normal forms is obvious, since they provide a unambiguous way of specifying the elements of M and, from there, for working with them in practice. As can be expected, the complexity of a normal form is a significant element. It can be defined either by considering the complexity of the language of normal words (regular, algebraic, etc.), or that of the associated normalisation map, that is, the procedure that transforms an arbitrary word into an equivalent normal word (linear, polynomial, etc.).

A huge number of normal forms appear in literature, based on quite different initial approaches, and it is certainly difficult to establish nontrivial results unifying all possible types. In this text, we concentrate on some families of normal forms that turn out to be simple in terms of complexity measures, and whose

© Springer International Publishing Switzerland 2015
I. Potapov (Ed.): DLT 2015, LNCS 9168, pp. 14–45, 2015.
DOI: 10.1007/978-3-319-21500-6_2

main specificity is to satisfy some *locality* assumptions, meaning that both the property of being normal and the procedure that transforms an arbitrary word into an equivalent normal word only involve factors of a bounded length, here factors of length two (and that is why we call them "quadratic"). As we shall see, several well-known classes of normalisation processes enter this framework, for instance the seminal example of the greedy normal form in Artin's braid monoids [1, 22, 23] but also the normal forms in Artin–Tits monoids based on rewriting systems as in [25] or, in a very different context, the normal form in plactic monoids based on Young tableaux and the Robinson–Schenstedt algorithm [5, 6].

The current introductory text is organised in four sections, going from the particular to the general. In Sec. 1, we analyze two motivating examples of greedy normal forms, involving free abelian monoids, a toy case that already contains the main ideas, and braid monoids, a more complicated case. Extending these examples, we describe in Sec. 2 the mechanism of the Δ-normal form in the now classical framework of Garside monoids. Next, in Sec. 3, we explain how most of the results can be generalized and, at the same time, simplified, using the notion of an S-normal form derived from a Garside family. Finally, in Sec. 4, we introduce quadratic normalisations, which provide a natural unifying framework for the normal forms we are interested in. Having defined a complexity measure called the class, we characterise Garside normalisations among quadratic normalisations, and mention (positive and negative) termination results for the rewriting systems naturally associated with quadratic normalisations.

Most proofs are omitted or only sketched. However, it turns out that some arguments, mainly in Sec. 2 and 3, are very elementary, and then we included them, hopefully making this text both more informative and thought-provoking.

Excepte in concluding remarks at the end of sections, we exclusively consider monoids. A number of results, in particular most of those involving Garside normalisation, can be extended to groups of fractions. Also, the whole approach extends to categories, viewed as monoids with a partial multiplication.

Our notation is standard. We use $\mathbb{Z}$ for the set of integers, $\mathbb{N}$ for the set of nonnegative integers. If S is a set, we denote by S^* the free monoid over S and call its elements S-*words*, or simply *words*. In this context, S is also called alphabet, and its elements letters. We write $\|w\|$ for the length of w, and $S^{[p]}$ for the set of all S-words of length p. We use $w|w'$ for the concatenation of two S-words w and w'. We say that w' is a *factor* of w if there exist u, v satisfying $w = u|w'|v$. If M is a monoid generated by a set S, we say that an S-word $s_1|\cdots|s_p$ is a S-*decomposition* for an element g of M if $g = s_1 \cdots s_p$ holds in M.

1   Two Examples

We shall describe a specific type of normal form often called the "greedy normal form". Before describing it in full generality, we begin here with two examples: a very simple one involving free abelian monoids, and then the seminal example of Artin's braid monoids as investigated after Garside.

1.1 Free Abelian Monoids

Our first example, free abelian monoids, is particularly simple, but it is fundamental as it can serve as a model for the sequel: our goal will be to obtain for more complicated monoids counterparts of the results that are trivial here.

Consider the free abelian monoid $(\mathbb{N}, +)^n$ with $n \geqslant 1$, simply denoted by $\mathbb{N}^n$. We shall see the elements of $\mathbb{N}^n$ as sequences of nonnegative integers indexed by $\{1, ..., n\}$, thus writing $g(k)$ for the kth entry of an element g, and use a multiplicative notation: $fg = h$ means $\forall k\,(f(k) + g(k) = h(k))$. Let A_n be the family $\{\mathsf{a}_1, ..., \mathsf{a}_n\}$, where a_i is defined by $\mathsf{a}_i(k) = 1$ for $k = i$, and 0 otherwise. Then A_n is a basis of $\mathbb{N}^n$ as an abelian monoid.

It is straightforward to obtain a normal form $\mathrm{NF}^{\mathrm{Lex}}$ for $\mathbb{N}^n$ with respect to A_n by fixing a linear ordering on A_n, for instance $\mathsf{a}_1 < \cdots < \mathsf{a}_n$, and, for g in $\mathbb{N}^n$, defining $\mathrm{NF}^{\mathrm{Lex}}(g)$ to be the lexicographically smallest word representing g.

We shall be interested here in another normal form, connected with another generating family. In this basic example, the construction may seem uselessly intricate, but we shall see that it nicely extends to less trivial cases, which is not the case of the above lexicographical normal form. Let us put

$$S_n := \{s \in \mathbb{N}^n \mid s(k) \in \{0, 1\} \text{ for } k = 1, ..., n\}. \tag{1.1}$$

For f, g in $\mathbb{N}^n$, define $f \preccurlyeq g$ to mean $\forall k\,(f(k) \leqslant g(k))$, and write $f \prec g$ for $f \preccurlyeq g$ with $f \neq g$. The relation $\preccurlyeq$ is a partial order, connected with the operation of $\mathbb{N}^n$, since $f \preccurlyeq g$ is equivalent to $\exists g' \in \mathbb{N}^n\,(fg' = g)$, that is, f *divides* g in $\mathbb{N}^n$. Then S_n consists of the 2^n divisors of the element Δ_n defined by

$$\Delta_n(k) := 1 \quad \text{for } k = 1, ..., n. \tag{1.2}$$

We recall that, if M is a (left-cancellative) monoid generated by a family S, the *Cayley graph of M with respect to S* is the S-labeled oriented graph with vertex set M and, for g, h in M and s in S, there is an s-labeled edge from g to h if, and only if, $gs = h$ holds in M. Then, the Cayley graph of $\mathbb{N}^n$ with respect to A_n is an n-dimensional grid, and S_n corresponds to the n-dimensional cube that is the elementary tile of the grid, see Fig. 1.

Proposition 1.3 *Every element of $\mathbb{N}^n$ admits a unique decomposition of the form $s_1 | \cdots | s_p$ with $s_1, ..., s_p$ in S_n satisfying $s_p \neq 1$, and, for every $i < p$,*

$$\forall s \in S_n\,(s_i \prec s \ \Rightarrow \ s \nmid s_i s_{i+1} \cdots s_p). \tag{1.4}$$

Condition (1.4) is a maximality statement. It says that s_1 contains as much of g as possible in order to remain in S_n and that, for every i, the entry s_i similarly contains as much of the right chunk $s_i \cdots s_p$ as possible to remain in S_n: when we try to replace s_i with a larger element s of S_n, then we quit the divisors of $s_i \cdots s_p$. This should make it clear why the decomposition $s_1 | \cdots | s_p$ of g is usually called *greedy*.

Example 1.5 Assume $n = 3$ and consider $g = \mathsf{a}^3\mathsf{b}\mathsf{c}^2$, that is, $g = (3,1,2)$ (we write $\mathsf{a}, \mathsf{b}, \mathsf{c}$ for $\mathsf{a}_1, \mathsf{a}_2, \mathsf{a}_3$). The maximal element of S_3 that divides g is Δ_3, with $g = \Delta_3 \cdot \mathsf{a}^2\mathsf{c}$. Then, the maximal element of S_3 that divides $\mathsf{a}^2\mathsf{c}$ is $\mathsf{a}\mathsf{c}$, with $\mathsf{a}^2\mathsf{c} = \mathsf{a}\mathsf{c} \cdot \mathsf{a}$. The latter element left-divides Δ_3. So the greedy decomposition of g as provided by Prop. 1.3 is the length-three S_3-word $\Delta_3|\mathsf{a}\mathsf{c}|\mathsf{a}$, see Fig. 1.

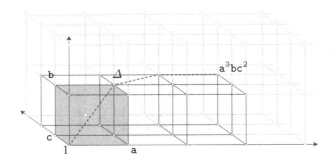

Fig. 1. The Cayley graph of $\mathbb{N}^n$ with respect to A_n, here for $n = 3$; we write $\mathsf{a}, \mathsf{b}, \mathsf{c}$ for $\mathsf{a}_1, \mathsf{a}_2, \mathsf{a}_3$, and Δ for Δ_3. The dark grey cube corresponds to the 8 elements of S_3. Then, the greedy decomposition of $\mathsf{a}^3\mathsf{b}\mathsf{c}^2$ corresponds to the dashed path: among the many possible ways of going from 1 to $\mathsf{a}^3\mathsf{b}\mathsf{c}^2$, we choose at each step the largest possible element of S_3 that divides the considered element, thus remaining in the light grey domain, which corresponds to the divisors of $\mathsf{a}^3\mathsf{b}\mathsf{c}^2$.

Prop. 1.3 is easy. It can be derived from the following (obvious) observation:

Lemma 1.6 *For every n, the divisibility relation of $\mathbb{N}^n$ is a lattice order, and S_n is a finite sublattice formed by the divisors of Δ_n, which are 2^n in number.*

We recall that a *lattice order* is a partial order in which every pair of elements admits a greatest lower bound and a lowest upper bound. When considering a divisibility relation, it is natural to use "least common multiple" (*lcm*) and "greatest common divisor" (*gcd*) for the least upper and greatest lower bounds.

Once Lemma 1.6 is available, Prop. 1.3 easily follows: indeed, starting from g, if g is not 1, there exists a maximal element s_1 of S_n dividing g, namely the left-gcd of g and Δ_n. So there exists g' satisfying $g = s_1g'$. If g' is not 1, we repeat with g', finding a maximal element s_2 of S_n dividing g', etc. The sequence $s_1|s_2|\cdots$ so obtained then satisfies the maximality condition of (1.4).

Remark 1.7 Let $\mathbb{Z}^n$ be the rank n free abelian group. Then $\mathbb{Z}^n$ is a group of (left) fractions for the monoid $\mathbb{N}^n$, meaning that every element of $\mathbb{Z}^n$ admits an expression $f^{-1}g$ with f, g in $\mathbb{N}^n$. It is easy to extend the greedy normal form of Prop. 1.3 into a unique normal form on the group $\mathbb{Z}^n$: indeed, every element of $\mathbb{Z}^n$ can be expressed as $\Delta_n^m g$ with m in $\mathbb{Z}$ and g in $\mathbb{N}^n$, hence it admits a decomposition $\Delta_n^m|s_1|\cdots|s_p$ with $s_1, ..., s_p$ in S_n satisfying (1.4). The latter is not unique in general, but it is if, in addition, one requires $s_1 \neq \Delta_n$.

1.2   Braid Monoids

Much less trivial, our second example involves braid monoids as investigated by F.A. Garside in [24]. For our current purpose, it is convenient to start with a presentation of the braid monoid B_n^+, namely

$$B_n^+ = \left\langle \sigma_1, ..., \sigma_{n-1} \ \middle| \ \begin{array}{l} \sigma_i\sigma_j = \sigma_j\sigma_i \quad \text{for } |i-j| \geq 2 \\ \sigma_i\sigma_j\sigma_i = \sigma_j\sigma_i\sigma_j \text{ for } |i-j| = 1 \end{array} \right\rangle^+. \tag{1.8}$$

The braid group B_n is the group which, as a group, admits the presentation (1.8). As shown by E. Artin in [3] (see, for instance, [7] or [18]), B_n interprets as the group of isotopy classes of n-strand braid diagrams, which are planar diagrams obtained by concatenating diagrams of the type

$$\sigma_i : \quad | \cdots | \times | \cdots |$$

$$\text{and} \quad \sigma_i^{-1} : \quad | \cdots | \times | \cdots |$$

with $1 \leq i \leq n-1$. When an n-strand braid diagram is viewed as the projection of n nonintersecting curves in a cylinder $D^2 \times \mathbb{R}$ as in Fig. 2, the relations of (1.8) correspond to the natural notion of a deformation, or *ambient isotopy*. Then, the monoid B_n^+ corresponds to isotopy classes of positive braid diagrams, meaning those diagrams in which all crossings have the orientation of σ_i.

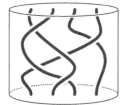 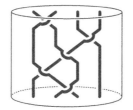 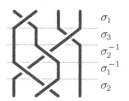

Fig. 2. Viewing an n-strand braid diagram (here $n = 4$) as the plane projection of a 3D-figure in a cylinder; on the right, by decomposing the diagram into elementary diagrams involving only one crossing, with two possible orientations, one obtains an encoding of an n-braid diagram by a word in the alphabet $\{\sigma_1^{\pm 1}, ..., \sigma_{n-1}^{\pm 1}\}$

   Defining unique normal forms for the elements of the monoid B_n^+ (called *positive n-strand braids*) is both easy and difficult. Indeed, by definition, every positive n-strand admits decompositions in terms of the letters $\sigma_1, ..., \sigma_{n-1}$, and, as in Subsec. 1.1, we obtain a distinguished expression by considering the lexicographically smallest expression. This, however, is *not* a good idea: the normal form so obtained is almost useless (nevertheless, see [2], building on unpublished work by Bronfman, for an application in combinatorics), mainly because there is no simple rule for obtaining the normal form of $\sigma_i g$ or $g\sigma_i$ from that of σ_i. In other words, one cannot *compute* the normal form easily.

A much better normal form can be obtained as follows. For each n, let Δ_n be the positive n-strand braid inductively defined by

$$\Delta_1 := 1, \quad \Delta_n := \Delta_{n-1}\sigma_{n-1}\cdots\sigma_2\sigma_1 \quad \text{for } n \geqslant 2, \tag{1.9}$$

corresponding to a (positive) half-turn of the whole family of n strands:

Next, let us call a positive braid *simple* if it can be represented by a positive diagram in which any two strands cross at most once. One shows that the latter property does not depend on the choice of the diagram. By definition, the trivial braid 1, and every braid σ_i is simple. We see above that Δ_n is also simple. Let S_n be the family of all simple n-strand braids.

As in Subsec. 1.1, let $\preccurlyeq$ be the *left*-divisibility relation of the monoid B_n^+: so $f \preccurlyeq g$ holds if, and only if, there exists g' (in B_n^+) satisfying $fg' = g$. For $n \geqslant 3$, the monoid B_n^+ is not abelian, so left-divisibility does not coincide in general with right-divisibility, defined symmetrically by $\exists g'\,(g'f = g)$.

Then, we have the following counterpart of Lemma 1.6:

Lemma 1.10 [24] *For every n, the left-divisibility relation of the monoid B_n^+ is a lattice order, and S_n is a finite sublattice formed by the left-divisors of Δ_n, which are $n!$ in number.*

Contrasting with Lemma 1.6, the proof of Lemma 1.10 is far from trivial. Why $n!$ appears here is easy to explain. Every n-strand braid g induces a well-defined permutation $\pi(g)$ of $\{1,...,n\}$, where $\pi(g)(i)$ is the initial position of the strand that finishes at position i in any diagram representing g. In this way, one obtains a surjective homomorphism from B_n to the symmetric group $\mathfrak{S}_n$. It turns out that, for every permutation f of $\{1,...,n\}$, there exists exactly one simple n-strand braid whose permutation is f: so simple braids (also called "permutation braids") provide a (set-theoretic) section for the projection of B_n to $\mathfrak{S}_n$, and they are $n!$ in number, see Fig. 3.

Once Lemma 1.10 is available, repeating the argument of Subsec. 1.1 (with some care) easily leads to

Proposition 1.11 [1,22] *Every element of B_n^+ admits a unique decomposition of the form $s_1|\cdots|s_p$ with $s_1,...,s_p$ in S_n satisfying $s_p \neq 1$, and, for every $i < p$,*

$$\forall s \in S_n\,(\,s_i \prec s \;\Rightarrow\; s \not\preccurlyeq s_i s_{i+1}\cdots s_p\,). \tag{1.12}$$

In other words, we obtain for every positive braid a unique greedy decomposition exactly similar to the one of Prop. 1.3.

Example 1.13 Consider $g = \sigma_2\sigma_3\sigma_2\sigma_2\sigma_1\sigma_2\sigma_3\sigma_3$ in B_4^+. First, by cutting when two strands that already crossed are about to cross for the second time, we obtain a decomposition into then three simple chunks $\sigma_2\sigma_3\sigma_2|\sigma_2\sigma_1\sigma_2\sigma_3|\sigma_3$. Next, we push

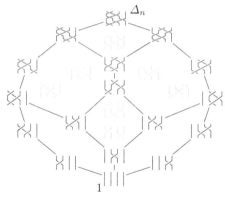

Fig. 3. The lattice $(\mathrm{Div}(\Delta_n), \preccurlyeq)$ formed by the $n!$ left-divisors of the braid Δ_n in the monoid B_n^+, here in the case $n = 4$: a copy of the n-permutahedron associated with the symmetric group $\mathfrak{S}_n$ equipped with what is called the weak order, see for instance [9]; topologically, this is an $n-2$-sphere tessellated by hexagons and squares which correspond to the relations of (1.8)

the crossings upwards as much as possible, we obtain $\sigma_2\sigma_3\sigma_2\sigma_1|\sigma_2\sigma_1\sigma_3|\sigma_3$, and finally $\sigma_1\sigma_2\sigma_3\sigma_2\sigma_1|\sigma_2\sigma_1\sigma_3$, as in the diagram below. with only two entries.

We cannot go farther, so the decomposition is greedy.

Remark 1.14 Here again, the greedy normal form extends from the monoid to the group. It turns out that B_n is a group of fractions for B_n^+, and that Δ_n is a sort of universal denominator for the elements of the group, meaning that every element of B_n can be expressed as $\Delta_n^m g$ with m in $\mathbb{Z}$ and g in B_n^+. As above, it follows that every element of B_n admits a unique decomposition $\Delta_n^m|s_1|\cdots|s_p$ with m in $\mathbb{Z}$, $s_1, \ldots, s_p$ in S_n satisfying (1.4) and, in addition, $s_1 \neq \Delta_n$.

2   The Δ-Normal Form in a Garside Monoid

In the direction of more generality, we now explain how to unify the examples of Sec. 1 into the notion of a Δ-normal form associated with a Garside element in what is now classically called a Garside monoid.

2.1   Garside Monoids

The greedy normal form of the braid monoid B_n^+ has been extended to other similar monoids a long time ago. Typically, an *Artin–Tits monoid*, which, by definition, is a monoid defined by relations of the form $stst... = tsts...$ where both sides have the same length, is called *spherical* if the Coxeter group obtained by making all generators involutive, that is, adding $s^2 = 1$ for each generator s, is finite. For instance, (1.8) shows that B_n^+ is an Artin–Tits monoid, whose associated Coxeter group is the finite group $\mathfrak{S}_n$, so B_n^+ is spherical. Then, building on [10], it was shown in [12] that all properties of the greedy normal form of braid monoids extend to spherical Artin–Tits monoids.

A further extension came with the notion of a *Garside monoid* (and of a *Garside group*) introduced in [14] and slightly generalised in [13]. Recall that a monoid M is *left-cancellative* (resp. *right-cancellative*) if $fg = fg'$ (resp. $gf = g'f$) implies $g = g'$, and *cancellative* if it is both left- and right-cancellative. As above, for f, g in M, we say that f is a *left-divisor* of g, or, equivalently, that g is a *right-multiple* of f, written $f \preccurlyeq g$, if there exists g' in M satisfying $fg' = g$. We write $f \prec g$ for $f \preccurlyeq g$ with $g \not\preccurlyeq f$ (which amounts to $f \neq g$ if M has no nontrivial invertible element). For $g'f = g$, we symmetrically say that f is a *right-divisor* of g, or, equivalently, that g is a *left-multiple* of f. Note that the set gM of all right-multiples of g is the right-ideal generated by g, and, similarly, the set Mg of all right-multiples of g is the left-ideal generated by g, as involved in the definition of Green's relations of M, see for instance [28].

Definition 2.1 A *Garside monoid* is a pair (M, Δ), where M is a cancellative monoid satisfying the following conditions:

(i) There exists $\lambda : M \to \mathbb{N}$ satisfying, for all f, g,

$$\lambda(fg) \geqslant \lambda(f) + \lambda(g) \qquad \text{and} \qquad g \neq 1 \Rightarrow \lambda(g) \neq 0.$$

(ii) Any two elements of M admit left- and right-lcms and gcds.

(iii) Δ is a *Garside element* of M, meaning that the left- and right-divisors of Δ coincide and generate M,

(iv) The family $\mathrm{Div}(\Delta)$ of all divisors of Δ in M is finite.

Note that condition (i) in Def. 2.1 implies in particular that the monoid M has no nontrivial invertible element (meaning: not equal to 1): indeed, $1 = 1 \cdot 1$ implies $\lambda(1) = \lambda(1 \cdot 1) \geqslant \lambda(1) + \lambda(1)$, hence $\lambda(1) = 0$, so $fg = 1$ implies $0 \geqslant \lambda(f) + \lambda(g)$, whence $f = g = 1$.

Example 2.2 For every n, the pair $(\mathbb{N}^n, \Delta_n)$, with Δ_n as defined in (1.2), is a Garside monoid. Indeed, $\mathbb{N}^n$ is cancellative, we can define $\lambda(g)$ to be the common length of all A_n-words representing an element g of $\mathbb{N}^n$ and, according to Lemma 1.6, the left- and right-divisibility relations (which coincide since $\mathbb{N}^n$ is abelian) are lattice orders. Finally, Δ_n is a Garside element, since its divisors include A_n, and the family $\mathrm{Div}(\Delta_n)$ is finite, since it has 2^n elements.

Similarly, the pairs (B_n^+, Δ_n), with Δ_n now defined by (1.9), is a Garside monoid as well. That B_n^+ is cancellative is proved in [24], for $\lambda(g)$ we can take again the common length of all braid words representing g, and Lemma 1.10 provides the remaining conditions.

In the same vein, it can be shown that, if M is a spherical Artin–Tits monoid and Δ is the lifting of the longest element w_0 of the associated finite Coxeter group, then (M, Δ) is a Garside monoid. Actually, many more examples are known. Let us mention two.

Example 2.3 First, let B_n^{*+}, the *dual braid monoid*, be the submonoid of the braid group B_n generated by the $n(n-1)/2$ braids

$$a_{i,j} := \sigma_{j-1} \cdots \sigma_{i+1} \sigma_i \sigma_{i+1}^{-1} \cdots \sigma_{j-1}^{-1} \qquad \text{with } 1 \leqslant i < j \leqslant n,$$

and let $\Delta_n^* := a_{1,2} a_{2,3} \cdots a_{n-1,n}$ [8]. Then (B_n^{*+}, Δ_n^*) is a Garside monoid, and B_n is its group of fractions (which shows that a group may be the group of fractions of several Garside monoids). Note that B_n^{*+} includes B_n^+, since $\sigma_i = a_{i,i+1}$ holds. The inclusion is strict for $n \geqslant 3$, since $a_{1,3}$ is not a positive braid. The lattice of the divisors of Δ_n^* has $\frac{1}{n+1}\binom{2n}{n}$ elements, which are in one–one correspondence with the noncrossing partitions of $\{1, ..., n\}$ [4].

Second, for $n \geqslant 1$ and $e_1, ..., e_n \geqslant 2$, let

$$T_{e_1,...,e_n}^+ := \langle a_1, ..., a_n \mid a_1^{e_1} = a_2^{e_2} = \cdots = a_n^{e_n} \rangle^+.$$

Define Δ to be the common value of $a_i^{e_i}$ for all i. Then $(T_{e_1,...,e_n}^+, \Delta)$ is a Garside monoid. The lattice $\mathrm{Div}(\Delta)$ has $e_1 + \cdots + e_n - n + 2$ elements and it consists of n disjoint chains of lengths $e_1, ..., e_n$ connecting 1 to Δ.

As one can expect, a greedy normal form exists in every Garside monoid:

Proposition 2.4 *Assume that (M, Δ) is a Garside monoid. Say that a $\mathrm{Div}(\Delta)$-word $s_1 | \cdots | s_p$ is Δ-normal if, for every $i < p$, we have*

$$\forall s \in \mathrm{Div}(\Delta) \ (s_i \prec s \ \Rightarrow \ s \nmid s_i s_{i+1} \cdots s_p), \tag{2.5}$$

and that it is strict if, in addition, $s_p \neq 1$ holds. Then every element of M admits a unique strict Δ-normal decomposition.

Proof. We show, using induction on ℓ, that every element g of M satisfying $\lambda(g) \leqslant \ell$ admits a strict Δ-normal decomposition. For $\ell = 0$, the only possibility is $g = 1$, and then the empty sequence is a Δ-normal decomposition of g. Assume $\ell \geqslant 1$, and let g satisfy $\lambda(g) \leqslant \ell$. The case $g = 1$ has already been considered, so we can assume $g \neq 1$. Let s_1 be the left-gcd of g and Δ. As $\mathrm{Div}(\Delta)$ generates M, some nontrivial divisor of Δ must left-divide g, so $s_1 \neq 1$ holds. As M is left-cancellative, there is a unique element g' satisfying $g = s_1 g'$. By assumption, one has $\lambda(g') \leqslant \lambda(g) - \lambda(s_1) < \ell$. Then, by the induction hypothesis, g' admits a strict Δ-normal decomposition $s_2 | \cdots | s_p$. Then one easily checks that $s_1 | s_2 | \cdots | s_p$ is a strict Δ-normal decomposition for g.

As for uniqueness, it is easy to see that the first entry of any greedy decomposition of g must be s_1, and then we apply the induction hypothesis again. □

Note that, by definition, if $s_1 | \cdots | s_p$ is a Δ-normal word, then so is every word of the form $s_1 | \cdots | s_p | 1 | \cdots | 1$: uniqueness is guaranteed only when we forbid final entries 1.

2.2 Computing the Δ-Normal Form

Prop. 2.4 is an existential statement, which does not directly solves the question of practically computing a Δ-normal decomposition of an element given by an arbitrary $\mathrm{Div}(\Delta)$-word. It turns out that simple incremental methods exist, which explains the interest of the Δ-normal form.

We begin with preliminary results. Their proofs are not very difficult and we give them as they are typical of what can be called the "Garside approach".

Lemma 2.6 *Assume that (M, Δ) is a Garside monoid. For g in M, define $H(g)$ to be the left-gcd of g and Δ. Then, for all $s_1, ..., s_p$ in $\mathrm{Div}(\Delta)$, the following conditions are equivalent:*

(i) The sequence $s_1|\cdots|s_p$ is Δ-normal.
(ii) For every $i < p$, one has $s_i = H(s_i s_{i+1} \cdots s_p)$.
(iii) For every $i < p$, one has $s_i = H(s_i s_{i+1})$.

Proof. Assume that $s_1|\cdots|s_p$ is Δ-normal. By definition, we have $s_i \preccurlyeq \Delta$ and $s_i \preccurlyeq s_i s_{i+} \cdots s_p$, whence $s_i \preccurlyeq H(s_i s_{i+1} \cdots s_p)$ by the definition of a left-gcd. Conversely, let s be the right-lcm of s_i and $H(s_i s_{i+1} \cdots s_p)$. As we have $s_i \preccurlyeq \Delta$ and $H(s_i s_{i+1} \cdots s_p) \preccurlyeq \Delta$, the definition of a right-lcm implies $s \preccurlyeq \Delta$. Hence, we have $s_1 \preccurlyeq s \in \mathrm{Div}(\Delta)$ and $s \preccurlyeq s_i s_{i+1} \cdots s_p$, so (2.5) implies that $s_i \prec s$ is impossible. Therefore, we must have $s = s_i$, meaning that $H(s_i s_{i+1} \cdots s_p)$ left-divides s_i. We deduce $s_i = H(s_i s_{i+1} \cdots s_p)$. Therefore, (i) implies (ii).

The proof that (i) implies (iii) is exactly similar, replacing $s_i \cdots s_p$ with $s_i s_{i+1}$.

Conversely, (ii) implies $\forall s \in \mathrm{Div}(\Delta)\, (s \preccurlyeq s_i s_{i+1} \cdots s_p \Rightarrow s \preccurlyeq s_i)$, whence a fortiori $\forall s \in \mathrm{Div}(\Delta)\, (s \preccurlyeq s_i s_{i+1} \cdots s_p \Rightarrow s_i \not\prec s)$, which is equivalent to (2.5). So (i) and (ii) are equivalent.

Next, by the definition of the left-gcd, $s_i \preccurlyeq H(s_i s_{i+1}) \preccurlyeq H(s_i s_{i+1} \cdots s_p)$ is always true, so (ii) implies (iii).

Finally, let us assume (iii) and prove (ii). We give the argument for $p = 3$, the general case then follows by an induction on p. So, we assume $s_1 = H(s_1 s_2)$ and $s_2 = H(s_2 s_3)$, and want to prove $s_1 = H(s_1 s_2 s_3)$. The nontrivial argument uses the *right-complement* operation. By assumption, any two elements f, g of M admit a right-lcm, say h. Let us denote by $f\backslash g$ and $g\backslash f$ the (unique) elements satisfying $f(f\backslash g) = g(g\backslash f) = h$. Using the associativity of the right-lcm operation, one checks that the right-complement operation $\backslash$ obeys the law

$$(gh)\backslash f = (g\backslash f)\backslash h. \tag{2.7}$$

Now, let s be an element of $\mathrm{Div}(\Delta)$ left-dividing $s_1 s_2 s_3$. Our aim is to show that s left-divides s_1. By assumption, the right-lcm of s and $s_1 s_2 s_3$ is $s_1 s_2 s_3$, so we have $(s_1 s_2 s_3)\backslash s = 1$. Applying (2.7) with $f = s$, $g = s_1$, and $h = s_2 s_3$, we deduce $(s_1\backslash s)\backslash(s_2 s_3) = 1$, which means that $s_1\backslash s$ left-divides $s_2 s_3$. Because s_1 and s divide Δ, so does their right-lcm t, hence so does also their right-complement $s_1\backslash s$, which is a right-divisor of t. So we have $s_1\backslash s \preccurlyeq s_2 s_3$, and the assumption $s_2 = H(s_2 s_3)$ then implies $s_1\backslash s \preccurlyeq s_2$. Arguing back, we deduce that $s_1 s_2$ is the right-lcm of s and $s_1 s_2$, that is, that s left-divides $s_1 s_2$. From there, the assumption $s_1 = H(s_1 s_2)$ implies that s left-divides s_1, as expected. □

Lemma 2.6 is not yet sufficient to compute a Δ-normal decomposition for an arbitrary element, but it already implies an important property:

Proposition 2.8 *If (M, Δ) is a Garside monoid, then Δ-normal words form a regular language.*

Proof. By assumption, the family $\mathrm{Div}(\Delta)$ is finite. Moreover, by Lemma 2.6, a word $s_1|\cdots|s_p$ is Δ-normal if, and only if, each length-two factor $s_i|s_{i+1}$ is Δ-normal. Hence, the language of all Δ-normal words is defined over the alphabet $\mathrm{Div}(\Delta)$ by the exclusion of finitely many patterns, namely the pairs $s_i|s_{i+1}$ satisfying $s_i \neq H(s_i s_{i+1})$. Hence it is a regular language.

We now specifically consider the normalisation of $\mathrm{Div}(\Delta)$-words of length two.

Lemma 2.9 *If (M, Δ) is a Garside monoid, then, for all s_1, s_2 in $\mathrm{Div}(\Delta)$, the element $s_1 s_2$ has a unique Δ-normal decomposition of length two.*

Proof. Let s_1, s_2 belong to $\mathrm{Div}(\Delta)$. Let $s_1' := H(s_1 s_2)$. As M is left-cancellative, there exists a unique element s_2' satisfying $s_1' s_2' = s_1 s_2$. By construction, s_1 left-divides s_1', which implies that s_2' right-divides s_2, hence a fortiori Δ. So s_2' lies in $\mathrm{Div}(\Delta)$. By construction, the $\mathrm{Div}(\Delta)$-word $s_1'|s_2'$ is Δ-normal, and it is a decomposition of $s_1 s_2$. Three cases are possible: if s_2' is not 1, then $s_1'|s_2'$ is a strict Δ-normal decomposition of $s_1 s_2$ (which is then said to have Δ-length two); otherwise, $s_1 s_2$ is a divisor of Δ, so s_1' is a strict Δ-normal decomposition of $s_1 s_2$ (which is then said to have Δ-length one), unless s_1' is also 1, corresponding to $s_1 = s_2 = 1$, where the strict Δ-normal decomposition is empty (and $s_1 s_2$, which is 1, is said to have Δ-length zero). □

The previous result and the subsequent arguments become (much) more easily understandable when illustrated with diagrams. To this end, we associate to every element g of the considered monoid a g-labeled edge $\xrightarrow{\ g\ }$. We then associate with a product the concatenation of the corresponding edges (which amounts to viewing the ambient monoid as a category), and naturally represent equalities in the ambient monoid using commutative diagrams: for instance, the square on the right illustrates an equality $fg = f'g'$.

Next, assuming that a set of elements S is given and some distinguished subset L of $S^{[2]}$ has been fixed, typically length-two normal words of some sort, we shall indicate that a length-two S-word $s_1|s_2$ belongs to L (that is, "is normal") by connecting the corresponding edges with a small arc, as in $\xrightarrow{\ s_1\ }\!\frown\!\xrightarrow{\ s_2\ }$. Then, Lemma 2.9 is illustrated in the diagram on the right: it says that, for all given s_1, s_2 in $\mathrm{Div}(\Delta)$ (solid arrows), there exist s_1', s_2' in $\mathrm{Div}(\Delta)$ (dashed arrows) such that $s_1'|s_2'$ is Δ-normal and the diagram is commutative.

The second ingredient needed for computing the normal form involves what the call the (left) domino rule.

Definition 2.10 Assume that M is a left-cancellative monoid, S is a subset of M, and L is a family of S-words of length two. We say that the *left domino rule is valid for* L if, whenever $s_1, s_2, s_1', s_2', t_0, t_1, t_2$ lie in S and $s_1' t_1 = t_0 s_1$ and $s_2' t_2 = t_1 s_2$ hold in M, then the assumption that $s_1|s_2$, $s_1'|t_1$, and $s_2'|t_2$ lie in L implies that $s_1'|s_2'$ lies in L as well.

The left domino rule corresponds to the diagram on the right: the solid arcs are the assumptions, namely that $s_1'|t_1$, $s_2'|t_2$ and $s_1|s_2$ lie in L, the dotted arc is the expected conclusion, namely that $s_1'|s_2'$ does.

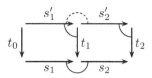

Lemma 2.11 *If* (M, Δ) *is a Garside monoid, then the left domino rule is valid for* Δ-*normal words of length two.*

Proof. Assume that $s_1, s_2, s_1', s_2', t_0, t_1, t_2$ lie in S and satisfy the assumptions of Def. 2.10 (with respect to Δ-normal words of length two). We want to show that $s_1'|s_2'$ is Δ-normal. In view of Lemma 2.6, assume $s \in \mathrm{Div}(\Delta)$ and $s \preccurlyeq s_1' s_2'$.
Then, we have $s \preccurlyeq s_1' s_2' t_2$, whence $s \preccurlyeq t_0 s_1 s_2$, see the diagram on the right. Arguing as in the proof of Lemma 2.6, we deduce $t_0 \backslash s \preccurlyeq s_1 s_2$. As $t_0 \backslash s$ lies in $\mathrm{Div}(\Delta)$, we deduce $t_0 \backslash s \preccurlyeq H(s_1 s_2) = s_1$, whence $s \preccurlyeq t_0 s_1 = s_1' t_1$. As s lies in $\mathrm{Div}(\Delta)$, we deduce $s \preccurlyeq H(s_1' t_1) = s_1'$. Therefore, we have $s_1' = H(s_1' s_2')$, and $s_1'|s_2'$ is Δ-normal.   □

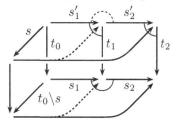

We can now easily compute a Δ-normal decomposition for every element. In order to describe the procedure (which can be translated into an algorithm directly), we work with $\mathrm{Div}(\Delta)$-words and, starting from an arbitrary $\mathrm{Div}(\Delta)$-word w, explain how to build a Δ-normal word that represents the same element. To this end, we first introduce notations that will be used throughout the paper.

Notation 2.12 (i) If S is a set and F is a map from $S^{[2]}$ to itself, then, for $i \geqslant 1$, we denote by F_i the (partial) map of S^* to itself that consists in applying F to the entries in position i and $i+1$. If $u = i_1|\cdots|i_n$ is a finite sequence of positive integers, we write F_u for the composite map $F_{i_n} \circ \cdots \circ F_{i_1}$ (so F_{i_1} is applied first).

(ii) If S is a set and N is a map from S^* to itself, we denote by $\overline{N}$ the restriction of N to $S^{[2]}$.

Then the main result about the Δ-normal form can be stated as follows:

Proposition 2.13 *Assume that* (M, Δ) *is a Garside monoid. Then, for every* $\mathrm{Div}(\Delta)$-*word* w *of length* p, *there exists a unique* Δ-*normal word* $N^\Delta(w)$ *of length* p *that represents the same element as* w. *Moreover, one has*

$$N^\Delta(w) = \overline{N^\Delta_{\delta_p}}(w), \qquad (2.14)$$

with δ_p *inductively defined by* $\delta_2 := 1$ *and* $\delta_p := \mathrm{sh}(\delta_{p-1})|1|2|\cdots|p-1$, *where* sh *is a shift of all entries by* $+1$.

Thus, for instance, (2.14) says that, in order to normalise a $\mathrm{Div}(\Delta)$-word of length four, we can successively normalise the length-two factors beginning at positions 3, 2, 3, 1, 2, and 3, thus in six steps.

Proof. We begin with an auxiliary result, namely finding a Δ-normal decomposition for a word of the form $t|s_1|\cdots|s_p$ where $s_1|\cdots|s_p$ is Δ-normal, that is, for multiplying a Δ-normal word by one more letter on the left. Then we claim that, putting $t_0 := t$ and, inductively, $s_i'|t_i := \overline{N^\Delta}(t_{i-1}|s_i)$ for $i = 1,\dots,p$, provides a Δ-normal decomposition of length $p+1$ for $ts_1\cdots s_p$. Indeed, the commutativity of the diagram in Fig. 4 gives the equality $ts_1\cdots s_p = s_1'\cdots s_p't_p$ in M, and the validity of the left domino rule implies that each pair $s_i'|s_{i+1}'$ is Δ-normal. In terms of $\overline{N^\Delta}$, we deduce the equality

$$N^\Delta(t|w) = \overline{N^\Delta}_{1|2|\cdots|p}(t|w). \qquad (2.15)$$

when w is a Δ-normal word of length p. From there, (2.5) follows by a straightforward induction. $\qquad\square$

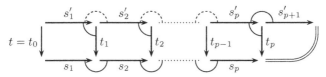

Fig. 4. Left-multiplying a Δ-normal word by an element of $\mathrm{Div}(\Delta)$: the left domino rule guarantees that the upper row is Δ-normal whenever the lower row is

The explicit description of Prop. 2.13 enables one to completely analyse the complexity of the Δ-normal form.

Corollary 2.16 *If (M, Δ) is a Garside monoid, then Δ-normal decompositions can be computed in linear space and quadratic time. The Word Problem for M with respect to $\mathrm{Div}(\Delta)$ lies in* DTIME(n^2).

Proof. By assumption, the set $\mathrm{Div}(\Delta)$ is finite, so the complete table of the map $\overline{N^\Delta}$ can be precomputed, and then each application of $\overline{N^\Delta}$ has time cost $O(1)$ and keeps the length unchanged. Then, as the sequence δ_p has length $p(p-1)/2$, Prop. 2.13 implies that a Δ-normal decomposition for an element represented by a $\mathrm{Div}(\Delta)$-word of length p can be obtained in time $O(p^2)$.

Computing Δ-normal decompositions yields a solution of the Word Problem, since two $\mathrm{Div}(\Delta)$-words w, w' represent the same element of M if, and only if, the Δ-normal words $N^\Delta(w)$ and $N^\Delta(w')$ only differ by the possible adjunction of final letters 1.

On the other hand, a direct application of (2.15) and Fig. 4 is the fact that, viewed as paths in the Cayley graph of M with respect to $\mathrm{Div}(\Delta)$, the Δ-normal forms of g and tg remain at a uniformly bounded distance, namely at most two. Thus, we can state (see [27]):

Corollary 2.17 *If (M, Δ) is a Garside monoid, then the Δ-normal words satisfy the 2-Fellow traveller Property on the left.*

2.3   The Right Counterpart

Owing to Prop. 2.13 and its normalisation recipe based on left-multiplication, the question naturally arises of a similar recipe based on right-multiplication, hence based on computing a Δ-normal decomposition of gt from one of g. Such a recipe does exist, but this is not obvious, because the definition of Δ-normality is not symmetric.

Proposition 2.18 *Assume that (M, Δ) is a Garside monoid. Then, for every* $\mathrm{Div}(\Delta)$-*word w of length p, one also has*

$$N^{\Delta}(w) = \overline{N^{\Delta}_{\widetilde{\delta}_p}}(w), \tag{2.19}$$

with $\widetilde{\delta}_p$ inductively defined by $\widetilde{\delta}_2 := 1$ and $\widetilde{\delta}_p := \widetilde{\delta}_{p-1}|p-1|\cdots|2|1$.

For instance, (2.19) says that, in order to normalise a $\mathrm{Div}(\Delta)$-word of length four, we can successively normalise the length-two factors beginning at positions 1, 2, 1, 3, 2, and 1, in six steps as in (2.14), but in a different order.

As can be expected, the proof of Prop. 2.18 relies on a symmetric counterpart of the left domino rule of Def. 2.10.

Definition 2.20 Assume that M is a left-cancellative monoid, S is a subset of M, and L is a family of S-words of length two. We say that the *right domino rule is valid for L* if, whenever $s_1, s_2, s_1', s_2', t_0, t_1, t_2$ lie in S and $t_0 s_1' = s_1 t_1$ and $t_1 s_2' = s_2 t_2$ hold in M, then the assumption that $s_1|s_2$, $t_0|s_1'$, and $t_1|s_2'$ lie in L implies that $s_1'|s_2'$ lies in L as well.

The right domino rule corresponds to the diagram on the right: the solid arcs are the assumptions, namely that $t_0|s_1'$, $t_1|s_2'$ and $s_1|s_2$ lie in L, and the dotted arc is the expected conclusion, namely that $s_1'|s_2'$ does.

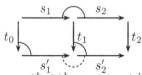

Then we have the counterpart of Lemma 2.11. Observe that the argument is totally different, reflecting the lack of symmetry in the definition of Δ-normality.

Lemma 2.21 *If (M, Δ) is a Garside monoid, then the right domino rule is valid for Δ-normal words of length two.*

Proof (sketch). For s in $\mathrm{Div}(\Delta)$, let ∂s be the element of $\mathrm{Div}(\Delta)$ satisfying $s\partial s = \Delta$, and let $\phi := \partial^2$. Then, $s\Delta = \Delta\phi(s)$ holds for every s in $\mathrm{Div}(\Delta)$, and one shows that ϕ extends to an automorphism of M. It follows, in particular, that $s_1|s_2$ being Δ-normal implies that $\phi(s_1)|\phi(s_2)$ is Δ-normal as well. By assumption, there exist t_0', t_1', t_2' satisfying $t_0 t_0' = t_1 t_1' = t_2 t_2' = \Delta$. By the above equality, we have $s_1\Delta = \Delta\phi(s_1)$ and $s_2\Delta = \Delta\phi(s_2)$, whence, by left-cancellation, $s_1' t_1' = t_0' \phi(s_1)$ and $s_2' t_2' = t_1' \phi(s_1)$. Thus the diagram below is commutative. Assume $s \preccurlyeq s_1' s_2'$ with s in $\mathrm{Div}(\Delta)$. Let $s' := s_1' \backslash s$. Our aim is to prove that s left-divides s_1', that is, that s' is 1.

The assumption that s left-divides $s_1' s_2'$ implies $s' \preccurlyeq s_2'$, whence $t_1 s' \preccurlyeq t_1 s_2'$. On the other hand, the assumption $s \preccurlyeq s_1' s_2'$ implies a fortiori $s \preccurlyeq s_1' s_2' t_2'$, that is, $s \preccurlyeq t_0' \phi(s_1) \phi(s_2)$ and, therefore, $t_0' \backslash s \preccurlyeq \phi(s_1) \phi(s_2)$. As $t_0' \backslash s$ lies in $\mathrm{Div}(\Delta)$ and $\phi(s_1) | \phi(s_2)$ is Δ-normal, we deduce $t_0' \backslash s \preccurlyeq \phi(s_1) \phi(s_2)$, whence $s \preccurlyeq t_0' \phi(s_1)$, which is also $s \preccurlyeq s_1' t_1'$.

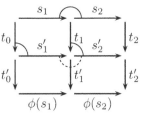

We deduce $s' \preccurlyeq t_1'$, and, therefore, $t_1 s' \preccurlyeq t_1 t_1' = \Delta$. Thus $t_1 s'$ lies in $\mathrm{Div}(\Delta)$ and it left-divides $t_1 s_2'$. By assumption, $t_1 | s_2'$ is Δ-normal, so we deduce that $t_1 s'$ left-divides t_1, implying $s' = 1$, as expected. Hence, $s_1' | s_2'$ is Δ-normal, the right domino rule is valid.   □

We can easily complete the argument.

Proof (of Prop. 2.18). The argument is symmetric of the one for Prop. 2.13. It consists in establishing for w a Δ-normal word of length p the equality

$$N^\Delta(w|t) = \overline{N^\Delta_{p|p-1|\cdots|1}}(w). \tag{2.22}$$

The latter immediately follows from the diagram

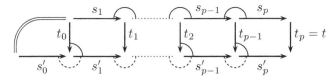

whose validity is guaranteed by the right domino rule.   □

We deduce a counterpart of Cor. 2.17.

Corollary 2.23 *Assume that (M, Δ) is a Garside monoid. Then Δ-normal words satisfy the 2-Fellow traveller Property on the right.*

Remark 2.24 As in Subsections 1.1 and 1.2, the Δ-normal decompositions associated with a Garside monoid (M, Δ) extend to the group of fractions of M. It directly follows from the definition that M satisfies the Ore conditions (cancellativity and existence of common right-multiples), hence embeds in a group of (left) fractions G (then called a *Garside group*). Then every element of G admits a unique decomposition of the form $\Delta^m | s_1 | \cdots | s_p$ where $s_1 | \cdots | s_p$ is Δ-normal and, in addition, we require $s_1 \neq \Delta$ (that is, m is maximal) and $s_p \neq 1$ (that is, p is minimal). It is easy to deduce from Prop. 2.8 and Cor. 2.17 and 2.23 that the Δ-normal form provides a biautomatic structure for G (in the sense of [23]).

3   The S-Normal Form Associated with a Garside Family

Looking at the mechanism of the Δ-normal form associated with a Garside monoid invites to a further extension. Indeed, one quickly sees that several assumptions in the definition of a Garside monoid are not used in the construction of a normal form that obeys the recipe of Prop. 2.13. This easy observation, and the need of using decompositions similar to Δ-normal ones in more general situations, led to introducing the notion of a *Garside family* [15], extensively developed in the book [17]. Also see [16] for the computational aspects.

3.1   The Notion of a Garside Family

Our aim is to define normal forms that work in the same way as the Δ-normal form of a Garside monoid, but in more general monoids (in fact, monoids can be extended into categories at no cost). So, we start with a monoid M equipped with a generating family S and try to define for the elements of M distinguished S-decompositions that resemble Δ-normal decompositions: in particular, if (M, Δ) is a Garside monoid and S is $\mathrm{Div}(\Delta)$, we should retrieve Δ-normal decompositions. Of course, we cannot expect to do that for an arbitrary generating family S, and this is where the notion of a Garside family will appear. First, if we try to just copy (2.5), problems arise. Therefore, we start from a new notion.

Definition 3.1 If M is a left-cancellative monoid and S is included in M, an S-word $s_1|s_2$ is called S-*normal* if the following condition holds:

$$\forall s \in S \; \forall f \in M \; (s \preccurlyeq f s_1 s_2 \Rightarrow s \preccurlyeq f s_1). \tag{3.2}$$

An S-word $s_1|\cdots|s_p$ is called S-*normal* if $s_i|s_{i+1}$ is S-normal for every $i < p$, and *strict* S-*normal* if it is S-normal with, in addition, $s_p \neq 1$.

Relation (3.2) is reminiscent of (iii) in Lemma 2.6, but with the important difference of the additional term f: we do not only consider the left-divisors of $s_1 s_2$ that lie in S, but, more generally, all elements of S that left-divide $f s_1 s_2$.

Example 3.3 Assume that (M, Δ) is a Garside monoid, and let $S := \mathrm{Div}(\Delta)$. Assume that $s_1|\cdots|s_p$ is S-normal in the sense of Def. 3.1. Then, for every i, (3.2) implies in particular $\forall s \in S \; (s \preccurlyeq s_1 s_2 \Rightarrow s \preccurlyeq s_1)$, whence $s_i = H(s_i s_{i+1})$. Hence, by Lemma 2.6, $s_1|\cdots|s_p$ is Δ-normal in the sense of Prop. 2.4.

The converse implication is also true, but less obvious. Indeed, assume that $s_1|\cdots|s_p$ is Δ-normal, and we have $s \preccurlyeq f s_i s_{i+1}$ for some s in $\mathrm{Div}(\Delta)$ and f in M. Then, using the right-complement operation $\backslash$ as in the proof of Lemma 2.6, we deduce that $f\backslash s$ left-divides $s_i s_{i+1}$, as illustrated in the diagram

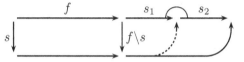

As $s_i|s_{i+1}$ is Δ-normal, we deduce $f\backslash s \preccurlyeq s_i$, whence $s \preccurlyeq f s_i$. Hence (3.2) is satisfied and $s_i|s_{i+1}$ is S-normal in the sense of Def. 3.1.

By definition, being S-normal is a local property only involving length-two subfactors. So we immediately obtain:

Proposition 3.4 *If M is a monoid and S is a finite subfamily of M, then S-normal words form a regular language.*

Hereafter we investigate S-normal decompositions. An easy, but important fact is that such decompositions are necessarily (almost) unique when they exist. We shall restrict to the case of monoids that admit no nontrivial invertible element (as Garside monoids do). This restriction is not necessary, but it makes statements more simple: essentially, one can cope with nontrivial invertible elements at the expense of replacing equality with the weaker equivalence relation $=^{\times}$, where $g =^{\times} g'$ means $g = ge$ for some invertible element e, see [17].

Lemma 3.5 *Assume that M is a left-cancellative monoid with no nontrivial invertible elements and S is included in M. Then every element of g admits at most one strict S-normal decomposition.*

Proof (sketch). Assume that $s_1|\cdots|s_p$ and $t_1|\cdots|t_q$ are S-normal decompositions of an element g. From the assumption that s_1 lies in S and left-divides $t_1\cdots t_q$, and that $t_1|\cdots|t_q$ is S-normal, one easily deduces $s_1 \preccurlyeq t_1$. By a symmetric argument, one deduces $t_1 \preccurlyeq s_1$, whence $s_1 = t_1$, because M has non nontrivial invertible element. Then use an induction. □

If we consider S-normal decompositions that are not strict, uniqueness is no longer true as, trivially, $s_1|\cdots|s_p$ being S-normal implies that $s_1|\cdots|s_p|1|\cdots|1$ is also S-normal (and represents the same element of the ambient monoid). Lemma 3.5 says that this is the only lack of uniqueness.

At this point, we are left with the question of the existence of S-normal decompositions, and this is where the central technical notion arises:

Definition 3.6 Assume that M is a left-cancellative monoid with no nontrivial invertible elements and S is a subset of M that contains 1. We say that S is a *Garside family in M* if every element g of M has an S-normal decomposition, that is, there exists an S-normal S-word $s_1|\cdots|s_p$ satisfying $s_1\cdots s_p = g$.

Example 3.7 It follows from the connection of Ex. 3.3 that, if $(M, \mathrm{Div}(\Delta)$ is a Garside monoid, then $\mathrm{Div}(\Delta)$ is a Garside family in M. So, in particular, the n-cube S_n of (1.1) is a Garside family in the abelian monoid $\mathbb{N}^n$ and, similarly, the family of all simple n-strand braids is a Garside family in the braid monoid B_n^+.

Many examples of a different flavour exist. For instance, *every* left-cancellative monoid M is a Garside family in itself, since every element g of M admits the length-one M-normal decomposition g (!). More interestingly, let K^+ be the "Klein bottle monoid"

$$K^+ := \langle \mathsf{a}, \mathsf{b} \mid \mathsf{a} = \mathsf{bab} \rangle^+,$$

which is the positive cone in the ordered group $\langle \mathsf{a}, \mathsf{b} \mid \mathsf{a} = \mathsf{bab} \rangle$, itself the fundamental group of the Klein bottle, and the nontrivial semidirect product $\mathbb{Z} \rtimes \mathbb{Z}$. Then K^+ cannot be made a Garside monoid since no function λ as in

Def. 2.1(i) may exist. However, if we put $\Delta := \mathsf{a}^2$, the left- and right-divisors of Δ coincide and the family $\mathrm{Div}(\Delta)$ is an (infinite) Garside family in K^+.

We refer to [17] for many examples of Garside families, and just mention the recent result [20] that every finitely generated Artin–Tits monoid admits a *finite* Garside family, independently of whether the associated Coxeter group is finite or not. In Fig. 5, we display such a finite Garside family for the monoid with presentation $\langle \sigma_1, \sigma_2, \sigma_3 \mid \sigma_1\sigma_2\sigma_1 = \sigma_2\sigma_1\sigma_2, \sigma_2\sigma_3\sigma_2 = \sigma_3\sigma_2\sigma_3, \sigma_3\sigma_1\sigma_3 = \sigma_1\sigma_3\sigma_1 \rangle^+$, that is, for what is called type $\widetilde{A}_2$.

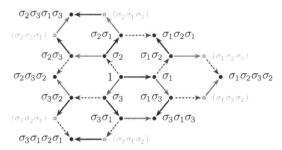

Fig. 5. A finite Garside family S in the Artin–Tits monoid of type $\widetilde{A}_2$: the sixteen right-divisors of the elements $\sigma_1\sigma_2\sigma_3\sigma_2$, $\sigma_2\sigma_3\sigma_1\sigma_3$, and $\sigma_3\sigma_1\sigma_3\sigma_1$. Attention! The family S is not closed under left-divisor, implying that some intermediate vertices (the six grey ones) do not belong to S

3.2   Computing S-Normal Decompositions

We postpone to the next subsection the question of recognising Garside families, and explain here how S-normal decompositions behave when they exist, that is, when S is a Garside family. To this end, the point is that the counterparts of Lemmas 2.9 and 2.11 are valid.

Lemma 3.8 *Assume that M is a left-cancellative monoid with no nontrivial invertible element and S is a Garside family in M. Then, for all s_1, s_2 in S, the element s_1s_2 has a unique S-normal decomposition of length two.*

Proof. Let s_1, s_2 belong to S. By assumption, s_1s_2 admits an S-normal decomposition, say $s_1'|\cdots|s_p'$. As s_1 belongs to S and $s_{p-1}'|s_p'$ is S-normal, the assumption $s_1 \preccurlyeq (s_1'\cdots s_{p-2}')s_{p-1}'s_p'$ implies $s_1 \preccurlyeq (s_1'\cdots s_{p-2}')s_{p-1}'$. Repeating the argument $p-1$ times, we conclude that s_1 left-divides s_1', say $s_1' = s_1t_1$. Left-cancelling s_1, we deduce $s_2 = t_1s_2'\cdots s_p'$ and, arguing as above, we conclude that s_2 must left-divide t_1s_2', say $t_1s_2' = s_2t_2$. Left-cancelling s_2, we deduce $1 = t_2s_3'\cdots s_p'$. As M has no nontrivial invertible element, the only possibility is $t_1 = s_3' = \cdots = s_p' = 1$, which implies that $s_1'|s_2'$ is an S-normal decomposition of s_1s_2. The argument is

illustrated in the diagram

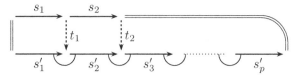

Thus, every element of $S^{[2]}$ has an S-normal decomposition of length two. Its uniqueness follows from Lemma 3.5.

Lemma 3.9 *Assume that M is a left-cancellative monoid with no nontrivial invertible element and S is a Garside family in M. Then the left domino rule is valid for S-normal words of length two.*

Proof. Assume that $s_1, s_2, s_1', s_2', t_0, t_1, t_2$ lie in S and satisfy the assumptions of Def. 2.10 (with respect to S-normal words of length two). We want to show that $s_1' | s_2'$ is S-normal. Assume $s \in S$ and $s \preccurlyeq fs_1's_2'$. First, $s \preccurlyeq s_1's_2'$ implies $s \preccurlyeq fs_1's_2't_2$, whence $s \preccurlyeq ft_0s_1s_2$. As $s_1|s_2$ is S-normal, we deduce $s \preccurlyeq ft_0s_1$, whence $s \preccurlyeq fs_1't_1$. As $s_1'|t_1$ is S-normal, we deduce $s \preccurlyeq fs_1'$ in turn. Therefore, $s_1'|s_2'$ is S-normal.   □

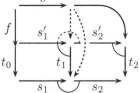

Arguing exactly as for Prop. 2.13 and using, in particular, Fig. 4, we obtain

Proposition 3.10 *Assume that M is a left-cancellative monoid with no nontrivial invertible element and S is a Garside family in M. Then, for every S-word w of length p, there exists a unique S-normal word $N^S(w)$ of length p that represents the same element as w. Moreover, with δ_p as in Prop. 2.13, one has*

$$N^S(w) = \overline{N^S_{\delta_p}}(w). \tag{3.11}$$

Thus, the recipe for computing the Δ-normal form associated with a Garside monoid extends without change to the S-normal form associated with an arbitrary Garside family S. As in Sec. 2, we deduce

Corollary 3.12 *Assume that M is a left-cancellative monoid with no nontrivial invertible element and S is a Garside family in M. Then S-normal decompositions can be computed in linear space and quadratic time. The Word Problem for M with respect to S lies in DTIME(n^2).*

On the other hand, as the diagram of Fig. 4 remains valid, we obtain

Corollary 3.13 *Assume that M is a left-cancellative monoid with no nontrivial invertible element and S is a Garside family in M. Then the S-normal words satisfy the 2-Fellow traveller Property on the left.*

In contrast to the particular case of Garside monoids, there is no symmetric counterpart involving right multiplication in the framework of an arbitrary Garside family. As can be expected, the existence of such a counterpart is equivalent to the validity of the right domino rule for S-normal words of length two. Now, the latter may fail, as the counterexample on the right shows: here S is the sixteen-element Garside family described in Fig. 5 in the Artin–Tits monoid of type $\widetilde{A}_2$.

For more results on the question, we refer to Chap. V of [17], where the notion of a *bounded* Garside family is introduced, and where it is proved that the right domino rule and the counterpart of Prop. 2.18 are valid, whenever S is a bounded Garside family.

3.3   Existence of S-Normal Decompositions

The notion of a Garside family is useful only if we can provide practical characterisations, which amounts to giving sufficient conditions for S-normal decompositions to exist. A number of such characterisations are known [17, Chap. IV], and we shall only mention a few of them.

Two types of characterisations exist, according to whether the ambient monoid satisfies or not additional conditions. Let us begin with the case of a monoid that is just assumed to be left-cancellative and, in this paper, to admit no nontrivial invertible element. In order to state the results, we need two definitions.

Definition 3.14 If M is a left-cancellative monoid, S is included in M, and g is an element of M, then an element s of S is said to be an S-*head* of g if we have $s \preccurlyeq g$ and $\forall t \in S \, (t \preccurlyeq g \Rightarrow t \preccurlyeq s)$.

In other words, an S-head of g is a greatest left-divisor of g lying in S. An S-head is unique whenever the ambient monoid M has no nontrivial invertible element: if s and s' are S-heads of g, the definition implies $s \preccurlyeq s'$ and $s' \preccurlyeq s$, whence $s' = s$. If (M, Δ) is a Garside monoid, the $\mathrm{Div}(\Delta)$-head of an element g exists and is simply the left-gcd of g and Δ, as considered in Lemma 2.6.

Definition 3.15 If M is a left-cancellative monoid and S is included in M, we say that S is *closed under right-comultiple* if the relation

$$\forall s, t \in S \; \forall g \in M \, ((s \preccurlyeq g \text{ and } t \preccurlyeq g)$$
$$\Rightarrow \exists r \in S \, (s \preccurlyeq r \text{ and } t \preccurlyeq r \text{ and } r \preccurlyeq g))$$

is satisfied in M, as illustrated on the right.

Thus, a family S is closed under right-comultiple if every common right-multiple of two elements s, t of S is a right-multiple of some common right-multiple of s and t that lies in S. Finally, we naturally say that a family S is *closed under right-divisor* if every right-divisor of an element of S belongs to S.

Proposition 3.16 [15, Prop. 3.9] or [17, Prop. IV.1.24] *Assume that M is a left-cancellative monoid with no nontrivial invertible element, and S is a generating subfamily of M that contains 1. Then S is a Garside family in M—that is, every element of M admits an S-normal decomposition—if, and only if, it satisfies one of the following equivalent conditions:*

(i) *Every nontrivial element of M admits an S-head, and S is closed under right-divisor.*

(ii) *Every element of S^2 admits a $\prec$-maximal left-divisor in S, and S is closed under right-comultiple and right-divisor.*

The conditions of Prop. 3.16 are not demanding: very little is required for the existence of S-normal decompositions. The difference between (i) and (ii) is that, in (ii), the existence of an S-head is relaxed twice: one considers elements of S^2 (that is, elements that can be expressed as the product of two elements of S) rather than arbitrary elements, and $\prec$-maximal left-divisors, which is weaker than $\preccurlyeq$-greatest left-divisors, since it amounts to replacing $s \preccurlyeq t$ with $t \not\prec s$.

Example 3.17 Prop. 3.16(i) makes it straightforward that, if (M, Δ) is a Garside monoid, $\mathrm{Div}(\Delta)$ is a Garside family: as noted above, the left-gcd of g and Δ is a $\mathrm{Div}(\Delta)$-head of g and, by definition, $\mathrm{Div}(\Delta)$ is closed under right-divisor.

The argument is similar for the family $\mathrm{Div}(\Delta)$ in the Klein bottle monoid K^+ of Example 3.7, but one easily finds examples of a completely different flavour. For instance, the reader can play with the family $\{\mathsf{b}^i \mid 0 \leqslant i \leqslant n + 1\} \cup \{\mathsf{a}\}$ in the monoid $\langle \mathsf{a}, \mathsf{b} \mid \mathsf{a}\mathsf{b}^n = \mathsf{b}^{n+1} \rangle^+$ with $n \geqslant 1$.

When the ambient monoid satisfies additional assumptions, the conditions of Prop. 3.16 can still be weakened.

Definition 3.18 A left-cancellative monoid is called *right-noetherian* if there is no infinite descending sequence with respect to strict right-divisibility.

Equivalently, a left-cancellative monoid is right-noetherian if, and only if, there is no infinite bounded ascending sequence with respect to strict left-divisibility, meaning that $g_1 \prec g_2 \prec \cdots \preccurlyeq g$ is impossible. When a monoid is right-noetherian, the existence of $\prec$-maximal elements is for free, and we deduce

Corollary 3.19 *Assume that M is a right-noetherian left-cancellative monoid with no nontrivial invertible element, and S is a generating subfamily of M that contains 1. Then S is a Garside family if, and only if, S is closed under right-comultiple and right-divisor.*

The criterion can be further improved as, for the ambient monoid to be right-noetherian, it is sufficient that the restriction of right-divisibility to the considered family S is, a trivial condition when S is finite, see [17, Prop. IV.2.18].

Finally, things become even more simple when the ambient monoid admits *conditional right-lcms*, meaning that any two elements that admit a common right-multiple admit a right-lcm. Then closure under right-comultiple boils down to closure under right-lcm (that is, the right-lcm of two elements of S belongs to S when it exists), and we obtain

Corollary 3.20 *Assume that M is a left-cancellative monoid that is right-noetherian, admits conditional right-lcms, and contains no nontrivial invertible element, and S is a generating subfamily of M that contains 1. Then S is a Garside family if, and only if, S is closed under right-lcm and right-divisor.*

Thus, in the context of Cor. 3.20, being a Garside family is a closure property. It follows that, for every generating set A, there exists a smallest Garside family S that includes A, namely the closure of A under right-lcm and right-divisor. When the ambient monoid is noetherian (meaning left- and right-noetherian), it admits a smallest generating family, namely the family of atoms (indecomposable elements), and therefore it admits a smallest Garside family, the closure of atoms under right-lcm and right-divisor. A typical example is the family $\mathrm{Div}(\Delta)$ in a Garside monoid (with Δ chosen minimal), but another example is the Garside family of Fig. 5 in the Artin–Tits monoid of type $\widetilde{A}_2$.

Let us mention a last result. We observed that the definition of an S-normal sequence in (3.2) is a priori more demanding than that of (2.5). It turns out that, when S satisfies convenient conditions, the conditions become equivalent:

Proposition 3.21 [17, Prop. IV.1.20] *Assume that M is a left-cancellative monoid with no nontrivial invertible element, and S is a generating family of M that is closed under right-comultiple and right-divisor. Then an S-word $s_1|\cdots|s_p$ is S-normal if, and only if, it satisfies the condition*

$$\forall s \in S \ (s_i \prec s \ \Rightarrow \ s \not\preccurlyeq s_i s_{i+1} \cdots s_p). \tag{3.22}$$

We already observed that the condition is necessary. That it is sufficient follows from arguments extending those of Ex. 3.3. Note that, by Prop. 3.16, every Garside family satisfies the assumptions of Prop. 3.21 and, therefore, the connection is valid in this case.

Remark 3.23 Once again, we can think of extending the results from the monoid to its enveloping group. Here, some care is needed as, in general, a left-cancellative monoid (even a cancellative one) need not embed in a group of left fractions: by the classical Ore theorem, this happens if, and only if, the monoid M is cancellative and any two elements of M admit a common left-multiple. But, even in this case, the existence of unique S-normal decompositions in M does not directly lead to unique distinguished decompositions for the elements of its group of fractions, because fractional decompositions need not be unique. However, when the monoid M admits left-lcms, a notion of irreducible fraction arises and one obtains unique decompositions (called "symmetric S-normal") for the elements of the group by using S-normal decompositions for the numerator and the denominator of an irreducible fractional decomposition, see [17, Sec. III.2].

4 Quadratic Normalisation

Proceeding one step further, we now consider more general normalisation processes that include those of the previous sections, but also new examples of a different flavour. However, we shall see that the mechanism of Garside normal-isation, as captured in Prop. 3.10, can be retrieved in the more general frame-work of what we shall call "quadratic normalisations of class $(4, 3)$". One of the benefits of such an extended approach is that some monoids that are not even left-cancellative, like plactic monoids, become in turn eligible.

4.1   Normalisations and Geodesic Normal Forms

We now restart from a general standpoint and consider (not necessarily cancella-tive) monoids equipped with a generating family. We are interested in normal forms in such monoids, according to the following abstract scheme:

Definition 4.1 Assume that M is a monoid and S is a generating subfamily of M. A *normal form on* (M, S) is a (set-theoretic) section of the canonical projection EV of S^* onto M. A normal form NF on (M, S) is called *geodesic* if, for every g in M, we have $\|\text{NF}(g)\| \leqslant \|w\|$ for every S-word w representing g.

Typically, we saw in Sec. 3 that, if M is left-cancellative with no nontriv-ial invertible element, every Garside family S of M provides a normal form on (M, S), associating with every element g of M the unique strict S-normal decompositon of g. This normal form is geodesic, since, by Prop. 3.10, the S-normal form of an element specified by an S-word of length p has length at most p.

As already done in Sec. 2 and 3, we shall rather work with words, and concen-trate on the normalisation maps that associate to an arbitrary word the unique equivalent normal word. This leads us to the following notion.

Definition 4.2 A *normalisation* is a pair (S, N), where S is a set and N is a map from S^* to itself satisfying, for all S-words u, v, w,

$$\|N(w)\| = \|w\|, \tag{4.3}$$

$$\|w\| = 1 \text{ implies } N(w) = w, \tag{4.4}$$

$$N(u|N(w)|v) = N(u|w|v). \tag{4.5}$$

An S-word w satisfying $N(w) = w$ is called *N-normal*. If M is a monoid, we say that (S, N) is a normalisation *for* M if M admits the presentation

$$\langle S \mid \{w = N(w) \mid w \in S^*\}\rangle^+. \tag{4.6}$$

Note that (4.5) implies that N is idempotent. We shall see below that the maps N^Δ and N^S considered in Sec. 2 and 3 are typical examples of normal-isations. Many others appear in [19]. The connection between normalisations and normal forms is easily described, especially in the case when all equivalent S-words have the same length.

Proposition 4.7 [19, Prop. 2.1.12] *If (S, N) is a normalisation for a mon-oid M, then putting* NF$(g) = N(w)$, *where w is any S-decomposition of g, pro-vides a normal form on (M, S).*

Conversely, if M is a monoid, S is a generating subfamily of M, and NF *is normal form on (M, S), and, moreover, any two S-decompositions of an element of M have the same length, then putting $N(w) = $ NF$($EV$(w))$ provides a normalisation for M.*

Moreover, it is easily seen that the two correspondences of Prop. 4.7 are inverses of one another.

When the elements of M may admit S-decompositions of different lengths (as in the case of a Garside family), more care is needed, but we can still merge unique normal forms and length-preserving normalisation maps at the expense of introducing a dummy letter that represents 1 and is eventually collapsed.

Definition 4.8 If (S, N) is a normalisation, an element e of S is called N-*neutral* if, for every S-word w, one has

$$N(w|e) = N(e|w) = N(w)|e. \qquad (4.9)$$

Then we write coll_e for the action of erasing e in an S-word. If M is a monoid, we say that (S, N) is a normalisation *mod e for M* if M admits the presentation

$$\langle S \mid \{w = N(w) \mid w \in S^*\} \cup \{e = 1\}\rangle^+. \qquad (4.10)$$

We invite the reader to check that, if S is a Garside family in a left-cancellative monoid M that admits no nontrivial invertible element, then (S, N^S) is a normalisation for M mod 1 and the S-normal words of Sec. 3 are the associated N^S-normal words. Then Prop. 4.7 extends in

Proposition 4.11 [19, Prop. 2.2.7] *If M is a monoid and (S, N) is a normalisation for M mod e, putting* NF$(g) = \mathrm{coll}_e(N(w))$, *where w is any S-decomposition of g, provides a geodesic normal form on $(M, S \setminus \{e\})$.*

Conversely, if M is a monoid, S generates M, and NF *is a geodesic normal form on (M, S), putting $N(w) = $ NF$($EV$(w))|e^m$, with e a new letter not in S evaluated to 1 in M and $m = \|w\| - \|$NF$($EV$(w))\|$, provides a normalisation for M mod e (with alphabet $S \cup \{e\}$).*

Again, the two correspondences of Prop. 4.11 are inverses of one another. Thus, investigating geodesic normal forms and investigating normalisations are one and the same question.

This general framework being set, we now turn to a more specific situation. By Prop. 3.4 and 3.10, subfactors of length two play a prominent rôle in Garside normalisation. This is the property we shall extend. We recall Notation 2.12, in particular that, for $N : S^* \to S^*$, we use $\overline{N}$ for the restriction of N to $S^{[2]}$.

Definition 4.12 A normalisation (S, N) is *quadratic* if the two conditions hold:
   (i) An S-word w is N-normal if, and only if, every length-two factor of w is.
   (ii) For every S-word w, there exists a finite sequence of positions u, depending on w, such that $N(w)$ is equal to $\overline{N}_u(w)$.

Example 4.13 By Prop. 3.4 and 3.10, if S is a Garside family, then the associated normalisation (S, N^S) is quadratic: Prop. 3.10 says that $u := \delta_p$ can be chosen for *every* S-word w of length p.

For a different example, as in Ex. 1.5, let (A_n, N^{Lex}) be the lexicographic normalisation for the free abelian monoid $\mathbb{N}^n$. Then (A_n, N^{Lex}) is quadratic. Indeed, an A_n-word is N^{Lex}-normal if, and only if, all its length-two subfactors are $\mathsf{a}_i|\mathsf{a}_j$ with $i \leqslant j$. On the other hand, every A_n-word can be transformed into a N^{Lex}-normal word by switching adjacent letters that are not in the due order.

Simple counterexamples show that none of the two conditions in Def. 4.12 implies the other. When a normalisation (S, N) is quadratic, then, by definition, the restriction $\overline{N}$ of N to $S^{[2]}$ is crucial and most properties can be read from $\overline{N}$. In particular, one shows that, if M is a monoid and (S, N) is a quadratic normalisation for M (*resp.* for M mod e), then M admits the presentation

$$\langle S \mid \{s|t = \overline{N}(s|t) \mid s, t \in S\}\rangle^+, \tag{4.14}$$

$$(resp.\ \langle S \setminus \{e\} \mid \{s|t = \mathrm{coll}_e(\overline{N}(s|t)) \mid s, t \in S \setminus \{e\}\}\rangle^+\ ). \tag{4.15}$$

So, the relations between words of length two bear all information.

Before turning to more elaborate results, let us immediately note the following direct consequence of Def. 4.12(i):

Proposition 4.16 *If (S, N) is a quadratic normalisation and S is finite, then N-normal words form a regular language.*

4.2   The Class of a Quadratic Normalisation

We now introduce a parameter, called the class, evaluating the complexity of the normalisation process associated with a quadratic normalisation.

If (S, N) is a quadratic normalisation and w is an S-word, $N(w)$ is obtained by successively applying the restriction $\overline{N}$ of N to $S^{[2]}$ at various positions. So, in particular, for $\|w\| = 3$, we have $N(w) = \overline{N}_u(w)$ for some finite sequence u of positions 1 and 2. As $\overline{N}$ is idempotent, repeating 1 or 2 in u is useless, and it is enough to consider sequences u of the form 121... or 212... (we omit separators).

Notation 4.17 For $m \geqslant 0$, we write 121...$[m]$ for the alternating sequence 121... of length m, and similarly for 212...$[m]$.

So, if (S, N) is a quadratic normalisation, then, for every S-word w of length three, there exists m such that $N(w)$ is $\overline{N}_{121...[m]}(w)$ or $\overline{N}_{212...[m]}(w)$.

Definition 4.18 We say that a quadratic normalisation (S, N) is *of left class c* (*resp. right-class c*) if $N(w) = \overline{N}_{121...[c]}(w)$ (*resp.* $N(w) = \overline{N}_{212...[c]}(w)$) holds for every w in $S^{[3]}$, and *of class (c, c')* if it is of left class c and right class c'.

Example 4.19 If (M, Δ) is a Garside monoid, Prop. 2.13 gives $N^\Delta(w) = \overline{N}^\Delta_{212}(w)$ for every $\mathrm{Div}(\Delta)$-word w of length three. Hence, $(\mathrm{Div}(\Delta), N^\Delta)$ is of right class 3. Symmetrically, Prop. 2.18 gives $N^\Delta(w) = \overline{N}^\Delta_{121}(w)$, so $(\mathrm{Div}(\Delta), N^\Delta)$ is also of left class 3. Hence, the normalisation $(\mathrm{Div}(\Delta), N^\Delta)$ is of class $(3, 3)$.

If S is a Garside family in a left-cancellative monoid with no nontrivial invertible element, then Prop. 3.10 implies that (S, N^S) is of right class 3, but, lacking in general a counterpart of Prop. 2.18, we have no hint for the left class.

The reader can check that the lexicographic normalisation (A_n, N^{Lex}) of Ex. 1.5 also has class $(3, 3)$. On the other hand, there exist examples of normalisations with an arbitrarily high minimal class: see [19, Ex. 3.3.9], where the minimal class is $(3 + \lfloor \log_2 n \rfloor, 3 + \lfloor \log_2 n \rfloor)$ for a size n alphabet.

Let us mention one more normalisation, very different from the previous examples. If X is a linearly ordered finite set, the *plactic* monoid over X is [6]

$$P_X = \left\langle X \; \middle| \; \begin{array}{l} \mathsf{xzy = zxy} \text{ for } \mathsf{x \leqslant y < z} \\ \mathsf{yxz = yzx} \text{ for } \mathsf{x < y \leqslant z} \end{array} \right\rangle^+ .$$

Then P_X is also generated by the family C_X of nonempty columns over X, defined to be strictly decreasing products of elements of X. Call a pair of columns $s_1|s_2$ *normal* if $\|s_1\| \geqslant \|s_2\|$ holds and, for every $1 \leqslant k \leqslant \|s_2\|$, the kth element of s_1 is at most the one of s_2. Then normal sequences $s_1|\cdots|s_p$ are in one-to-one correspondence with Young tableaux, and every element of P_X is represented by a unique tableau of minimal length (in terms of columns). Thus, mapping a C_X-word to the unique corresponding tableau defines a geodesic normal form on (P_X, C_X). Writing $\widehat{C}_X$ for C_X enriched with one empty column $\emptyset$ and using Prop. 4.11, we obtain a normalisation $(\widehat{C}_X, N)$ for P_X mod $\emptyset$. Then, condition (i) in Def. 4.12 is satisfied by the definition of tableaux. Moreover, for every $\widehat{C}_X$-word w, the normal tableau $N(w)$ can be computed by resorting to the Robinson–Schensted's insertion algorithm, progressively replacing each pair $s_1|s_2$ of subsequent columns by $\overline{N}(s_1|s_2)$, which is a tableau with two columns (if the algorithm returns a tableau with one column, we insert an empty column to keep the length unchanged). So, the normalisation $(\widehat{C}_X, N)$ also satisfies condition (ii) in Def. 4.12 and, therefore, it is quadratic. Then, the computations of [5, §§3.2–3.4 and §§4.2–4.4] show that $(\widehat{C}_X, N)$ is of class $(3, 3)$.

There exists an easy connection between the left and the right class.

Lemma 4.20 *If a quadratic normalisation is of left class c, then it is of left class c' for every c' with $c' \geqslant c$, and of right class c'' for every c'' with $c'' \geqslant c+1$.*

Proof. Assume that (S, N) is of left class c. First, for w in $S^{[3]}$, the equality $N(w) = \overline{N}_{121\ldots[c]}(w)$ implies $N(w) = \overline{N}_{121\ldots[c+1]}(w)$, since $N(w)$ is N-invariant. So (S, N) is of left class $c+1$ as well and, from there, it is of left class c' for $c' \geqslant c$.

On the other hand, we have $N(w) = \overline{N}_{121\ldots[c]}(\overline{N}_2(w)) = \overline{N}_{212\ldots[c+1]}(w)$ by the assumption and by (4.5). Hence (S, N) is of right class $c+1$ and, from there, of right class c'' for every c'' with $c'' \geqslant c+1$.

Hence, the minimal class of a quadratic normalisation (S, N) is either (c, c') with $|c'-c| \leqslant 1$, or (∞, ∞), the latter being excluded for S finite. By Lemma 4.20, a Garside normalisation is of right class 4, and we can state:

Proposition 4.21 *If M is a left-cancellative monoid with no nontrivial invertible element and S is a Garside family in M, then the normalisation (S, N^S) is of class $(4, 3)$.*

4.3 Quadratic Normalisations of Class $(4, 3)$

We shall now show that many properties of Garside normalisations extend to all quadratic normalisations of class $(4, 3)$. The extension comes from the following observation:

Lemma 4.22 *A quadratic normalisation (S, N) is of class $(4, 3)$ if, and only if, the left domino rule is valid for the family of all N-normal words of length two.*

Proof. Assume that (S, N) is of right class 3, and let L be the family of all N-normal words of length two. Let $s_1, s_2, s_1', s_2', t_0, t_1, t_2$ be elements of S satisfying the assumptions of Def. 2.10. By the definition of the right class, we have $N(t_0|s_1|s_2) = \overline{N}_{212}(t_0|s_1|s_2)$. As, by assumption, $s_1|s_2$ is N-normal, we obtain $N(t_0|s_1|s_2) = \overline{N}_{12}(t_0|s_1|s_2) = \overline{N}_2(s_1'|t_1|s_2) = s_1'|s_2'|t_2$. So $s_1'|s_2'|t_2$ is N-normal, hence so is $s_1'|s_2'$. Therefore, the left domino rule is valid for L.

Conversely, assume that the left domino rule is valid for L. Let $t_0|r_1|r_2$ belong to $S^{[3]}$. Put $s_1|s_2 = \overline{N}(r_1|r_2)$, $s_1'|t_1 = \overline{N}(t_0|s_1)$, and $s_2'|t_2 = \overline{N}(t_1|s_2)$. Then $s_2'|t_2$ is N-normal by construction, and $s_1'|s_2'$ is N-normal by the left domino rule, so $s_1'|s_2'|t_2$ is N-normal. Hence we have $N(w) = \overline{N}_{212}(w)$ for every w in $S^{[3]}$. Therefore, (S, N) is of right class 3 and, therefore, of class $(4, 3)$.   □

Using the left domino rule exactly as in Sec. 2 and 3, we deduce

Proposition 4.23 *If (S, N) is a quadratic normalisation of class $(4, 3)$, then, for every word w of length p, we have*

$$N(w) = \overline{N}_{\delta_p}(w). \tag{4.24}$$

So the universal recipe given by the sequence of positions δ_p is valid for every normalisation of class $(4, 3)$. Of course, a similar recipe associated with the sequence of positions $\widetilde{\delta}_p$ as in Prop. 2.18 is valid for every normalisation of class $(3, 4)$, with the right domino rule replacing the left one. In the case of a normalisation of class $(3, 3)$, both recipes are valid, as in the case of the Garside normalisation associated with a Garside monoid or, more generally, with a bounded Garside family.

Arguing exactly as in the previous sections, we deduce

Corollary 4.25 *If (S, N) is a quadratic normalisation of class $(4, 3)$ for a monoid M, then N-normal decompositions can be computed in linear space and quadratic time. The Word Problem for M with respect to S lies in DTIME(n^2), and, if e is a N-neutral element of S and M_e is the quotient of M obtained by collapsing e, so does the Word Problem for M_e with respect to $S \setminus \{e\}$.*

Corollary 4.26 *If (S, N) is a quadratic normalisation of class $(4, 3)$ (resp. of class $(3, 4)$), then N-normal words satisfy the 2-Fellow traveller Property on the left (resp. on the right).*

Let us turn to another question, and mention (without proof) one further result. We start from the (easy) observation that the class of a normalisation (S, N) can be characterised by algebraic relations satisfied by the map $\overline{N}$ and its translated copy:

Lemma 4.27 [19, Prop. 3.3.5] *A quadratic normalisation (S, N) is of left class c if, and only if, the map $\overline{N}$ satisfies $\overline{N}_{121...[c]} = \overline{N}_{121...[c+1]} = \overline{N}_{212...[c+1]}$; it is of class (c, c) if, and only if, the map $\overline{N}$ satisfies $\overline{N}_{121...[c]} = \overline{N}_{212...[c]}$.*

So, in particular, if a normalisation (S, N) is of class $(4, 3)$, the map $\overline{N}$, which, by definition, is idempotent, satisfies $\overline{N}_{212} = \overline{N}_{2121} = \overline{N}_{1212}$. The next result provides an axiomatisation of class $(4, 3)$ normalisations: it shows that, conversely, every idempotent map satisfying the above relation necessarily stems from such a normalisation.

Proposition 4.28 [19, Prop. 4.3.1] *If S is a set and F is a map from $S^{[2]}$ to itself satisfying*

$$F_{212} = F_{2121} = F_{1212}, \tag{4.29}$$

there exists a quadratic normalisation (S, N) of class $(4, 3)$ satisfying $F = \overline{N}$.

The problem is to extend F into a map F^* on S^* such that (S, F^*) is a quadratic normalisation of class $(4, 3)$. The idea of the proof is to take the recipe given by (4.24) as a definition, and to show that the resulting map has the expected properties. The result is not trivial, and there is no counterpart for higher classes. The specific reasons why the result works for class $(4, 3)$ are the algebraic properties of the monoid that admits the presentation

$$\left\langle \sigma_1, ..., \sigma_{p-1} \mid \sigma_i^2 = \sigma_i \text{ for } i \geqslant 1, \begin{array}{ll} \sigma_i \sigma_j = \sigma_j \sigma_i & \text{for } j - i \geqslant 2 \\ \sigma_j \sigma_i \sigma_j = \sigma_i \sigma_j \sigma_i \sigma_j = \sigma_j \sigma_i \sigma_j \sigma_i \text{ for } j = i+1 \end{array} \right\rangle^+.$$

This monoid is a sort of asymmetric version of a symmetric group (or rather of the corresponding Hecke algebra at $q = 0$), which is considered and used by A. Hess and V. Ozornova in [26], and investigated by D. Krammer in [30].

4.4 Characterising Garside Normalisations

We observed in Prop. 4.21 that every Garside normalisation is of class $(4, 3)$, a result that is optimal in general, since the right domino rule fails for the finite Garside family of Fig. 5, implying that the associated normalisation is not of class $(3, 3)$. Conversely, it is easy to see that the lexicographic normalisation of Ex. 4.13, which is of class $(3, 3)$, hence a fortiori $(4, 3)$, does not stem from a Garside family. So the question arises of characterising Garside normalisations among all normalisations of class $(4, 3)$. The answer is simple.

Definition 4.30 Assume that (S, N) is a (quadratic) normalisation for a monoid M. We say that (S, N) is *left-weighted* if, for all s, t, s', t' in S, the equality $s'|t' = N(s|t)$ implies $s \preccurlyeq s'$ in M.

Thus, a normalisation (S, N) is left-weighted if, for every s in S, the first entry of $N(s|t)$ is always a right-multiple of s in the associated monoid.

Proposition 4.31 [19, Prop. 5.4.3] *Assume that (S, N) is a quadratic normalisation mod 1 for a monoid M that is left-cancellative and contains no nontrivial invertible element. Then the following are equivalent:*
(i) *The family S is a Garside family in M and $N = N^S$ holds.*
(ii) *The normalisation (S, N) is of class $(4, 3)$ and is left-weighted.*

The implication (i) $\Rightarrow$ (ii) is almost straightforward. Indeed, if S is a Garside family and $s'_1|s'_2 = N^S(s_1|s_2)$ holds, we have $s_1 \preccurlyeq s'_1 s'_2$ with $s \in S$, so the assumption that $s'_1|s'_2$ is S-normal implies $s_1 \preccurlyeq s'_1$. Hence N^S is left-weighted.

The converse implication is much more delicate. The main point is to show that S is a Garside family in M, which is proved by establishing that S is closed under right-divisor and every element of the ambient monoid M has an S-head, and then using Prop. 3.16(i).

4.5   Connection with Rewriting Systems

There exists a simple connection between normalisations as introduced above and rewriting systems. We refer to [21] or [29] for basic terminology.

Lemma 4.32 *If (S, N) is a quadratic normalisation for a monoid M, then putting $R = \{s|t \rightarrow \overline{N}(s|t) \mid s, t \in S, s|t \neq \overline{N}(s|t)\}$ provides a rewriting system (S, R) that is quadratic, reduced, normalising, confluent, and presents M.*

Conversely, if (S, R) is a quadratic, reduced, normalising, and confluent rewriting system presenting a monoid M, putting $N(w) = w'$, where w' is the R-normal form of w, provides a quadratic normalisation (S, N) for M.

The above correspondences are inverses of one another.

Example 4.33 If (A_n, N^{Lex}) is the lexicographic normalisation for the free commutative monoid $\mathbb{N}^n$, the associated quadratic rewriting system (A_n, R_n) consists of the $n(n-1)/2$ rules $a_i|a_j \rightarrow a_j|a_i$ for $1 \leqslant j < i \leqslant n$.

The correspondence of Lemma. 4.32 extends to a normalisation mod a neutral letter e at the expense of defining a new system R_e by replacing $s|t \rightarrow \overline{N}(s|t)$ with $s|t \rightarrow \mathrm{coll}_e(\overline{N}(s|t))$, that is, of erasing the involved N-neutral letter.

By Lemma 4.32, a quadratic normalisation (S, N) yields a reduced quadratic rewriting system (S, R) that is normalising and confluent, meaning that, from every S-word, there is a rewriting sequence leading to a N-normal word. This however does not rule out the possible existence of infinite rewriting sequences: the system (S, R) need not a priori be terminating. Here again, class $(4, 3)$ is the point where transition occurs.

Proposition 4.34 [19, Prop. 5.7.1] *If (S, N) is a quadratic normalisation of class $(4, 3)$, then the associated rewriting system (S, R) is terminating, and so is $(S \setminus \{e\}, R_e)$ if e is a N-neutral element of S. More precisely, every rewriting sequence from a length-p word has length at most $2^p - p - 1$.*

The (delicate) proof consists in showing that every sequence of R-rewritings inevitably approaches a N-normal word: because of the left domino rule, in whatever order the rewritings are operated, the distance between the current word and its image under N cannot increase, and it must even decrease at some predictible intervals.

Either by taking into account the influence of the right domino rule in the proof of the above result, or by an alternative direct argument based on the classical Matsumoto lemma for the symmetric group $\mathfrak{S}_p$, one can show that, in the case of a normalisation of class $(3, 3)$, the upper bound $2^p - p - 1$ drops to $p(p-1)/2$.

Owing to Prop. 4.21, we obtain as a direct application of Prop. 4.34:

Corollary 4.35 *Assume that M is a left-cancellative monoid with no nontrivial invertible element and S is a Garside family in M. Then the associated rewriting system is terminating. More precisely, every rewriting sequence from a length-p word has length at most $2^p - p - 1$.*

By contrast, we have:

Proposition 4.36 *There exists a quadratic normalisation of class $(4, 4)$ such that the associated rewriting system is not terminating.*

Proof (sketch). Let $S := \{a, b, b', b'', c, c', c'', d\}$ and let R consist of the five rules $ab \to ab'$, $b'c' \to bc$, $bc' \to b''c''$, $b'c \to b''c''$, $cd \to c'd$. Then (S, R) is quadratic by definition, and the diagram

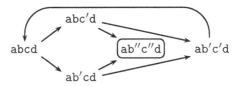

in which $ab''c''d$ is R-normal, shows that (S, R) is not terminating, since it admits the length-3 cycle $abcd \to ab'cd \to ab'c'd \to abcd$. However, one can show (with some care) that (S, R) is normalising and confluent, and that the associated normalisation is of class $(4, 4)$. □

It can be noted that terminating rewriting systems may also arise when the minimal class is $(4, 4)$: a beautiful example is provided by the Chinese monoid based on a set of size 3, see [11].

References

1. Adyan, S.I.: Fragments of the word Delta in a braid group. Mat. Zam. Acad. Sci. SSSR **36–1**, 25–34 (1984); translated Math. Notes of the Acad. Sci. USSR **36–1**, 505–510 (1984)
2. Albenque, M., Nadeau, P.: Growth function for a class of monoids 21st International Conference on Formal Power Series and Algebraic Combinatorics (FPSAC 2009), 2538 Discrete Math. Theor. Comput. Sci. Proc., AK, Assoc. Discrete Math. Theor. Comput. Sci., Nancy (2009)
3. Artin, E.: Theory of braids. Ann. of Math. **48**, 101–126 (1947)
4. Bessis, D., Digne, F., Michel, M.: Springer theory in braid groups and the Birman-Ko-Lee monoid. Pacific J. Math. **205**, 287–309 (2002)
5. Bokut, L., Chen, Y., Chen, W., Li, J.: New approaches to plactic monoid via Gröbner-Shirshov bases. J. Algebra **423**, 301–317 (2015)
6. Cain, A., Gray, R., Malheiro, A.: Finite Gröbner-Shirshov bases for Plactic algebras and biautomatic structures for Plactic monoids. J. Algebra **423**, 37–52 (2015)
7. Birman, J.: Braids, Links, and Mapping Class Groups. Ann. of Math. Studies, **82** (1975). Princeton Univ. Press
8. Birman, J., Ko, K.H., Lee, S.J.: A new approach to the word problem in the braid groups. Adv. in Math. **139–2**, 322–353 (1998)
9. Bjorner, A., Brenti, F.: Combinatorics of Coxeter Groups. Graduate Texts in Mathematics, vol. 231, Springer (2005)
10. Brieskorn, E., Saito, K.: Artin-Gruppen und Coxeter-Gruppen. Invent. Math. **17**, 245–271 (1972)
11. Cassaigne, J., Espie, M., Krob, D., Novelli, J.C., Hivert, F.: The Chinese monoid. Internat. J. Algebra Comput. **11**, 301–334 (2001)
12. Charney, R.: Artin groups of finite type are biautomatic. Math. Ann. **292–4**, 671–683 (1992)
13. Dehornoy, P.: Groupes de Garside. Ann. Sci. Éc. Norm. Supér. **35**, 267–306 (2002)
14. Dehornoy, P., Paris, L.: Gaussian groups and Garside groups, two generalizations of Artin groups. Proc. London Math. Soc. **79–3**, 569–604 (1999)
15. Dehornoy, P., Digne, F., Michel, J.: Garside families and Garside germs. J. Algebra **380**, 109–145 (2013)
16. Dehornoy, P., Gebhardt, V.: Algorithms for Garside calculus. J. Symbolic Comput. **63**, 68–116 (2014)
17. Dehornoy, P., Digne, F., Godelle, E., Krammer, D., Michel, J.: Foundations of Garside Theory. EMS Tracts in Mathematics, **22** (2015)
18. Dehornoy, P., Dynnikov, I., Rolfsen, D., Wiest, B.: Ordering Braids Mathematical Surveys and Monographs. Amer. Math. Soc. **148** (2008)
19. Dehornoy, P., Guiraud, Y.: Quadratic normalisation in monoids. arXiv:1504.02717
20. Dehornoy, P., Dyer, M., Hohlweg, C.: Garside families in Artin-Tits monoids and low elements in Coxeter groups. Comptes-Rendus Math. **353**, 403–408 (2015)
21. Dershowitz, N., Jouannaud, J.P.: Rewrite Systems. In: van Leeuwen, J. (ed.) Handbook of Theoretical Computer Science B: Formal Methods and Semantics, Chap. 6, pp. 243–320. North-Holland (1990)
22. El-Rifai, E.A., Morton, H.R.: Algorithms for positive braids. Quart. J. Math. Oxford **45–2**, 479–497 (1994)
23. Epstein, D., Cannon, J., Holt, D., Levy, S., Paterson, M., Thurston, W.: Word Processing in Groups. Jones and Bartlett Publishers (1992)

24. Garside, F.A.: The braid group and other groups. Quart. J. Math. Oxford **20**, 235–254 (1969)
25. Gaussent, S., Guiraud, Y., Malbos, P.: Coherent presentations of Artin monoids Compos. Math. (to appear). arXiv:1203.5358
26. Hess, A., Ozornova, V.: Factorability, string rewriting and discrete Morse theory. arXiv:1412.3025
27. Hoffmann, M., Thomas, R.M.: A geometric characterisation of automatic semigroups. Theoret. Comput. Sci. **369**, 300–313 (2006)
28. Howie, J.M.: Fundamentals of Semigroup Theory. Clarendon, Oxford (1995)
29. Klop, J.W.: Term Rewriting Systems. In: Abramsky, S., Gabbay, D.M., Maibaum, T.S.E. (eds.) Handbook of Logic in Computer Science, vol. 2, Chap. 1, pp. 1–117. Oxford University Press (1992)
30. Krammer, D.: An asymmetric generalisation of Artin monoids Groups Complex. Cryptol. **5**, 141–168 (2013)

Grammar-Based Tree Compression

Markus Lohrey[✉]

Universität Siegen, Siegen, Germany
lohrey@eti.uni-siegen.de

Abstract. This paper gives a survey on recent progress in grammar-based compression for trees. Also algorithms that directly work on grammar-compressed trees will be surveyed.

1 Introduction

Trees are an omnipresent data structure in computer science. Large trees occur for instance in XML processing or automated deduction. For certain appliations it is important to work with compact tree representations. A widely studied standard compact tree representation is the *dag* (*directed acyclic graph*), see e.g. [5,10–12,38]. A dag is basially a folded tree, where nodes may share children. The tree represented by a dag is obtained by unfolding the dag. One of the nice things about dags is that every tree has a unique minimal (or smallest) dag that can be computed in linear time [10]. In the minimal dag of a tree t, isomorphic subtrees of t are represented only once. Figure 1 shows a tree and its minimal dag (we consider ranked ordered trees, where every node is labelled with a symbol, whose rank determines the number of children of the node). In [38], dags are used to obtain a universal (in the information theoretic sense) compressor for binary trees under certain distributions. Dags can achieve exponential compression in the best case: The minimal dag of a full binary tree of height n is a linear chain of length n.

In recent years, another compact tree representation that generalizes dags has been studied: Tree straight-line programs, briefly TSLPs. Whereas dags can only share repeated complete subtrees, TSLPs can also share repeated occurrences of subtrees with gaps (i.e., subtrees, where some smaller subtrees are removed). A TSLP can be seen as a very restricted context-free tree grammar that produces exactly one tree. It consists of rewrite rules (productions) of the form $A(x_1, \ldots, x_k) \rightarrow t(x_1, \ldots, x_k)$. Here, A is a nonterminal of rank k and $x_1, \ldots, x_k$ are parameters that are replaced by concrete trees in the application of this rule. The nodes of the tree $t(x_1, \ldots, x_k)$ are labelled with terminal symbols (the node labels of the tree produced by the TSLP), nonterminal symbols and the parameters $x_1, \ldots, x_k$. There is a distinguished start nonterminal S of rank 0. To produce a single tree, it is required that (i) for every nonterminal A there is exactly one rule with A on the left-hand side, and (ii) that from a nonterminal A one cannot reach A by more than one rewrite step. Finally, it is

---

This research is supported by the DFG-project LO 748/10-1.

I. Potapov (Ed.): DLT 2015, LNCS 9168, pp. 46–57, 2015.
DOI: 10.1007/978-3-319-21500-6_3

required that every parameter x_i appears at most once in the right-hand side of A (linearity). Dags can be seen as TSLPs, where every nonterminal has rank zero. As for dags, TSLPs allow exponential compression in the best case, but due to the ability to share also internal patterns, one can easily come up with examples where the minimal dag is exponentially larger than the smallest TSLP, see Section 2.

TSLPs generalize straight-line programs for words (SLPs). These are context-free grammars that produce a single word. There exist several grammar-based string compressors that produce (a suitable encoding of) an SLP for an input word. Prominent examples are LZ78, RePair, Sequitur, and BiSection. Theoretical results on the compression ratio of these algorithmis can be found in [8]. Over the last couple of years, the idea of grammar-based compression has been extended from words to trees. In Section 3 we will discuss several grammar-based tree compressors based on TSLPs.

SLPs and TSLPs are a simple and mathematically clean data structure. These makes them well-suited for the development of efficient algorithms on compressed objects. The goal of such algorithms is to manipulate and analyze compressed objects and thereby beat a naive decompress-and-compute strategy, where the uncompressed object is first computed and then analyzed. A typical example for this is pattern matching. Here, we have a large text, which is stored in compressed form and want to locate occurrences of a pattern (which is usually given in explicit form) in the text. But algorithms on compressed objects can be also useful for problems, where we do not directly deal with compression. In many algorithms huge intermediate data structures have to be stored, which are the main bottleneck in the computation. An obvious potential solution in such a situation is to store these intermediate data structures in a compressed way.

A survey on algorithms that work on SLP-compressed words can be found in [24], which contains a section on algorithms on TSLP-compressed trees as well. In Section 4 we give a more detailed and up-to-date survey on algorithms for trees that are represented by TSLPs.

2 Tree Straight-Line Programs

For background on trees and tree grammars see [9]. Here, trees are rooted, ordered and node-labelled. Every node has a label from a finite alphabet Σ. Moreover, with every symbol $a \in \Sigma$ a natural number (the rank of a) is associated. Symbols of rank zero are called *constants* and symbols of rank one are *unary*. If a tree node v is labelled with a symbol of rank n, then v has exactly n children, which are linearly ordered. Such trees can conveniently be represented as terms. The size $|t|$ of a tree t is the number of nodes of t. Here is an example:

Example 1. Let f be a symbol of rank 2, h a symbol of rank 1, and a a symbol of rank 0 (a constant). Then the term $h(f(h(f(h(h(a)),a)),h(f(h(h(a)),a))))$ corresponds to the tree of size 14, shown in Figure 1.

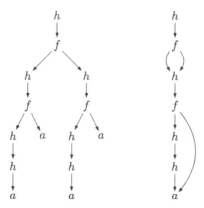

Fig. 1. A node-labelled tree and its minimal dag

A tree straight-line program (TSLP for short and also called SLCF tree grammar in [25,28] or SLT grammar in [27]) over the terminal alphabet Σ (which is a ranked alphabet in the above sense) is a tuple $\mathcal{G} = (N, \Sigma, S, P)$, such that

- N is a finite set of ranked symbols (the nonterminals) with $N \cap \Sigma = \emptyset$,
- $S \in N$ has rank 0 (the initial nonterminal),
- and P is a finite set of productions of the form $A(x_1, \ldots, x_n) \to t$ where A is a nonterminal of rank n and t is a tree built up from the ranked symbols in $\Sigma \cup N$ and the parameters $x_1, \ldots, x_n$ which are considered as symbols of rank 0 (i.e., constants). Every x_i is required to appear exactly once in t. Moreover, it is required that every nonterminal occurs on the left-hand side of exactly one production, and that the relation $\{(A, B) \in N \times N \mid (A(x_1, \ldots, x_n) \to t) \in P, B$ occurs in $t\}$ is acyclic.

A TSLP $\mathcal{G}$ generates a tree val($\mathcal{G}$) in the natural way. During the derivation process, the parameters $x_1, \ldots, x_n$ are instantiated with concrete trees. Instead of giving a formal definition, let us consider an example.

Example 2. Let S, A, B, C be nonterminals, let S be the start nonterminal and let the TSLP $\mathcal{G}$ consist of the following productions:

$$S \to A(B(a), B(a))$$
$$A(x_1, x_2) \to C(C(x_1, a), C(x_2, a))$$
$$C(x_1, x_2) \to h(f(x_1, x_2))$$
$$B(x) \to h(h(x))$$

Then val(G) is the tree from Example 1. It can be derived as follows:

$$S \rightarrow A(B(a), B(a))$$
$$\rightarrow C(C(B(a), a), C(B(a), a))$$
$$\rightarrow C(C(h(h(a)), a), C(h(h(a)), a))$$
$$\rightarrow h(f(C(h(h(a)), a), C(h(h(a)), a)))$$
$$\rightarrow h(f(h(f(h(h(a)), a)), h(f(h(h(a)), a))))$$

The size of a TSLP $\mathcal{G} = (N, \Sigma, S, P)$ is defined as the total number of all nodes in right-hand sides of P, where nodes labelled with a parameter are not counted (see [17] for a discussion of this). Hence, the size of the TSLP in Example 2 is 14. It is easy to show that the size of tree val$(\mathcal{G})$ is bounded by $2^{O(|\mathcal{G}|)}$.

The following result from [28] turned out to be very useful for algorithmic problems on trees that are represented by TSLPs:

Theorem 1 ([28]). *From a given TSLP $\mathcal{G} = (N, \Sigma, S, P)$, where every $A \in N$ has rank at most k and every $\sigma \in \Sigma$ has rank at most r, one can compute in time $O(k \cdot r \cdot |\mathcal{G}|)$ a TSLP $\mathcal{H}$ of size $O(r \cdot |\mathcal{G}|)$ such that (i) val$(\mathcal{G}) = $ val$(\mathcal{H})$ and every nonterminal of $\mathcal{H}$ has rank at most one.*

This result is sharp in the sense that transforming a TSLP into an equivalent TSLP where every nonterminal has rank zero involves an exponential blow-up in the size of the TSLP. For instance, the tree $t_n = f^{2^n}(a)$ with 2^n many occurrences of the unary symbol f can be produced by a TSLP of size $O(n)$ ($S \rightarrow A_n(a)$, $A_i(x) \rightarrow A_{i-1}(A_{i-1}(x))$ for $1 \leq i \leq n$, and $A_0(x) \rightarrow f(x)$) but the minimal dag for t_n is t_n itself. Note that a dag can be transformed into a TSLP, where every nonterminal has rank zero: The nodes of the dag are the nonterminals of the TSLP, and if a σ-labelled node v has the children $v_1, \ldots, v_n$ in the dag (from left to right), then we introduce the production $v \rightarrow \sigma(v_1, \ldots, v_n)$. Similarly, a TSLP, where every nonterminal has rank zero, can be transformed into a dag of the same size.

In [25, 27, 28], also *non-linear* TSLPs were studied. A non-linear TSLP may contain productions of the form $A(x_1, \ldots, x_k) \rightarrow t$, where a parameter x_i occurs several times in t. Non-linear TSLPs can achieve double exponential compression: The non-linear TSLP with the productions $S \rightarrow A_n(a)$, $A_i(x) \rightarrow A_{i-1}(A_{i-1}(x))$ for $1 \leq i \leq n$, and $A_0(x) \rightarrow f(x, x)$ produces a full binary tree of height 2^n and hence has $2^{2^n+1} - 1$ many nodes.

TSLPs generalize SLPs, which produce words instead of trees. In SLPs, symbols do not have a rank, and the productions are simply of the form $A \rightarrow w$, where w consists of terminal symbols and nonterminals. It is required again that every nonterminal occurs on the left-hand side of exactly one production, and that the relation $\{(A, B) \mid (A \rightarrow w)$ is a production and nonterminal B occurs in $w\}$ is acyclic. More details on SLPs can be found in [24].

3 Constructing Small TSLPs

Efficient algorithms that generate for a given input tree t a linear TSLP $\mathcal{G}$ with val$(\mathcal{G}) = t$ are described in [6,26]. The algorithm from [26], called *TreeRePair*, is an extension of the grammar-based string compressor RePair [20] to trees. On a collection of XML skeleton trees (where the data values were removed) the compression ratio of TreeRePair (measured in the size of the computed TSLP divided by the number of edges of the input tree) was about 2.8 %, whereas for the same data set the compression ratio achieved by the minimal dag (number of edges of the minimal dag divided by the number of edges of the input tree) is about 12%, see [26].

Altough TreeRePair works very well on real XML data, its performance is quite poor from a theoretical viewpoint: In [26], a familiy of (binary) trees t_n ($n \geq 1$) is constructed, such that (i) t_n has size $O(n)$, (ii) a TSLP of size $O(\log n)$ for t_n exists, but (iii) the TSLP for t_n computed by TreeRePair has size $\Omega(n)$.

For a tree t let opt(t) be the size of a smallest TSLP for the tree t. Similarly, for a word s let opt(s) be the size of a smallest SLP for the word s. It was shown in [8] that unless $\mathsf{P} = \mathsf{NP}$ there is no polynomial time algorithm that computes for a given word s an SLP of size less than $8569/8568 \cdot$ opt(s). The same result holds also for trees: Simply encode a word by the tree consisting of unary nodes and a single leaf. A TSLP for this tree is basically an SLP for the original word. For SLPs the best known polynomial time grammar-based compressors achieve an approximation ratio of $O(\log(\frac{n}{\mathsf{opt}(s)}))$, i.e., the size of the computed SLP for an input word s of length n is bounded by $O(\mathsf{opt}(s) \cdot \log(\frac{n}{\mathsf{opt}(s)}))$ [8,16,32,33]. Recently, this bound has been also shown for trees:

Theorem 2 ([17]). *From a given tree t of size n, one can compute in linear time a TSLP $\mathcal{G}$ of size $O(r \cdot g + r \cdot g \cdot \log(\frac{n}{r \cdot g}))$ such that* val$(\mathcal{G}) = t$. *Here, $g = $ opt(t) and r is the maximal rank of a node label in t.*

The algorithm from [17] uses three different types of compression operations that are executed repeatedly on the current tree in the following order as long as the tree has size at least two. At the same time we build up the TSLP for the input tree.

Chain Compression: For every unary symbol a, we replace every maximal occurrence of a pattern $a^n(x)$ (maximal means that the parent node of the topmost a-node is not labelled with a and also the unique child of the deepest a-node is not labelled with a) by a single tree that is labelled with a fresh unary symbol a_n. We call such maximal patterns *maximal a-chains*. Moreover, we add to the TSLP productions that generate from the nonterminal $a_n(x)$ the a-chain $a^n(x)$. These productions basically form an SLP for a^n. If $a^{n_1}(x), a^{n_2}(x), \ldots, a^{n_k}(x)$ are all maximal a-chains in the current tree with $n_1 < n_2 < \cdots < n_k$, then the total size of all productions needed to produce these chains can be bounded by $O(k + \sum_{i=1}^{k} \log(n_i - n_{i-1}))$ with $n_0 = 0$.

Pair Compression: After the chain compression step, there do not exist occurrences of a pattern $a(a(x))$ in the current tree for a unary symbol a. Let Σ_1 be the set of all unary symbols that appear in the current tree. We first compute a partition $\Sigma_1 = \Sigma_{0,1} \cup \Sigma_{1,1}$. Then, every occurrence of a pattern $a(b(x))$ with $a \in \Sigma_{0,1}$ and $b \in \Sigma_{1,1}$ is replaced by a single node labelled with the fresh unary symbol $c_{a,b}$. Moreover, we introduce the TSLP-production $c_{a,b}(x) \to a(b(x))$. The partition $\Sigma_1 = \Sigma_{0,1} \cup \Sigma_{1,1}$ is chosen such that the number of occurrences of a pattern $a(b(x))$ with $a \in \Sigma_{0,1}$ and $b \in \Sigma_{1,1}$ is large. More precisely, one can choose the partition such that there are at most $(n_1 - c + 2)/4$ such occurrences, where n_1 is the number of unary nodes in the current tree and c is the number of maximal chains consisting of unary nodes.

Leaf Compression: We eliminate all leaves of the current tree as follows: Let v be an f-labelled node such that v has at least one leaf among its children. Let $n \geq 1$ be the rank of f, and let $1 \leq i_1 < i_2 < \cdots < i_k \leq n$ be the positions of the leaves among the children of v. Let a_j be the label (a constant) of the i_j^{th} child of v. Then we remove all children of v, which are leaves, and replace the label f of v by the fresh symbol $f_{i_1,a_1,\ldots,i_k,a_k}$, which has rank $n - k$. Moreover, we add to the TSLP the production $f_{i_1,a_1,\ldots,i_k,a_k} \to f(x_1,\ldots,x_{i_1-1}, a_1, x_{i_1+1}, \ldots x_{i_2-1}, a_2, x_{i_2+1}, \ldots, x_{i_k-1}, a_k, x_{i_k+1}, \ldots, x_n)$.

Chain compression and pair compression are the two compression steps in Jeż's string compressor from [16]. They allow to shrink chains of unary nodes. Intuitively, if there are no long chains of unary nodes in the tree, then there must be many leaves and leaf compression will shrink the size of the tree substantially. More precisely, it can be shown that in a single phase, consisting of chain compression, followed by pair compression, followed by leaf compression, the size of the tree drops by a constant factor. This allows to come up with a linear bound on the running time. To bound the size of the produced TSLP and, in particular, to compare it with the size of a smallest TSLP for the input tree, Jeż's recompression technique is used in [17].

To the knowledge of the author, there is no algorithm for computing a small non-linear TSLP for a given input tree and thereby achieves a reasonable approximation ratio. This raises the question of whether the size of a smallest non-linear TSLP can be approximated in polynomial time up to a factor of say $\log n$ (assuming reasonable assumptions from complexity theory).

It is well known that for every word $w \in \Sigma^*$ there exists an SLP for w of size $O(\frac{n}{\log_\sigma n})$, where $\sigma = |\Sigma|$. Examples of grammar-based compressors that achieve this bound are for instance LZ78 or BiSection [18]. A simple information theoretic argument shows that the bound $O(\frac{n}{\log_\sigma n})$ is optimal. By the following result from [15] the same bound holds also for binary trees and TSLPs.

Theorem 3 ([15]). *From a given tree t of size n, where every terminal symbol has rank at most 2, one can compute in linear time a TSLP $\mathcal{G}$ of size $O(\frac{n}{\log_\sigma n})$ such that $\mathrm{val}(\mathcal{G}) = t$. Here, σ is the number of different node labels that appear in t.*

In [15], only the bound $O(n \log n)$ on the running time is stated. The linear time algorithm will appear in a long version of [15]. Let us briefly sketch the linear time algorithm for a binary trees t with node labels from a set Σ ($|\Sigma| = \sigma$). The algorithm works in two steps:

Step 1. We decompose the tree t into $O(\frac{n}{\log_\sigma n})$ many clusters (connected subgraphs) of size at most $c \cdot \log_\sigma n$ for a constant c that will be chosen later. Each cluster is a full subtree of t with at most two full subtrees of t removed from it. Hence, we can write such a cluster as a tree $u(x_1, \ldots, x_k)$ with $k \geq 2$, where every parameter x_i appears exactly once. We replace each cluster $u(x_1, \ldots, x_k)$ in t by a single node labelled with a nonterminal A_u of rank at most two, and introduce the production $A_u(x_1, \ldots, x_k) \to u(x_1, \ldots, x_k)$. Note that the resulting tree s has size $O(\frac{n}{\log_\sigma n})$. We add the production $S \to s$, where S is the start nonterminal. With some care, this first step can be done in linear time.

Step 2. The TSLP we obtain from the previous step has size $O(n)$, so nothing is gained. We now compute in linear time (using [10]) the minimal dag for the forest consisting of all cluster trees $u(x_1, \ldots, x_k)$. Recall that each such tree has size at most $c \cdot \log_\sigma n$. Hence, to bound the size of the minimal dag of this forest, one only has to count the number of binary trees of size at most $c \cdot \log_\sigma n$, where every node is labelled with a symbol from $\Sigma \cup \{x_1, x_2\}$. By choosing the constant c suitably, we can (using the formula for the number of binary trees of size m, which is given by the Catalan numbers) bound this number by $\sqrt{n}$. The minimal dag for the cluster trees together with the start production $S \to s$ translates into a TSLP for t of size $\sqrt{n} + O(\frac{n}{\log_\sigma n}) = O(\frac{n}{\log_\sigma n})$.

Theorem 3 can be generalized to trees of higher rank. Then the constant hidden in the big-O-notation depends on the maximal rank of a terminal symbol, but the precise dependence is not analyzed in [15].

A simple information theoretic argument shows that the average size of a minimal TSLP for a uniformly chosen tree of size n with labels from an alphabet of size σ is $\Omega(\frac{n}{\log_\sigma n})$ and hence, by Theorem 3, $\Theta(\frac{n}{\log_\sigma n})$. In [11] it is shown that the average size of the minimal dag of a uniformly chosen binary tree with n unlabelled nodes is $\Theta(\frac{n}{\sqrt{\log n}})$. In [3] this result is extended to node-labelled unranked trees.

With some additional effort, one can ensure that the TSLP $\mathcal{G}$ in Theorem 3 has height $O(\log n)$. This has an interesting application for the problem of transforming arithmetical expressions into circuits (i.e., dags). Let $\mathbb{S} = (S, +, \cdot)$ be a (not necessarily commutative) semiring. Thus, $(S, +)$ is a commutative monoid with identity element 0, $(S, \cdot)$ is a monoid with identity element 1, and $\cdot$ left and right distributes over $+$. An *arithmetical expression* is just a labelled binary tree where internal nodes are labelled with the semiring operations $+$ and $\cdot$, and leaf nodes are labelled with variables $y_1, y_2, \ldots$ or the constants 0 and 1. An *arithmetical circuit* is a (not necessarily minimal) dag whose internal nodes are labelled with $+$ and $\cdot$ and whose leaf nodes are labelled with variables or the constants 0 and 1. The *depth* of a circuit is the length of a longest path from the root node to a leaf. An arithmetical circuit evaluates to a multivariate noncommutative polynomial $p(y_1, \ldots, y_n)$ over $\mathbb{S}$, where $y_1, \ldots, y_n$ are the variables occurring at the leaf nodes. Two arithmetical circuits are equivalent if they evaluate to the same polynomial. Brent [4] has shown that every arithmetical expression of size n over a commutative ring can be transformed into an equivalent circuit of depth $O(\log n)$ and size $O(n)$ (the proof easily generalizes to semirings). Using Theorem 3 one can refine the size bound to $O(\frac{n \cdot \log m}{\log n})$, where m is the number of different variables in the formula:

Theorem 4 ([15]). *A given arithmetical expression F of size n having m different variables can be transformed in time $O(n)$ into an arithmetical circuit C of depth $O(\log n)$ and size $O(\frac{n \cdot \log m}{\log n})$ such that over every semiring, C and F evaluate to the same noncommutative polynomial (in m variables).*

To show Theorem 4 one first transforms the arithmetical expression into a TSLP of size $O(\frac{n}{\log_m n}) = O(\frac{n \cdot \log m}{\log n})$. Then one transforms the TSLP into a circuit that evaluates to the same polynomial (over any semiring) as the TSLP. Only for this second step, one has to use the semiring structure.

There are also some other tree compressors that use grammar formalisms slightly different from TSLPs. In [1] so called elementary ordered tree grammars are used, and a polynomial time compressor with an approximation ratio of $O(n^{5/6})$ is presented. Also the *top dags* from [2] can be seen as a variation of TSLPs for unranked trees. In [2] it was shown that for every tree t of size n the top dag has size $O(\frac{n}{\log^{0.19} n})$. An extension of TSLPs to higher order tree grammars was proposed in [19].

4   Algorithmic Problems for TSLP-Compressed Trees

Let us now consider algorithmic problems for TSLP-compressed trees. Probably the most basic question is whether two trees, both given by TSLPs, are equal.

Theorem 5 ([6, 34]). *For two given TSLPs $\mathcal{G}$ and $\mathcal{H}$ it can be checked in polynomial time, whether $\operatorname{val}(\mathcal{G}) = \operatorname{val}(\mathcal{H})$.*

For the proof of Theorem 5 one constructs in polynomial time from a TSLP $\mathcal{G}$ an SLP $\mathcal{G}'$ such that $\operatorname{val}(\mathcal{G}')$ represents a depth-first left-to-right transversal of the tree $\operatorname{val}(\mathcal{G})$. For this, $\mathcal{G}'$ contains $k+1$ nonterminals $A_{0,1}, A_{1,2}, A_{2,3}, \ldots, A_{k-1,k}$, $A_{k,0}$ for a rank-k nonterminal A of $\mathcal{G}$. Intuitively, $A_{0,1}$ produces the part of the traversal of $\operatorname{val}_{\mathcal{G}}(A)$ from the root of $\operatorname{val}_{\mathcal{G}}(A)$ to the position of the first parameter, $A_{i,i+1}$ $(1 \leq i \leq k-1)$ produces the part of the traversal from the position of the ith parameter to the position of the $(i+1)^{\text{th}}$ parameter, and $A_{k,0}$ produces the part of the traversal from the position of the kth parameter back to the root. For the TSLP from Example 2 we obtain the following SLP:

$$S_{0,0} \rightarrow A_{0,1} B_{0,1} a B_{1,0} A_{1,2} B_{0,1} a B_{1,0} A_{2,0}$$

$$A_{0,1} \rightarrow C_{0,1} C_{0,1}, \quad A_{1,2} \rightarrow C_{1,2} a C_{2,0} C_{1,2} C_{0,1}, \quad A_{2,0} \rightarrow C_{1,2} a C_{2,0} C_{2,0}$$

$$C_{0,1} \rightarrow hf, \quad C_{1,2} \rightarrow \varepsilon, \quad C_{2,0} \rightarrow \varepsilon$$

$$B_{0,1} \rightarrow hh, \quad B_{1,0} \rightarrow \varepsilon$$

For a ranked tree, its depth-first left-to-right transversal uniquely represents the tree. Therefore, for two TSLPs $\mathcal{G}$ and $\mathcal{H}$ we have val($\mathcal{G}$) = val($\mathcal{H}$) if and only if val($\mathcal{G}'$) = val($\mathcal{H}'$). Hence, equality of trees that are represented by linear TSLPs can be reduced to checking equality of SLP-compressed words, which can be checked in polynomial time by a famous result of Plandowski [31] (which has been indendently shown in [14,30]). In [34], Theorem 5 is shown by a direct extension of Plandowski's algorithm for SLPs.

It is open whether Theorem 5 can be extended to non-linear TSLPs. For these, the best upper bound on the equivalence problem is PSPACE [6] and no good lower bound is known.

In [13], Theorem 5 has been extended to the unification problem. Unification is a classical problem in logic and deduction. One considers trees s and t with distinguished variables (these variables should be not confused with the parameters in TSLPs), which label leaf nodes. The trees s and t are unifiable if there exists a substitution that maps every variable x appearing in s or t to a variable-free tree (also called ground term) such that $\sigma(s) = \sigma(t)$. Here, $\sigma(s)$ (resp., $\sigma(t)$) denotes the tree that is obtained by replacing every x-labelled leaf of s (resp., t) by the tree $\sigma(x)$. The following result has been shown in [13]:

Theorem 6 ([13]). *For two given TSLPs $\mathcal{G}$ and $\mathcal{H}$ (where some of the terminal symbols of rank 0 are declared as variables) it can be checked in polynomial time, whether* val($\mathcal{G}$) *and* val($\mathcal{H}$) *are unifiable.*

In fact, the representation of the most general unifier of val($\mathcal{G}$) and val($\mathcal{H}$) in terms of TSLPs for the variables is computed in [13] in polynomial time.

In [36], the authors studied the compressed submatching problem: The input consists of TSLPs $\mathcal{G}$ (the pattern TSLP) and $\mathcal{H}$, where some of the terminal symbols of rank 0 appearing in $\mathcal{G}$ are declared as variables, and it is asked whether there exists a substitution σ such that $\sigma(\text{val}(\mathcal{G}))$ is a subtree of val($\mathcal{H}$). Whereas the complexity of the general compressed submatching problem is still open (the best upper bound is NP), Schmidt-Schauß proved in [36]:

Theorem 7 ([36]). *Compressed submatching can be solved in polynomial time, if (i) no variable appears more than once in the tree produced by the pattern TSLP (i.e., this tree is linear) or (ii) all nonterminals in the pattern TSLP have rank zero (i.e., the pattern TSLP is in fact a dag).*

So far, we considered ordered trees, where the children of a node are linearly ordered. Deciding isomorphism of unordered trees, where the children are not ordered is more difficult than for ordered trees. For explicitly given unordered trees, isomorphism can be decided in logspace by a result of Lindell [22]. For unordered trees that are given by dags, one can solve the isomorphism problem by a simple partition refinement algorithm [29]. Recently this result has been extended to unordered trees that are represented by TSLPs [27]:

Theorem 8 ([27]). *For two given TSLPs $\mathcal{G}$ and $\mathcal{H}$ it can be checked in polynomial time, whether* val($\mathcal{G}$) *and* val($\mathcal{H}$) *are isomorphic as unordered trees.*

For non-linear TSLPs it was shown in [27] that the problem whether val($\mathcal{G}$) and val($\mathcal{H}$) are isomorphic as unordered trees is PSPACE-hard and in EXPTIME.

In [25, 28], the problem of evaluating tree automata over TSLP-compressed input trees was considered. A tree automaton runs on a ranked input tree bottom-up and thereby assigns states to tree nodes. Transitions are of the form $(q_1, \ldots, q_n, f, q)$, where f is a node label of rank n and $q_1, \ldots, q_n, q$ are states of the tree automaton. Then a run of the tree automaton is a mapping ρ from the tree nodes to states that is consistent with the set of transitions in the following sense: If a tree node is labelled with the symbol f (of rank n) and $v_1, \ldots, v_n$ are the children of v in that order, then $(\rho(v_1), \ldots, \rho(v_n), f, \rho(v))$ must be a transition of the tree automaton. A tree automaton accepts a tree if there is a run that assigns a final state to the root of the tree (every tree automaton has a distinguished set of final states). The problem of checking whether an explicitly given tree is accepted by a tree automaton that is part of the input is complete for the class LogCFL (which is contained in the parallel class NC^2) [23]. For a fixed tree automaton this problem belongs to NC^1 [23]. For TSLP-compressed trees we have:

Theorem 9 ([28]). *It is* P-*complete to check for a given TSLP $\mathcal{G}$ and a given tree automaton $\mathcal{A}$, whether $\mathcal{A}$ accepts* val($\mathcal{G}$).

The polynomial time algorithm works in two steps:

Step 1. Using Theorem 1 the input TSLP $\mathcal{G}$ is transformed in polynomial time into a TSLP $\mathcal{H}$ such that val($\mathcal{G}$) = val($\mathcal{H}$) and every nonterminal of $\mathcal{H}$ has rank at most one.

Step 2. For a TSLP $\mathcal{G}$, where every nonterminal has rank 0 or 1, it is easy to evaluate a tree automaton $\mathcal{A}$ on val($\mathcal{G}$). Bottom-up on the structure of the TSLP, one computes for every nonterminal A of rank 0 the set of states to which $\text{val}_{\mathcal{G}}(A)$ can evaluate (i.e., those states that may appear in a run at the root), whereas for a nonterminal A of rank 1 one computes a binary relation on the set of states of $\mathcal{A}$. This relation contains a pair (q_1, q_2) if and only if the following holds: There is a mapping from the nodes of $\text{val}_{\mathcal{G}}(A)$ to the states of $\mathcal{A}$ such that (i) the above condition of a run is satisfied, (ii) to the unique parameter-labelled node of $\text{val}_{\mathcal{G}}(A)$ the state q_1 is assigned, and (iii) to the root the state q_2 is assigned. It is easy to compute this information for a nonterminal A assuming it has been computed for all nonterminals in the right-hand side of A.

In [28], also a generalization of Theorem 9 to tree automata with sibling constraints is shown. In this model, transitions can depend on (dis)equalities between the children of the node to which the transition is applied to.

The problem, whether a given tree automaton accepts the tree val($\mathcal{G}$), where $\mathcal{G}$ is a given non-linear TSLP was shown to be PSPACE-complete in [25]. In fact, PSPACE-hardness already holds for a fixed tree automaton.

Several other algorithmic problems for TSLP-compressed input trees are studied in [7, 13, 21, 35–37].

References

1. Akutsu, T.: A bisection algorithm for grammar-based compression of ordered trees. Information Processing Letters **110**(18–19), 815–820 (2010)
2. Bille, P., Gørtz, I.L., Landau, G.M., Weimann, O.: Tree compression with top trees. In: Fomin, F.V., Freivalds, R., Kwiatkowska, M., Peleg, D. (eds.) ICALP 2013, Part I. LNCS, vol. 7965, pp. 160–171. Springer, Heidelberg (2013)
3. Bousquet-Mélou, M., Lohrey, M., Maneth, S., Noeth, E.: XML compression via DAGs. Theory of Computing Systems (2014). doi:10.1007/s00224-014-9544-x
4. Brent, R.P.: The parallel evaluation of general arithmetic expressions. Journal of the Association for Computing Machinery **21**(2), 201–206 (1974)
5. Buneman, P., Grohe, M., Koch, C.: Path queries on compressed XML. In: Proceedings of VLDB 2003, pp. 141–152. Morgan Kaufmann (2003)
6. Busatto, G., Lohrey, M., Maneth, S.: Efficient memory representation of XML document trees. Information Systems **33**(4–5), 456–474 (2008)
7. Carles Creus, A.G., Godoy, G.: One-context unification with STG-compressed terms is in NP. In: Proceedings of RTA 2012, vol. 15 of LIPIcs, pp. 149–164. Schloss Dagstuhl - Leibniz-Zentrum für Informatik (2012)
8. Charikar, M., Lehman, E., Lehman, A., Liu, D., Panigrahy, R., Prabhakaran, M., Sahai, A., Shelat, A.: The smallest grammar problem. IEEE Transactions on Information Theory **51**(7), 2554–2576 (2005)
9. Comon, H., Dauchet, M., Gilleron, R., Jacquemard, F., Lugiez, D., Löding, C., Tison, S., Tommasi, M.: Tree automata techniques and applications (2007). http://tata.gforge.inria.fr/
10. Downey, P.J., Sethi, R., Tarjan, R.E.: Variations on the common subexpression problem. Journal of the Association for Computing Machinery **27**(4), 758–771 (1980)
11. Flajolet, P., Sipala, P., Steyaert, J.-M.: Analytic variations on the common subexpression problem. In: Paterson, M. (ed.) ICALP 1990. LNCS, vol. 443, pp. 220–234. Springer, Heidelberg (1990)
12. Frick, M., Grohe, M., Koch, C.: Query evaluation on compressed trees (extended abstract). In: Proceedings of LICS 2003, pp. 188–197. IEEE Computer Society Press (2003)
13. Gascón, A., Godoy, G., Schmidt-Schauß, M.: Unification and matching on compressed terms. ACM Transactions on Computational Logic **12**(4), 26 (2011)
14. Hirshfeld, Y., Jerrum, M., Moller, F.: A polynomial algorithm for deciding bisimilarity of normed context-free processes. Theoretical Computer Science **158**(1&2), 143–159 (1996)
15. Hucke, D., Lohrey, M., Noeth, E.: Constructing small tree grammars and small circuits for formulas. In: Proceedings of FSTTCS 2014, vol. 29 of LIPIcs, pp. 457–468. Schloss Dagstuhl - Leibniz-Zentrum für Informatik (2014)
16. Jeż, A.: Approximation of grammar-based compression via recompression. In: Fischer, J., Sanders, P. (eds.) CPM 2013. LNCS, vol. 7922, pp. 165–176. Springer, Heidelberg (2013)
17. Jeż, A., Lohrey, M.: Approximation of smallest linear tree grammars. In: Proceedings of STACS 2014, vol. 25 of LIPIcs, pp. 445–457. Schloss Dagstuhl - Leibniz-Zentrum für Informatik (2014)
18. Kieffer, J.C., Yang, E.H.: Grammar-based codes: A new class of universal lossless source codes. IEEE Transactions on Information Theory **46**(3), 737–754 (2000)

19. Kobayashi, N., Matsuda, K., Shinohara, A.: Functional programs as compressed data. In: Proceedings of PEPM 2012, pp. 121–130. ACM Press (2012)
20. Larsson, N.J., Moffat, A.: Offline dictionary-based compression. In: Proceedings of DCC 1999, pp. 296–305. IEEE Computer Society Press (1999)
21. Levy, J., Schmidt-Schauß, M., Villaret, M.: The complexity of monadic second-order unification. SIAM Journal on Computing **38**(3), 1113–1140 (2008)
22. Lindell, S.: A logspace algorithm for tree canonization (extended abstract). In: Proceedings of STOC 1992, pp. 400–404. ACM Press (1992)
23. Lohrey, M.: On the parallel complexity of tree automata. In: Middeldorp, A. (ed.) RTA 2001. LNCS, vol. 2051, pp. 201–215. Springer, Heidelberg (2001)
24. Lohrey, M.: Algorithmics on SLP-compressed strings: A survey. Groups Complexity Cryptology **4**(2), 241–299 (2012)
25. Lohrey, M., Maneth, S.: The complexity of tree automata and XPath on grammar-compressed trees. Theoretical Computer Science **363**(2), 196–210 (2006)
26. Lohrey, M., Maneth, S., Mennicke, R.: XML tree structure compression using RePair. Information Systems **38**(8), 1150–1167 (2013)
27. Lohrey, M., Maneth, S., Peternek, F.: Compressed tree canonization.Technical report, arXiv.org (2015). http://arxiv.org/abs/1502.04625. An extended abstract will appear in Proceedings of ICALP 2015
28. Lohrey, M., Maneth, S., Schmidt-Schauß, M.: Parameter reduction and automata evaluation for grammar-compressed trees. Journal of Computer and System Sciences **78**(5), 1651–1669 (2012)
29. Lohrey, M., Mathissen, C.: Isomorphism of regular trees and words. Information and Computation **224**, 71–105 (2013)
30. Mehlhorn, K., Sundar, R., Uhrig, C.: Maintaining dynamic sequences under equality-tests in polylogarithmic time. In: Proceedings of SODA 1994, pp. 213–222. ACM/SIAM (1994)
31. Plandowski, W.: Testing equivalence of morphisms on context-free languages. In: van Leeuwen, J. (ed.) ESA 1994. LNCS, vol. 855, pp. 460–470. Springer, Heidelberg (1994)
32. Rytter, W.: Application of Lempel-Ziv factorization to the approximation of grammar-based compression. Theoretical Computer Science **302**(1–3), 211–222 (2003)
33. Sakamoto, H.: A fully linear-time approximation algorithm for grammar-based compression. Journal of Discrete Algorithms **3**(2–4), 416–430 (2005)
34. Schmidt-Schauß, M.: Polynomial equality testing for terms with shared sub-structures. Technical Report Report 21, Institut für Informatik, J. W. Goethe-Universität Frankfurt am Main (2005)
35. Schmidt-Schauß, M.: Matching of compressed patterns with character-variables. In: Proceedings of RTA 2012, vol. 15 of LIPIcs, pp. 272–287. Schloss Dagstuhl - Leibniz-Zentrum für Informatik (2012)
36. Schmidt-Schauß, M.: Linear compressed pattern matching for polynomial rewriting (extended abstract). In: Proceedings of TERMGRAPH 2013, vol. 110 of EPTCS, pp. 29–40 (2013)
37. Schmidt-Schauss, M., Sabel, D., Anis, A.: Congruence closure of compressed terms in polynomial time. In: Tinelli, C., Sofronie-Stokkermans, V. (eds.) FroCoS 2011. LNCS, vol. 6989, pp. 227–242. Springer, Heidelberg (2011)
38. Zhang, J., Yang, E.-H., Kieffer, J.C.: A universal grammar-based code for lossless compression of binary trees. IEEE Transactions on Information Theory **60**(3), 1373–1386 (2014)

On Decidability of Intermediate Levels
of Concatenation Hierarchies

Jorge Almeida[1], Jana Bartoňová[2], Ondřej Klíma[2], and Michal Kunc[2 (✉)]

[1] CMUP, Dep. Matemática, Faculdade de Ciências,
Universidade do Porto, Rua do Campo Alegre 687, 4169-007 Porto, Portugal
jalmeida@fc.up.pt
[2] Department of Mathematics and Statistics, Masaryk University,
Kotlářská 2, 611 37 Brno, Czech Republic
{xbartonovaj,klima,kunc}@math.muni.cz

Abstract. It is proved that if definability of regular languages in the Σ_n fragment of the first-order logic on finite words is decidable, then it is decidable also for the Δ_{n+1} fragment. In particular, the decidability for Δ_5 is obtained. More generally, for every concatenation hierarchy of regular languages, it is proved that decidability of one of its half levels implies decidability of the intersection of the following half level with its complement.

1 Introduction

A remarkable connection between finite automata and regular languages on one hand and logic on the other hand was found by McNaughton and Papert [6], who proved that star-free languages are exactly those languages that are definable in first-order logic FO[<] on finite words. The decidability of star-freeness of a regular language is in turn known thanks to Schützenberger [15], who provided a key connection with algebra by showing that star-free languages are those whose syntactic monoids are aperiodic.

Within the class of all star-free languages, Brzozowski and Cohen [5] defined in 1971 the so-called *dot-depth hierarchy*, based on the polynomial closure and Boolean closure operators. A long standing open question about this hierarchy is to algorithmically determine the minimum level in the hierarchy to which a given star-free language belongs. The logical significance of this problem was discovered by Thomas [17], who proved that levels of the natural variant of the dot-depth hierarchy known as the Straubing–Thérien hierarchy correspond to levels of the quantifier-alternation hierarchy within FO[<] (see Figure 1). More precisely, a regular language is definable in the Σ_n fragment if and only if it belongs to the polynomial closure of the $(n-1)$th level of the Straubing–Thérien hierarchy; this polynomial closure is often referred to as the $(n-1/2)$th

The first author was partially supported by CMUP (UID/MAT/00144/2013), which is funded by FCT (Portugal) with national (MEC) and European structural funds through the programs FEDER, under the partnership agreement PT2020. The last two authors were supported by grant 15-02862S of the Czech Science Foundation.

I. Potapov (Ed.): DLT 2015, LNCS 9168, pp. 58–70, 2015.
DOI: 10.1007/978-3-319-21500-6_4

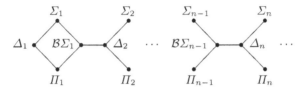

Fig. 1. The quantifier-alternation hierarchy of sentences of FO[<]

level of the hierarchy. Furthermore, the nth level of the Straubing–Thérien hierarchy contains precisely $\mathcal{B}\Sigma_n$-definable languages, that is, languages definable by any sentence with n alternations of quantifiers. Finally, the Δ_n fragment of FO[<] defines exactly the unambiguous polynomial closure of the $(n-1)$th level of the Straubing–Thérien hierarchy, which is equal to the intersection of the $(n-1/2)$th level with its complement. It is also worth mentioning that the connection between the original dot-depth hierarchy and the Straubing–Thérien hierarchy is well understood through the work of Straubing [16], which provides an algebraic transformation from the latter to the former which preserves decidability in both directions.

The Straubing–Thérien hierarchy is the most prominent example of a concatenation hierarchy of regular languages; each concatenation hierarchy is obtained in the same way, once its lowest level is suitably chosen. Pin, Straubing and Thérien [8] and Pin and Weil [10] described the algebraic counterpart of the unambiguous polynomial closure and polynomial closure operators using the Mal'cev product of pseudovarieties of finite (ordered) monoids. These results (together with [14, Theorem 4.6.50]) show that decidability of any integer level in an arbitrary concatenation hierarchy implies decidability of the intersection of the next half level with its complement.

The decidability of the Straubing–Thérien hierarchy turned out to be very difficult. The problem was eventually solved for levels up to $3/2$ (see Subsection 8.1 of [7]), but the first algorithm for deciding definability in $\mathcal{B}\Sigma_2$ has been announced only as late as 2014 by Place and Zeitoun [12]. In the same paper, it is also shown that definability in Σ_3 is decidable (to which, for shortness, we will refer by saying Σ_3 is decidable), and decidability of Σ_4 was recently proved by Place [11]. Place and Zeitoun [12] further provided, for each n, a non-effective description of languages definable in Σ_{n+1} and Δ_{n+1} based on inequalities valid in syntactic ordered monoids of languages definable in Σ_n. Their proof is specific for the Straubing–Thérien hierarchy, as it is based on manipulations with first-order formulas.

Thus, only a finite number of levels of the Straubing–Thérien hierarchy have been proved to be decidable. The main result of this paper, whose proof employs techniques of profinite monoids, implies the following statement:

For each n, the problem of definability in Δ_{n+1} polynomially reduces to the problem of definability in Σ_n. In particular, if Σ_n is decidable, then so is Δ_{n+1}.

More generally, we provide, for an arbitrary concatenation hierarchy, a polynomial time reduction of the membership problem for the intersection of any half level with its complement to the membership problem for the previous half level. Combining with the results of Place and Zeitoun [12], one obtains that our result implies decidability of Δ_4, a fact that was independently discovered by Place [11]. Furthermore, from the decidability of Σ_4 also proved by Place [11], our result yields the decidability of Δ_5.

In the following section, the definition of concatenation hierarchies and basic concepts of the algebraic theory of regular languages are recalled, in order to fix terminology and notation; for a more comprehensive overview of results on these hierarchies and for a general introduction to the theory, we refer to a handbook chapter by Pin [7]. Section 3 describes the relationship between intermediate levels in concatenation hierarchies, while Sections 4 and 5 are devoted to turning this relationship into a polynomial time reduction of the membership problems.

2   Basic Concepts

2.1   Concatenation Hierarchies of Regular Languages

A *class* of regular languages consists of a set of regular languages over each finite alphabet. A *positive variety* $\mathcal{V}$ is a class of regular languages such that languages of $\mathcal{V}$ over each alphabet are closed under finite intersections, finite unions and quotients, and additionally, for every homomorphism $f \colon A^* \to B^*$ and every $L \subseteq B^*$ belonging to $\mathcal{V}$, the language $f^{-1}(L)$ belongs to $\mathcal{V}$ as well. A positive variety $\mathcal{V}$ is a *variety* if it is also closed under complementation. For a positive variety $\mathcal{V}$, we denote by Co-$\mathcal{V}$ the positive variety consisting precisely of complements of languages in $\mathcal{V}$.

For a set of languages T over an alphabet A, its *polynomial closure* is the set of all languages over A, which are finite unions of languages of the form $L_0 a_1 L_1 \ldots a_n L_n$, where $n \geqslant 0$, $a_i \in A$, and $L_i \in T$. If $\mathcal{V}$ is a variety of languages, then we denote by Pol$\mathcal{V}$ the class of languages obtained by performing, over every alphabet, the polynomial closure of languages from $\mathcal{V}$. The resulting class Pol$\mathcal{V}$ is a positive variety of languages (see Theorem 7.1 of [7]). Moreover, we denote by BPol$\mathcal{V}$ the variety of languages consisting of Boolean combinations of languages from Pol$\mathcal{V}$. In other words, BPol$\mathcal{V}$ is the join of Pol$\mathcal{V}$ and Co-Pol$\mathcal{V}$ in the lattice of all positive varieties. A widely studied variant of the polynomial closure is the unambiguous polynomial closure UPol, which is known to satisfy UPol$\mathcal{V}$ = Pol$\mathcal{V} \cap$ Co-Pol$\mathcal{V}$ for every variety $\mathcal{V}$; for details, see Subsection 7.2 of [7].

For a variety of languages $\mathcal{V}_0$, the *concatenation hierarchy* of basis $\mathcal{V}_0$ is a hierarchy of classes of languages defined by the rules $\mathcal{V}_{n+1/2} = \text{Pol}\,\mathcal{V}_n$ and $\mathcal{V}_{n+1} = \text{BPol}\,\mathcal{V}_n$, for every integer $n \geqslant 0$. Each half level in the hierarchy is a positive variety and each integer level is a variety. Such a hierarchy, together with complements of half levels, is depicted in Figure 2. The *Straubing–Thérien hierarchy* is the concatenation hierarchy whose basis $\mathcal{V}_0$ is formed by the smallest variety of languages, which contains only the languages $\emptyset$ and A^* over each

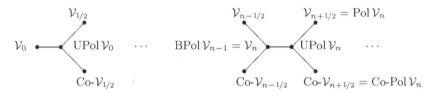

Fig. 2. A concatenation hierarchy of languages

alphabet A. In this hierarchy, the equality $\mathcal{V}_0 = \mathrm{UPol}\,\mathcal{V}_0$ holds, and all other inclusions in Figure 2 are proper. In the case of the Straubing–Thérien hierarchy, Figure 2 represents the language counterpart of Figure 1, except for $\mathcal{V}_0$.

2.2 Ordered Monoids

A binary relation R on a set M is said to be a *quasiorder* if it is reflexive and transitive. Every quasiorder R on M determines an equivalence relation R^{e} on M, consisting of all pairs $(s, t) \in R$ such that $(t, s) \in R$. Then R induces a partial order on the quotient set M/R^{e}, and we denote the resulting partially ordered set by M/R. The *transitive closure* of a binary relation R on M is denoted $\mathrm{T}(R)$.

A binary relation R on a monoid M is called *stable* if for all $(s, t) \in R$ and $z \in M$, both pairs (sz, tz) and (zs, zt) belong to R. If R is a stable and reflexive relation on M, then its transitive closure $\mathrm{T}(R)$ is a stable quasiorder on M.

An *ordered monoid* $(M, \leqslant)$ is a monoid M equipped with a stable partial order $\leqslant$. A homomorphism between ordered monoids $(M, \leqslant)$ and $(N, \leqslant)$ is a mapping $\varphi \colon M \to N$ which is a monoid homomorphism and at the same time monotone, *i.e.*, for all $s, t \in M$ satisfying $s \leqslant t$, the inequality $\varphi(s) \leqslant \varphi(t)$ holds in N. Given a stable quasiorder R on an unordered monoid M, the ordered set M/R can be turned into an ordered monoid, and the natural projection $\pi \colon M \to M/R$ becomes a homomorphism of ordered monoids, with M ordered by the equality relation.

2.3 Pseudovarieties and Pseudoidentities

There is a one-to-one correspondence between varieties of languages and *pseudovarieties* of finite monoids, which are classes of finite monoids closed under homomorphic images, submonoids and finite direct products. Similarly, positive varieties of languages are in one-to-one correspondence with pseudovarieties of finite ordered monoids. For instance, a positive variety $\mathcal{V}$ corresponds to the pseudovariety V generated by syntactic ordered monoids of languages in $\mathcal{V}$. Then a language $L \subseteq A^*$ belongs to $\mathcal{V}$ if and only if its syntactic ordered monoid belongs to V, while an ordered monoid M belongs to V if and only if all preimages of upper sets in M under homomorphisms $\varphi \colon A^* \to M$ belong to $\mathcal{V}$. In particular, the decidability of the membership problems for $\mathcal{V}$ and for V are equivalent.

For an arbitrary pseudovariety V of ordered monoids, with the corresponding positive variety $\mathcal{V}$, the pseudovariety corresponding to the positive variety Co-$\mathcal{V}$

is $V^d = \{(M, \geqslant) \mid (M, \leqslant) \in V\}$. We denote by $\mathcal{B}V$ the join of V and its dual V^d in the lattice of all pseudovarieties of ordered monoids; it is the pseudovariety corresponding to the Boolean closure of $\mathcal{V}$. The pseudovariety $\mathcal{B}V$ can be equivalently characterized as the pseudovariety of ordered monoids generated by the class $\{(M, =) \mid (M, \leqslant) \in V\}$. We call a pseudovariety V of ordered monoids *selfdual* if $V = V^d$; these are precisely pseudovarieties of the form $\mathcal{B}W$ for some pseudovariety W, or equivalently, those which satisfy $V = \mathcal{B}V$. There is a one-to-one correspondence between selfdual pseudovarieties of ordered monoids and pseudovarieties of monoids; we will not distinguish between a selfdual pseudovariety V and the corresponding pseudovariety of monoids $\{M \mid (M, \leqslant) \in V\}$.

Reiterman [13] (see [7, Section 4.1]) proved that every pseudovariety of monoids V can be characterized by some set of so-called *pseudoidentities*; each such set is usually called a *basis of pseudoidentities* for V. Pseudoidentities are generalizations of identities, that is, pairs $u = v$ of words $u, v \in A^*$, which determine equational axioms for classes of finite monoids. Since not all pseudovarieties of monoids are equational in this sense, Reiterman's idea was to consider identities of generalized words. For this purpose, one takes the completion $\widehat{A^*}$ of the free monoid A^* with respect to the metric induced by homomorphisms from A^* into finite monoids, in which, roughly speaking, two distinct words are very close if it takes a homomorphism from A^* into a very large finite monoid to distinguish them. The metric monoid $\widehat{A^*}$ is *profinite*, which means that it is compact and each pair of its elements can be distinguished by a continuous homomorphism to a finite monoid endowed with the discrete topology. Moreover, the monoid $\widehat{A^*}$ is characterized by a universal property among profinite monoids, namely, for every mapping $\alpha \colon A \to S$ to a profinite monoid S, there exists a unique continuous homomorphism $\overline{\alpha} \colon \widehat{A^*} \to S$ extending α (see [2, Proposition 3.4]). This property can be in particular used if S is a finite monoid.

A pseudoidentity $u = v$ is a pair of elements $u, v \in \widehat{A^*}$, and this pseudoidentity is satisfied by a finite monoid M if $\overline{\alpha}(u) = \overline{\alpha}(v)$ holds for every mapping $\alpha \colon A \to M$. In the case of pseudovarieties of ordered monoids, the previous definition must be modified to use the so-called inequalities. An *inequality* $u \leqslant v$ is a pair of elements of $\widehat{A^*}$, and it is satisfied by an ordered monoid $(M, \leqslant)$ if $\overline{\alpha}(u) \leqslant \overline{\alpha}(v)$ holds for every mapping $\alpha \colon A \to M$. Note that in this paper, as usual in recent literature, the ordering of monoids corresponding to a positive variety of languages is dual to the one used by Pin [7], and thus all inequalities in [7] characterizing pseudovarieties of ordered monoids have to be reversed. This stems from the fact that the reverse of the syntactic order which used to appear in the literature has come to be preferred.

An element e of a monoid M is called an *idempotent* if $ee = e$. For an arbitrary element s of a profinite monoid S, the sequence $s^{n!}$ converges in S to an idempotent, which is denoted by s^ω. In particular, if S is a finite monoid, then s^ω is the unique idempotent which is a power of s. We also write $s^{\omega-1}$ for the limit of the sequence $s^{n!-1}$. For every continuous homomorphism $\gamma \colon S \to T$ of profinite monoids and every $s \in S$, it is clear from the definition that $\gamma(s^\omega) = \gamma(s)^\omega$. A finite monoid is *aperiodic* if it satisfies the pseudoidentity $x^{\omega+1} = x^\omega$.

The pseudovariety of aperiodic monoids corresponds to the variety of star-free languages (see [7, Theorem 5.2]).

3 Bases of Inequalities for Polynomial Closure

The aim of this section is to obtain such descriptions of various levels of concatenation hierarchies by inequalities, that can be turned into algorithms for testing membership.

We will rely on a result that gives the inequalities of the pseudovarieties corresponding to $\mathrm{Pol}\,\mathcal{V}$ and $\mathrm{UPol}\,\mathcal{V}$ as a function of those of the pseudovariety corresponding to $\mathcal{V}$:

Proposition 1 ([8–10], [7, Theorems 6.5, 7.1 and 7.3]). *Let* V *be a pseudovariety of monoids and* $\mathcal{V}$ *be the corresponding variety of languages. Then*

(i) *the pseudovariety of ordered monoids corresponding to the positive variety of languages* $\mathrm{Pol}\,\mathcal{V}$ *is defined by the set of all inequalities* $v^\omega \leqslant v^\omega u v^\omega$, *where* u, v *belong to* $\widehat{A^*}$ *for some finite set* A *and* V *satisfies* $u = v$ *and* $v = v^2$,

(ii) *the pseudovariety of monoids corresponding to the variety* $\mathrm{UPol}\,\mathcal{V}$ *is defined by the set of all pseudoidentities* $v^\omega = v^\omega u v^\omega$, *where* u, v *belong to* $\widehat{A^*}$ *for some finite set* A *and* V *satisfies* $u = v$ *and* $v = v^2$.

In the case of the Straubing–Thérien hierarchy, Place and Zeitoun [12] described a basis of inequalities for $\mathsf{V}_{n+1/2}$ very similar to the basis obtained from Proposition 1(i) by taking V equal to V_n. The main difference is that instead of assuming that the pseudoidentity $u = v$ is valid in V_n, it is required that the inequality $u \leqslant v$ holds in $\mathsf{V}_{n-1/2}$. A general relationship between such characterizations will now be explained using a new operation on pseudovarieties.

The inequalities of Proposition 1 may be interpreted in terms of Mal'cev products, an algebraic operation on pseudovarieties of monoids. Here we follow another route by defining the new operation solely in terms of inequalities. For an arbitrary pseudovariety of ordered monoids V, let V^{m} denote the pseudovariety of ordered monoids with basis of inequalities consisting of all inequalities $v^{\omega+1} \leqslant v^\omega u v^\omega$, where $u, v \in \widehat{A^*}$ for some finite set A and the inequality $u \leqslant v$ holds in V. An alternative definition for V^{m} is obtained by taking the inequalities of the form $v^\omega \leqslant v^\omega u v^\omega$, where $u, v \in \widehat{A^*}$ for some finite set A and both the inequality $u \leqslant v$ and the pseudoidentity $v^2 = v$ hold in V. To obtain the inequalities of the latter definition from the former, assume that V satisfies $u \leqslant v = v^2$. Then the inequality $v^\omega u \leqslant v^\omega$ is also valid in V, and thus $(v^\omega)^{\omega+1} \leqslant v^\omega u v^\omega$ holds in V^{m} by the former definition, which is the same inequality as $v^\omega \leqslant v^\omega u v^\omega$. Conversely, assuming the latter definition, take any inequality $u \leqslant v$ valid in V. Then obviously $v^{\omega-1} u \leqslant v^\omega = (v^\omega)^2$ also holds in V, and so the latter definition gives validity of the inequality $v^\omega \leqslant v^{\omega-1} u v^\omega$ in V^{m}, which immediately implies validity of $v^{\omega+1} \leqslant v^\omega u v^\omega$.

The operator $(\_)^{\mathrm{m}}$ is monotone, that is, $\mathsf{V} \subseteq \mathsf{W}$ implies $\mathsf{V}^{\mathrm{m}} \subseteq \mathsf{W}^{\mathrm{m}}$, because every inequality from the basis for W^{m} is included in the basis for V^{m}. Additionally, denote by V^{M} the monoid pseudovariety $\mathsf{V}^{\mathrm{m}} \cap (\mathsf{V}^{\mathrm{m}})^{\mathrm{d}}$. Then V^{M} is defined by pseudoidentities of the form $v^{\omega+1} = v^{\omega}uv^{\omega}$, where $u, v \in \widehat{A}^*$ for some finite set A are such that the inequality $u \leqslant v$ holds in V. An alternative basis can be given as in the case of V^{m}.

Remark. The alternative definition of the pseudovariety V^{m} in particular shows that if V is selfdual, then V^{m} and V^{M} are defined by the inequalities and pseudoidentities of Proposition 1, and thus can be expressed in terms of the Mal'cev product as $\mathsf{V}^{\mathrm{m}} = \mathsf{W} \,\textcircled{m}\, \mathsf{V}$ and $\mathsf{V}^{\mathrm{M}} = \mathsf{LI} \,\textcircled{m}\, \mathsf{V}$, with W and LI pseudovarieties of finite semigroups defined by the inequality $x^{\omega} \leqslant x^{\omega}yx^{\omega}$ and the pseudoidentity $x^{\omega} = x^{\omega}yx^{\omega}$, respectively (see [7, Theorem 6.5]). However, if V is not selfdual, such characterizations cannot be formulated, as the corresponding Mal'cev products are not defined.

The following result is immediate from the definition of V^{m}, since validity of $v \leqslant u$ in an ordered monoid implies validity of $v^{\omega+1} \leqslant v^{\omega}uv^{\omega}$.

Lemma 2. *Every pseudovariety of ordered monoids* V *satisfies the inclusion* $\mathsf{V}^{\mathrm{d}} \subseteq \mathsf{V}^{\mathrm{m}}$. *In particular, if* V *is selfdual, then* $\mathsf{V} \subseteq \mathsf{V}^{\mathrm{m}}$.   □

The next lemma provides a condition on a pseudovariety V which guarantees that the pseudovariety $(\mathcal{B}\mathsf{V})^{\mathrm{m}}$ can be obtained by applying the operator $(\_)^{\mathrm{m}}$ directly to V.

Lemma 3. *Let* $\mathsf{V} = \mathsf{W}^{\mathrm{m}}$, *with* W *a selfdual pseudovariety of ordered monoids. Then* $\mathsf{V}^{\mathrm{m}} = (\mathcal{B}\mathsf{V})^{\mathrm{m}}$ *and* $\mathsf{V}^{\mathrm{M}} = (\mathcal{B}\mathsf{V})^{\mathrm{M}}$.

Proof. Clearly, the first equality implies the second one. Since the operator $(\_)^{\mathrm{m}}$ is monotone, we have $\mathsf{V}^{\mathrm{m}} \subseteq (\mathcal{B}\mathsf{V})^{\mathrm{m}}$. It remains to prove that every inequality from the basis for V^{m} is valid in $(\mathcal{B}\mathsf{V})^{\mathrm{m}}$ as well. Let $u, v \in \widehat{A}^*$ be such that $u \leqslant v$ holds in V. Then obviously $v^{\omega}uv^{\omega} \leqslant v^{\omega+1}$ also holds in V. On the other hand, because W is selfdual by assumption, the inclusion $\mathsf{W} \subseteq \mathsf{V}$ holds by Lemma 2, and consequently $u \leqslant v$ is valid in W. Hence, the pseudovariety $\mathsf{V} = \mathsf{W}^{\mathrm{m}}$ satisfies $v^{\omega+1} \leqslant v^{\omega}uv^{\omega}$. Together, this shows that the pseudoidentity $v^{\omega}uv^{\omega} = v^{\omega+1}$ is valid in V; therefore, it is also valid in $\mathcal{B}\mathsf{V}$. Now consider elements $\overline{u} = v^{\omega}uv^{\omega}$ and $\overline{v} = v^{\omega+1}$ of $\widehat{A}^*$. Then $\overline{u} = \overline{v}$ holds in $\mathcal{B}\mathsf{V}$. This implies that the inequality $\overline{v}^{\omega+1} \leqslant \overline{v}^{\omega}\overline{u}\,\overline{v}^{\omega}$ is valid in $(\mathcal{B}\mathsf{V})^{\mathrm{m}}$. However, this inequality is equivalent to $v^{\omega+1} \leqslant v^{\omega}uv^{\omega}$, because $(v^{\omega+1})^{\omega+1} = v^{\omega+1}$ and $(v^{\omega+1})^{\omega} = v^{\omega}$ hold in $\widehat{A}^*$. Hence, the inequality $v^{\omega+1} \leqslant v^{\omega}uv^{\omega}$ is valid in $(\mathcal{B}\mathsf{V})^{\mathrm{m}}$.   □

Lemma 3 has the following consequence for concatenation hierarchies.

Proposition 4. *Let* $(\mathcal{V}_k)_{k\in\mathbb{N}/2}$ *be an arbitrary concatenation hierarchy and let* $(\mathsf{V}_k)_{k\in\mathbb{N}/2}$ *be the corresponding hierarchy of pseudovarieties of ordered monoids. Then, for each positive integer n, the following equalities hold:*

$$\mathsf{V}_{n+1/2} = (\mathsf{V}_n)^{\mathrm{m}} = (\mathsf{V}_{n-1/2})^{\mathrm{m}}.$$

Proof. The equality $\mathsf{V}_{n+1/2} = (\mathsf{V}_n)^{\mathrm{m}}$ comes from Proposition 1(i) via the alternative definition of the operator $(\_)^{\mathrm{m}}$. The equality $(\mathsf{V}_n)^{\mathrm{m}} = (\mathsf{V}_{n-1/2})^{\mathrm{m}}$ follows from Lemma 3, because of $\mathsf{V}_{n-1/2} = (\mathsf{V}_{n-1})^{\mathrm{m}}$ and $\mathsf{V}_n = \mathcal{B}\mathsf{V}_{n-1/2}$.   □

4 Testing the Inequalities in an Ordered Monoid

It is not clear that the inequalities defining V^m may be effectively tested in a given ordered monoid $(M, \leqslant)$. The aim of this section is to identify for which pairs (s, t) of elements of M the inequality $t^{\omega+1} \leqslant t^\omega st^\omega$ should be tested in order to assure that $(M, \leqslant)$ belongs to V^m. For this purpose, we define a binary relation on M determined by inequalities that hold in V.

Recall that for a mapping $\alpha \colon A \to M$, there is a unique continuous homomorphism $\overline{\alpha} \colon (\widehat{A^*}, =) \to (M, \leqslant)$ that extends α. We define a relation $\sigma_V(M)$ on M as the set of all pairs $(\overline{\alpha}(u), \overline{\alpha}(v))$, where $\alpha \colon A \to M$ is a mapping from an arbitrary finite alphabet A, and $u, v \in \widehat{A^*}$ are such that the inequality $u \leqslant v$ holds in V. Thus, $\sigma_V(M)$ is an order analog of the 2-pointlike pair relation on a finite monoid, which in turn may be viewed as a topological separation problem on regular languages within the free profinite monoid $\widehat{A^*}$ [1]. In the work of Place and Zeitoun [12] an element of $\sigma_{V_{3/2}}(M)$ is called a two-element Σ_2-chain of M, and may also be viewed as a topological separation property.

The relation $\sigma_V(M)$ is obviously reflexive. It is also stable, since for given $(\overline{\alpha}(u), \overline{\alpha}(v)) \in \sigma_V(M)$ and $z \in M$, the pairs $(\overline{\alpha}(u)z, \overline{\alpha}(v)z)$ and $(z\overline{\alpha}(u), z\overline{\alpha}(v))$ can be shown to belong to $\sigma_V(M)$ by choosing a new letter $x \notin A$, observing that $ux \leqslant vx$ and $xu \leqslant xv$ hold in V, and extending α by setting $\alpha(x) = z$. However, it need not be transitive. The following lemma shows that instead of using all possible mappings α, it is sufficient to use an arbitrary surjective mapping.

Lemma 5. *Let V be a pseudovariety of ordered monoids and $(M, \leqslant)$ a finite ordered monoid. Let $\alpha \colon A \to M$ be an arbitrary surjective mapping. Then, for every $s, t \in M$, the pair (s, t) belongs to $\sigma_V(M)$ if and only if $(s, t) = (\overline{\alpha}(u), \overline{\alpha}(v))$ for some $u, v \in \widehat{A^*}$ such that the inequality $u \leqslant v$ holds in V.*

Proof. The "if" statement is trivial. In order to prove the "only if" part, assume that $(s, t) \in \sigma_V(M)$, that is, there exists a mapping $\beta \colon B \to M$ such that $(s, t) = (\overline{\beta}(u), \overline{\beta}(v))$ for some $u, v \in \widehat{B^*}$, with $u \leqslant v$ valid in V. The universal property of $\widehat{B^*}$ and surjectivity of α imply that there exists a continuous homomorphism $\gamma \colon \widehat{B^*} \to \widehat{A^*}$ such that $\overline{\beta} = \overline{\alpha} \circ \gamma$. Then $\gamma(u) \leqslant \gamma(v)$ holds in V, and $(s, t) = (\overline{\alpha}(\gamma(u)), \overline{\alpha}(\gamma(v)))$. □

An alternative characterization of ordered monoids $(M, \leqslant)$ belonging to the pseudovariety V^m can be formulated using the relation $\sigma_V(M)$.

Lemma 6. *Let V be a pseudovariety of ordered monoids and $(M, \leqslant)$ a finite ordered monoid. Then $(M, \leqslant) \in V^m$ if and only if $t^{\omega+1} \leqslant t^\omega st^\omega$ holds for every pair $(s, t) \in \sigma_V(M)$.*

An analogous characterization of the pseudovariety V^M follows directly from the characterizations of V^m and $(V^m)^d$ given by Lemma 6.

Corollary 7. *Let V be a pseudovariety of ordered monoids and M a finite monoid. Then $M \in V^M$ if and only if $t^{\omega+1} = t^\omega st^\omega$ holds for every pair $(s, t) \in \sigma_V(M)$.*

We return now to concatenation hierarchies with an application of Lemma 6 that is obtained directly from Proposition 4:

Proposition 8. *Let $(\mathcal{V}_k)_{k \in \mathbb{N}/2}$ be an arbitrary concatenation hierarchy of languages and let $(\mathsf{V}_k)_{k \in \mathbb{N}/2}$ be the corresponding hierarchy of pseudovarieties of ordered monoids. Let $(M, \leqslant)$ be a finite ordered monoid. Then, for every positive integer n, the following conditions are equivalent:*

(i) $(M, \leqslant) \in \mathsf{V}_{n+1/2}$;
(ii) $t^{\omega+1} \leqslant t^{\omega} s t^{\omega}$ *holds for all* $(s, t) \in \sigma_{\mathsf{V}_n}(M)$;
(iii) $t^{\omega+1} \leqslant t^{\omega} s t^{\omega}$ *holds for all* $(s, t) \in \sigma_{\mathsf{V}_{n-1/2}}(M)$.

In the case of star-free languages, Proposition 8 can be formulated in a slightly different way, since then one can assume that the monoid M is aperiodic, and thus t^{ω} can be used in place of $t^{\omega+1}$; this in particular shows that Proposition 8 generalizes Theorem 7 of Place and Zeitoun [12] from the case of the Straubing–Thérien hierarchy to an arbitrary concatenation hierarchy.

One of the contributions of [12] is an algorithm for computing the relation $\sigma_{\mathsf{V}_{3/2}}(M)$, which directly implies decidability of $\mathsf{V}_{5/2}$ and $\mathsf{V}_{5/2} \cap (\mathsf{V}_{5/2})^{\mathrm{d}}$. In this paper, rather than computing directly the relation $\sigma_{\mathsf{V}}(M)$, we attempt to prove decidability of pseudovarieties by computing the transitive closure of $\sigma_{\mathsf{V}}(M)$. The following section shows that this approach can be used to verify validity of all equalities $t^{\omega+1} = t^{\omega} s t^{\omega}$ in Corollary 7. Although one needs to be able to verify all inequalities $t^{\omega+1} \leqslant t^{\omega} s t^{\omega}$ in order to decide membership in the pseudovariety V^{m}, dealing with equalities is sufficient to obtain decidability of the pseudovariety V^{M}.

5   Polynomial Reduction of the Membership Problem

Let M be a finite monoid and let V be a pseudovariety of ordered monoids. We say that a stable quasiorder ρ on M is a V-quasiorder if the quotient ordered monoid M/ρ belongs to V. For a pair of stable quasiorders ρ and τ, the relation $\rho \cap \tau$ is a stable quasiorder as well, and the quotient $M/(\rho \cap \tau)$ is isomorphic to a submonoid of the ordered monoid $M/\rho \times M/\tau$. This shows that the set of all V-quasiorders of the monoid M is closed under intersection, and consequently there exists the smallest V-quasiorder on M, denoted $\rho_{\mathsf{V}}(M)$. If the pseudovariety V is decidable, then the relation $\rho_{\mathsf{V}}(M)$ is computable, since there are only finitely many binary relations on the finite set M.

Lemma 9. *Let M be a finite monoid and let V be a pseudovariety of ordered monoids. Then $\rho_{\mathsf{V}}(M)$ is equal to the transitive closure of $\sigma_{\mathsf{V}}(M)$.*

Proof. For every mapping $\alpha \colon A \to M$, the composition of $\overline{\alpha}$ with the natural projection $\pi \colon M \to M/\rho_{\mathsf{V}}(M) \in \mathsf{V}$ satisfies $\pi \overline{\alpha}(u) \leqslant \pi \overline{\alpha}(v)$ for every inequality $u \leqslant v$ valid in V. This shows that $\sigma_{\mathsf{V}}(M) \subseteq \rho_{\mathsf{V}}(M)$, and consequently $\mathrm{T}(\sigma_{\mathsf{V}}(M)) \subseteq \rho_{\mathsf{V}}(M)$ holds, as $\rho_{\mathsf{V}}(M)$ is transitive.

Conversely, because $\sigma_V(M)$ is a stable reflexive relation on M, its transitive closure $\tau = \mathrm{T}(\sigma_V(M))$ is a stable quasiorder on M. Therefore, in order to prove that $\tau \supseteq \rho_V(M)$, it suffices to verify that the ordered monoid M/τ belongs to V. So, let $u \leqslant v$ be an arbitrary inequality that holds in V, with $u, v \in \widehat{A^*}$, and consider any mapping $\beta \colon A \to M/\tau$. Let $\alpha \colon A \to M$ be any mapping such that $\beta = \pi \circ \alpha$, where $\pi \colon M \to M/\tau$ denotes the natural projection. Then the unique extensions of the mappings α and β to the monoid $\widehat{A^*}$ satisfy $\overline{\beta} = \pi \circ \overline{\alpha}$. By definition of the relation $\sigma_V(M)$, the pair $(\overline{\alpha}(u), \overline{\alpha}(v))$ belongs to $\sigma_V(M)$, and consequently also to τ. Altogether, we obtain $\overline{\beta}(u) = \pi(\overline{\alpha}(u)) \leqslant \pi(\overline{\alpha}(v)) = \overline{\beta}(v)$, thus showing that $u \leqslant v$ is satisfied by M/τ. □

The following lemma, analogous to Lemma 4.1 of [3], shows that when verifying the condition of Corollary 7, that is, the condition

$$\bigl(\forall s, t \in M\bigr)\bigl((s, t) \in R \implies t^{\omega+1} = t^\omega s t^\omega\bigr), \tag{1}$$

one can take $R = \rho_V(M)$ instead of $R = \sigma_V(M)$.

Lemma 10. *Let M be a finite monoid and let R be a reflexive and stable binary relation on M. Then R satisfies (1) if and only if the transitive closure of R satisfies (1).*

Proof. Assume that $t^{\omega+1} = t^\omega s t^\omega$ holds for each $(s, t) \in R$. We show by induction with respect to n that $t^{\omega+1} = t^\omega s t^\omega$ holds also for each pair (s, t) from R^n. For $n = 1$, this is the assumption. Let $n > 1$ and let $(s, t) \in R^n$. Thus, there exists $z \in M$ such that $(s, z) \in R$ and $(z, t) \in R^{n-1}$. Since R is stable, we have $(t^\omega s t^\omega, t^\omega z t^\omega) \in R$. The induction assumption gives $t^{\omega+1} = t^\omega z t^\omega$, which means that $(t^\omega s t^\omega, t^{\omega+1}) \in R$. Then $t^{\omega+1} = (t^{\omega+1})^{\omega+1} = (t^{\omega+1})^\omega (t^\omega s t^\omega)(t^{\omega+1})^\omega = t^\omega s t^\omega$ by the assumption on R, as required. As the transitive closure of R is the union of all relations R^n, the statement is proved. □

In order to construct a polynomial time reduction of the membership problem for the pseudovariety V^{M} to that of V, we proceed in a way similar to [14, Subsection 4.6.2], where the membership problem for some Mal'cev products is solved by constructing certain congruences on a monoid M, and testing whether the quotient monoid belongs to V. However, in our case, as V is a variety of ordered monoids, instead of a congruence, an appropriate stable quasiorder has to be constructed.

Let M be a finite monoid. Denote by $\kappa(M)$ the union of all stable relations R on M satisfying (1). Then $\kappa(M)$ is obviously also a stable relation satisfying (1). Moreover, it is reflexive, as the identity relation has these properties, and it is transitive, as the transitive closure of a stable relation satisfying (1) is such a relation as well, by Lemma 10. This shows that $\kappa(M)$ is a quasiorder on M.

For a given pseudovariety of ordered monoids V, let V^κ be the class of all monoids M such that $M/\kappa(M) \in \mathsf{V}$.

Lemma 11. *Let V be a pseudovariety of ordered monoids. Then $\mathsf{V}^\kappa = \mathsf{V}^{\mathrm{M}}$. In particular, V^κ is a pseudovariety of monoids.*

Proof. Corollary 7 shows that M belongs to V^{M} if and only if $\sigma_{\mathsf{V}}(M)$ satisfies (1), which holds precisely when $\rho_{\mathsf{V}}(M)$ satisfies (1), according to Lemmata 9 and 10. This is equivalent to the requirement $\rho_{\mathsf{V}}(M) \subseteq \kappa(M)$, which holds if and only if $M/\kappa(M) \in \mathsf{V}$, by definition of $\rho_{\mathsf{V}}(M)$. The latter condition is exactly the definition of membership of M in V^{κ}. □

The algorithm for deciding membership in V^{M} is based on calculating the relation $\kappa(M)$.

Lemma 12. *For an arbitrary finite monoid M, the quasiorder $\kappa(M)$ can be calculated in polynomial time.*

Proof. For each element $m \in M$, let κ_m be the binary relation on M defined by the rule

$$(a, b) \in \kappa_m \iff (\forall p, q \in M)(pbq = m \implies m^{\omega} \cdot paq \cdot m^{\omega} = m^{\omega+1}).$$

Each of the relations κ_m can be calculated in polynomial time. It will be proved that the relation $\kappa(M)$ is equal to the intersection of all relations κ_m, which means that it can be calculated in polynomial time as well.

In order to show that $\kappa(M) \subseteq \kappa_m$ for every $m \in M$, let $(a, b) \in \kappa(M)$ and $p, q \in M$ be such that $pbq = m$. The stability of $\kappa(M)$ implies that $(paq, pbq) \in \kappa(M)$, and since $pbq = m$, condition (1) gives $m^{\omega} \cdot paq \cdot m^{\omega} = m^{\omega+1}$.

Conversely, in order to prove that the relation $\bigcap_{m \in M} \kappa_m$ is contained in $\kappa(M)$, it is sufficient to verify that $\bigcap_{m \in M} \kappa_m$ is a stable relation satisfying (1). The stability of the intersection follows directly from the stability of each κ_m, which is obvious from the definition. In order to verify condition (1), let (s, t) belong to all relations κ_m, with $m \in M$. In particular, it belongs to κ_t, which gives $t^{\omega} s t^{\omega} = t^{\omega+1}$ by choosing $p = q = 1$. □

The desired polynomial reduction of the membership problem for V^{M} is obtained directly by combining Lemmata 11 and 12 and the definition of V^{κ}.

Proposition 13. *For every pseudovariety of ordered monoids V, the membership problem for the pseudovariety V^{M} can be reduced in polynomial time to the membership problem for V.*

Note that for an arbitrary variety of languages $\mathcal{V}$, with the corresponding pseudovariety of monoids V, Proposition 1(ii) states that the pseudovariety corresponding to the variety $\mathrm{UPol}\,\mathcal{V}$ is precisely V^{M}. Therefore, as a special case of Proposition 13 we obtain the known result that decidability of the variety $\mathcal{V}$ implies decidability of the variety $\mathrm{UPol}\,\mathcal{V}$ (see [14, Theorem 4.6.50]).

Propositions 4 and 13 together give the main result of this paper:

Theorem 14. *Let $(\mathcal{V}_k)_{k \in \mathbb{N}/2}$ be an arbitrary concatenation hierarchy of regular languages and let $(\mathsf{V}_k)_{k \in \mathbb{N}/2}$ be the corresponding hierarchy of pseudovarieties of ordered monoids. If $\mathsf{V}_{n-1/2}$ is a decidable pseudovariety, then $\mathsf{V}_{n+1/2} \cap (\mathsf{V}_{n+1/2})^{\mathrm{d}}$ is also decidable, that is, the variety $\mathrm{UPol}\,\mathcal{V}_n$ is decidable.*

Using decidability of the variety $\mathcal{V}_{7/2}$ proved by Place [11], this theorem gives

Corollary 15. *It is algorithmically decidable whether a given regular language is definable in Δ_5.*

6   Conclusion

We have shown how structural techniques of the theory of profinite monoids may be used to climb some steps of the decidable part of concatenation hierarchies, which is achieved through a polynomial time reduction. In the case of the Straubing-Thérien hierarchy, this corresponds to step up from a Σ_n fragment of FO[<] to the next Δ_{n+1} fragment.

Our method consists in studying an operator $(\_)^M$ on pseudovarieties of ordered monoids defined by the pseudoidentities used by Pin, Straubing and Thérien [8] to give an algebraic characterization of unambiguous polynomial closure. To check whether a given finite monoid M belongs to the pseudovariety V^M, one needs to verify that the equality $t^{\omega+1} = t^\omega s t^\omega$ holds in M whenever (s,t) is a member of the binary relation $\sigma_\mathsf{V}(M)$ on M whose elements are obtained by evaluating inequalities $u \leqslant v$ valid in V. Decidability of $\sigma_\mathsf{V}(M)$, that is of the separation property considered by Place and Zeitoun [12], then immediately gives decidability of V^M. We managed to show that the weaker hypothesis of decidability of V suffices for the same conclusion by establishing the following key steps: (i) the relation $\sigma_\mathsf{V}(M)$ may be replaced by its transitive closure $\rho_\mathsf{V}(M)$, which is the smallest stable quasiorder such that $M/\rho_\mathsf{V}(M) \in \mathsf{V}$; (ii) the pseudovariety V^M consists of all finite monoids M such that $M/\kappa(M) \in \mathsf{V}$, where $\kappa(M)$ is the largest stable relation on M such that $(s,t) \in \kappa(M)$ implies $t^\omega s t^\omega = t^{\omega+1}$; (iii) the relation $\kappa(M)$ may be computed in polynomial time.

Since our main result steps up from Σ_n to Δ_{n+1}, not to Σ_{n+1}, it cannot be simply iterated. The approach of Place and Zeitoun [11,12] has been to try to show that one may climb from Σ_n to Σ_{n+1} by proving decidability of the separation property for Σ_n. If such a condition were inherited by Σ_{n+1}, then all Σ_n and Δ_n fragments would be decidable. However, proving that condition turns out to be very complicated and has so far only been achieved by a deep but casuistic combinatorial analysis up to Σ_4. If here also one could relax such a condition to plain decidability to be able to climb the decidable part of the hierarchy, then again all Σ_n and Δ_n fragments would be decidable. Thus, it would be very interesting to adapt our algebraic approach to V^M to the pseudovariety V^m.

In a less ambitious route, it should be possible to extend our approach to regular languages of infinite words. More general classes of regular languages may also be amenable to our methods, taking into account that the polynomial closure of lattices of regular languages have already been investigated by Branco and Pin [4]. A further natural question is to characterize the language counterpart of the operator $(\_)^m$.

Acknowledgments. The authors thank Jean-Éric Pin and the anonymous referees for their comments and suggestions, which significantly contributed to improve the presentation of this paper.

References

1. Almeida, J.: Some algorithmic problems for pseudovarieties. Publ. Math. Debrecen **54**(Suppl), 531–552 (1999)
2. Almeida, J.: Profinite semigroups and applications. In: Kudryavtsev, V.B., Rosenberg, I.G. (eds.) Structural Theory of Automata, Semigroups and Universal Algebra, pp. 1–45. Springer (2005)
3. Almeida, J., Klíma, O.: New decidable upper bound of the second level in the Straubing-Thérien concatenation hierarchy of star-free languages. Discrete Math. Theor. Comput. Sci. **12**, 41–58 (2010)
4. Branco, M.J.J., Pin, J.-É.: Equations defining the polynomial closure of a lattice of regular languages. In: Albers, S., Marchetti-Spaccamela, A., Matias, Y., Nikoletseas, S., Thomas, W. (eds.) ICALP 2009, Part II. LNCS, vol. 5556, pp. 115–126. Springer, Heidelberg (2009)
5. Brzozowski, J.A., Cohen, R.S.: Dot-depth of star-free events. J. Comput. System Sci. **5**, 1–15 (1971)
6. McNaughton, R., Papert, S.: Counter-Free Automata. MIT Press (1971)
7. Pin, J.-É.: Syntactic semigroups. In: Rozenberg, G., Salomaa, A. (eds.) Handbook of Formal Languages, Chapter 10. Springer (1997)
8. Pin, J.-É., Straubing, H., Thérien, D.: Locally trivial categories and unambiguous concatenation. J. Pure Appl. Algebra **52**, 297–311 (1988)
9. Pin, J.-É., Weil, P.: Profinite semigroups, Mal'cev products and identities. J. Algebra **182**, 604–626 (1996)
10. Pin, J.-É., Weil, P.: Polynomial closure and unambiguous product. Theory Comput. Systems **30**, 383–422 (1997)
11. Place, T.: Separating regular languages with two quantifier alternations. In: Proc. LICS (2015), to appear
12. Place, T., Zeitoun, M.: Going higher in the first-order quantifier alternation hierarchy on words. In: Esparza, J., Fraigniaud, P., Husfeldt, T., Koutsoupias, E. (eds.) ICALP 2014, Part II. LNCS, vol. 8573, pp. 342–353. Springer, Heidelberg (2014)
13. Reiterman, J.: The Birkhoff theorem for finite algebras. Algebra Universalis **14**, 1–10 (1982)
14. Rhodes, J., Steinberg, B.: The q-theory of Finite Semigroups. Springer (2009)
15. Schützenberger, M.-P.: On finite monoids having only trivial subgroups. Inform. and Control **8**, 190–194 (1965)
16. Straubing, H.: Finite semigroup varieties of the form $V * D$. J. Pure Appl. Algebra **36**, 53–94 (1985)
17. Thomas, W.: Classifying regular events in symbolic logic. J. Comput. System Sci. **25**, 360–376 (1982)

Ergodic Infinite Permutations of Minimal Complexity

Sergey V. Avgustinovich[1], Anna E. Frid[2], and Svetlana Puzynina[1,3](✉)

[1] Sobolev Institute of Mathematics, Novosibirsk, Russia
avgust@math.nsc.ru
[2] Aix-Marseille Université, Marseille, France
anna.e.frid@gmail.com
[3] LIP, ENS de Lyon, Université de Lyon, Lyon, France
s.puzynina@gmail.com

Abstract. An *infinite permutation* can be defined as a linear ordering of the set of natural numbers. Similarly to infinite words, a *complexity* $p(n)$ of an infinite permutation is defined as a function counting the number of its factors of length n. For infinite words, a classical result of Morse and Hedlund, 1940, states that if the complexity of an infinite word satisfies $p(n) \leq n$ for some n, then the word is ultimately periodic. Hence minimal complexity of aperiodic words is equal to $n + 1$, and words with such complexity are called *Sturmian*. For infinite permutations this does not hold: There exist aperiodic permutations with complexity functions of arbitrarily slow growth, and hence there are no permutations of minimal complexity.

In the paper we introduce a new notion of *ergodic permutation*, i.e., a permutation which can be defined by a sequence of numbers from $[0, 1]$, such that the frequency of its elements in any interval is equal to the length of that interval. We show that the minimal complexity of an ergodic permutation is $p(n) = n$, and that the class of ergodic permutations of minimal complexity coincides with the class of so-called Sturmian permutations, directly related to Sturmian words.

1 Introduction

In this paper, we continue the study of combinatorial properties of infinite permutations analogous to those of words. In this approach, infinite permutations are interpreted as equivalence classes of real sequences with distinct elements, such that only the order of elements is taken into account. In other words, an infinite permutation is a linear order in $\mathbb{N}$. We consider it as an object close to an infinite word, but instead of symbols, we have transitive relations $<$ or $>$ between each pair of elements.

S. Puzynina—Supported by the LABEX MILYON (ANR-10-LABX-0070) of Université de Lyon, within the program "Investissements d'Avenir" (ANR-11-IDEX-0007) operated by the French National Research Agency (ANR).

I. Potapov (Ed.): DLT 2015, LNCS 9168, pp. 71–84, 2015.
DOI: 10.1007/978-3-319-21500-6_5

Infinite permutations in the considered sense were introduced in [10]; see also a very similar approach coming from dynamics [6] and summarised in [3]. Since then, they were studied in two main directions: first, permutations directly constructed with the use of words are studied to reveal new properties of words used for their construction [8,16–18,20–22]. In the other approach, properties of infinite permutations are compared with those of infinite words, showing some resemblance and some difference.

In particular, both for words and permutations, the (factor) complexity is bounded if and only if the word or the permutation is ultimately periodic [10,19]. However, the minimal complexity of an aperiodic word is $n + 1$, and the words of this complexity are well-studied Sturmian words; as for the permutations, there is no "minimal" complexity function for the aperiodic case. By contrary, if we modify the definition to consider the *maximal pattern complexity* [13,14], the result for permutations is more classifying than that for words: in both cases, there is a minimal complexity for aperiodic objects, but for permutations, unlike for words, the cases of minimal complexity are characterised [4]. All the permutations of lowest maximal pattern complexity are closely related to Sturmian words, whereas words may have lowest maximal pattern complexity even if they have another structure [14].

Other results on the comparison of words and permutations include an attempt to define automatic permutations [12] analogously to automatic words [1], a discussion [11] of the Fine and Wilf theorem and a study of square-free permutations [5].

In this paper, we return to the initial definition of factor complexity and prove a result on permutations of minimal complexity analogous to that for words. To do it, we restrict ourselves to *ergodic* permutations. This new notion means that a permutation can be defined by a sequence of numbers from $[0, 1]$ such that the frequency of its elements in any interval is equal to the length of that interval. We show that this class of permutations is natural and wide, and the ergodic permutations of minimal complexity are exactly Sturmian permutations in the sense of Makarov [18].

The paper is organized as follows. After general basic definitions and a necessary section on the properties of Sturmian words (and permutations), we introduce ergodic permutations and study their basic properties. The main result of the paper, Theorem 5.2, is proved in Section 5.

2 Basic Definitions

In what concerns words, in this paper we mostly follow the terminology and notation from [15]. We consider finite and infinite words over a finite alphabet Σ; here we consider $\Sigma = \{0, 1\}$. A *factor* of an infinite word is any sequence of its consecutive letters. The factor $u[i] \cdots u[j]$ of an infinite word $u = u[0]u[1] \cdots u[n] \cdots$, with $u_k \in \Sigma$, is denoted by $u[i..j]$; *prefixes* of a finite or an infinite word are as usual defined as starting factors. A factor s of a right infinite word u is called *right* (resp., *left*) *special* if sa, sb (resp., as, bs) are both factors of u for distinct letters $a, b \in \Sigma$. A word which is both left and right special is called *bispecial*.

The length of a finite word s is denoted by $|s|$. An infinite word $u = vwwww \cdots = vw^{\omega}$ for some non-empty word w is called *ultimately ($|w|$-)periodic*; otherwise it is called *aperiodic*.

The *complexity* $p_u(n)$ of an infinite word u is a function counting the number of its factors of length n; see [7] for a survey. Due to a classical result of Morse and Hedlund [19], if the complexity $p_u(n)$ of an infinite word u satisfies $p_u(n) \leq n$ for some n, then u is ultimately periodic. Therefore, the minimal complexity of aperiodic words is $n + 1$; such words are called *Sturmian* and are discussed in the next section.

When considering words on the binary alphabet $\{0, 1\}$, we refer to the *order* on finite and infinite words meaning lexicographic (partial) order: $0 < 1$, $u < v$ if $u[0..i] = v[0..i]$ and $u[i + 1] < v[i + 1]$ for some i. For words such that one of them is the prefix of the other the order is not defined.

A *conjugate* of a finite word w is any word of the form vu, where $w = uv$. Clearly, conjugacy is an equivalence, and in particular, all the words from the same conjugate class have the same number of occurrences of each symbol.

Analogously to a factor of a word, for a sequence $(a[n])_{n=0}^{\infty}$ of real numbers, we denote by $a[i..j]$ and call a *factor* of $(a[n])$ the finite sequence of numbers $a[i], a[i+1], \ldots, a[j]$. We need sequences of numbers to correctly define an *infinite permutation* α as an equivalence class of real infinite sequences with pairwise distinct elements under the following equivalence $\sim$: we have $(a[n])_{n=0}^{\infty} \sim (b[n])_{n=0}^{\infty}$ if and only if for all i, j the conditions $a[i] < a[j]$ and $b[i] < b[j]$ are equivalent. Since we consider only sequences of pairwise distinct real numbers, the same condition can be defined by substituting $(<)$ by $(>)$: $a[i] > a[j]$ if and only if $b[i] > b[j]$. So, an infinite permutation is a linear ordering of the set $\mathbb{N}_0 = \{0, \ldots, n, \ldots\}$. We denote it by $\alpha = (\alpha[n])_{n=0}^{\infty}$, where $\alpha[i]$ are abstract elements equipped by an order: $\alpha[i] < \alpha[j]$ if and only if $a[i] < a[j]$ or, which is the same, $b[i] < b[j]$ of every *representative* sequence $(a[n])$ or $(b[n])$ of α. So, one of the simplest ways to define an infinite permutation is by a representative, which can be any sequence of distinct real numbers.

Example 2.1. Both sequences $(a[n]) = (1, -1/2, 1/4, \ldots)$ with $a[n] = (-1/2)^n$ and $(b[n])$ with $b[n] = 1000 + (-1/3)^n$ are representatives of the same permutation $\alpha = \alpha[0], \alpha[1], \ldots$ defined by

$$\alpha[2n] > \alpha[2n + 2] > \alpha[2k + 3] > \alpha[2k + 1]$$

for all $n, k \geq 0$.

A *factor* $\alpha[i..j]$ of an infinite permutation α is a finite sequence $(\alpha[i], \alpha[i + 1], \ldots, \alpha[j])$ of abstract elements equipped by the same order than in α. Note that a factor of an infinite permutation can be naturally interpreted as a finite permutation: for example, if in a representative $(a[n])$ we have a factor $(2.5, 2, 7, 1.6)$, that is, the 4th element is the smallest, followed by the 2nd, 1st and 3rd, then in the permutation, it will correspond to a factor $\begin{pmatrix} 1\ 2\ 3\ 4 \\ 3\ 2\ 4\ 1 \end{pmatrix}$, which we will denote simply as (3241). Note that in general, we number elements of infinite objects (words, sequences or permutations) starting with 0 and elements of finite objects starting with 1.

A factor of a sequence (permutation) should not be confused with its subsequence $a[n_0], a[n_1], \ldots$ (subpermutation $\alpha[n_0], \alpha[n_1], \ldots$) which is indexed with a growing subsequence (n_i) of indices.

Note, however, that in general, an infinite permutation cannot be defined as a permutation of $\mathbb{N}_0$. For instance, the permutation from Example 2.1 has all its elements between the first two ones.

Analogously to words, the *complexity* $p_\alpha(n)$ of an infinite permutation α is the number of its distinct factors of length n. As for words, this is a non-decreasing function, but it was proved in [10] that contrary to words, we cannot distinguish permutations of "minimal" complexity:

Example 2.2. For each unbounded non-decreasing function $f(n)$, we can find a permutation α on $\mathbb{N}_0$ such that ultimately, $p_\alpha(n) < f(n)$. The needed permutation can be defined by the inequalities $\alpha[2n-1] < \alpha[2n+1]$ and $\alpha[2n] < \alpha[2n+2]$ for all $n \geq 1$, and $\alpha[2n_k - 2] < \alpha[2k - 1] < \alpha[2n_k]$ for a sequence $\{n_k\}_{k=1}^\infty$ which grows sufficiently fast (see [10] for more details).

In this paper, we prove that by contrary, as soon as we restrict ourselves to *ergodic* permutations defined below, the minimal complexity of an ergodic permutation is n. The ergodic permutations of complexity n are directly related to Sturmian words which we discuss in the next section.

3   Sturmian Words and Sturmian Permutations

Definition 3.1. An aperiodic infinite word u is called *Sturmian* if its factor complexity satisfies $p_u(n) = n + 1$ for all $n \in \mathbb{N}$.

Sturmian words are by definition binary and they have the lowest possible factor complexity among aperiodic infinite words. This extremely popular class of words admits various types of characterizations of geometric and combinatorial nature (see, e. g., Chapter 2 of [15]). In this paper, we need their characterization via irrational rotations on the unit circle found already in the seminal paper [19].

Definition 3.2. *For an irrational* $\sigma \in (0, 1)$, *the rotation* by slope σ is the mapping R_σ from $[0, 1)$ (identified with the unit circle) to itself defined by $R_\sigma(x) = \{x + \sigma\}$, where $\{x\} = x - \lfloor x \rfloor$ is the fractional part of x.

Considering a partition of $[0, 1)$ into $I_0 = [0, 1 - \sigma)$, $I_1 = [1 - \sigma, 1)$, define an infinite word $s_{\sigma,\rho}$ by

$$s_{\sigma,\rho}[n] = \begin{cases} 0 & \text{if } R_\sigma^n(\rho) = \{\rho + n\sigma\} \in I_0, \\ 1 & \text{if } R_\sigma^n(\rho) = \{\rho + n\sigma\} \in I_1. \end{cases}$$

We can also define $I_0' = (0, 1 - \sigma]$, $I_1' = (1 - \sigma, 1]$ and denote the corresponding word by $s_{\sigma,\rho}'$. As it was proved by Morse and Hedlund, Sturmian words on $\{0, 1\}$ are exactly words $s_{\sigma,\rho}$ or $s_{\sigma,\rho}'$.

The same irrational rotation R_σ can be used to define Sturmian permutations:

Definition 3.3. *A Sturmian permutation $\beta = \beta(\sigma, \rho)$ is defined by its representative $(b[n])$, where $b[n] = R_\sigma^n(\rho) = \{\rho + n\sigma\}$.*

These permutations are obviously related to Sturmian words: indeed, $\beta[i+1] > \beta[i]$ if and only if $s[i] = 0$, where $s = s_{\sigma,\rho}$. Strictly speaking, the case of s' corresponds to a permutation β' defined with the upper fractional part.

Sturmian permutations have been studied in [18]; in particular, it is known that their complexity is $p_\beta(n) \equiv n$.

To continue, we now need a series of properties of a Sturmian word $s = s(\sigma, \rho)$. They are either trivial or classical, and the latter can be found, in particular, in [15].

1. The frequency of ones in s is equal to the slope σ.
2. In any factor of s of length n, the number of ones is either $\lfloor n\sigma \rfloor$, or $\lceil n\sigma \rceil$. In the first case, we say that the factor is *light*, in the second case, it is *heavy*.
3. The factors of s from the same conjugate class are all light or all heavy.
4. Let the continued fraction expansion of σ be $\sigma = [0, 1 + d_1, d_2, \ldots]$. Consider the sequence of *standard* finite words s_n defined by

$$s_{-1} = 1, s_0 = 0, s_n = s_{n-1}^{d_n} s_{n-2} \text{ for } n > 0.$$

Then
   - The set of bispecial factors of s coincides with the set of words obtained by erasing the last two symbols from the words $s_n^k s_{n-1}$, where $0 < k \le d_{n+1}$.
   - For each n, we can decompose s as a concatenation

$$s = p \prod_{i=1}^{\infty} s_n^{k_i} s_{n-1}, \tag{1}$$

   where $k_i = d_{n+1}$ or $k_i = d_{n+1}+1$ for all i, and p is a suffix of $s_n^{d_{n+1}+1} s_{n-1}$.
   - For all $n \ge 0$, if s_n is light, then all the words $s_n^k s_{n-1}$ for $0 < k \le d_{n+1}$ (including s_{n+1}) are heavy, and vice versa.
5. A *Christoffel word* can be defined as a word of the form $0b1$ or $1b0$, where b is a bispecial factor of a Sturmian word s. For a given b, both Christoffel words are also factors of s and are conjugate of each other. Moreover, they are conjugates of all but one factors of s of that length.
6. The lengths of Christoffel words in s are exactly the lengths of words $s_n^k s_{n-1}$, where $0 < k \le d_{n+1}$. Such a word is also conjugate of both Christoffel words of the respective length obtained from one of them by sending the first symbol to the end of the word.

The following statement will be needed for our result.

Proposition 3.4. *Let n be such that $\{n\alpha\} < \{i\alpha\}$ for all $0 < i < n$. Then the word $s_{\alpha,0}[0..n-1]$ is a Christoffel word. The same assertion holds if $\{n\alpha\} > \{i\alpha\}$ for all $0 < i < n$.*

PROOF. We will prove the statement for the inequality $\{n\alpha\} < \{i\alpha\}$; the other case is symmetric. First notice that there are no elements $\{i\alpha\}$ in the interval $[1 - \alpha, 1 - \alpha + \{n\alpha\})$ for $0 \leq i < n$. Indeed, assuming that for some i we have $1 - \alpha \leq \{i\alpha\} < 1 - \alpha + \{n\alpha\}$, we get that $0 \leq \{(i + 1)\alpha\} < \{n\alpha\}$, which contradicts the conditions of the claim.

Next, consider a word $s_{\alpha,1-\varepsilon}[0..n-1]$ for $0 < \varepsilon < \{n\alpha\}$, i.e., the word obtained from the previous one by rotating by ε clockwise. Clearly, all the elements except for $s[0]$ stay in the same interval, so the only element which changes is $s[0]$: $s_{\alpha,0}[0] = 0$, $s_{\alpha,1-\varepsilon}[0] = 1$, $s_{\alpha,0}[1..n - 1] = s_{\alpha,1-\varepsilon}[1..n - 1]$. This means that the factor $s_{\alpha,0}[1..n - 1]$ is left special.

Now consider a word $s_{\alpha,1-\varepsilon'}[0..n - 1]$ for $\{n\alpha\} < \varepsilon' < \min_{i \in \{0<i<n\}}\{i\alpha\}$, i.e., the word obtained from $s_{\alpha,0}[0..n - 1]$ by rotating by ε' (i.e., we rotate a bit more). Clearly, all the elements except for $s[0]$ and $s[n - 1]$ stay in the same interval, so the only elements which change are $s[0]$ and $s[n - 1]$: $s_{\alpha,0}[0] = 0$, $s_{\alpha,1-\varepsilon'}[0] = 1$, $s_{\alpha,0}[n-1] = 1$, $s_{\alpha,1-\varepsilon'}[n-1] = 0$, $s_{\alpha,0}[1..n-2] = s_{\alpha,1-\varepsilon'}[1..n-2]$. This means that the factor $s_{\alpha,0}[1..n - 2]$ is right special.

So, the factor $s_{\alpha,0}[1..n - 2]$ is both left and right special and hence bispecial. By the construction, $s_{\alpha,0}[0..n - 1]$ is a Christoffel word.

The proof is illustrated by Fig. 1, where all the numbers on the circle are denoted modulo 1.   □

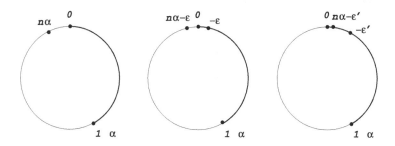

Fig. 1. Intervals for a bispecial word

Note also that in the Sturmian permutation $\beta = \beta(\sigma, \rho)$, we have $\beta[i] < \beta[j]$ for $i < j$ if and only if the respective factor $s[i..j - 1]$ of s is light (and, symmetrically, $\beta[i] > \beta[j]$ if and only if the factor $s[i..j - 1]$ is heavy).

4   Ergodic Permutations

In this section, we define a new notion of an ergodic permutation.

Let $(a[i])_{i=1}^{\infty}$ be a sequence of real numbers from the interval $[0, 1]$, representing an infinite permutation, a and p also be real numbers from $[0, 1]$. We say that the *probability* for any element $a[j]$ to be less than a exists and is equal to p if

$$\forall \varepsilon > 0 \; \exists N \in \mathbb{N} \; \forall n > N \; \forall j \in \mathbb{N} \; \left| \frac{\#\{a[j+k] | 0 \le k < n, a[j+k] < a\}}{n} - p \right| < \varepsilon.$$

In other words, if we substitute all the elements from $(a[i])$ which are smaller than a by 1, and those which are bigger by 0, the above condition means that the uniform frequency of the letter 1 exists and equals p. So, the probability to be smaller than a is the uniform frequency of the elements which are less than a. For more on uniform frequencies of letters in words we refer to [9].

We note that this is not exactly probability on the classical sense, since we do not have a random sequence. But we are interested in permutations where this "probability" behaves in certain sense like probability of a random sequence uniformly distributed on $[0, 1]$:

Definition 4.1. A sequence $(a[i])_{i=1}^{\infty}$ of real numbers is *canonical* if and only if

- all the numbers are pairwise distinct;
- for all i we have $0 \le a[i] \le 1$;
- and for all a, the probability for any element $a[i]$ to be less than a is well-defined and equal to a.

More formally, the last condition should be rephrased as

$$\forall a \in [0, 1], \varepsilon > 0 \; \exists N \in \mathbb{N} \; \forall n > N, j \in \mathbb{N} \; \left| \frac{\#\{a[j+k] | 0 \le k < n, a[j+k] < a\}}{n} - a \right| < \varepsilon.$$

Remark 4.2. The set $\{a[i] | i \in \mathbb{N}\}$ for a canonical sequence $(a[i])$ is dense on $[0, 1]$.

Remark 4.3. In a canonical sequence, the frequency of elements which fall into any interval $(t_1, t_2) \subseteq [0, 1]$ exists and is equal to $t_2 - t_1$.

Remark 4.4. Symmetrically to the condition "the probability to be less than a is a" we can consider the equivalent condition "the probability to be greater than a is $1 - a$".

Definition 4.5. An infinite permutation $\alpha = (\alpha_i)_{i=1}^{\infty}$ is called *ergodic* if it has a canonical representative.

Example 4.6. Since for any irrational σ and for any ρ the sequence of fractional parts $\{\rho + n\sigma\}$ is uniformly distributed in $[0, 1)$, a Sturmian permutation $\beta_{\sigma,\rho}$ is ergodic.

Example 4.7. Consider the sequence

$$\frac{1}{2}, 1, \frac{3}{4}, \frac{1}{4}, \frac{5}{8}, \frac{1}{8}, \frac{3}{8}, \frac{7}{8}, \dots$$

defined as the fixed point of the morphism

$$\varphi_{tm} : [0,1] \mapsto [0,1]^2, \varphi_{tm}(x) = \begin{cases} \frac{x}{2} + \frac{1}{4}, \frac{x}{2} + \frac{3}{4}, & \text{if } 0 \leq x \leq \frac{1}{2}, \\ \frac{x}{2} + \frac{1}{4}, \frac{x}{2} - \frac{1}{4}, & \text{if } \frac{1}{2} < x \leq 1. \end{cases}$$

If can be proved that this sequence is canonical and thus the respective permutation is ergodic. In fact, this permutation, as well as its construction [17], are closely related to the famous Thue-Morse word [2], and thus it is reasonable to call it the *Thue-Morse permutation*.

Proposition 4.8. *The canonical representative $(a[n])$ of an ergodic permutation α is unique.*

PROOF. Given α, for each i we define

$$a[i] = \lim_{n \to \infty} \frac{\#\{\alpha[k] | 0 \leq k < n, \alpha[k] < \alpha[i]\}}{n}$$

and see that, first, this limit must exist since α is ergodic, and second, $a[i]$ is the only possible value of an element of a canonical representative of α.   □

Note, however, that even if all the limits exist, it does not imply the existence of the canonical representative. Indeed, there is another condition to fulfill: for different i the limits must be different.

Consider a growing sequence $(n_i)_{i=1}^{\infty}$, $n_i \in \mathbb{N}$, $n_{i+1} > n_i$. The respective subpermutation $(\alpha[n_i])$ will be called *N-growing* (*N-decreasing*) if $n_{i+1} - n_i \leq N$ and $\alpha[n_{i+1}] > \alpha[n_i]$ ($\alpha[n_{i+1}] < \alpha[n_i]$) for all i. A subpermutation which is N-growing or N-decreasing is called *N-monotone*.

Proposition 4.9. *If a permutation has a N-monotone subpermutation for some N, then it is not ergodic.*

PROOF. Suppose the opposite and consider a subsequence $(a[n_i])$ of the canonical representative a corresponding to the N-monotone (say, N-growing) subpermutation $(\alpha[n_i])$. Consider $b = \lim_{i \to \infty} a[n_i]$ (which exists) and an $\varepsilon < 1/N$. Let M be the number such that $a[n_m] > b - \varepsilon$ for $m \geq M$. Then the probability for an element $a[i]$ to be in the interval $[a[n_M], b]$ must be equal to $b - a[n_M] < \varepsilon$ due to Remark 4.3. On the other hand, since all $a[n_m]$ for $m > M$ are in this interval, and $n_{i+1} - n_i \leq N$, this probability is at least $1/N > \varepsilon$. A contradiction.   □

An element $\alpha[i]$, $i > N$, of a permutation α is called *N-maximal* (resp., *N-minimal*) if $\alpha[i]$ is greater (resp., less) than all the elements at the distance at most N from it: $\alpha[i] > \alpha[j]$ (resp., $\alpha[i] < \alpha[j]$) for all $j = i - N, i - N + 1, \ldots, i - 1, i + 1, \ldots, i + N$.

Proposition 4.10. *In an ergodic permutation α, for each N there exists an N-maximal and an N-minimal element.*

PROOF. Consider a permutation α without N-maximal elements and prove that it is not ergodic. Suppose first that there exists an element $\alpha[n_1]$, $n_1 > N$, in α which is greater than any of its N left neighbours: $\alpha[n_1] > \alpha[n_1 - i]$ for all i from 1 to N. Since $\alpha[n_1]$ is not N-maximal, there exist some $i \in \{1, \ldots, N\}$ such that $\alpha[n_1 + i] > \alpha[n_1]$. If such i are several, we take the maximal $\alpha[n_1 + i]$ and denote $n_2 = n_1 + i$. By the construction, $\alpha[n_2]$ is also greater than any of its N left neighbours, and we can continue the sequence of elements $\alpha[n_1] < \alpha[n_2] < \cdots < \alpha[n_k] < \cdots$. Since for all k we have $n_{k+1} - n_k \leq N$, it is an N-growing subpermutation, and due to the previous proposition, α is not ergodic.

Now suppose that there are no elements in α which are greater than all their N left neighbours:

For all $n > N$, there exists some $i \in \{1, \ldots, N\}$ such that $\alpha[n - i] > \alpha[n]$. (2)

We take $\alpha[n_1]$ to be the greatest of the first N elements of α and $\alpha[n_2]$ to be the greatest among the elements $\alpha[n_1 + 1], \ldots, \alpha[n_1 + N]$. Then due to (2) applied to n_2, $\alpha[n_1] > \alpha[n_2]$. Moreover, $n_2 - n_1 \leq N$ and for all $n_1 < k < n_2$ we have $\alpha[k] < \alpha[n_2]$.

Now we take n_3 such that $\alpha[n_3]$ is the maximal element among $\alpha[n_2 + 1], \ldots, \alpha[n_2 + N]$, and so on. Suppose that we have chosen $n_1, \ldots, n_i$ such that $\alpha[n_1] > \alpha[n_2] > \cdots > \alpha[n_i]$, and

For all $j \leq i$ and for all k such that $n_{j-1} < k < n_j$, we have $\alpha[k] < \alpha[n_j]$. (3)

For each new $\alpha[n_{i+1}]$ chosen as the maximal element among $\alpha[n_i+1], \ldots, \alpha[n_i + N]$, we have $n_{i+1} - n_i \leq N$. Due to (2) applied to n_{i+1} and by the construction, $\alpha[n_{i+1}] < \alpha[l]$ for some l from $n_{i+1} - N$ to n_i. Because of (3), without loss of generality we can take $l = n_j$ for some $j \leq i$. Moreover, we cannot have $\alpha[n_i] < \alpha[n_{i+1}]$ and thus $j < i$: otherwise n_{i+1} would have been chosen as n_{j+1} since it fits the condition of maximality better.

So, we see that $\alpha[n_i] > \alpha[n_{i+1}]$, (3) holds for $i+1$ as well as for i, and thus by induction the subpermutation $\alpha[n_1] > \cdots > \alpha[n_i] > \cdots$ is N-decreasing. Again, due to the previous proposition, α is not ergodic. □

5 Minimal Complexity of Ergodic Permutations

Proposition 5.1. *For any ergodic permutation α, we have $p_\alpha(n) \geq n$.*

PROOF. Due to Proposition 4.10, there exists an n-maximal element α_i, $i > n$. All the n factors of α of length n containing it are different: in each of them, the maximal element is at a different position. □

The complexity of Sturmian permutations considered in Section 3 is known to be $p_\alpha(n) = n$ [18]. In what follows, we are going to prove that these are the only ergodic examples of this minimal complexity, and thus the Sturmian construction remains a natural "simplest" example if we restrict ourselves to ergodic permutations. So, the rest of the section is devoted to the proof of

Theorem 5.2. *The minimal complexity of an ergodic permutation α is $p_\alpha(n) \equiv n$. The set of ergodic permutations of minimal complexity coincides with the set of Sturmian permutations.*

Since the complexity of ergodic permutations satisfies $p_\alpha(n) \geq n$ due to Proposition 5.1; and the complexity of Sturmian permutations is $p_\alpha(n) \equiv n$, it remains to prove just that if $p_\alpha(n) \equiv n$ for an ergodic permutation α, then α is Sturmian.

Definition 5.3. *Given an infinite permutation $\alpha = \alpha[1] \cdots \alpha[n] \cdots$, consider its underlying infinite word $s = s[1] \cdots s[n] \cdots$ over the alphabet $\{0, 1\}$ defined by*

$$s[i] = \begin{cases} 0, & \text{if } \alpha[i] < \alpha[i+1], \\ 1, & \text{otherwise.} \end{cases}$$

Note that in some papers the word s was denoted by γ and considered directly as a word over the alphabet $\{<, >\}$.

It is not difficult to see that a factor $s[i + 1..i + n - 1]$ of s contains only a part of information on the factor $\alpha[i + 1..i + n]$ of α, i.e., does not define it uniquely. Different factors of length $n - 1$ of s correspond to different factors of length n of α. So,

$$p_\alpha(n) \geq p_s(n - 1).$$

Together with the above mentioned result of Morse and Hedlund [19], it gives the following

Proposition 5.4. *If $p_\alpha(n) = n$, then the underlying sequence s of α is either ultimately periodic or Sturmian.*

Now we consider different cases separately.

Proposition 5.5. *If $p_\alpha(n) \equiv n$ for an ergodic permutation α, then its underlying sequence s is aperiodic.*

PROOF. Suppose the converse and let p be the minimal period of s. If $p = 1$, then the permutation α is monotone, increasing or decreasing, so that its complexity is always 1, a contradiction. So, $p \geq 2$. There are exactly p factors of s of length $p - 1$: each residue modulo p corresponds to such a factor and thus to a factor of α of length p. The factor $\alpha[kp + i..(k + 1)p + i - 1]$, where $i \in \{1, \ldots, p\}$, does not depend on k, but for all the p values of i, these factors are different.

Now let us fix i from 1 to p and consider the subpermutation $\alpha[i], \alpha[p + i], \ldots, \alpha[kp+i], \ldots$. It cannot be monotone due to Proposition 4.9, so, there exist k_1 and k_2 such that $\alpha[k_1 p + i] < \alpha[(k_1 + 1)p + i]$ and $\alpha[k_2 p + i] > \alpha[(k_2 + 1)p + i]$. So, $\alpha[k_1 p + i..(k_1 + 1)p + i] \neq \alpha[k_2 p + i..(k_2 + 1)p + i]$. We see that each of p factors of α of length p, uniquely defined by the residue i, can be extended to the right to a factor of length $p + 1$ in two different ways, and thus $p_\alpha(p + 1) \geq 2p$. Since $p > 1$ and thus $2p > p + 1$, it is a contradiction. $\square$

So, Propositions 5.4 and 5.5 imply that the underlying word s of an ergodic permutation α of complexity n is Sturmian. Let $s = s(\sigma, \rho)$, that is,

$$s[n] = \lfloor \sigma(n + 1) + \rho \rfloor - \lfloor \sigma n + \rho \rfloor.$$

In the proofs we will only consider $s(\sigma, \rho)$, since for $s'(\sigma, \rho)$ the proofs are symmetric.

It follows directly from the definitions that the Sturmian permutation $\beta = \beta(\sigma, \rho)$ defined by its canonical representative b with $b[n] = \{\sigma n + \rho\}$ has s as the underlying word.

Suppose that α is a permutation whose underlying word is s and whose complexity is n. We shall prove the following statement concluding the proof of Theorem 5.2:

Lemma 5.6. *Let α be a permutation of complexity $p_\alpha(n) \equiv n$ whose underlying word is $s(\sigma, \rho)$. If α is ergodic, then $\alpha = \beta(\sigma, \rho)$.*

PROOF. Assume the converse, i.e., that α is not equal to β. We will prove that hence α is not ergodic, which is a contradiction.

Recall that in general, $p_\alpha(n) \geq p_s(n-1)$, but here we have the equality since $p_\alpha(n) \equiv n$ and $p_s(n) \equiv n + 1$. It means that a factor u of s of length $n - 1$ uniquely defines a factor of α of length n which we denote by α^u. Similarly, there is a unique factor β^u of β.

Clearly, if u is of length 1, we have $\alpha^u = \beta^u$: if $u = 0$, then $\alpha^0 = \beta^0 = (12)$, and if $u = 1$, then $\alpha^1 = \beta^1 = (21)$. Suppose now that $\alpha^u = \beta^u$ for all u of length up to $n - 1$, but there exists a word v of length n such that $\alpha^v \neq \beta^v$.

Since for any factor $v' \neq v$ of v we have $\alpha^{v'} = \beta^{v'}$, the only difference between α^v and β^v is the relation between the first and last element: $\alpha^v[1] < \alpha^v[n + 1]$ and $\beta^v[1] > \beta^v[n + 1]$, or vice versa. (Note that we number elements of infinite objects starting with 0 and elements of finite objects starting with 1.)

Consider the factor b^v of the canonical representative b of β corresponding to an occurrence of β^v. We have $b^v = (\{\tau\}, \{\tau + \sigma\}, \dots, \{\tau + n\sigma\})$ for some τ.

Proposition 5.7. *All the numbers $\{\tau + i\sigma\}$ for $0 < i < n$ are situated outside of the interval whose ends are $\{\tau\}$ and $\{\tau + n\sigma\}$.*

PROOF. Consider the case of $\beta^v[1] < \beta^v[n + 1]$ (meaning $\{\tau\} < \{\tau + n\sigma\}$) and $\alpha^v[1] > \alpha^v[n + 1]$; the other case is symmetric. Suppose by contrary that there is an element $\{\tau + i\sigma\}$ such that $\{\tau\} < \{\tau + i\sigma\} < \{\tau + n\sigma\}$ for some i. It means that $\beta^v[1] < \beta^v[i] < \beta^v[n + 1]$. But the relations between the 1st and the ith elements, as well as between the ith and $(n + 1)$st elements, are equal in α^v and in β^v, so, $\alpha^v[1] < \alpha^v[i]$ and $\alpha^v[i] < \alpha^v[n + 1]$. Thus, $\alpha^v[1] < \alpha^v[n + 1]$, a contradiction. □

Proposition 5.8. *The word v belongs to the conjugate class of a Christoffel factor of s, or, which is the same, of a factor of the form $s_n^k s_{n-1}$ for $0 < k \leq d_{n+1}$.*

PROOF. The condition "For all $0 < i < n$, the number $\{\tau + i\sigma\}$ is not situated between $\{\tau\}$ and $\{\tau + n\sigma\}$" is equivalent to the condition "$\{n\alpha\} < \{i\alpha\}$ for all $0 < i < n$" considered in Proposition 3.4 and corresponding to a Christoffel word of the same length. The set of factors of s of length n is exactly the set

$\{s_{\alpha,\tau}[0..n-1]|\tau \in [0,1]\}$. These words are n conjugates of the Christoffel word plus one singular factor corresponding to $\{\tau\}$ and $\{\tau + n\sigma\}$ situated in the opposite ends of the interval $[0,1]$ ("close" to 0 and "close" to 1), so that all the other points $\{\tau + i\sigma\}$ are between them.

Example 5.9. Consider a Sturmian word s of the slope $\sigma \in (1/3, 2/5)$. Then the factors of s of length 5 are 01001, 10010, 00101, 01010, 10100, 00100. Fig. 2 depicts permutations of length 6 with their underlying words. In the picture the elements of the permutations are denoted by points; the order between two elements is defined by which element is "higher" on the picture. We see that in the first five cases, the relation between the first and the last elements can be changed, and in the last case, it cannot since there are other elements between them. Indeed, the first five words are exactly the conjugates of the Christoffel word 1 010 0, where the word 010 is bispecial.

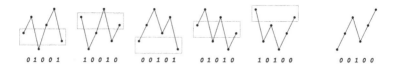

01001   10010   00101   01010   10100   00100

Fig. 2. Five candidates for v and a non-candidate word

Note also that due to Proposition 5.8, the shortest word v such that $\alpha^v \neq \beta^v$ is a conjugate of some $s_n^k s_{n-1}$ for $0 < k \leq d_{n+1}$.

In what follows without loss of generality we suppose that the word s_n is heavy and thus s_{n-1} and $s_n^k s_{n-1}$ for all $0 < k \leq d_{n+1}$ are light.

Consider first the easiest case: $v = s_n^{d_{n+1}} s_{n-1} = s_{n+1}$. This word is light, so, $\beta^{s_{n+1}}[1] < \beta^{s_{n+1}}[|s_{n+1}| + 1]$. Since the first and the last elements of $\alpha^{s_{n+1}}$ must be in the other relation, we have $\alpha^{s_{n+1}}[1] > \alpha^{s_{n+1}}[|s_{n+1}| + 1]$. At the same time, since s_n is shorter than s_{n+1}, we have $\alpha^{s_n} = \beta^{s_n}$ and in particular, since s_n is heavy, $\alpha^{s_n}[1] > \alpha^{s_n}[|s_n| + 1]$.

Due to (1), the word s after a finite prefix can be represented as an infinite concatenation of occurrences of s_{n+1} and s_n: $s = p \prod_{i=1}^{\infty} s_n^{t_i} s_{n+1}$, where $t_i = k_i - d_{n+1} = 0$ or 1. But both α^{s_n} and $\alpha^{s_{n+1}}$ are permutations with the last elements less than the first ones. Moreover, if we have a concatenation uw of factors u and w of s, we see that the first symbol of α^w is the last symbol of α^u: $\alpha^u[|u| + 1] = \alpha^w[1]$. So, an infinite sequence of factors s_n and s_{n+1} of s gives us a chain of the first elements of respective factors of the permutation α, and each next elements is less than the previous one. This chain is a $|s_{n+1}|$-monotone subpermutation, and thus α is not ergodic.

Now let us consider the general case: v is from the conjugate class of $s_n^t s_{n-1}$, where $0 < t \leq d_{n+1}$. We consider two cases: the word $s_n^t s_{n-1}$ can be cut either in one of the occurrences of s_n, or in the suffix occurrence of s_{n-1}.

In the first case, $v = r_1 s_n^l s_{n-1} s_n^{t-l-1} r_2$, where $s_n = r_2 r_1$ and $0 \leq l < t$. Then

$$s = p \prod_{i=1}^{\infty} s_n^{k_i} s_{n-1} = p r_2 (r_1 r_2)^{k_1 - l - 1} \prod_{i=2}^{\infty} v(r_1 r_2)^{k_i - t}. \tag{4}$$

We see that after a finite prefix, the word s is an infinite catenation of words v and $r_1 r_2$. The word $r_1 r_2$ is shorter than v and heavy since it is a conjugate of s_n. So, $\alpha^{r_1 r_2} = \beta^{r_1 r_2}$ and in particular, $\alpha^{r_1 r_2}[1] > \alpha^{r_1 r_2}[|r_1 r_2| + 1]$. The word v is light since it is a conjugate of $s_n^t s_{n-1}$, but the relation between the first and the last elements of α^v is different than between those in β^v, that is, $\alpha^v[1] > \alpha^v[|v| + 1]$. But as above, in a concatenation uw, we have $\alpha^u[|u| + 1] = \alpha^w[1]$, so, we see a $|v|$-decreasing subpermutation in α. So, α is not ergodic.

Analogous arguments work in the second case, when $s_n^t s_{n-1}$ is cut somewhere in the suffix occurrence of s_{n-1}: $v = r_1 s_n^t r_2$, where $s_{n-1} = r_2 r_1$. Note that s_{n-1} is a prefix of s_n, and thus $s_n = r_2 r_3$ for some r_3. In this case,

$$s = p \prod_{i=1}^{\infty} s_n^{k_i} s_{n-1} = p r_2 (r_3 r_2)^{k_1} \prod_{i=2}^{\infty} v(r_3 r_2)^{k_i - t}. \tag{5}$$

As above, we see that after a finite prefix, s is an infinite catenation of the heavy word $r_3 r_2$, a conjugate of s_n, and the word v. For both words, the respective factors of α have the last element less than the first one, which gives a $|v|$-decreasing subpermutation. So, α is not ergodic.

The case when s_n is not heavy but light is considered symmetrically and gives rise to $|v|$-increasing subpermutations. This concludes the proof of Theorem 5.2.

References

1. Allouche, J.-P., Shallit, J.: Automatic sequences – theory, applications, generalizations. Cambridge University Press (2003)
2. Allouche, J.-P., Shallit, J.: The ubiquitous Prouhet-Thue-Morse sequence. In: Sequences and Their Applications, Discrete Mathematics and Theoretical Computer Science, pp. 1–16. Springer, London (1999)
3. Amigó, J.: Permutation Complexity in Dynamical Systems - Ordinal Patterns. Permutation Entropy and All That, Springer Series in Synergetics (2010)
4. Avgustinovich, S.V., Frid, A., Kamae, T., Salimov, P.: Infinite permutations of lowest maximal pattern complexity. Theoretical Computer Science **412**, 2911–2921 (2011)
5. Avgustinovich, S.V., Kitaev, S., Pyatkin, A., Valyuzhenich, A.: On square-free permutations. J. Autom. Lang. Comb. **16**(1), 3–10 (2011)
6. Bandt, C., Keller, G., Pompe, B.: Entropy of interval maps via permutations. Nonlinearity **15**, 1595–1602 (2002)
7. Cassaigne, J., Nicolas, F.: Factor complexity. Combinatorics, automata and number theory, Encyclopedia Math. Appl. **135**, 163–247 (2010). Cambridge Univ. Press
8. Elizalde, S.: The number of permutations realized by a shift. SIAM J. Discrete Math. **23**, 765–786 (2009)

9. Ferenczi, S., Monteil, T.: Infinite words with uniform frequencies, and invariant measures. Combinatorics, automata and number theory. Encyclopedia Math. Appl. **135**, 373–409 (2010). Cambridge Univ. Press
10. Fon-Der-Flaass, D.G., Frid, A.E.: On periodicity and low complexity of infinite permutations. European J. Combin. **28**, 2106–2114 (2007)
11. Frid, A.: Fine and Wilf's theorem for permutations. Sib. Elektron. Mat. Izv. **9**, 377–381 (2012)
12. Frid, A., Zamboni, L.: On automatic infinite permutations. Theoret. Inf. Appl. **46**, 77–85 (2012)
13. Kamae, T., Zamboni, L.: Sequence entropy and the maximal pattern complexity of infinite words. Ergodic Theory and Dynamical Systems **22**, 1191–1199 (2002)
14. Kamae, T., Zamboni, L.: Maximal pattern complexity for discrete systems. Ergodic Theory and Dynamical Systems **22**, 1201–1214 (2002)
15. Lothaire, M.: Algebraic combinatorics on words. Cambridge University Press (2002)
16. Makarov, M.: On permutations generated by infinite binary words. Sib. Elektron. Mat. Izv. **3**, 304–311 (2006)
17. Makarov, M.: On an infinite permutation similar to the Thue-Morse word. Discrete Math. **309**, 6641–6643 (2009)
18. Makarov, M.: On the permutations generated by Sturmian words. Sib. Math. J. **50**, 674–680 (2009)
19. Morse, M., Hedlund, G.: Symbolic dynamics II: Sturmian sequences. Amer. J. Math. **62**, 1–42 (1940)
20. Valyuzhenich, A.: On permutation complexity of fixed points of uniform binary morphisms. Discr. Math. Theoret. Comput. Sci. **16**, 95–128 (2014)
21. Widmer, S.: Permutation complexity of the Thue-Morse word. Adv. Appl. Math. **47**, 309–329 (2011)
22. Widmer, S.: Permutation complexity related to the letter doubling map, WORDS (2011)

Diverse Palindromic Factorization Is NP-complete

Hideo Bannai[1], Travis Gagie[2,3]([✉]), Shunsuke Inenaga[1],
Juha Kärkkäinen[2], Dominik Kempa[2,3], Marcin Piątkowski[4],
Simon J. Puglisi[2,3], and Shiho Sugimoto[1]

[1] Department of Informatics, Kyushu University, Fukuoka, Japan
[2] Department of Computer Science, University of Helsinki, Helsinki, Finland
[3] Helsinki Institute for Information Technology, Espoo, Finland
travis.gagie@gmail.com
[4] Faculty of Mathematics and Computer Science,
Nicolaus Copernicus University, Toruń, Poland

Abstract. We prove that it is NP-complete to decide whether a given string can be factored into palindromes that are each unique in the factorization.

1 Introduction

Several papers have appeared on the subject of palindromic factorization. The palindromic length of a string is the minimum number of palindromic substrings into which the string can be factored. Notice that, since a single symbol is a palindrome, the palindromic length of a string is always defined and at most the length of the string. Ravsky [8] proved a tight bound on the maximum palindromic length of a binary string in terms of its length. Frid, Puzynina, and Zamboni [4] conjectured that any infinite string in which the palindromic length of any finite substring is bounded, is ultimately periodic. Their work led other researchers to consider how to efficiently compute a string's palindromic length and give a minimum palindromic factorization. It is not difficult to design a quadratic-time algorithm that uses linear space, but doing better than that seems to require some string combinatorics.

Alatabbi, Iliopoulos and Rahman [1] first gave a linear-time algorithm for computing a minimum factorization into maximal palindromes, if such a factorization exists. Notice that *abaca* cannot be factored into maximal palindromes, for example, because its maximal palindromes are *a*, *aba*, *a*, *aca* and *a*. Fici, Gagie, Kärkkäinen and Kempa [3] and I, Sugimoto, Inenaga, Bannai and Takeda [6] independently then described essentially the same $\mathcal{O}(n \log n)$-time

T. Gagie and S.J. Puglisi—Supported by grants 268324, 258308 and 284598 from the Academy of Finland.

M. Piątkowski—Supported by a research fellowship within the project "Enhancing Educational Potential of Nicolaus Copernicus University in the Disciplines of Mathematical and Natural Sciences" (project no. POKL.04.01.01-00-081/10).

© Springer International Publishing Switzerland 2015
I. Potapov (Ed.): DLT 2015, LNCS 9168, pp. 85–96, 2015.
DOI: 10.1007/978-3-319-21500-6_6

algorithm for computing a minimum palindromic factorization. Shortly there-after, Kosolobov, Rubinchik and Shur [7] gave an algorithm for recognizing strings with a given palindromic length. Their result can be used to compute the palindromic length ℓ of a string of length n in $\mathcal{O}(n\ell \log \ell)$ time. We also note that Gawrychowski and Uznański [5] used similar techniques as Fici et al. and I et al., for finding approximately the longest palindrome in a stream.

We call a factorization *diverse* if each of the factors is unique. Some well-known factorizations, such as the LZ77 [10] and LZ78 [11] parses, are diverse (except that the last factor may have appeared before). Fernau, Manea, Mercaş and Schmid [2] very recently proved that it is NP-complete to determine whether a given string has a diverse factorization of size at least k. It seems natural to consider the problem of determining whether a given string has a diverse factorization into palindromes. For example, *bgikkpps* and *bgikpspk* each have exactly one such factorization — i.e., (*b*, *g*, *i*, *kk*, *pp*, *s*) and (*b*, *g*, *i*, *kpspk*), respectively — but *bgkpispk* has none. This problem is obviously in NP and in this paper we prove that it is NP-hard and, thus, NP-complete. Some people might dismiss as doubly useless a lower bound for a problem with no apparent application; nevertheless, we feel the proof is pretty (albeit somewhat intricate) and we would like to share it. We conjecture that it is also NP-complete to determine whether a given string has a palindromic factorization in which each factor appears at most a given number $k > 1$ times.

2   Outline

The circuit satisfiability problem was one of the first to be proven NP-complete and is often the first taught in undergraduate courses. It asks whether a given Boolean circuit C is satisfiable, i.e., has an assignment to its inputs that makes its single output true. We will show how to build, in time linear in the size of C, a string that has a diverse palindromic factorization if and only if C is satisfiable. It follows that diverse palindromic factorization is also NP-hard. Our construction is similar to the Tseitin Transform [9] from Boolean circuits to CNF formulas.

Because AND, OR and NOT gates can be implemented with a constant number of NAND gates, we assume without loss of generality that C is composed only of NAND gates with two inputs and one output each, and splitters that each divide one wire into two. Furthermore, we assume each wire in C is labelled with a unique symbol (considering a split to be the end of an incoming wire and the beginning of two new wires, so all three wires have different labels). For each such symbol a, and some auxiliary symbols we introduce during our construction, we use as characters in our construction three related symbols: a itself, $\bar{a}$ and x_a. We indicate an auxiliary symbol related to a by writing a' or a''. We write x_a^j to denote j copies of x_a. We emphasize that, despite their visual similarity, a and $\bar{a}$ are separate characters, which play complementary roles in our reduction. We use \$ and # as generic separator symbols, which we consider to be distinct for each use; to prevent confusion, we add different superscripts to their different uses within the same part of the construction.

We can build a sequence $C_0, \ldots, C_t$ of subcircuits such that C_0 is empty, $C_t = C$ and, for $1 \leq i \leq t$, we obtain C_i from C_{i-1} by one of the following operations:

- adding a new wire (which is both an input and an output in C_i),
- splitting an output of C_{i-1} into two outputs,
- making two outputs of C_{i-1} the inputs of a new NAND gate.

We will show how to build in time linear in the size of C, inductively and in turn, a sequence of strings $S_1, \ldots, S_t$ such that S_i represents C_i according to the following definitions:

Definition 1. *A diverse palindromic factorization P of a string S_i encodes an assignment τ to the inputs of a circuit C_i if the following conditions hold:*

- *if τ makes an output of C_i labelled a true, then a, $x_a \bar{a} x_a$ and $x_a \bar{a} x_a$ are complete factors in P but $\bar{a}$, $x_a a x_a$ and x_a^j are not for $j > 1$;*
- *if τ makes an output of C_i labelled a false, then $\bar{a}$, x_a and $x_a a x_a$ are complete factors in P but a, $x_a \bar{a} x_a$ and x_a^j are not for $j > 1$;*
- *if a is a label in C but not in C_i, then none of a, $\bar{a}$, $x_a a x_a$, $x_a \bar{a} x_a$ and x_a^j for $j \geq 1$ are complete factors in P.*

Definition 2. *A string S_i represents a circuit C_i if each assignment to the inputs of C_i is encoded by some diverse palindromic factorization of S_i, and each diverse palindromic factorization of S_i encodes some assignment to the inputs of C_i.*

Once we have S_t, we can easily build in constant time a string S that has a diverse palindromic factorization if and only if C is satisfiable. To do this, we append $\$ \# x_a a x_a$ to S_t, where $\$$ and $\#$ are symbols not occurring in S_t and a is the label on C's output. Since $\$$ and $\#$ do not occur in S_t and occur as a pair of consecutive characters in S, they must each be complete factors in any palindromic factorization of S. It follows that there is a diverse palindromic factorization of S if and only if there is a diverse palindromic factorization of S_t in which $x_a a x_a$ is not a factor, which is the case if and only if there is an assignment to the inputs of C that makes its output true.

3   Adding a Wire

Suppose C_i is obtained from C_{i-1} by adding a new wire labelled a. If $i = 1$ then we set $S_i = x_a a x_a \bar{a} x_a$, whose two diverse palindromic factorizations $(x_a, a, x_a \bar{a} x_a)$ and $(x_a a x_a, \bar{a}, x_a)$ encode the assignments true and false to the wire labelled a, which is both the input and output in C_i. If $i > 1$ then we set

$$S_i = S_{i-1} \$ \# x_a a x_a \bar{a} x_a \,,$$

where $\$$ and $\#$ are symbols not occurring in S_{i-1} and not equal to a', $\overline{a'}$ or $x_{a'}$ for any label a' in C.

Since \$ and # do not occur in S_{i-1} and occur as a pair of consecutive characters in S_i, they must each be complete factors in any palindromic factorization of S_i. Therefore, any diverse palindromic factorization of S_i is the concatenation of a diverse palindromic factorization of S_{i-1} and either ($, #, x_a, a, $x_a\bar{a}x_a$) or ($, #, x_aax_a, $\bar{a}$, x_a). Conversely, any diverse palindromic factorization of S_{i-1} can be extended to a diverse palindromic factorization of S_i by appending either ($, #, x_a, a, $x_a\bar{a}x_a$) or ($, #, x_aax_a, $\bar{a}$, x_a).

Assume S_{i-1} represents C_{i-1}. Let τ be an assignment to the inputs of C_i and let P be a diverse palindromic factorization of S_{i-1} encoding τ restricted to the inputs of C_{i-1}. If τ makes the input (and output) of C_i labelled a true, then P concatenated with ($, #, x_a, a, $x_a\bar{a}x_a$) is a diverse palindromic factorization of S_i that encodes τ. If τ makes that input false, then P concatenated with ($, #, x_aax_a, $\bar{a}$, x_a) is a diverse palindromic factorization of S_i that encodes τ. Therefore, each assignment to the inputs of C_i is encoded by some diverse palindromic factorization of S_i.

Now let P be a diverse palindromic factorization of S_i and let τ be the assignment to the inputs of C_{i-1} that is encoded by a prefix of P. If P ends with ($, #, x_a, a, $x_a\bar{a}x_a$) then P encodes the assignment to the inputs of C_i that makes the input labelled a true and makes the other inputs true or false according to τ. If P ends with ($, #, x_aax_a, $\bar{a}$, x_a) then P encodes the assignment to the inputs of C_i that makes the input labelled a false and makes the other inputs true or false according to τ. Therefore, each diverse palindromic factorization of S_i encodes some assignment to the inputs of C_i.

Lemma 1. *We can build a string S_1 that represents C_1. If we have a string S_{i-1} that represents C_{i-1} and C_i is obtained from C_{i-1} by adding a new wire, then in constant time we can append symbols to S_{i-1} to obtain a string S_i that represents C_i.*

4   Splitting a Wire

Now suppose C_i is obtained from C_{i-1} by splitting an output of C_{i-1} labelled a into two outputs labelled b and c. We set

$$S_i' = S_{i-1}\,\$\#\,x_a^3b'x_aax_ac'x_a^5\,\$'\#'\,x_a^7\bar{b}'x_a\bar{a}x_a\bar{c}'x_a^9\,,$$

where \$, \$', #, #', b', $\bar{b}'$, c' and $\bar{c}'$ are symbols not occurring in S_{i-1} and not equal to a', $\bar{a}'$ or $x_{a'}$ for any label a' in C.

Since \$, \$', # and #' do not occur in S_{i-1} and occur as pairs of consecutive characters in S_i', they must each be complete factors in any palindromic factorization of S_i'. Therefore, a simple case analysis shows that any diverse palindromic factorization of S_i' is the concatenation of a diverse palindromic factorization of S_{i-1} and one of

$$(\$,\ \#,\ x_a^3,\ b',\ x_aax_a,\ c',\ x_a^5,\ \$',\ \#',\ x_a^2,\ x_a^4,\ x_a\overline{b}'x_a,\ \bar{a},\ x_a\overline{c}'x_a,\ x_a^8),$$
$$(\$,\ \#,\ x_a^3,\ b',\ x_aax_a,\ c',\ x_a^5,\ \$',\ \#',\ x_a^4,\ x_a^2,\ x_a\overline{b}'x_a,\ \bar{a},\ x_a\overline{c}'x_a,\ x_a^8),$$
$$(\$,\ \#,\ x_a^3,\ b',\ x_aax_a,\ c',\ x_a^5,\ \$',\ \#',\ x_a^6,\ x_a\overline{b}'x_a,\ \bar{a},\ x_a\overline{c}'x_a,\ x_a^8),$$
$$(\$,\ \#,\ x_a^2,\ x_ab'x_a,\ a,\ x_ac'x_a,\ x_a^4,\ \$',\ \#',\ x_a^7,\ \overline{b}',\ x_a\bar{a}x_a,\ \overline{c}',\ x_a^3,\ x_a^6),$$
$$(\$,\ \#,\ x_a^2,\ x_ab'x_a,\ a,\ x_ac'x_a,\ x_a^4,\ \$',\ \#',\ x_a^7,\ \overline{b}',\ x_a\bar{a}x_a,\ \overline{c}',\ x_a^6,\ x_a^3),$$
$$(\$,\ \#,\ x_a^2,\ x_ab'x_a,\ a,\ x_ac'x_a,\ x_a^4,\ \$',\ \#',\ x_a^7,\ \overline{b}',\ x_a\bar{a}x_a,\ \overline{c}',\ x_a^9).$$

In any diverse palindromic factorization of S_i', therefore, either b' and c' are complete factors but $\overline{b}'$ and $\overline{c}'$ are not, or vice versa.

Conversely, any diverse palindromic factorization of S_{i-1} in which a, x_a and $x_a\bar{a}x_a$ are complete factors but $\bar{a}$, x_aax_a and x_a^j are not for $j > 1$, can be extended to a diverse palindromic factorization of S_i' by appending either of

$$(\$,\ \#,\ x_a^3,\ b',\ x_aax_a,\ c',\ x_a^5,\ \$',\ \#',\ x_a^2,\ x_a^4,\ x_a\overline{b}'x_a,\ \bar{a},\ x_a\overline{c}'x_a,\ x_a^8),$$
$$(\$,\ \#,\ x_a^3,\ b',\ x_aax_a,\ c',\ x_a^5,\ \$',\ \#',\ x_a^6,\ x_a\overline{b}'x_a,\ \bar{a},\ x_a\overline{c}'x_a,\ x_a^8);$$

any diverse palindromic factorization of S_{i-1} in which $\bar{a}$, x_a and x_aax_a are complete factors but a, $x_a\bar{a}x_a$ and x_a^j are not for $j > 1$, can be extended to a diverse palindromic factorization of S_i' by appending either of

$$(\$,\ \#,\ x_a^2,\ x_ab'x_a,\ a,\ x_ac'x_a,\ x_a^4,\ \$',\ \#',\ x_a^7,\ \overline{b}',\ x_a\bar{a}x_a,\ \overline{c}',\ x_a^3,\ x_a^6),$$
$$(\$,\ \#,\ x_a^2,\ x_ab'x_a,\ a,\ x_ac'x_a,\ x_a^4,\ \$',\ \#',\ x_a^7,\ \overline{b}',\ x_a\bar{a}x_a,\ \overline{c}',\ x_a^9).$$

We set

$$S_i = S_i'\,\$''\#''\,x_bbx_bb'x_b\overline{b}'x_b\bar{b}x_b\,\$'''\#'''\,x_ccx_cc'x_c\overline{c}'x_c\bar{c}x_c,$$

where $\$''$, $\$'''$, $\#''$ and $\#'''$ are symbols not occurring in S_i' and not equal to a', $\overline{a}'$ or $x_{a'}$ for any label a' in C. Since $\$''$, $\$'''$, $\#''$ and $\#'''$ do not occur in S_i' and occur as pairs of consecutive characters in S_i', they must each be complete factors in any palindromic factorization of S_i. Therefore, any diverse palindromic factorization of S_i is the concatenation of a diverse palindromic factorization of S_i' and one of

$$(\$'',\ \#'',\ x_b,\ b,\ x_bb'x_b,\ \overline{b}',\ x_b\bar{b}x_b,\ \$''',\ \#''',\ x_c,\ c,\ x_cc'x_c,\ \overline{c}',\ x_c\bar{c}x_c),$$
$$(\$'',\ \#'',\ x_bbx_b,\ b',\ x_b\overline{b}'x_b,\ \bar{b},\ x_b,\ \$''',\ \#''',\ x_ccx_c,\ c',\ x_c\overline{c}'x_c,\bar{c},x_c).$$

Conversely, any diverse palindromic factorization of S_i' in which b' and c' are complete factors but $\overline{b}'$ and $\overline{c}'$ are not, can be extended to a diverse palindromic factorization of S_i by appending

$$(\$'',\ \#'',\ x_b,\ b,\ x_bb'x_b,\ \overline{b}',\ x_b\bar{b}x_b,\ \$''',\ \#''',\ x_c,\ c,\ x_cc'x_c,\ \overline{c}',\ x_c\bar{c}x_c);$$

any diverse palindromic factorization of S_i' in which $\overline{b}'$ and $\overline{c}'$ are complete factors but b' and c' are not, can be extended to a diverse palindromic factorization of S_i by appending

$$(\$'',\ \#'',\ x_bbx_b,\ b',\ x_b\overline{b}'x_b,\ \bar{b},\ x_b,\ \$''',\ \#''',\ x_ccx_c,\ c',\ x_c\overline{c}'x_c,\ \bar{c},\ x_c).$$

Assume S_{i-1} represents C_{i-1}. Let τ be an assignment to the inputs of C_{i-1} and let P be a diverse palindromic factorization of S_{i-1} encoding τ. If τ makes the output of C_{i-1} labelled a true, then P concatenated with, e.g.,

$$(\$, \#, x_a^3, b', x_aax_a, c', x_a^5, \$', \#', x_a^2, x_a^4, x_a\overline{b'}x_a, \bar{a}, x_a\overline{c'}x_a, x_a^8,$$
$$\$'', \#'', x_b, b, x_bb'x_b, \bar{b'}, x_b\bar{b}x_b, \$''', \#''', x_c, c, x_cc'x_c, \bar{c'}, x_c\bar{c}x_c)$$

is a diverse palindromic factorization of S_i. Notice b, c, x_b, x_c, $x_b\bar{b}x_b$ and $x_c\bar{c}x_c$ are complete factors but $\bar{b}$, $\bar{c}$, x_bbx_b, x_ccx_c, x_b^j and x_c^j for $j > 1$ are not. Therefore, this concatenation encodes the assignment to the inputs of C_i that makes them true or false according to τ.

If τ makes the output of C_{i-1} labelled a false, then P concatenated with, e.g.,

$$(\$, \#, x_a^2, x_ab'x_a, a, x_ac'x_a, x_a^4, \$', \#', x_a^7, \bar{b'}, x_a\bar{a}x_a, \bar{c'}, x_a^3, x_a^6,$$
$$\$'', \#'', x_bbx_b, b', x_b\bar{b'}x_b, \bar{b}, x_b, \$''', \#''', x_ccx_c, c', x_c\bar{c'}x_c, \bar{c}, x_c)$$

is a diverse palindromic factorization of S_i. Notice $\bar{b}$, $\bar{c}$, x_b, x_c, x_bbx_b and x_ccx_c are complete factors but b, c, $x_b\bar{b}x_b$, $x_c\bar{c}x_c$, x_b^j and x_c^j for $j > 1$ are not. Therefore, this concatenation encodes the assignment to the inputs of C_i that makes them true or false according to τ. Since C_{i-1} and C_i have the same inputs, each assignment to the inputs of C_i is encoded by some diverse palindromic factorization of S_i.

Now let P be a diverse palindromic factorization of S_i and let τ be the assignment to the inputs of C_{i-1} that is encoded by a prefix of P. If P ends with any of

$$(\$, \#, x_a^3, b', x_aax_a, c', x_a^5, \$', \#', x_a^2, x_a^4, x_a\overline{b'}x_a, \bar{a}, x_a\overline{c'}x_a, x_a^8),$$
$$(\$, \#, x_a^3, b', x_aax_a, c', x_a^5, \$', \#', x_a^4, x_a^2, x_a\overline{b'}x_a, \bar{a}, x_a\overline{c'}x_a, x_a^8),$$
$$(\$, \#, x_a^3, b', x_aax_a, c', x_a^5, \$', \#', x_a^6, x_a\overline{b'}x_a, \bar{a}, x_a\overline{c'}x_a, x_a^8)$$

followed by

$$(\$'', \#'', x_b, b, x_bb'x_b, \bar{b'}, x_b\bar{b}x_b, \$''', \#''', x_c, c, x_cc'x_c, \bar{c'}, x_c\bar{c}x_c),$$

then a must be a complete factor in the prefix of P encoding τ, so τ must make the output of C_{i-1} labelled a true. Since b, c, x_b, x_c, $x_b\bar{b}x_b$ and $x_c\bar{c}x_c$ are complete factors in P but $\bar{b}$, $\bar{c}$, x_bbx_b, x_ccx_c, x_b^j and x_c^j for $j > 1$ are not, P encodes the assignment to the inputs of C_i that makes them true or false according to τ.

If P ends with any of

$$(\$, \#, x_a^2, x_ab'x_a, a, x_ac'x_a, x_a^4, \$', \#', x_a^7, \bar{b'}, x_a\bar{a}x_a, \bar{c'}, x_a^3, x_a^6),$$
$$(\$, \#, x_a^2, x_ab'x_a, a, x_ac'x_a, x_a^4, \$', \#', x_a^7, \bar{b'}, x_a\bar{a}x_a, \bar{c'}, x_a^6, x_a^3),$$
$$(\$, \#, x_a^2, x_ab'x_a, a, x_ac'x_a, x_a^4, \$', \#', x_a^7, \bar{b'}, x_a\bar{a}x_a, \bar{c'}, x_a^9)$$

followed by

$$(\$'', \#'', x_bbx_b, b', x_b\bar{b'}x_b, \bar{b}, x_b, \$''', \#''', x_ccx_c, c', x_c\bar{c'}x_c, \bar{c}, x_c),$$

then $\bar{a}$ must be a complete factor in the prefix of P encoding τ, so τ must make the output of C_{i-1} labelled a false. Since $\bar{b}$, $\bar{c}$, x_b, x_c, $x_b b x_b$ and $x_c c x_c$ are complete factors but b, c, $x_b \bar{b} x_b$, $x_c \bar{c} x_c$, x_b^j and x_c^j for $j > 1$ are not, P encodes the assignment to the inputs of C_i that makes them true or false according to τ.

Since these are all the possibilities for how P can end, each diverse palindromic factorization of S_i encodes some assignment to the inputs of C_i. This gives us the following lemma:

Lemma 2. *If we have a string S_{i-1} that represents C_{i-1} and C_i is obtained from C_{i-1} by splitting an output of C_{i-1} into two outputs, then in constant time we can append symbols to S_{i-1} to obtain a string S_i that represents C_i.*

5   Adding a NAND Gate

Finally, suppose C_i is obtained from C_{i-1} by making two outputs of C_{i-1} labelled a and b the inputs of a new NAND gate whose output is labelled c. Let C'_{i-1} be the circuit obtained from C_{i-1} by splitting the output of C_{i-1} labelled a into two outputs labelled a_1 and a_2, where a_1 and a_2 are symbols we use only here. Assuming S_{i-1} represents C_{i-1}, we can use Lemma 2 to build in constant time a string S'_{i-1} representing C'_{i-1}. We set

$$S'_i = S'_{i-1} \$\# x_{c'}^3 a'_1 x_{c'} a_1 x_{c'} \overline{a_1} x_{c'} \overline{a'_1} x_{c'}^5$$
$$\$'\#' x_{c'}^7 a'_2 x_{c'} a_2 x_{c'} \overline{a_2} x_{c'} \overline{a'_2} x_{c'}^9$$
$$\$''\#'' x_{c'}^{11} b' x_{c'} b x_{c'} \overline{b} x_{c'} \overline{b'} x_{c'}^{13} ,$$

where all of the symbols in the suffix after S'_{i-1} are ones we use only here.

Since $\$$, $\$'$, $\$''$, $\$'''$, $\#$ and $\#'$ do not occur in S_{i-1} and occur as pairs of consecutive characters in S'_i, they must each be complete factors in any palindromic factorization of S'_i. Therefore, any diverse palindromic factorization of S'_i consists of

1. a diverse palindromic factorization of S'_{i-1},
2. $(\$, \#)$,
3. a diverse palindromic factorization of $x_{c'}^3 a'_1 x_{c'} a_1 x_{c'} \overline{a_1} x_{c'} \overline{a'_1} x_{c'}^5$,
4. $(\$', \#')$,
5. a diverse palindromic factorization of $x_{c'}^7 a'_2 x_{c'} a_2 x_{c'} \overline{a_2} x_{c'} \overline{a'_2} x_{c'}^9$,
6. $(\$'', \#'')$,
7. a diverse palindromic factorization of $x_{c'}^{11} b' x_{c'} b x_{c'} \overline{b} x_{c'} \overline{b'} x_{c'}^{13}$.

If a_1 is a complete factor in the factorization of S'_{i-1}, then the diverse palindromic factorization of

$$x_{c'}^3 a'_1 x_{c'} a_1 x_{c'} \overline{a_1} x_{c'} \overline{a'_1} x_{c'}^5$$

must include either

$$(a'_1, \ x_{c'} a_1 x_{c'}, \ \overline{a_1}, \ x_{c'} \overline{a'_1} x_{c'}) \qquad \text{or} \qquad (a'_1, \ x_{c'} a_1 x_{c'}, \ \overline{a_1}, \ x_{c'}, \ \overline{a'_1}).$$

Notice that in the former case, the factorization need not contain $x_{c'}$. If $\overline{a_1}$ is a complete factor in the factorization of S'_{i-1}, then the diverse palindromic factorization of

$$x^3_{c'} a'_1 x_{c'} a_1 x_{c'} \overline{a_1} x_{c'} \overline{a'_1} x^5_{c'}$$

must include either

$$(x_{c'} a'_1 x_{c'},\ a_1,\ x_{c'} \overline{a_1} x_{c'},\ \overline{a'_1}) \qquad \text{or} \qquad (a'_1,\ x_{c'},\ a_1,\ x_{c'} \overline{a_1} x_{c'},\ \overline{a'_1}).$$

Again, in the former case, the factorization need not contain $x_{c'}$. A simple case analysis shows analogous propositions hold for a_2 and b; we leave the details for the full version of this paper.

We set

$$S''_i = S'_i\ \$^{\dagger}\#^{\dagger}\ x^{15}_{c'} \overline{a'_1} x_{c'} c' x_{c'} \overline{b'} x^{17}_{c'}\ \$^{\dagger\dagger}\#^{\dagger\dagger}\ x^{19}_{c'} \overline{a'_2} x_{c'} d x_{c'} b' x^{21}_{c'},$$

where $\$^{\dagger}$, $\#^{\dagger}$, $\$^{\dagger\dagger}$, $\#^{\dagger\dagger}$, c' and d are symbols we use only here. Any diverse palindromic factorization of S''_i consists of

1. a diverse palindromic factorization of S'_i,
2. $(\$^{\dagger},\ \#^{\dagger})$,
3. a diverse palindromic factorization of $x^{15}_{c'} \overline{a'_1} x_{c'} c' x_{c'} \overline{b'} x^{17}_{c'}$,
4. $(\$^{\dagger\dagger},\ \#^{\dagger\dagger})$,
5. a diverse palindromic factorization of $x^{19}_{c'} \overline{a'_2} x_{c'} d x_{c'} b' x^{21}_{c'}$.

Since a_1 and a_2 label outputs in C'_{i-1} split from the same output in C_{i-1}, it follows that a_1 is a complete factor in a diverse palindromic factorization of S'_{i-1} if and only if a_2 is. Therefore, we need consider only four cases:

- The factorization of S'_{i-1} includes a_1, a_2 and b as complete factors, so the factorization of S'_i includes as complete factors either $x_{c'} \overline{a'_1} x_{c'}$, or $\overline{a'_1}$ and $x_{c'}$; either $x_{c'} \overline{a'_2} x_{c'}$, or $\overline{a'_2}$ and $x_{c'}$; either $x_{c'} \overline{b'} x_{c'}$, or $\overline{b'}$ and $x_{c'}$; and b'. Trying all the combinations — there are only four, since $x_{c'}$ can appear as a complete factor at most once — shows that any diverse palindromic factorization of S''_i includes one of

$$(\overline{a'_1},\ x_{c'} c' x_{c'},\ \overline{b'},\ \ldots,\ \overline{a'_2},\ x_{c'},\ d,\ x_{c'} b' x_{c'}),$$
$$(\overline{a'_1},\ x_{c'} c' x_{c'},\ \overline{b'},\ \ldots,\ x_{c'} \overline{a'_2} x_{c'},\ d,\ x_{c'} b' x_{c'}),$$

  with the latter only possible if $x_{c'}$ appears earlier in the factorization.
- The factorization of S'_{i-1} includes a_1, a_2 and $\overline{b}$ as complete factors, so the factorization of S'_i includes as complete factors either $x_{c'} \overline{a'_1} x_{c'}$, or $\overline{a'_1}$ and $x_{c'}$; either $x_{c'} \overline{a'_2} x_{c'}$, or $\overline{a'_2}$ and $x_{c'}$; $\overline{b'}$; and either $x_{c'} b' x_{c'}$, or b' and $x_{c'}$. Trying all the combinations shows that any diverse palindromic factorization of S''_i includes one of

$$(\overline{a'_1},\ x_{c'},\ c',\ x_{c'} \overline{b'} x_{c'},\ \ldots,\ \overline{a'_2},\ x_{c'} d x_{c'},\ b'),$$
$$(x_{c'} \overline{a'_1} x_{c'},\ c',\ x_{c'} \overline{b'} x_{c'},\ \ldots,\ \overline{a'_2},\ x_{c'} d x_{c'},\ b'),$$

  with the latter only possible if $x_{c'}$ appears earlier in the factorization.

– The factorization of S'_{i-1} includes $\overline{a_1}$, $\overline{a_2}$ and b as complete factors, so the factorization of S'_i includes as complete factors $\overline{a'_1}$; $\overline{a'_2}$; either $x_{c'}\overline{b'}x_{c'}$, or $\overline{b'}$ and $x_{c'}$; and b'. Trying all the combinations shows that any diverse palindromic factorization of S''_i includes one of

$$(x_{c'}\overline{a'_1}x_{c'},\ c',\ x_{c'},\ \overline{b'},\ \ldots,\ x_{c'}\overline{a'_2}x_{c'},\ d,\ x_{c'}b'x_{c'}),$$
$$(x_{c'}\overline{a'_1}x_{c'},\ c',\ x_{c'}\overline{b'}x_{c'},\ \ldots,\ x_{c'}\overline{a'_2}x_{c'},\ d,\ x_{c'}b'x_{c'}),$$

with the latter only possible if $x_{c'}$ appears earlier in the factorization.

– The factorization of S'_{i-1} includes $\overline{a_1}$, $\overline{a_2}$ and $\overline{b}$ as complete factors, so the factorization of S'_i includes as complete factors $\overline{a'_1}$; $\overline{a'_2}$; $\overline{b'}$; and either $x_{c'}b'x_{c'}$, or b' and $x_{c'}$. Trying all the combinations shows that any diverse palindromic factorization of S''_i that extends the factorization of S'_i includes one of

$$(x_{c'}\overline{a'_1}x_{c'},\ c',\ x_{c'}\overline{b'}x_{c'},\ \ldots,\ x_{c'}\overline{a'_2}x_{c'},\ d,\ x_{c'},\ b),$$
$$(x_{c'}\overline{a'_1}x_{c'},\ c',\ x_{c'}\overline{b'}x_{c'},\ \ldots,\ x_{c'}\overline{a'_2}x_{c'},\ d,\ x_{c'}b'x_{c'}),$$

with the latter only possible if $x_{c'}$ appears earlier in the factorization.

Summing up, any diverse palindromic factorization of S''_i always includes $x_{c'}$ and includes either $x_{c'}c'x_{c'}$ if the factorization of S'_{i-1} includes a_1, a_2 and b as complete factors, or c' otherwise.

We set

$$S'''_i = S''_i \,\$^{\dagger\dagger\dagger}\#^{\dagger\dagger\dagger}\, x^{23}_{c'}c''x_{c'}c'x_{c'}\overline{c'}x_{c'}\overline{c''}x^{25}_{c'}\,,$$

where $\$^{\dagger\dagger\dagger}$ and $\#^{\dagger\dagger\dagger}$ are symbols we use only here. Any diverse palindromic factorization of S'''_i consists of

1. a diverse palindromic factorization of S''_i,
2. $(\$^{\dagger\dagger\dagger},\ \#^{\dagger\dagger\dagger})$,
3. a diverse palindromic factorization of $x^{23}_{c'}c''x_{c'}c'x_{c'}\overline{c'}x_{c'}\overline{c''}x^{25}_{c'}$.

Since $x_{c'}$ must appear as a complete factor in the factorization of S''_i, if c' is a complete factor in the factorization of S''_i, then the factorization of

$$x^{23}_{c'}\overline{c''}x_{c'}c'x_{c'}\overline{c'}x_{c'}c''x^{25}_{c'}$$

must include

$$(c'',\ x_{c'}c'x_{c'},\ \overline{c'},\ x_{c'}\overline{c''}x_{c'});$$

otherwise, it must include

$$(x_{c'}c''x_{c'},\ c',\ x_{c'}\overline{c'}x_{c'},\ \overline{c''}).$$

That is, the factorization of $x^{23}_{c'}\overline{c''}x_{c'}c'x_{c'}\overline{c'}x_{c'}c''x^{25}_{c'}$ includes c'', $x_{c'}$ and $x_{c'}\overline{c''}x_{c'}$ but not $\overline{c''}$ or $x_{c'}c''x_{c'}$, if and only if the factorization of S''_i includes c'; otherwise, it includes $\overline{c''}$, $x_{c'}$ and $x_{c'}c''x_{c'}$ but not c'' or $x_{c'}\overline{c''}x_{c'}$.

We can slightly modify and apply the results in Section 4 to build in constant time a string T such that in any diverse palindromic factorization of

$$S_i = S'''_i \,\$^{\ddagger}\#^{\ddagger}\, T\,,$$

if c'' is a complete factor in the factorization of S'''', then c, x_c and $x_c\bar{c}x_c$ are complete factors in the factorization of T but $\bar{c}$, x_ccx_c and x_c^j are not for $j > 1$; otherwise, $\bar{c}$, x_c and x_ccx_c are complete factors but c, $x_c\bar{c}x_c$ and x_c^j are not for $j > 1$. Again, we leave the details for the full version of this paper.

Assume S_{i-1} represents C_{i-1}. Let τ be an assignment to the inputs of C_{i-1} and let P be a diverse palindromic factorization of S_{i-1} encoding τ. By Lemma 2 we can extend P to P' so that it encodes the assignment to the inputs of C'_{i-1} that makes them true or false according to τ. Suppose τ makes the output of C_{i-1} labelled a true but the output labelled b false. Then P' concatenated with, e.g.,

$$(\$, \ \#, \ x_{c'}^3, \ a_1', \ x_{c'}a_1x_{c'}, \ \overline{a_1}, \ x_{c'}\overline{a_1'}x_{c'}, \ x_c^4,$$
$$\$', \ \#', \ x_{c'}^7, \ a_2', \ x_{c'}a_2x_{c'}, \overline{a_2}, \ x_{c'}\overline{a_2'}x_{c'}, \ x_{c'}^8,$$
$$\$'', \ \#'', \ x_{c'}^{10}, \ x_{c'}b'x_{c'}, \ b, \ x_{c'}\bar{b}x_{c'}, \ \bar{b'}, \ x_{c'}^{13})$$

is a diverse palindromic factorization P'' of S_i' which, concatenated with, e.g.,

$$(\$^\dagger, \ \#^\dagger, \ x_{c'}^{15}, \ \overline{a_1'}, \ x_{c'}, \ c', \ x_{c'}\overline{b'}x_{c'}, \ x_{c'}^{16},$$
$$\$^\ddagger, \ \#^\ddagger, \ x_{c'}^{19}, \ \overline{a_2'}, \ x_{c'}dx_{c'}, \ b', \ x_{c'}^{21})$$

is a diverse palindromic factorization P''' of S_i'' which, concatenated with, e.g.,

$$(\$^{\dagger\dagger\dagger}, \ \#^{\dagger\dagger\dagger}, \ x_{c'}^{23}, \ \overline{c''}, \ x_{c'}c'x_{c'}, \ \overline{c'}, \ x_{c'}c''x_{c'}, \ x_{c'}^{24})$$

is a diverse palindromic factorization $P^\dagger$ of S_i'''. Since $P^\dagger$ does not contain c'' as a complete factor, it can be extended to a diverse palindromic factorization $P^\ddagger$ of S_i in which $\bar{c}$, x_c and x_ccx_c are complete factors but c, $x_c\bar{c}x_c$ and x_c^j are not for $j > 1$. Notice $P^\ddagger$ encodes the assignment to the inputs of C_i that makes them true or false according to τ. The other three cases — in which τ makes the outputs labelled a and b both false, false and true, and both true — are similar and we leave them for the full version of this paper. Since C_{i-1} and C_i have the same inputs, each assignment to the inputs of C_i is encoded by some diverse palindromic factorization of S_i.

Now let P be a diverse palindromic factorization of S_i and let τ be the assignment to the inputs of C_{i-1} that is encoded by a prefix of P. Let P' be the prefix of P that is a diverse palindromic factorization of S_i''' and suppose the factorization of

$$x_{c'}^{23}c''x_{c'}c'x_{c'}\overline{c'}x_{c'}\overline{c''}x_{c'}^{25}$$

in P' includes $\overline{c''}$ as a complete factor, which is the case if and only if P includes $\bar{c}$, x_c and x_ccx_c as complete factors but not c, $x_c\bar{c}x_c$ and x_c^j for $j > 1$. We will show that τ must make the outputs of C_{i-1} labelled a and b true. The other case — in which the factorization includes c'' as a complete factor and we want to show τ makes at least one of the inputs labelled a and b false — is similar but longer, and we leave it for the full version of this paper.

Let P'' be the prefix of P' that is a diverse palindromic factorization of S_i''. Since $\overline{c''}$ is a complete factor in the factorization of

$$x_{c'}^{23}c''x_{c'}c'x_{c'}\overline{c'}x_{c'}\overline{c''}x_{c'}^{25}$$

in P', so is c'. Therefore, c' is not a complete factor in the factorization of

$$x_{c'}^{15}\overline{a_1'}x_{c'}c'x_{c'}\overline{b'}x_{c'}^{17}$$

in P'', so $\overline{a_1'}$ and $\overline{b'}$ are.

Let P''' be the prefix of P'' that is a diverse palindromic factorization of S_i'. Since $\overline{a_1'}$ and $\overline{b'}$ are complete factors later in P'', they are not complete factors in P'''. Therefore, $\overline{a_1}$ and $\overline{b}$ are complete factors in the factorizations of

$$x_{c'}^3 a_1' x_{c'} a_1 x_{c'},\ \overline{a_1}x_{c'}\overline{a_1'}x_{c'}^5 \qquad \text{and} \qquad x_{c'}^{11}b'x_{c'}bx_{c'}\overline{b}x_{c'}\overline{b'}x_{c'}^{13}$$

in P''', so they are not complete factors in the prefix $P^\dagger$ of P that is a diverse palindromic factorization of S_{i-1}'. Since we built S_{i-1}' from S_{i-1} with Lemma 2, it follows that a_1 and b are complete factors in the prefix of P that encodes τ. Therefore, τ makes the outputs of C_{i-1} labelled a and b true.

Going through all the possibilities for how P can end, which we will do in the full version of this paper, we find that each diverse palindromic factorization of S_i encodes some assignment to the inputs of C_i. This gives us the following lemma:

Lemma 3. *If we have a string S_{i-1} that represents C_{i-1} and C_i is obtained from C_{i-1} by making two outputs of C_{i-1} the inputs of a new NAND gate, then in constant time we can append symbols to S_{i-1} to obtain a string S_i that represents C_i.*

6   Conclusion

By Lemmas 1, 2 and 3 and induction, given a Boolean circuit C composed only of splitters and NAND gates with two inputs and one output, in time linear in the size of C we can build, inductively and in turn, a sequence of strings $S_1, \ldots, S_t$ such that S_i represents C_i. As mentioned in Section 2, once we have S_t we can easily build in constant time a string S that has a diverse palindromic factorization if and only if C is satisfiable. Therefore, diverse palindromic factorization is NP-hard. Since it is obviously in NP, we have the following theorem:

Theorem 1. *Diverse palindromic factorization is NP-complete.*

Acknowledgments. Many thanks to Gabriele Fici for his comments on a draft of this paper, and to the anonymous referee who pointed out a gap in the proof of Lemma 3.

References

1. Alitabbi, A., Iliopoulos, C.S., Rahman, M.S.: Maximal palindromic factorization. In: Proceedings of the Prague Stringology Conference (PSC), pp. 70–77 (2013)

2. Fernau, H., Manea, F., Mercaş, R., Schmid, M.L.: Pattern matching with variables: fast algorithms and new hardness results. In: Proceedings of the 32nd Symposium on Theoretical Aspects of Computer Science (STACS), pp. 302–315 (2015)
3. Fici, G., Gagie, T., Kärkkäinen, J., Kempa, D.: A subquadratic algorithm for minimum palindromic factorization. Journal of Discrete Algorithms **28**, 41–48 (2014)
4. Frid, A.E., Puzynina, S., Zamboni, L.: On palindromic factorization of words. Advances in Applied Mathematics **50**(5), 737–748 (2013)
5. Gawrychowski, P., Uznański, P.: Tight tradeoffs for approximating palindromes in streams. Technical Report 1410.6433, arxiv.org (2014)
6. I, T., Sugimoto, S., Inenaga, S., Bannai, H., Takeda, M.: Computing palindromic factorizations and palindromic covers on-line. In: Kulikov, A.S., Kuznetsov, S.O., Pevzner, P. (eds.) CPM 2014. LNCS, vol. 8486, pp. 150–161. Springer, Heidelberg (2014)
7. Kosolobov, D., Rubinchik, M., Shur, A.M.: Pal^k is linear recognizable online. In: Italiano, G.F., Margaria-Steffen, T., Pokorný, J., Quisquater, J.-J., Wattenhofer, R. (eds.) SOFSEM 2015-Testing. LNCS, vol. 8939, pp. 289–301. Springer, Heidelberg (2015)
8. Ravsky, O.: On the palindromic decomposition of binary words. Journal of Automata, Languages and Combinatorics **8**(1), 75–83 (2003)
9. Tseitin, G.S.: On the complexity of derivation in propositional calculus. In: Slisenko, A.O. (ed.) Structures in Constructive Mathematics and Mathematical Logic, Part II, pp. 115–125 (1968)
10. Ziv, J., Lempel, A.: A universal algorithm for sequential data compression. IEEE Transactions on Information Theory **22**(3), 337–343 (1977)
11. Ziv, J., Lempel, A.: Compression of individual sequences via variable-rate coding. IEEE Transactions on Information Theory **24**(5), 530–536 (1978)

Factorization in Formal Languages

Paul C. Bell[1], Daniel Reidenbach[1], and Jeffrey Shallit[2]([⊠])

[1] Department of Computer Science, Loughborough University, Loughborough, Leicestershire LE11 3TU, UK
{P.Bell,D.Reidenbach}@lboro.ac.uk
[2] School of Computer Science, University of Waterloo, Waterloo, ON N2L 3G1, Canada
shallit@cs.uwaterloo.ca

Abstract. We consider several language-theoretic aspects of unique factorization in formal languages. We reprove the familiar fact that the set $\mathrm{uf}(L)$ of words having unique factorization into elements of L is regular if L is regular, and from this deduce an quadratic upper and lower bound on the length of the shortest word not in $\mathrm{uf}(L)$. We observe that $\mathrm{uf}(L)$ need not be context-free if L is context-free.

Next, we consider some variations on unique factorization. We define a notion of "semi-unique" factorization, where every factorization has the same number of terms, and show that, if L is regular or even finite, the set of words having such a factorization need not be context-free. Finally, we consider additional variations, such as unique factorization "up to permutation" and "up to subset". Although all these variations have been considered before, it appears that the languages of words having these properties have not been positioned in the Chomsky hierarchy up to now. We also consider the length of the shortest word *not* having the desired property.

1 Introduction

Let L be a formal language. We say $x \in L^*$ has *unique factorization* if whenever

$$x = y_1 y_2 \cdots y_m = z_1 z_2 \cdots z_n$$

for $y_1, y_2, \ldots, y_m, z_1, z_2, \ldots, z_n \in L$ then $m = n$ and $y_i = z_i$ for $1 \le i \le m$. If every element of L^* has unique factorization into elements of L, then L is called a *code*.

Although codes have been studied extensively (see, for example, [1,8]), in this paper we look at some novel aspects of unique factorization. Namely, we look at some variations of unique factorization, consider the language of words possessing this type of unique factorization, and position the resulting language in the Chomsky hierarchy. We also consider the length of the shortest word *not* having the desired property, if it exists.

© Springer International Publishing Switzerland 2015
I. Potapov (Ed.): DLT 2015, LNCS 9168, pp. 97–107, 2015.
DOI: 10.1007/978-3-319-21500-6_7

2 Unique Factorizations

Given L, we define $\mathrm{uf}(L)$ to be the set of all elements of L^* having unique factorization into elements of L. So L is a code iff $L^* = \mathrm{uf}(L)$. We recall the following familiar fact:

Proposition 1. *If L is regular, then so is $\mathrm{uf}(L)$.*

Proof. If L contains the empty word ϵ then no elements of L^* have unique factorization, and so $\mathrm{uf}(L) = \emptyset$. So, without loss of generality we can assume $\epsilon \notin L$.

To prove the result, we show that the relative complement $L^* - \mathrm{uf}(L)$ is regular. Let L be accepted by a deterministic finite automaton (DFA) M. On input $x \in L^*$, we build a nondeterministic finite automaton (NFA) M' to guess two different factorizations of x and verify they are different. The machine M' maintains the single state of the DFA M for L as it scans the elements of x, until M' reaches a final state q. At this point M' moves, via an ϵ-transition, to a new kind of state that records pairs. Transitions on these "doubled" states still follow M's transition function in both coordinates, with the exception that if either state is in F, we allow a "reset" implicitly to q_0. Each implicit return to q_0 marks, in a factorization, the end of a term. The final states of M' are the "doubled" states with both elements in F.

More precisely, assume $M = (Q, \Sigma, \delta, q_0, F)$. Since $\epsilon \notin L(M)$, we know $q_0 \notin F$. We create the machine $M' = (Q', \Sigma, \delta', q_0, F')$ as follows:

$$\delta'(q, a) = \begin{cases} \{\delta(q, a)\}, & \text{if } q \notin F; \\ \{\delta(q_0, a), \ [\delta(q_0, a), \delta(q, a)]\}, & \text{if } q \in F. \end{cases}$$

Writing $r = \delta(p, a)$, $s = \delta(q, a)$, $t = \delta(q_0, a)$, we also set

$$\delta'([p, q], a) = \begin{cases} \{[r, s]\}, & \text{if } p \notin F, q \notin F; \\ \{[r, s], [t, s]\}, & \text{if } p \in F, q \notin F; \\ \{[r, s], [r, t]\}, & \text{if } p \notin F, q \in F; \\ \{[r, s], [t, s], [r, t], [t, t]\}, & \text{if } p \in F, q \in F. \end{cases}$$

Finally, we set $F' = F \times F$. To see that the construction works, suppose that $x \in L^*$ has two different factorizations

$$x = y_1 y_2 \cdots y_j y_{j+1} \cdots y_k = y_1 y_2 \cdots y_j z_{j+1} \cdots z_\ell$$

with y_{j+1} a proper prefix of z_{j+1}. Then an accepting path starts with singleton sets until the end of y_j. The next transition goes to a pair having first element $\delta(q_0, a)$ with a the first letter of y_{j+1}. Subsequent transitions eventually lead to a pair in $F \times F$.

On the other hand, if x is accepted, then two different factorizations are traced out by the accepting computation in each coordinate. The factorizations are guaranteed to be different because of the transition to $[\delta(q_0, a), \delta(q, a)]$. $\square$

Remark 2. There is a shorter and more transparent proof of this result, as follows. Given a DFA for L, create an NFA A for L^* by adding ϵ-transitions from every final state back to the initial state, and then removing the ϵ-transitions using the familiar method (e.g., [6, Theorem2.2]). Next, using the Boolean matrix interpretation of finite automata (e.g., [15] and [12, §3.8]), we can associate an adjacency matrix M_a with the transitions of A on the letter a. Then, on input $x = a_1 a_2 \cdots a_i$, a DFA can compute the matrix $M_x = M_{a_1} M_{a_2} \cdots M_{a_i}$ using ordinary integer matrix multiplication, with the proviso that any entry that is 2 or more is changed to 2 after each matrix multiplication. This can be done by a DFA since the number of such matrices is at most 3^{n^2} where n is the number of states of M. Then, accepting if and only if the entry in the row and column corresponding to the initial state of A is 1, we get a DFA accepting exactly those x having unique factorization into elements of L. While this proof is much simpler, the state bound it provides is quite extravagant compared to our previous proof.

Corollary 3. *Suppose L is accepted by a DFA with n states. If L is not a code, then there exists a word $x \in L^*$ with at least two distinct factorizations into elements of L, with $|x| < n^2 + n$.*

Proof. Our construction in the proof of Proposition 1 gives an NFA M' accepting all words with at least two different factorizations, and it has $n^2 + n$ states. If M' accepts anything at all, it accepts a word of length at most $n^2 + n - 1$. ☐

Proposition 4. *For all $n \geq 2$, there exists an $O(n)$-state DFA accepting a language L that is not a code, such that the shortest word in L^* having two factorizations into elements of L is of length $\Omega(n^2)$.*

Proof. Consider the language $L_n = b(a^n)^* \cup (a^{n+1})^* b$. It is easy to see that L_n can be accepted by a DFA with $2n + 5$ states, but the shortest word in L_n^* having two distinct factorizations into elements of L_n is $b\,a^{n(n+1)}\,b$, of length $n^2 + n + 2$. ☐

In fact, there are even examples of finite languages with the same property.

Proposition 5. *For all $n \geq 2$, there exists an $O(n)$-state DFA accepting a finite language L that is not a code, such that the shortest word in L^* having two factorizations is of length $\Omega(n^2)$.*

Proof. Let $\Sigma = \{b, a_1, a_2, \ldots, a_n\}$ be an alphabet of size $n + 1$, and let L_n be the language of $2n$ words

$$\{a_1, a_n\} \cup \{b^i a_{i+1} : 1 \leq i < n\} \cup \{a_i b^i : 1 \leq i < n\}$$

defined over Σ.

Then it is easy to see that L_n can be accepted with a DFA of $2n + 2$ states, while the shortest word having two distinct factorizations is

$$a_1 b a_2 b^2 a_3 b^3 \cdots a_{n-1} b^{n-1} a_n,$$

which is of length $n(n+1)/2$. ☐

Remark 6. The previous example can be recoded over a three-letter alphabet by mapping each a_i to the base-2 representation of i, padded, if necessary, to make it of length ℓ, where $\ell = \lceil \log_2 n \rceil$. With some reasonably obvious reuse of states this can still be accepted by a DFA using $O(n)$ states, and the shortest word with two distinct factorizations is still of length $\Omega(n^2)$.

Theorem 7. *If L is a context-free language, then* $\mathrm{uf}(L)$ *need not be context-free.*

Proof. Our example is based on two languages (see, for example, [10]):

(a) PALSTAR, the set of all strings over the alphabet $\Sigma = \{0, 1\}$ that are the concatenation of one or more even-length palindromes; and
(b) PRIMEPALSTAR, the set of all elements of PALSTAR that *cannot* be written as the concatenation of two or more elements of PALSTAR.

Clearly PALSTAR is a context-free language (CFL). We see that $\mathrm{uf}(\text{PALSTAR}) = $ PRIMEPALSTAR, which was proven in [10] to be non-context-free.   □

3   Semi-unique Factorizations

We now consider a variation on unique factorization. We say that $x \in L^*$ has *semi-unique factorization* if all factorizations of x into elements of L consist of the same number of factors. More precisely, x has semi-unique factorization if whenever

$$x = y_1 y_2 \cdots y_m = z_1 z_2 \cdots z_n$$

for $y_1, y_2, \ldots, y_m, z_1, z_2, \ldots, z_n \in L$, then $m = n$.

Given a language L, we define $\mathrm{su}(L)$ to be the set of all elements of L^* having semi-unique factorization over L. This concept was previously studied by Weber and Head [14], where a language L was called *numerically decipherable* if $L = \mathrm{su}(L)$, and an efficient algorithm was proposed for testing this property.

Example 8. Let $L = \{a, ab, aab\}$. Then $\mathrm{su}(L) = (ab)^* a^*$.

Theorem 9. *If L is regular, then* $\mathrm{su}(L)$ *is a co-CFL (and hence a context-sensitive language).*

Proof. To see that $\mathrm{su}(L)$ is a co-CFL, mimic the proof of Proposition 1. We use a stack to keep track of the difference between the number of terms in the two guessed factorizations, and another flag in the state to say which, the "top", or the "bottom" state, has more terms (since the stack can't hold negative counters). We accept if we guess two factorizations having different numbers of terms.

It now follows immediately that $\mathrm{su}(L)$ is a context-sensitive language (CSL), by the Immerman-Szelepcsényi theorem [7,13].   □

Corollary 10. *Given a regular language L, it is decidable if there exist elements $x \in L^*$ lacking semi-unique factorization.*

Proof. Given L, we can construct a pushdown automaton (PDA) accepting $L^* -$ su(L). We convert this PDA to a context-free grammar G generating the same language (e.g., [6, Theorem 5.4]). Finally, we use well-known techniques (e.g., [6, Theorem 6.6]) to determine whether $L(G)$ is empty. $\square$

Theorem 11. *If L is regular then* su(L) *need not be a CFL.*

Proof. Let

$$L = a0^+b + 1 + c(23)^+ + 23d + a + 0 + b1^+c(23)^+ + a0^+b1^+c2 + 32 + 3d.$$

Consider su(L) and intersect with the regular language $a0^+b1^+c(23)^+d$.

Then there are only three possible factorizations for a given word here. They look like (using parens to indicate factors)

$(a0^ib)1 \cdot 1 \cdot 1 \cdots 1(c(23)^k)(23d)$, which has $j + 3$ terms if j is the number of 1's;

$(a)0 \cdot 0 \cdots 0(b1^jc(23)^k)(23d)$, which has $i + 3$ terms if i is the number of 0's; and

$(a0^ib1^jc2)(32)(32) \cdots (32)(3d)$, which has $k + 2$ terms, if k is the number of (32)'s.

So if all three factorizations have the same number of terms we must have $i = j = k - 1$, giving us

$$\{a0^nb1^nc(23)^{n-1}d : n \geq 1\},$$

which is not a CFL. $\square$

There are even examples, as in Theorem 11, where L is finite. For expository purposes, we give an example over the 21-letter alphabet

$$\Sigma = \{0, 1, 2, 3, 4, 5, 6, 7, 8, a, b, c, d, e, f, g, h, i, j, k, l\}.$$

Theorem 12. *If L is finite, then* su(L) *need not be a CFL.*

Proof. Define

$$L_1 = \{0ab, cd, ab, cd127, efgh, efgh3, 4ijkl, ijkl, 5, 68\}$$
$$L_2 = \{0abc, dabc, d1, 27e, fg, he, h34ij, klij, kl568\}$$
$$L_3 = \{0a, bcda, bcd12, 7ef, ghef, gh34i, jk, li, jkl56, 8\}$$

and set $L := L_1 \cup L_2 \cup L_3$.

Consider possible factorizations of words of the form

$$0(abcd)^m127(efgh)^n34(ijkl)^p568$$

for some integers $m, n, p \geq 1$. Any factorization of such a word into elements of L must begin with either $0ab$, $0abc$, or $0a$. There are three cases to consider:

Case 1: the first word is $0ab$. Then the next word must begin with c, and there are only two possible choices: cd and $cd127$. If the next word is cd then since no word begins with 1 the only choice is to pick a word starting with a, and there is only one: ab. After picking this, we are back in the same situation, and can only choose between cd followed by ab, or $cd127$. Once $cd127$ is picked we must pick a word that begins with e. However, there are only two: $efgh$ and $efgh3$. If we pick $efgh$ we are left in the same situation. Once we pick $efgh3$ we must pick a word starting with 4, but there is only one: $4ijkl$. After this we can either pick 5 and then 68, or we can pick $ijkl$ a number of times, followed by 568.

This gives the factorization

$$(0ab)((cd)(ab))^{m-1}(cd127)(efgh)^{n-1}(efgh3)(4ijkl)(ijkl)^{p-1}(5)(68)$$

having $1 + 2(m-1) + 1 + (n-1) + 1 + 1 + (p-1) + 1 + 1 = 2m + n + p + 2$ terms.

Case 2: the first word is $0abc$. Then the next word must begin with d, and there are only two choices: $dabc$ and $d1$. If we pick $dabc$ we are back in the same situation. If we pick $d1$ then the next word must begin with 2, but there is only one such word: $27e$. Then the next word must begin with f, but there is only one: fg. Then the next word must begin with h, but there are only two: he and $h34ij$. If we pick he we are back in the same situation. Otherwise we must have a word beginning with k, but there are only two: $klij$ and $kl568$. This gives the factorization

$$(0abc)(dabc)^{m-1}(d1)(27e)((fg)(he))^{n-1}(fg)(h34ij)(klij)^{p-1}(kl568)$$

having $1 + (m-1) + 2 + 2(n-1) + 1 + 1 + (p-1) + 1 = m + 2n + p + 2$ terms.

Case 3: the first word is $0a$. Then only $bcda$ and $bcd12$ start with b, so we must choose $bcda$ over and over until we choose $bcd12$. Only one word starts with 7 so we must choose $7ef$. Now we must choose $ghef$ again and again until we choose $gh34i$. We now choose jk and li alternately until $jkl56$. Finally, we pick 8.

This gives us a factorization

$$(0a)(bcda)^{m-1}(bcd12)(7ef)(ghef)^{n-1}(gh34i)((jk)(li))^{p-1}(jkl56)(8)$$

with $1 + (m-1) + 2 + (n-1) + 1 + 2(p-1) + 2 = m + n + 2p + 2$.

So for all these three factorizations to have the same number of terms, we must have

$$2m + n + p + 2 = m + 2n + p + 2 = m + n + 2p + 2.$$

Eliminating variables we get that $m = n = p$. So when we compute $su(L)$ and intersect with the regular language $0(abcd)^{+}127(efgh)^{+}34(ijkl)^{+}568$ we get

$$\{0(abcd)^{n}127(efgh)^{n}34(ijkl)^{n}568 \ : \ n \geq 1\},$$

which is clearly a non-CFL. □

Remark 13. The previous two examples can be recoded over a binary alphabet, by mapping the i'th letter to the string ba^ib.

4 Permutationally Unique Factorization

In this section we consider yet another variation on unique factorization, which are factorizations that are unique up to permutations of the factors. This concept was introduced by Lempel [9] under the name "multiset decipherable codes". For other work on these codes, see [2,5,11].

Formally, given a language L we say $x \in L^*$ has *permutationally unique factorization* if whenever $x = y_1 y_2 \cdots y_m = z_1 z_2 \cdots z_n$ for

$$y_1, y_2, \ldots, y_m, z_1, z_2, \ldots, z_n \in L,$$

then $m = n$ and there exists a permutation σ of $\{1, \ldots, n\}$ such that $y_i = z_{\sigma(i)}$ for $1 \le i \le n$. In other words, we consider two factorizations that differ only in the order of the factors to be the same. We define $\mathrm{ufp}(L)$ to be the set of $x \in L^*$ having permutationally unique factorization.

Example 14. Consider $L = \{a^3, a^4\}$. Then

$$\mathrm{ufp}(L) = \{a^3, a^4, a^6, a^7, a^8, a^9, a^{10}, a^{11}, a^{13}, a^{14}, a^{17}\}.$$

Theorem 15. *If L is finite then $\mathrm{ufp}(L)$ is a co-CFL and hence a CSL.*

Proof. We sketch the construction of a PDA accepting $\overline{\mathrm{ufp}(L)}$. If a word is in L^* but has two permutationally distinct factorizations, then there has to be some factor appearing in the factorizations a different number of times. Our PDA nondeterministically guesses two different factorizations and a factor $t \in L$ that appears a different number of times in the factorizations, then verifies the factorizations and checks the number. It uses the stack to hold the absolute value of the difference between the number of times t appears in the first factorization and the second. It accepts if both factorizations end properly and the stack is nonempty. □

Theorem 16. *If L is finite then $\mathrm{ufp}(L)$ need not be a CFL.*

Proof. Let $\Sigma = \{a, b, c\}$. Define $L = \{A, B, S_1, S_2, T_1, T_2\} \subseteq \Sigma^+$ as follows:

$$A = aa, \; B = aaa, \; S_1 = ab, \; S_2 = ac, \; T_1 = ba, \; T_2 = ca.$$

Let $R = aa(ab)^+(ac)^+aa(ba)^+(ca)^+aaa$, and consider words of the form

$$w := aa(ab)^r(ac)^s aa(ba)^t(ca)^q aaa \in \mathrm{ufp}(L) \cap R$$

with $r, s, t, q \ge 1$ and the following two factorizations of w:

$$AS_1^r S_2^s A T_1^t T_2^q B = aa \cdot (ab)^r \cdot (ac)^s \cdot aa \cdot (ba)^t \cdot (ca)^q \cdot aaa \tag{1}$$
$$BT_1^r T_2^s S_1^t S_2^q AA = aaa \cdot (ba)^r \cdot (ca)^s \cdot (ab)^t \cdot (ac)^q \cdot aa \cdot aa \tag{2}$$

It is not difficult to see that w must be of one of these two forms. Since w has prefix $aaab$, it must start with either AS_1 or BT_1. If it starts with $AS_1 = aa \cdot ab$, the next factors must be S_1^{r-1} to match $(ab)^r$, so we have AS_1^r. We then see $(ac)^s$, which can only match with S_2^s. Next, we see '$aaba$', thus we must choose $AT_1 = aa \cdot ba$. We then have $(ba)^{t-1}$, which can only match with T_1^{t-1}, and then $(ca)^q$, matching only with T_2^q. Finally the suffix is 'aaa' which can only match with B as required.

If w starts with $BT_1 = aaa \cdot ba$, the next part is $(ba)^{r-1}$, which only matches with T_1^{r-1}. Then we see $(ca)^s$, so we must use factors T_2^s. We then see $(ab)^t$ and $(ac)^q$, matching with S_1^t and S_2^q respectively. Finally we have '$aaaa$' matching only with AA as required.

If $r = t$ and $s = q$, then the number of each factor $(A, B, S_1, S_2, T_1, T_2)$ in factorizations (1) and (2) is identical. Therefore, w always has more than one factorization (of type (1) or (2)); however, that factorization is only non-permutationally equivalent if $r \neq t$ or $s \neq q$. Therefore

$$\mathrm{ufp}(L) \cap R = \{aa \cdot (ab)^r \cdot (ac)^s \cdot aa \cdot (ba)^t \cdot (ca)^q \cdot aaa \mid (r = t) \wedge (s = q)\}$$
$$= \{AS_1^r S_2^s AT_1^r T_2^s B \; : \; r, s \geq 1\},$$

which is not a context-free language.   □

5   Subset-Invariant Factorization

In this section we consider yet another variation on unique factorization, previously studied under the name "set-decipherable code" by Blanchet-Sadri and Morgan [2].

We say a word $x \in L^*$ has *subset-invariant factorization* (into elements of L) if there exists a subset $S \subseteq L$ with the property that every factorization of x into elements of L uses exactly the elements of S – no more, no less – although each element may be used a different number of times. More precisely, x has subset-invariant factorization if there exists $S = S(x)$ such that whenever $x = y_1 y_2 \cdots y_m$ with $y_1, y_2, \ldots, y_m \in L$, then $S = \{y_1, y_2, \ldots, y_m\}$. We let $\mathrm{ufs}(L)$ denote the set of those $x \in L^*$ having such a factorization.

Theorem 17. *If L is finite then $\mathrm{ufs}(L)$ is regular.*

Proof. The proof is similar to the proof of Theorem 15 above. On input x we nondeterministically attempt to construct two different factorizations into elements of L, recording which elements of L we have seen so far. We accept if we are successful in constructing two different factorizations (which will be different if and only if some element was chosen in one factorization but not the other). This NFA accepts $L^* - \mathrm{ufs}(L)$. So if L is finite, it follows that $\mathrm{ufs}(L)$ is regular.

In more detail, here is the construction. States of our NFA are 6-tuples of the form $[w_1, s_1, v_1, w_2, s_2, v_2]$ where w_1, w_2 are the words of L we are currently trying to match; s_1, s_2 are, respectively, the suffixes of w_1, w_2 we have yet to see, and v_1, v_2 are binary characteristic vectors of length $|L|$, specifying which elements of L have been seen in the factorization so far (including w_1 and w_2, although technically they may not have been seen yet). Letting $C(z)$ denote the vector with all 0's except a 1 in the position corresponding to the word $z \in L$, the initial states are $[w, w, C(w), x, x, C(x)]$ for all words $w, x \in L$. The final states are of the form $[w, \epsilon, v_1, x, \epsilon, v_2]$ where $v_1 \neq v_2$. Transitions on a letter a look like $\delta([w_1, as_1, v_1, w_2, as_2, v_2], a) = [w_1, s_1, v_1, w_2, s_2, v_2]$. In addition there are ϵ-transitions that update the corresponding vectors if s_1 or s_2 equals ϵ, and that "reload" the new w_1 and w_2 we are expecting to see:

$$\delta([w_1, \epsilon, v_1, w_2, s_2, v_2], \epsilon) = \{[w, w, v_1 \vee C(w), w_2, s_2, v_2] \ : \ w \in L\}$$
$$\delta([w_1, s_1, v_1, w_2, \epsilon, v_2], \epsilon) = \{[w_1, s_1, v_1, w, w, v_2 \vee C(w)] \ : \ w \in L\}.$$

$\square$

The preceding proof also shows that the shortest word failing to have subset-invariant factorization is bounded polynomially:

Corollary 18. *Suppose $|L| = n$ and the length of the longest word of L is m. Then if some word of L^* fails to have subset-invariant factorization, there is a word with this property of length $\leq 2m^2n^2$.*

Proof. Let $u \in L^+$ be a minimal length word such that $u \in L^+ - \mathrm{ufs}(L)$. Consider the states of the NFA traversed in processing u. Let $S_0 := [w, w, C(w), x, x, C(x)]$ be the initial state and $S_F := [w_F, \epsilon, v_F, x_F, \epsilon, v'_F]$ the final state, where $v_F \neq v'_F$ and $C(w), C(x)$ are defined as in the proof of Theorem 17. By definition, there must exist some $z \in L$ such that v_F and v'_F differ on $C(z)$, i.e., $v_F^T \cdot C(z) + v'_F{}^T \cdot C(z) = 1$.

Initially the characteristic vectors have a single 1, and once an element is set to 1 in a characteristic vector in the NFA, it is never reset to 0. Thus, there exists some $1 \leq k \leq |u|$ such that $u = u_1 \cdots u_{k-1} \cdot u_k \cdot u_{k+1} \cdots u_{|u|}$ where $S_{k-1} = \delta(S_0, u_1 \cdots u_{k-1})$ has a 0 in the characteristic vectors at position z, and $\delta(S_{k-1}, u_k)$ has a 1 in exactly one of the two characteristic vectors at position z. We shall now prove that $|u_1 \cdots u_{k-1}|, |u_{k+1} \cdots u_{|u|}| \leq m^2n^2$, which proves the result.

We prove the result for the word $v = u_1 \cdots u_{k-1}$; a similar analysis holds for $u_{k+1} \cdots u_{|u|}$. Let $S_0, S_1, \ldots S_{k-1}$ be the states of the NFA visited as we process v. We prove that there does not exist $0 \leq i < j \leq k - 1$ such that $S_i = [w_1, s_1, v_1, w_2, s_2, v_2]$ and $S_j = [w_1, s_1, v'_1, w_2, s_2, v'_2]$. We proceed by contradiction. Assume that such an i and j exist. Then $u_{i+1} \cdots u_j$ is such that $\delta(S_i, u_{i+1} \cdots u_j) = S_j$. However, $\delta(S_i, u_{j+1} \cdots u_k)$ and $\delta(S_j, u_{j+1} \cdots u_k)$ can only differ in their binary characteristic vectors, since the transition function does not depend upon the characteristic vectors when we update the words w_1, s_1, w_2, s_2. Thus, we can remove the factor $u_{i+1} \cdots u_j$ from u and still reach a final state of the form $S_{F_2} := [w_F, \epsilon, v_{F_2}, x_F, \epsilon, v'_{F_2}]$, for which we still have that $v_{F_2} \neq v'_{F_2}$, since they differ on element z due to letter u_k. Continuing this idea iteratively, the maximal number of states k is bounded by m^2n^2. Doubling this bound gives the result. $\square$

The next result shows that we can achieve a quadratic lower bound.

Proposition 19. *There exist examples with $|L| = 2n$ and longest word of length n for which the shortest word of L^* failing to have subset-invariant factorization is of length $n(n + 1)/2$.*

Proof. We just use the example of Proposition 5.   □

Theorem 20. *If L is regular then* $\mathrm{ufs}(L)$ *need not be a CFL.*

Proof. We use a variation of the construction in the proof of Theorem 16. Let $L = (ab)^+(ac)^+aa + (ba)^+(ca)^+ + aa + aaa$. Then (using the notation in the proof of Theorem 16), if

$$w := aa(ab)^r(ac)^s aa(ba)^t(ca)^q aaa \in \mathrm{ufs}(L) \cap R$$

with $r, s, t, q \geq 1$ then there are two different factorizations of w:

$$w = aa \cdot (ab)^r(ac)^s aa \cdot (ba)^t(ca)^q \cdot aaa$$
$$= aaa \cdot (ba)^r(ca)^s \cdot (ab)^t(ac)^q aa \cdot aa$$

which are subset-invariant if and only if $r = t$ and $s = q$. So

$$\mathrm{ufs}(L) \cap R = \{aa(ab)^r(ac)^s aa(ba)^r(ca)^s aaa \ : \ r, s \geq 1\},$$

which is not a CFL.   □

Acknowledgments. The idea of considering semi-unique factorization was inspired by a talk of Nasir Sohail at the University of Waterloo in April 2014.

We are very grateful to the referees for pointing out relevant citations to the literature that we did not know about.

References

1. Berstel, J., Perrin, D., Reutenauer, C.: Codes and automata. In: Encyclopedia of Mathematics and its Applications, vol. 129. Cambridge University Press (2010)
2. Blanchet-Sadri, F., Morgan, C.: Multiset and set decipherable codes. Computers and Mathematics with Applications **41**, 1257–1262 (2001)
3. Burderi, F., Restivo, A.: Coding partitions. Discrete Mathematics and Theoretical Computer Science **9**, 227–240 (2007)
4. Head, T., Weber, A.: Deciding code related properties by means of finite transducers. In: Capocelli, R., De Santis, A., Vaccaro, U. (Eds.) Sequences II: Methods in Communication, Security, and Computer Science, pp. 260–272. Springer (1993)
5. Head, T., Weber, A.: Deciding multiset decipherability. IEEE Trans. Info. Theory **41**, 291–297 (1995)
6. Hopcroft, J.E., Ullman, J.D.: Introduction to Automata Theory, Languages, and Computation. Addison-Wesley (1979)

7. Immerman, N.: Nondeterministic space is closed under complementation. SIAM J. Comput. **17**, 935–938 (1988)
8. Jürgensen, H., Konstantinidis, S.: Codes. In: Rozenberg, G., Salomaa, A. (Eds.) Handbook of Formal Languages, Word, Language, Grammar, vol. 1, pp. 511–607. Springer (1991)
9. Lempel, A.: On multiset decipherable codes. IEEE Trans. Info. Theory **32**, 714–716 (1986)
10. Rampersad, N., Shallit, J., Wang, M.-W.: Inverse star, borders, and palstars. Info. Proc. Letters **111**, 420–422 (2011)
11. Restivo, A.: A note on multiset decipherable codes. IEEE Trans. Info. Theory **35**, 662–663 (1989)
12. Shallit, J.: A Second Course in Formal Languages and Automata Theory. Cambridge University Press (2009)
13. Szelepcsényi, R.: The method of forcing for nondeterministic automata. Bull. EATCS **33**, 96–100 (1987)
14. Weber, A., Head, T.: The finest homophonic partition and related code concepts. In: Privara, I., Ružička, P., Rovan, B. (eds.) MFCS 1994. LNCS, vol. 841, pp. 618–628. Springer, Heidelberg (1994)
15. Zhang, G.-Q.: Automata, Boolean matrices, and ultimate periodicity. Inf. Comput. **152**, 138–154 (1999)

Consensus Game Acceptors

Dietmar Berwanger[(✉)] and Marie van den Bogaard[(✉)]

LSV, ENS Cachan, CNRS and University of Paris-Saclay, Paris-Saclay, France
{dwb,mvdb}@lsv.fr

Abstract. We study a game for recognising formal languages, in which
two players with imperfect information need to coordinate on a common
decision, given private input strings correlated by a finite graph. The
players have a joint objective to avoid an inadmissible decision, in spite
of the uncertainty induced by the input.

We show that the acceptor model based on consensus games characterises context-sensitive languages, and conversely, that winning strategies in such games can be described by context-sensitive languages. We
also discuss consensus game acceptors with a restricted observation pattern that describe nondeterministic linear-time languages.

1 Introduction

The idea of viewing computation as an interactive process has been at the origin of many enlightening developments over the past three decades. With the
concept of alternation, introduced around 1980 by Chandra and Stockmeyer,
and independently by Kozen [6], computation steps are attributed to conflicting
players seeking to reach or avoid certain outcome states. This approach relies
on determined games with perfect information, and it lead to important and
elegant results, particularly in automata theory. Around the same time, Peterson and Reif [18] initiated a study on computation via games with imperfect
information, also involving teams of players. This setting turned out to be even
more expressive, but also overwhelmingly difficult to comprehend. (See [3,10],
for more recent accounts.)

In this paper, we propose a game model of a language acceptor based on
coordination games between two players with imperfect information. Compared
to the model of Reif and Peterson, our setting is extremely simple: the games are
played on a finite graph, plays are of finite duration, they involve only one yes/no
decision, and the players have no means to communicate. Moreover, they are
bound to take their decisions in consensus. Given an input word that may yield
different observations to each of the players, they have to settle simultaneously
and independently on a common decision, otherwise they lose.

We model such systems as consensus game acceptors, a particular case of
coordination games with perfect recall, also described as multiplayer concurrent
games or synchronous distributed games with incomplete information in the
computer-science literature. Our motivation for studying the acceptor model
is to obtain lower bounds on the complexity of basic computational problems

© Springer International Publishing Switzerland 2015
I. Potapov (Ed.): DLT 2015, LNCS 9168, pp. 108–119, 2015.
DOI: 10.1007/978-3-319-21500-6_8

regarding these more general games, specifically (1) solvability: whether a winning strategy exists, for a given game, and (2) implementability: which computational resources are needed to implement a winning strategy, if any exists.

Without the restrictions to consensus and to a single decision per play, the solvability problem for coordination games with safety winning conditions is known to be undecidable for two or more players [18,19]. Furthermore, Janin [11] points out that there exist two-player safety games that admit a winning strategy but none that can be implemented by a Turing machine.

Our first result establishes a correspondence between context-sensitive languages and winning strategies in consensus games: We prove that every context-sensitive language L corresponds to a consensus game in which the characteristic function of L describes a winning strategy, and conversely, every consensus game that admits a joint winning strategy also admits one characterised by a context-sensitive language. Games with imperfect information for one player against the environment (which, here we call Input) display a similar correspondence with regular languages; they correspond to consensus games where the two player receive identical (or functionally dependent) observations. In extension, we consider consensus games where the observations of the two players can be ordered, and we show that the resulting acceptor model subsumes context-free languages and moreover allows to describe languages decidable in nondeterministic time.

The correspondence has several consequences in terms of game complexity. On the one hand, it reveals that consensus games preserve surprisingly much of the computational complexity found in games with imperfect information, in spite of the restriction to a single decision and to consensus. Consensus games are relevant because they represent coordination games in the limiting case where signalling is impossible. The classical constructions for proving undecidability of synchronous distributed games typically simulate a communication channel that may lose one message and involve an unbounded number of non-trivial decisions by which the players describe configurations of a Turing machine [2, 19,22]. In contrast, our undecidability argument for acceptor games relies on the impossibility to attain common knowledge when knowledge hierarchies can grow unboudedly, and this can be witnessed by a single decision. Apart of this, we obtain a simple game family in which winning strategies are PSPACE-hard to run, in the length of the play, or, in a positive perspective, where winning strategies can be implemented by a linear-bounded automata whenever they exist.

2   Preliminaries

For classical notions of formal languages, in particular context-sensitive languages, we refer, e.g., to the textbook of Salomaa [21]. We use the characterisation of context-sensitive languages in terms of nondeterministic linear-bounded automata, given by Kuroda [12] and the following well-known results from the same article: (1) For a fixed context-sensitive language L over an alphabet Σ, the problem whether a given word $w \in \Sigma^*$ belongs to L is PSPACE-hard. (2) The problem of determining whether a given context-sensitive language represented by a linear-bounded automaton contains any non-empty word is undecidable.

2.1   Consensus Game Acceptors

Consensus acceptors are games between two cooperating players 1 and 2, and an additional agent called Input. Given a finite *observation alphabet* Γ common to both players, a *consensus game acceptor* $G = (V, E, (\beta^1, \beta^2), v_0, \Omega)$ is described by a finite set V of *states*, a *transition* relation $E \subseteq V \times V$, and a pair of *observation* functions $\beta^i : V \rightarrow \Gamma$ that label every state with an observation, for each player $i = 1, 2$. There is a distinguished initial state $v_0 \in V$ with no incoming transition. States with no outgoing transitions are called final states; the admissibility condition $\Omega : V \rightarrow \mathcal{P}(\{0,1\})$ maps every final state $v \in V$ to a nonempty subset of admissible decisions $\Omega(v) \subseteq \{0,1\}$. The observations at the initial and the final states do not matter, we may assume that they are labelled with the same observation $\#$ for both players.

The game is played as follows: Input chooses a finite path $\pi = v_0, v_1, \ldots, v_{n+1}$ in G from the initial state v_0, following transitions $(v_\ell, v_{\ell+1}) \in E$, for all $\ell \leq n$, to a final state v_{n+1}. Then, each player i receives a private sequence of observations $\beta^i(\pi) := \beta^i(v_1), \beta^i(v_1), \ldots, \beta^i(v_n)$ and is asked to take a *decision* $a^i \in \{0,1\}$, independently and simultaneously. The players win if they agree on an admissible decision, that is, $a^1 = a^2 \in \Omega(v_{n+1})$; otherwise they lose. Without risk of confusion we sometimes write $\Omega(\pi)$ for $\Omega(v_{n+1})$.

We say that two plays π, π' are *indistinguishable* to a player i, and write $\pi \sim^i \pi'$, if $\beta^i(\pi) = \beta^i(\pi')$. This is an equivalence relation, and its classes, called the *information sets* of Player i, correspond to observation sequences $\beta^i(\pi)$. A *strategy* for Player i is a mapping $s^i : V^* \rightarrow \{0,1\}$ from plays π to decisions $s^i(\pi) \in \{0,1\}$ such that $s^i(\pi) = s^i(\pi')$, for any pair $\pi \sim^i \pi'$ of indistinguishable plays. A joint strategy is a pair $s = (s^1, s^2)$; it is *winning*, if $s^1(\pi) = s^2(\pi) \in \Omega(\pi)$, for all plays π. In this case, the components s^1 and s^2 are equal, and we use the term *winning strategy* to refer to the joint strategy or either of its components. Finally, a game is *solvable*, if there exists a (joint) winning strategy.

In the terminology of distributed systems, consensus game acceptors correspond to *synchronous* systems with *perfect recall* and *known initial state*. They are a particular case of distributed games with safety objectives [16], coordination games with imperfect information [4], or multi-player concurrent games [1].

Strategies and Knowledge. We say that two plays π and π' are *connected*, and write $\pi \sim^* \pi'$, if there exists a sequence of plays $\pi_1, \ldots, \pi_k$ such that $\pi \sim^1 \pi_1 \sim^2 \cdots \sim^1 \pi_k \sim^2 \pi'$. Then, a mapping $f : V^* \rightarrow \{0,1\}$ from plays to decisions is a strategy that satisfies the consensus condition if, and only if, $f(\pi) = f(\pi')$, for all $\pi \sim^* \pi'$. In terms of distributed knowledge, this means that, for every play π, the events $\{\pi \in V^* \mid f(\pi) = 1\}$ and $\{\pi \in V^* \mid f(\pi) = 0\}$ are common knowledge among the players. (For an introduction to knowledge in distributed systems, see the book of Fagin, Halpern, Moses, and Vardi [9, Ch.10,11].) Such a consensus strategy — or, more precisely, the pair (f, f)— may still fail, due to prescribing inadmissible decisions. We say that a decision $a \in \{0, 1\}$ is *safe* at a play π if $a \in \Omega(\pi')$, for all $\pi' \sim^* \pi$. Then, a consensus strategy f is winning, if and only if, it prescribes a safe decision $f(\pi)$, for every play π.

It is sometimes convenient to represent a strategy for a player i as a function $f^i : \Gamma^* \to \{0, 1\}$. Every such function describes a valid strategy, because observation sequences identify information sets; we refer to an *observation-based* strategy in contrast to the *state-based* representation $s^i : V^* \to \{0, 1\}$. Note that the components of a joint winning strategy need no longer be equal in the observation-based representation. However, once the strategy for one player is fixed, the strategy of the other player is determined by the consensus condition, so there is no risk of confusion in speaking of a winning strategy rather than a joint strategy pair.

As an example, consider the game depicted in Figure 1, with observation alphabet $\Gamma = \{a, b, \lhd, \rhd, \Box\}$. States v at which the two players receive different observations are split, with $\beta^1(v)$ written in the upper part and $\beta^2(v)$ in the lower part; states at which the players receive the same observation carry only one symbol. The admissible decisions at final states are indicated on the outgoing arrow. Notice that upon receiving the observation sequence $a^2 b^2$, for instance, the first player is constrained to choose decision 1, due to the following sequence of indistinguishable plays that leads to a play where deciding 0 is not admissible.

$$\begin{pmatrix} a, a \\ a, \lhd \\ b, \rhd \\ b, b \end{pmatrix} \sim^2 \begin{pmatrix} a, a \\ \lhd, \lhd \\ \rhd, \rhd \\ b, b \end{pmatrix} \sim^1 \begin{pmatrix} a, \lhd \\ \lhd, \rhd \\ \rhd, \lhd \\ b, \rhd \end{pmatrix} \sim^2 \begin{pmatrix} \lhd, \lhd \\ \rhd, \rhd \\ \lhd, \lhd \\ \rhd, \rhd \end{pmatrix} \sim^1 \begin{pmatrix} \lhd, \Box \\ \rhd, \Box \\ \lhd, \Box \\ \rhd, \Box \end{pmatrix} \sim^2 \begin{pmatrix} \Box, \Box \\ \Box, \Box \\ \Box, \Box \\ \Box, \Box \end{pmatrix}$$

In contrast, decision 0 may be safe when Player 1 receives input $a^3 b^2$, for instance. Actually, the strategy $s^1(w)$ that prescribes 1 if, and only if, $w \in \{a^n b^n : n \in \mathbb{N}\}$ determines a joint winning strategy. Next, we shall make the relation between games and languages more precise.

3 Describing Languages by Games

We consider languages L over a letter alphabet Σ. The empty word ε is excluded from the language, and also from its complement $\bar{L} := (\Sigma^* \setminus \{\varepsilon\}) \setminus L$. As acceptors for such languages, we consider games over an observation alphabet $\Gamma \supseteq \Sigma$, and we assume that no observation sequence in Σ^+ is omitted: for every word $w \in \Sigma^+$, and each player i, there exists a play π that yields the observation sequence $\beta^i(\pi) = w$. Every consensus game acceptor can be modified to satisfy this condition without changing the winning strategies.

Given an acceptor game G, we associate to every observation-based strategy $s \in S^1$ of the first player, the language $L(s) := \{w \in \Sigma^* : s(w) = 1\}$. We say that the game G *covers* a language $L \subseteq \Sigma^*$, if G is solvable and

- $L = L(s)$, for *some* winning strategy $s \in S^1$, and
- $L \subseteq L(s)$, for *every* winning strategy $s \in S^1$.

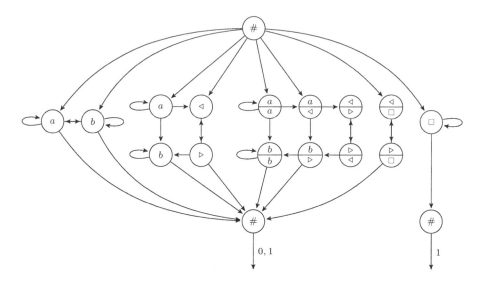

Fig. 1. A consensus game acceptor

If, moreover, $L = L(s)$ for *every* winning strategy in G, we say that G *charac-terises* L. In this case, all winning strategies map L to 1 and $\bar{L}$ to 0.

As suggested above, the consensus game acceptor represented in Figure 1 covers the language $\{a^n b^n \ : \ n \in \mathbb{N}\}$. To characterise a language rather than covering it, we need to add constraints that require to reject inputs.

Given two games G, G', we define the *conjunction* $G \wedge G'$ as the acceptor game obtained by taking the disjoint union of G and G' and identifying the initial states. Then, winning strategies of the component games can be turned into winning strategies of the composite game, if they agree on the observation sequences over the common alphabet.

Lemma 1. *Let G, G' be two acceptor games over observation alphabets Γ, Γ'. Then, an observation-based strategy r is winning in $G \wedge G'$ if, and only if, there exist observation-based winning strategies s, s' in G, G' that agree with r on Γ^* and on Γ'^*, respectively.*

Whenever a language and its complement are covered by two acceptor games, we can construct a new game that characterises the language. The construction involves *inverting* the decisions in a game, that is, replacing the admissible deci-sions for every final state $v \in V$ with $\Omega(v) = \{0\}$ by $\Omega(v) := \{1\}$ and vice versa; final states v with $\Omega(v) = \{0, 1\}$ remain unchanged.

Lemma 2. *Suppose two acceptor games G, G' cover a language $L \subseteq \Sigma^*$ and its complement $\bar{L}$, respectively. Let G'' be the game obtained from G' by inverting the admissible decisions. Then, the game $G \wedge G''$ characterises L.*

3.1 Domino Frontier Languages

We use domino systems as an alternative to encoding machine models and formal grammars (See [23] for a survey.). A *domino system* $\mathcal{D} = (D, E_h, E_v)$ is described by a finite set of *dominoes* together with a horizontal and a vertical compatibility relation $E_h, E_v \subseteq D \times D$. The generic domino tiling problem is to determine, for a given system $\mathcal{D}$, whether copies of the dominoes can be arranged to tile a given space in the discrete grid $\mathbb{Z} \times \mathbb{Z}$, such that any two vertically or horizontally adjacent dominoes are compatible. Here, we consider finite rectangular grids $Z(\ell, m) := \{0, \ldots, \ell + 1\} \times \{0, \ldots, m\}$, where the first and last column, and the bottom row are distinguished as border areas. Then, the question is whether there exists a *tiling* $\tau : Z(\ell, m) \to D$ that assigns to every point $(x, y) \in Z(\ell, m)$ a domino $\tau(x, y) \in D$ such that:

- if $\tau(x, y) = d$ and $\tau(x + 1, y) = d'$ then $(d, d') \in E_h$, and
- if $\tau(x, y) = d$ and $\tau(x, y + 1) = d'$ then $(d, d') \in E_v$.

The *Border-Constrained Corridor* tiling problem takes as input a domino system $\mathcal{D}$ with two distinguished border dominoes # and □, together with a sequence $w = w_1, \ldots, w_\ell$ of dominoes $w_i \in D$, and asks whether there exists a height m such that the rectangle $Z(\ell, m)$ allows a tiling τ with w in the top row, # in the first and last column, and □ in the bottom row:

- $\tau(i, 0) = w_i$, for all $i = 1, \ldots, \ell$;
- $\tau(0, y) = \tau(\ell + 1, y) = \#$, for all $y = 0, \ldots, m - 1$;
- $\tau(x, m) = \Box$, for all $x = 1, \ldots, \ell$.

Domino systems can be used to recognise formal languages. For a domino system $\mathcal{D}$ with side and bottom border dominoes as above, the *frontier language* $L(\mathcal{D})$ is the set of words $w \in D^*$ that yield positive instances of the border-constrained corridor tiling problem. We use the following correspondence between context-sensitive languages and domino systems established by Latteux and Simplot.

Theorem 3 ([13, 14]). *For every context-sensitive language $L \subseteq \Sigma^*$, we can effectively construct a domino system $\mathcal{D}$ over a set of dominoes $D \supseteq \Sigma$ with frontier language $L(\mathcal{D}) = L$.*

Figure 2 describes a domino system for recognising the language $a^n b^n$ also covered by the game in Figure 1. In the following, we show that domino systems can generally be described in terms of consensus game acceptors.

3.2 Uniform Encoding of Domino Problems in Games

Game formulations of domino tiling problems are standard in complexity theory, going back to the early work of Chlebus [7]. However, these reductions are typically non-uniform: they construct, for every input instance consisting of a domino system together with a border constraint, a different game which depends, in particular, on the size of the constraint. Here, we use imperfect information to construct a *uniform* reduction that associates to a fixed domino system $\mathcal{D}$ a game $G(\mathcal{D})$, such that for every border constraint w, the question whether $\mathcal{D}, w$ allows a correct tiling is reduced to the question of whether decision 1 is safe in a certain play associated to w in $G(\mathcal{D})$.

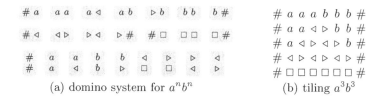

(a) domino system for $a^n b^n$ (b) tiling $a^3 b^3$

Fig. 2. Characterising a language with dominoes

Proposition 4. *For every domino system D, we can construct, in polynomial time, a consensus game acceptor that covers the frontier language of D.*

Proof. Let us fix a domino system $\mathcal{D} = (D, E_h, E_v)$ with a left border domino $\#$ and a bottom domino $\square$. We construct an acceptor game G for the alphabet $\Sigma := D \setminus \{\#, \square\}$ to cover the frontier language $L(\mathcal{D})$.

The game is built as follows. There are domino states of two types: singleton states d for each $d \in D \setminus \{\#\}$ and pair states (d, b) for each $(d, b) \in E_v$. At singleton states d, the two players receive the same observation d.

At states (d, b), the first player observes d and the second player b. The domino states are connected by moves $d \to d'$ for every $(d, d') \in E_h$, and $(d, b) \to (d', b')$ whenever (d, d') and (b, b') are in E_h. There is an initial state v_0 and two final states $\hat{z}$ and z, all associated to the the observation $\#$ for the border domino. From v_0 there are moves to all compatible domino states d with $(\#, d) \in E_h$, and all pair states (d, b) with $(\#, d)$ and $(\#, b) \in E_h$. Conversely, the final state z is reachable from all domino states d with $(d, \#) \in E_h$, and all pair states (d, b) with $(d, \#)$ and $(b, \#) \in E_h$; the final $\hat{z}$ is reachable only from the singleton bottom domino state $\square$. Finally, admissible decisions are $\Omega(z) = \{0, 1\}$ and $\Omega(\hat{z}) = \{1\}$. Clearly, G is an acceptor game, and the construction can be done in polynomial time.

Note that any sequence $x = d_1, d_2, \ldots, d_\ell \in D^\ell$ that forms a horizontally consistent row in a tiling by $\mathcal{D}$ corresponds in the game to a play $\pi_x = v_0, d_1, d_2, \ldots, d_\ell, z$ or $\pi_x = v_0, \square^\ell, \hat{z}$. Conversely, every play in G corresponds either to one possible row, in case Input chooses a single domino in the first move, or to two rows, in case it chooses a pair. Moreover, a row x can appear on top of a row $y = b_1, b_2, \ldots, b_\ell \in D^\ell$ in a tiling if, and only if, there exists a play ρ in G such that $\pi_x \sim^1 \rho \sim^2 \pi_y$, namely $\rho = v_0, (d_1, b_1), (d_2, b_2), \ldots (d_\ell, b_\ell), z$.

Now, we claim that, at an observation sequence $\pi = w$ for $w \in \Sigma^\ell$ the decision 0 is safe if, and only if, there exists no correct corridor tiling by $\mathcal{D}$ with w in the top row. According to our remark, there exists a correct tiling of the corridor with top row w, if and only if, there exists a sequence of rows corresponding to plays $\pi_1, \ldots, \pi_m$, and a sequence of witnessing plays $\rho_1, \ldots, \rho_{m-1}$ such that $w = \pi_1 \sim^1 \rho_1 \sim^2 \pi_2 \cdots \sim^1 \rho_{m-1} \sim^2 \pi_m = \Box^\ell$. However, the decision 0 is unsafe in the play $\Box^\ell$ and therefore at w as well. Hence, every winning strategy s for G must prescribe $s(w) = 1$, for every word w in the frontier language of $\mathcal{D}$, meaning that $L(s) \subseteq L(\mathcal{D})$.

Finally, consider the mapping $s : D^* \to A$ that prescribes $s(w) = 1$ if, and only, if $w \in L(\mathcal{D})$. The observation-based strategy s in the acceptor game G is winning since $s(\Box^*) = 1$, and it witnesses the condition $L(s) = L(\mathcal{D})$. This concludes the proof that the constructed acceptor game G covers the frontier language of $\mathcal{D}$. $\qquad\square$

4   Characterising Context-Sensitive Languages

Theorem 5. *For every context-sensitive language $L \subseteq \Sigma^*$, we can construct effectively a consensus game acceptor that characterises L.*

Proof. Let $L \subseteq \Sigma^*$ be an arbitrary context-sensitive language, represented, e.g., by a linear-bounded automaton. By Theorem 3, it is possible to construct a domino system $\mathcal{D}$ with frontier language L. Then, by Proposition 4, we can construct an acceptor game G that covers $L(\mathcal{D}) = L$. Due to the Immerman-Szelepcsényi Theorem, context-sensitive languages are effectively closed under complement, so we construct an acceptor game G' that covers $\bar{L}$ following the same procedure. Finally, we combine the games G and G' as described in Lemma 2 to obtain an acceptor game that characterises L. $\qquad\square$

One interpretation of the characterisation is that, for every context-sensitive language, there exists a consensus game that is as hard to play as it is to decide membership in the language. On the one hand, this implies that winning strategies for consensus games are in general PSPACE-hard. Indeed, there are instances of acceptor games that admit winning strategies, however, any machine that computes the decision to take in a play requires space polynomial in the length of the play.

Theorem 6. *There exists a solvable consensus game acceptor for which every winning strategy is PSPACE-hard.*

Proof. There exist context-sensitive languages with a PSPACE-hard word problem [12]. Let us fix such a language $L \subseteq \Sigma^*$ together with a consensus game G that characterises it, according to Theorem 5. This is a solvable game, and every winning strategy can be represented as an observation-based strategy s for the first player. Then, the membership problem in L reduces (in linear time) to the problem of deciding the value of s in a play in G: For any input word $w \in \Sigma^*$, we have $w \in L$ if, and only if, $s(w) = 1$. In conclusion, it is PSPACE-hard to decide whether $s(\pi) = 1$, for every winning strategy s in G. $\qquad\square$

On the other hand, it follows that determining whether a consensus game admits a winning strategy is no easier than solving the emptiness problem of context-sensitive languages, which is well known to be undecidable.

Theorem 7. *The question whether an acceptor game admits a winning strategy is undecidable.*

Proof. We reduce the emptiness problem for a context-sensitive grammar to the solvability problem for a acceptor game.

For an arbitrary context-sensitive language $L \in \Sigma^*$ given as a linear bounded automaton, we construct an acceptor game G that characterises L, in polynomial time, according to Theorem 5. Additionally, we construct an acceptor game G' that characterises the empty language over Σ^*: this can be done, for instance, by connecting a clique over letters in Σ observable for both players to a final state at which only the decision 0 is admissible. Now, for any word $w \in \Sigma^*$, the game G' requires decision 0 at every observation sequences $w \in \Sigma^*$, whereas G requires decision 1 whenever $w \in L$. Accordingly, the acceptor game $G \wedge G'$ is solvable if, and only if, L is empty. As the emptiness problem for context-sensitive languages is undecidable [12], it follows that the solvability problem is undecidable for consensus game acceptors. □

We have seen that every context-sensitive language corresponds to a consensus game acceptor such that language membership tests reduce to winning strategy decisions in a play. Conversely, every solvable game admits a winning strategy that is the characteristic function of some context-sensitive language. Intuitively, a strategy should prescribe 0 at a play π whenever there exists a connected play π' at which 0 is the only admissible decision. Whether this is the case can be verified by a nondeterministic machine using space linear in the size of π.

Theorem 8. *Every solvable acceptor game admits a winning strategy that is implementable by a nondeterministic linear bounded automaton.*

5   Games for Weaker Language Classes

The relation between the observation sequences received by the players in a synchronous game on a finite graph can also be explained in terms of letter-to-letter transducers, that is, finite-state automata where the transitions are labelled with input and output letters (See, e.g., [20, Ch. IV]). For a game G, the relation $\{(\beta^1(\pi), \beta^2(\pi)) \in \Gamma^* \times \Gamma^* : \pi \text{ a play in } G\}$ and its inverse are recognised by letter-to-letter transducers with the same transition structure as G. Conversely, every transducer τ can be turned into a game by letting one player observe the input and the other player the output letter of every transition. The consensus condition requires decisions to be invariant under the transitive closure τ^* of the described relation over Γ^*, which corresponds to iterating letter-to-letter transductions. Denoting by $L_{acc} \subseteq \Gamma^*$ the language of observation sequences for plays in which only the decision 1 is admissible, the Σ-language covered by G is $L := \Sigma^* \tau^* L_{\mathrm{acc}}$. To characterise L, we additionally need to ensure $\bar{L} = \Sigma^* \tau^* L_{\mathrm{rej}}$, for the language L_{rej} of observation sequences for plays in which only decision 0 is admissible. Thus, every consensus game acceptor can be described as a collection of three finite-state devices: two automata recognising the accepting and rejecting *seed* languages L_{acc} and L_{rej}, and a (nondeterministic) letter-to-letter transducer τ relating the observation sequences of the players.

Properties of iterated letter-to-letter transductions, or equivalently, length-preserving transductions, have been investigated in [15], also leading to the conclusion that iterated transducers capture context-sensitive languages. In the following, we investigate restrictions of consensus game acceptors towards capturing weaker language classes.

Firstly, we remark that regular languages correspond to games where the two players receive the same observation at every node.

Proposition 9. *Every regular language $L \subseteq \Sigma^*$ is characterised by a consensus game acceptor with identical observations for the players.*

Here, the consensus condition is ineffective, the model reduces to one-player games with imperfect information. To characterise a regular language L, we can build a game from a deterministic automaton for L, by moving symbols from transitions into the target states and allowing Input to go from every accepting state in the automaton to a final game state v_{acc} with $\Omega(v_{\mathrm{acc}}) = \{1\}$, and from every rejecting state to a final state v_{rej} with $\Omega(v_{\mathrm{rej}}) = \{0\}$. Conversely, given a consensus game acceptor G with identical observations, the accepting seed language L_{acc} mentioned above yields the language characterised by G. Clearly, winning strategies in such games are regular.

We say that a consensus game acceptor has *ordered observations* if its alphabet Γ can be ordered so that $\beta^1(v) \geq \beta^2(v)$, for every state $v \in V$. One consequence of this restriction is that the implementation complexity of winning strategy drops from PSPACE to NP.

Proposition 10. *Every solvable acceptor game with ordered observations admits a winning strategy that is characterised by a language in NP.*

Without loss of generality we can assume that the symbols occurring in L_{rej} or L_{acc} are disjoint from the input alphabet Σ and order them below. Then, given a sequence of observations $\pi \in \Sigma^*$, any sequence of indistinguishable plays that starts with observation π and leads to L_{rej} or L_{acc} is of length at most $|\Gamma| \times |\pi|$. To decide whether to prescribe 0 or 1 at π, a nondeterministic machine can guess and verify such a sequence in at most cubic time.

Despite this drop of complexity, games with ordered observations are sufficiently expressive to cover context-free languages.

Lemma 11. *Every context-free language is covered by a consensus game acceptor with ordered observations.*

Firstly, any Dyck language over a finite alphabet Σ of parentheses, possibly with extra neutral symbols, can be covered by a consensus game acceptor over

Σ extended with one additional $\square$ symbol ordered below Σ. The accepting seed language is $L_{\mathrm{acc}} = \square^*$, and the plays allow to project either a neutral symbol or an innermost pair of parentheses from the observation sequence of the first player by replacing them with $\square$ observations for the second player. Next, we observe that the class of languages covered by consensus game acceptors with ordered observations is effectively closed under intersection and letter-to-letter transductions, and thus particularly under letter-to-letter homomorphisms. The statement then follows by the Chomsky-Schützenberger [8] representation theorem, in the non-erasing variant proved by Okhotin [17]: every context-free game is the letter-to-letter homomorphic image of a Dyck language with neutral symbols with a regular language.

Since it is undecidable whether two context-free languages have non-empty intersection [21], the above lemma also implies that the solvability problem for consensus game acceptors is undecidable, already when observations are ordered. Concretely, we can represent the standard formulation of Posts Correspondence Problem as a solvability problem for such restricted consensus games.

Corollary 12. *The question whether a consensus game acceptor with ordered observations admits a winning strategy is undecidable.*

Due to the characterisation of nondeterministic linear-time languages as homomorphic images of intersections of three context-free languages due to Book and Greibach [5], we can draw the following conclusion.

Theorem 13. *For every language L decidable in nondeterministic linear time, we can effectively construct a consensus game acceptor with ordered observations that covers L.*

Acknowledgments. This work was supported by the European Union Seventh Framework Programme under Grant Agreement 601148 (CASSTING).

References

1. Alur, R., Henzinger, T.A., Kupferman, O.: Alternating-time temporal logic. J. ACM **49**, 672–713 (2002)
2. Arnold, A., Walukiewicz, I.: Nondeterministic controllers of nondeterministic processes. In: Logic and Automata, vol. 2. Amsterdam University Press (2007)
3. Azhar, S., Peterson, G., Reif, J.: Lower bounds for multiplayer non-cooperative games of incomplete information. Journal of Computers and Mathematics with Applications **41**, 957–992 (2001)
4. Berwanger, D., Kaiser, Ł.: Information tracking in games on graphs. Journal of Logic, Language and Information **19**, 395–412 (2010)
5. Book, R.V., Greibach, S.A.: Quasi-realtime languages. Mathematical Systems Theory **4**, 97–111 (1970)
6. Chandra, A.K., Kozen, D., Stockmeyer, L.J.: Alternation. Journal of the ACM **28**, 114–133 (1981)
7. Chlebus, B.S.: Domino-tiling games. Journal of Computer and System Sciences **32**, 374–392 (1986)

8. Chomsky, N., Schützenberger, M.: The algebraic theory of context-free languages. In: Computer Programming and Formal Systems. Studies in Logic and the Foundations of Mathematics, vol. 35, pp. 118–161. Elsevier (1963)

9. Fagin, R., Halpern, J.Y., Moses, Y., Vardi, M.Y.: Reasoning about Knowledge. MIT Press (1995)

10. Hearn, R.A., Demaine, E.D.: Games, Puzzles, and Computation. A. K. Peters Ltd., Natick (2009)

11. Janin, D.: On the (high) undecidability of distributed synthesis problems. In: van Leeuwen, J., Italiano, G.F., van der Hoek, W., Meinel, C., Sack, H., Plášil, F. (eds.) SOFSEM 2007. LNCS, vol. 4362, pp. 320–329. Springer, Heidelberg (2007)

12. Kuroda, S.-Y.: Classes of languages and linear-bounded automata. Information and Control **7**, 207–223 (1964)

13. Latteux, M., Simplot, D.: Context-sensitive string languages and recognizable picture languages. Information and Computation **138**, 160–169 (1997)

14. Latteux, M., Simplot, D.: Recognizable picture languages and domino tiling. Theoretical Computer Science **178**, 275–283 (1997)

15. Latteux, M., Simplot, D., Terlutte, A.: Iterated length-preserving rational transductions. In: Brim, L., Gruska, J., Zlatuška, J. (eds.) MFCS 1998. LNCS, vol. 1450, pp. 286–295. Springer, Heidelberg (1998)

16. Mohalik, S., Walukiewicz, I.: Distributed games. In: Pandya, P.K., Radhakrishnan, J. (eds.) FSTTCS 2003. LNCS, vol. 2914, pp. 338–351. Springer, Heidelberg (2003)

17. Okhotin, A.: Non-erasing variants of the Chomsky–Schützenberger theorem. In: Yen, H.-C., Ibarra, O.H. (eds.) DLT 2012. LNCS, vol. 7410, pp. 121–129. Springer, Heidelberg (2012)

18. Peterson, G.L., Reif, J.H.: Multiple-person alternation. In: Proc. 20th Annual Symposium on Foundations of Computer Science (FOCS 1979), pp. 348–363. IEEE (1979)

19. Pnueli, A., Rosner, R.: Distributed reactive systems are hard to synthesize. In: Proceedings of the 31st Annual Symposium on Foundations of Computer Science, FoCS 1990, pp. 746–757. IEEE Computer Society Press (1990)

20. Sakarovitch, J.: Elements of Automata Theory. Cambridge University Press (2009)

21. Salomaa, A.: Formal Languages. Academic Press, New York (1973)

22. Schewe, S.: Distributed synthesis is simply undecidable. Inf. Process. Lett. **114**, 203–207 (2014)

23. van Emde Boas, P.: The convenience of tilings. In: Complexity, Logic, and Recursion Theory. Lecture Notes in Pure and Applied Mathematics, vol. 18, pp. 331–363. Marcel Dekker Inc. (1997)

On the Size of Two-Way Reasonable Automata for the Liveness Problem

Maria Paola Bianchi[1]([✉]), Juraj Hromkovič[1], and Ivan Kováč[2]

[1] Department of Computer Science, ETH Zurich, Zürich, Switzerland
{maria.bianchi,juraj.hromkovic}@inf.ethz.ch
[2] Department of Computer Science, Comenius University, Bratislava, Slovakia
ikovac@dcs.fmph.uniba.sk

Abstract. The existence of a size gap between deterministic and nondeterministic two-way automata is one of the most famous open problems in automata theory. This problem is also related to the famous DLOG vs. NLOG question. An exponential gap between the number of states of two-way nondeterministic automata (2NFAs) and their deterministic counterparts (2DFAs) has been proved only for some restrictions of 2DFAs up to now. It seems that the hardness of this problem lies in the fact that, when trying to prove lower bounds, we must consider every possible automaton, without imposing any particular structure or meaning to the states, while when designing a specific automaton we always assign an unambiguous interpretation to the states. In an attempt to capture the concept of *meaning of states*, a new model of two-way automata, namely *reasonable automaton* (RA), was introduced in [6]. In a RA, each state is associated with a logical formula expressing some properties of the input word, and transitions are designed to maintain consistency within this setting. In this paper we extend the study, started in [6], of the descriptional complexity of RAs solving the liveness problem, showing several lower and upper bounds for different choices of allowed atomic predicates and connectors.

1 Introduction and Preliminaries

The relationship between determinism and nondeterminism is one of the predominant topics of research in theoretical computer science. Such a comparison for polynomial-time Turing machines is the well-known P vs. NP question, and is considered to be the hardest open problem in computer science. Thus, a lot of effort has been aimed to study the problem on simpler computational models, such as finite automata.

In the case of one-way finite automata it has been shown that nondeterminism, although it does not add any computational power [16], can reduce the number of states exponentially with respect to determinism [13,15]. The

---

This work was partially supported by grants SNF 200021_146372/1 and VEGA 1/0979/12.

I. Potapov (Ed.): DLT 2015, LNCS 9168, pp. 120–131, 2015.
DOI: 10.1007/978-3-319-21500-6_9

intuitive reason of this explosion in the number of states of one-way deterministic automata (1DFAs) with respect to their equivalent nondeterministic version (1NFAs) is that once a letter is read, the information given by this letter must be either stored in the current state or forgotten. Thus, as the amount of potentially useful information grows, so does the number of states in the 1DFA. On the other hand, 1NFAs can avoid this problem by storing only the part of information that will be actually relevant later.

It is not as intuitive to understand what happens in the two-way model [12,16], where the automaton has the ability of moving back on the input tape to retrieve some information, once known it is required. Although adding such an ability does not improve the computational power [12], both deterministic and nondeterministic two-way finite automata (2DFAs and 2NFAs, respectively) can be exponentially more succinct than their one-way counterparts [13,15,17]. For the exact bound for the simulation of 2DFAs/2NFAs by 1DFAs/1NFAs, see [8].

On the other hand, the costs of the simulations of 1NFAs by 2DFAs and of 2NFAs by 2DFAs are still unknown. The problem of stating them was raised in 1978 by Sakoda and Sipser [17], with the conjecture that they are not polynomial. To support such a conjecture, they presented a complete analogy with the P vs. NP question, by introducing a notion of reducibility between families of regular languages which allows to obtain families of complete languages for these simulations (see Section 1.1). In spite of all attempts to solve it, this problem is still open. The hardness of this question is emphasized by the direct connection to the DLOG vs. NLOG problem: in fact, if DLOG=NLOG, then there exists a polynomial p such that, for every n-state 2NFA A, there exists a $p(n)$-state 2DFA A' which is equivalent to A, when restricted to inputs of maximal length $p(n)$ [2]. As a direct consequence of this result, if we were able to show an exponential gap between the number of states of a 2DFA and a 2NFA by using only languages with words of polynomial length, we would immediately prove DLOG $\neq$ NLOG.

Actually, the complexity theory for finite automata can be developed as a part of standard complexity theory for Turing machines, with classes, reductions, complete problems and so on. This approach was first suggested in [17] and later followed in [9,11], under the name of *minicomplexity theory*.

Due to the hardness of the problem of comparing the size of 2DFAs and 2NFAs, a lot of attention has been given to restricted versions of two-way automata, in order to obtain a deeper understanding of the issue and isolate the critical features of the model. In 1980, Sipser proved that if the resulting machine is required to be *sweeping* (deterministic and reversing the direction of its input head only at the endmarkers, two special symbols used to mark the left and right ends of the input), the simulation of a 2NFA is indeed exponential [18]. This, however, does not solve the general problem: in fact the simulation of unrestricted 2DFAs by sweeping 2DFAs also requires an exponential number of states [1,14]. Sipser's result was generalized by Hromkovič and Schnitger [7], who considered *oblivious* machines (following the same trajectory of input head movements along all inputs of equal length) and, recently, by Kapoutsis [10], considering 2DFAs where the number of input head reversals is sublinear in the length of the input. However, even the last condition gives a machine provably less succinct than unrestricted 2DFAs, hence the general problem remains open.

Other approaches were based not on the restriction of the automaton itself, but on the form of the input, i.e. from a unary alphabet. The situation for the unary languages differs from the general case: for any 1NFA, as well as any 2DFA, it is possible to find an equivalent 1DFA of subexponential size [3,4]. Also the relationship between 2DFAs and 2NFAs is in this case subexponential [5].

Hardness of proving lower bounds for the 2DFA vs. 2NFA problem is connected to the fact that one has to reason about all possible 2DFAs, including automata without any clear structure or meaning of states. When proving that an automaton solves some problem, we usually argue about the meaning of the states, about possible places in a word where it can be while in some state, or about reasons to change the state. So, there is a gap between the set of automata we reason about when trying to show an upper bound and the set of all possible automata we have to consider when proving lower bounds. Therefore, a possible step towards solving the general problem might be to focus on automata with clear meaning of states. To achieve this goal, Hromkovič et al. defined in [6] the model of *Reasonable Automaton*, in which each state is mapped to a logical formula, composed by (a combination of) some atomic propositions, which represent some property of the current input word.

1.1 Liveness Problem and Reasonable Automata

A complete problem for the question of the number of states needed to simulate a 2NFA by a 2DFA is the *two-way liveness* (2-LIV for short), defined in [17] as an infinite family of languages $\{C_n\}_{n\geq 1}$. The n-th language C_n is defined over an alphabet whose letters are graphs consisting of n left nodes and n right nodes. Directed arcs may join any distinct pair of nodes, and the input word is interpreted as a graph, which arises by identifying the adjacent columns of each two consecutive letters. A word belongs to C_n iff there exists a path from some vertex in the leftmost column to some vertex in the rightmost column.

It is possible to create a 2NFA with $O(n)$ states accepting C_n by nondeterministically choosing one of the leftmost vertices and then iteratively guessing the next vertex of the path. However, the existence of a 2DFA for this language with a number of states polynomial in n would imply that the same is true for any language family whose n-th member can be accepted by a 2NFA with n states.

A special case of 2-LIV is the *one-way liveness* problem (1-LIV for short) where, in each letter, only oriented edges from the left column to the right column are allowed. This problem turns out to be complete with respect to the question of simulation of 1NFAs by 2DFAs. The restriction of 1-LIV where the n-th language contains only words of length $f(n)$ will be denoted as 1-LIV$_{f(n)}$.

We define here formally the model of reasonable automaton introduced in [6]. In what follows, for any word z, we denote by z_i its i-th symbol.

Definition 1. *Let $\mathcal{F}$ be a set of propositional expressions over some set of atoms. A* Reasonable Automaton *(RA for short) $\mathcal{A}$ over $\mathcal{F}$ for words of fixed length m is a tuple $\mathcal{A} = (\Sigma, Q, m, q_s, Q_F, Q_R, \delta, \tau, \kappa)$, such that*

- Σ is the input alphabet,
- Q is the finite set of states, $Q_F, Q_R \subseteq Q$ are the sets of accepting and rejecting states, respectively, for which it holds $Q_F \cap Q_R = \emptyset$,
- $m \in \mathbb{N}$ is a positive integer,
- $q_s \in Q$ is the initial state,
- $\delta : Q \setminus (Q_F \cup Q_R) \times \Sigma \to Q$ is the (total) transition function,
- $\tau : Q \setminus (Q_F \cup Q_R) \to \{1, 2, \dots, m\}$ is a function that assigns to every state q which is neither accepting nor rejecting its focus $\tau(q)$, i.e. the position on the input tape that $\mathcal{A}$ scans while in the state q.
- $\kappa : Q \to \mathcal{F}$ is a mapping such that the following properties hold:
  1. If $\mathcal{A}$ is in a state q while processing a word z, then the condition $\kappa(q)$ must be valid for z (in the given interpretation of the atoms).
  2. If $\delta(q, a) = p$ holds for a triple $p, q \in Q$ and $a \in \Sigma$, then, for each $z \in \Sigma^m$ such that $z_{\tau(q)} = a$ and the condition $\kappa(q)$ is valid for z, then the condition $\kappa(p)$ must be valid for z as well.
  3. For any $q \in Q_F$, the condition $\kappa(q)$ must not be valid for any $w \notin L(\mathcal{A})$.
  4. For any $q \in Q_R$, the condition $\kappa(q)$ must not be valid for any $w \in L(\mathcal{A})$.
  5. $\mathcal{A}$ never enters an infinite loop in the computation on any input word.

The computation starts in the state q_s on the $\tau(q_s)$-th symbol of the input word z. In each computation step, the next state is given by $\delta(q, a)$, where q is the current state and a is the symbol placed on the $\tau(q)$-th position. The automaton works on inputs of fixed length m, and each computation ends whenever the current state is either accepting or rejecting.

By imposing restrictions on the allowed propositional expressions $\mathcal{F}$, we obtain classes of RAs with very different succinctness. In [6,19] are proposed restrictions on the set of allowed logical connections and on the set of allowed atomic statements. Namely, two ways of the choice of logical connections are taken into consideration: in the first case, the stored information is only accumulated, which is represented as a conjunction of the predicates and their negations. In the second case, the information may have a more complicated form, as all well-formed propositional formulæ are allowed.

The chosen predicates represent different kinds of information one can gather about the input graph in the liveness problem families. In particular, the predicates $e(a, b)$, $p(a, b)$, $p(a, b, c)$, and $r(a)$ have been considered. The predicate $e(a, b)$ is true iff there is an edge from a to b in the input graph. Similarly, the predicate $p(a, b)$ is true iff there is a path from a to b in the given graph, and $p(a, b, c)$ is true iff there is a path from a to b passing through the vertex c. Finally, $r(a)$ is true iff the vertex a is reachable from the left side, i.e. if there exists a path from some vertex in the leftmost column to a.

In [6,19] the descriptional complexity of RAs solving 1-LIV in the case of words of fixed length has been studied, leading to the bounds shown in Table 1. In this paper, we extend such analysis by showing further bounds for several choices of predicates and connectives. We first prove that the linear upper bound for RAs for 1-LIV$_2$ in which any formula on predicates $e(a, b)$ is allowed is actually tight. Then we show an exponential upper bound on RAs allowing only conjunctions of either edge or path predicates, for words of constant length k, which is tight for $k = 2$. In the case of edge variables we also show a lower bound which is still exponential. Finally, we show that the choice of predicates of the form $r(a)$ leads to very weak expressive power since, even when allowing any propositional formula on them, the number of states needed for such a RA to solve the n-th language from 1-LIV$_2$ is $\Theta(2^{\frac{n}{2}})$.

Table 1. Known bounds for any constant k and polynomial r, such that $r(n) \leq dn^c$ for some constants c, d. The ones displayed in bold are contained in this paper. In the case of formulæ constructed through conjunctions and variables $p(a, b)$, an upper bound of $O(n^2 2^n)$ for 1-LIV$_2$ was previously stated in [6].

Predicates	Connectives	Problem	Lower bound	Upper bound
$e(a, b)$	full logic	1-LIV$_2$	$\Omega(n)$	$O(n)$
		1-LIV$_3$, 1-LIV$_4$		$O(n^2)$
		1-LIV$_{r(n)}$		$O(dn^{2+c+\log_2(dn^c)})$
$e(a, b)$	conjunction	1-LIV$_2$	$\Omega(2^n)$	$\mathbf{O(2^n)}$
		1-LIV$_k$	$\mathbf{\Omega(2^{(k-1)n})}$	$\mathbf{O(k \cdot 2^{(k-1)n})}$
$p(a, b)$	conjunction	1-LIV$_2$	$\Omega(2^n)$	$\mathbf{O(2^n)}$
		1-LIV$_k$		$\mathbf{O(k \cdot n \cdot 2^n)}$
$p(a, b, c)$	conjunction	1-LIV$_2$, 1-LIV$_3$	$\Omega(n^2)$	$O(n^2)$
		1-LIV$_4$		$O(n^3)$
		1-LIV$_{r(n)}$		$O(dn^{2+c+\log_2(dn^c)})$
$r(a)$	full logic	1-LIV$_2$	$\mathbf{\Omega(2^{\frac{n}{2}})}$	$\mathbf{O(2^{\frac{n}{2}})}$
$r(a)$	conjunction	1-LIV$_2$, 1-LIV$_3$, 1-LIV$_4$	$\Omega(2^n)$	$O(2^n)$
$r(a), p(a, b, c)$	conjunction	1-LIV$_2$	$\Omega(n)$	$O(n)$
		1-LIV$_3$, 1-LIV$_4$		$O(n^2)$

2 Main Results

We start by considering the 1-LIV problem on words of length 2. In [19] it was shown an upper bound of $O(n)$ states when considering RAs where the allowed formulæ are any propositional formula over predicates defined by the existence of edges. Here, we complement this result with a matching lower bound.

Theorem 1. *Consider propositional variables $e(a, b)$ ($\neg e(a, b)$, resp.) with the interpretation that there exists (does not exist, resp.) an edge between vertices a and b, and let $\mathcal{F}$ be the set of all propositional formulæ on such variables. Then any reasonable automaton solving the n-th language from the 1-LIV$_2$ family has at least n states.*

Proof. Let a be the first vertex in the first column, c the first vertex in the third column and B the set of all vertices in the second column of the graph defined by words of length 2 over the alphabet Σ_n. We define a subset of words $L = \{z^{(1)}, z^{(2)}, \ldots, z^{(2^n)}\} \subset \Sigma_n^2$ not in the n-th language from 1-LIV$_2$ as follows: for any $\emptyset \subseteq D \subseteq B$ there is a word in L such that it contains only edges (a, d), for every $d \in D$, and edges (d', c), for $d' \notin D$.

Suppose there exists a reasonable automaton $\mathcal{A}$ with $m < n$ states solving the n-th language from 1-LIV$_2$. For each $x, y \in L$, let Q_x and Q_y be the set of states encountered during the computation of $\mathcal{A}$ on the word x and y, respectively. Since $m < n$, there must exist two different words, say v and w, in L such that $Q_v = Q_w$. Then, consider the word $z = v_1 w_2$, which should be accepted by $\mathcal{A}$. However, we claim that the computation of $\mathcal{A}$ on z produces a set of states Q_z which is a subset of Q_v. We show this claim by induction on the length of the computation: (i) the first state of the computation is the starting state of $\mathcal{A}$, which is also in Q_v, (ii) if the i-th state encountered in the computation is $q_i \in Q_v = Q_w$ such that $\tau(q_i) = 1$ ($= 2$, resp.), then the next state is $\delta(q_i, v_1)$ ($\delta(q_i, w_2)$, resp.) which is one of the states reached through the computation of v (w, resp.), so it belongs to Q_v. Thus the claim is settled.

However, since there are no accepting states in Q_v, the word z cannot be accepted by $\mathcal{A}$, which is a contradiction. $\qquad\square$

In [6] it is proved that, when restricting the allowed connectors to only conjunctions, i.e. allowing the state to only *accumulate* information on edge existence, RAs need an exponential number of states to solve 1-LIV$_2$. Here we give a matching upper bound, improving the one given in [6] by a factor of n^2.

Theorem 2. *Consider propositional variables $e(a, b)$ ($\neg e(a, b)$, resp.) with the interpretation that there exists (does not exist, resp.) an edge between vertices a and b, and let $\mathcal{F}$ be propositional formulæ created by conjunctions of such variables. For each $n > 0$, there exists a RA $\mathcal{A}$ over $\mathcal{F}$ accepting the n-th language from 1-LIV$_2$ by using $O(2^n)$ states.*

Proof. Let us call the vertices on left side $A = \{a_1, \ldots, a_n\}$, the vertices in the middle $B = \{b_1, \ldots, b_n\}$ and the vertices on the right side $C = \{c_1, \ldots, c_n\}$. The idea behind the automaton solving the problem is the following. First, the automaton tries to reject the word, showing that only vertices (in the middle column) not reachable from the left side are connected to the right side. If this approach does not work, then it chooses one edge from the second letter and find its match in the first letter. (It does not matter which edge as long as the vertex incident with it is reachable from the left side. This information is however remembered in the state.) Formally, the automaton $\mathcal{A}$ has five types of states:

1. One initial state q_s with focus 1, corresponding to the formula $\top$.
2. The states $q_D^{(1)}$, for $D \subseteq B$, with focus 2, corresponding to formulæ $\bigwedge_{d \in D} \bigwedge_{i=1}^n \neg e(a_i, d)$.
3. The rejecting states $q_D^{(2)}$, for $D \subseteq B$, corresponding to formulæ $(\bigwedge_{d \in D} \bigwedge_{i=1}^n \neg e(a_i, d)) \wedge (\bigwedge_{d \notin D} \bigwedge_{k=1}^n \neg e(d, c_k)),$.
4. The states $q_{b_j, c_k}^{(3)}$ with focus 1, corresponding to formulæ $e(b_j, c_k)$.
5. The accepting states $q_{a_i, b_j, c_k}^{(4)}$ corresponding to formulæ $e(a_i, b_j) \wedge e(b_j, c_k)$.

The automaton changes its state from q_s to $q_D^{(1)}$ if D is the *maximal* set of vertices in the middle column which are not reachable from the left side. If in the second symbol there are no edges from the complement of D to the right side, then $\mathcal{A}$ changes it state from $q_D^{(1)}$ to the rejecting state $q_D^{(2)}$. Otherwise, it chooses one edge (the minimum according to some fixed total ordering of edges) $\{b_j, c_k\}$ that appears in the second letter such that $b_j \notin D$ and changes its state to $q_{b_j,c_k}^{(3)}$. When in the state $q_{b_j,c_k}^{(3)}$, $\mathcal{A}$ checks if the edge $\{a_i, b_j\}$ exists in the first symbol for some a_i on the left side. If it exists, then $\mathcal{A}$ goes to the accepting state $q_{a_i,b_j,c_k}^{(4)}$, otherwise it goes back to q_s.

The total number of states is therefore $(2^{n+1} + n^3 + n^2 + 1) \in O(2^n)$. To show that $\mathcal{A}$ is a RA, we need to ensure the validity of all five properties of κ in Definition 1. Properties 1, 3 and 4 are straightforward. To prove property 2, we notice that, in transitions from states of type 2 to states of type 4, the information of the preceding state is forgotten, and a new information is based only on the knowledge of the symbol under the head. For all other transitions, the information of the preceding state is accumulated together with new information about the symbol under the head. Therefore, if the formula of the preceding state is valid, the formula of the new state is valid as well. To show property 5 we notice that the only possible loop in the automaton is of the form $q_s, q_C^{(1)}, q_{b_j,c_k}^{(3)}, q_s$. Since states of type 4 are only reachable by words in the accepted language, this may only be part of a computation on a word $\omega \in L(\mathcal{A})$. However, in this case the set C chosen after the starting state would be the maximal set of vertices not connected to the left side, which implies that for any possible b_j in the choice of the next state $q_{b_j,c_k}^{(3)}$, there must exist an edge $\{a_i, b_j\}$ in the first symbol. Therefore the state after $q_{b_j,c_k}^{(3)}$ cannot be q_s. In other words, the automaton never goes from a state of type 4 back to the starting state if the computation started in q_s.   □

Since edges can be trivially represented by paths of length 1, we obtain the same upper bound of Theorem 2 for the case of variable representing paths.

Corollary 1. *Consider propositional variables $p(a, b)$ ($\neg p(a, b)$, resp.) with the interpretation that there exists (does not exist, resp.) a path between vertices a and b, and let $\mathcal{F}$ be propositional formulæ created by conjunctions of such variables. There exists a RA over $\mathcal{F}$ accepting the n-th language from 1-LIV$_2$ by using $O(2^n)$ states.*

We now extend the above reasoning to words of constant length.

Theorem 3. *Consider propositional variables $p(a, b)$ ($\neg p(a, b)$, resp.) with the interpretation that there exists (does not exist, resp.) a path between vertices a and b, and let $\mathcal{F}$ be propositional formulæ created by conjunctions of such variables. There exists a RA $\mathcal{A}$ over $\mathcal{F}$ accepting the n-th language from 1-LIV$_k$ by using $O(k \cdot n \cdot 2^n)$ states.*

Proof (Sketch). We call the vertices in the i-th column $V^{(i)} = \{a_1^{(i)}, \ldots, a_n^{(i)}\}$, and we build $\mathcal{A}$ for 1-LIV$_k$ as an extension of the one described in Theorem 2 for 1-LIV$_2$. Again, the automaton works in two phases using 5 types of states:

1. the starting state $q_{\emptyset}^{(1)}$, which has focus 1 and is associated to the formula $\top$.
2. states of the form $q_C^{(i)}$, for $2 \leq i \leq k$ and $C \subsetneq V^{(i)}$ have focus i and correspond to the formula $\bigcup_{a \in V^{(1)}, c \in C} \neg p(a, c)$, meaning that no vertex in the set C is reachable from the left side.
3. the rejecting state r, corresponding to the formula $\bigcup_{a \in V^{(1)}, b \in V^{(k+1)}} \neg p(a, b)$.
4. states of the form $p_{C, b}^{(i)}$, for $1 \leq i \leq k - 1$, $C \subseteq V^{(i+1)}$, $b \in V^{(k+1)}$ have focus i and correspond to the formula $\bigcup_{c \in C} p(c, b)$, meaning that b is reachable from all the vertices in C.
5. the accepting states of the form $p_{a, b}$ for $a \in V^{(1)}$ and $b \in V^{(k+1)}$, which correspond to the formula $p(a, b)$.

Hence, the total number of states is $1 + (k - 1)(2^n - 1) + 1 + (k - 1)n2^n + n^2$.

The automaton works as follows: in the first phase it scans the input left-to-right, moving through states of type 2, and tries to prove that the word does not belong to the language: when in a state $q_C^{(i)}$, it looks for the *maximal* set $C' \subseteq V^{(i+1)}$ reachable from $V^{(i)} \setminus C$. If $C' = V^{(i+1)}$, then it rejects the input, otherwise it moves to $q_{C'}^{(i+1)}$. If $i = k$, $\mathcal{A}$ chooses one vertex b in the right side reachable from $V^k \setminus C$ and moves to $p_{C', b}^{(k-1)}$, where C' is the *maximal* set reachable from b. In the second phase, $\mathcal{A}$ tries to find a path from b to the left side by scanning the input right-to-left through states of type 4: from $p_{C, b}^{(i)}$ it moves to $p_{C', b}^{(i-1)}$, where $C' \subseteq V^{(i)}$ is the *maximal* nonempty set reachable from C (in the last step it simply chooses a single vertex a instead of a set). If such set does not exist, it moves back to $q_0^{(1)}$. Notice that this last transition is never actually used in a computation starting from $q_0^{(1)}$, however it is necessary to ensure Property 2 from Definition 1. □

We can use a similar idea for proving an upper bound in the case of variables defined by edges: in this case, however, since variables representing the existence of a path are not allowed, in order to store in a state the information that some subset C of the i-th column is not reachable from the right column, we store the information that the set $C_{k-1} \subseteq V^{(k-1)}$ is not reachable from the right side, and $C_j \subseteq V^{(j)}$ is not reachable from the set $V^{(j+1)} \setminus C_j$ for all $i \leq j < k - 2$. This causes an increase in the number of states exponential in k.

Theorem 4. *Consider propositional variables $e(a, b)$ ($\neg e(a, b)$, resp.) with the interpretation that there exists (does not exist, resp.) an edge between vertices a and b, and let $\mathcal{F}$ be propositional formulæ created by conjunctions of such variables. There exists a RA over $\mathcal{F}$ accepting the n-th language from 1-LIV$_k$ by using $O(2^{(k-1)n})$ states.*

We complement Theorem 4 with an almost matching lower bound.

Theorem 5. *Consider propositional variables $e(a, b)$ ($\neg e(a, b)$, resp.) with the interpretation that there exists (does not exists, resp.) an edge between vertices a and b, and let $\mathcal{F}$ be propositional formulæ created by conjunctions of such variables. For any integer $k > 0$ any reasonable automaton solving the n-th language from the 1-LIV$_k$ family has $\Omega(2^{(k-1)n})$ states.*

Proof. Suppose there exists a reasonable automaton $\mathcal{A}$ recognizing the n-th language from the 1-LIV_k family. Let us call the vertices in the i-th column $V^{(i)} = \{a_1^{(i)}, \ldots, a_n^{(i)}\}$. We consider the language $L = \{z^{(1)}, z^{(2)}, \ldots, z^{((2^n-2)^{k-1})}\} \subset \Sigma_n^k$ such that, for every sequence of nonempty sets $U^{(2)}, U^{(3)}, \ldots, U^{(k)}$, such that $U^{(i)} \subsetneq V^{(i)}$, there exists a word $w \in L$ such that it consists of the edges $(a_1^{(1)}, b)$, for each $b \in U^{(2)}$, (b, c) for each $b \in U^{(i)}$ and $c \in U^{(i+1)}$, where $2 \le i < k$, (c', d') for each $c' \in V^{(i)} \setminus U^{(i)}$ and $d' \in V^{(i+1)} \setminus U^{(i+1)}$, where $2 \le i < k$, and $(d', a_1^{(k+1)})$ for each $d' \in V^{(k)} \setminus U^{(k)}$ (see Figure 1 for an example).

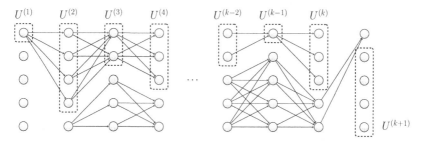

Fig. 1. The word in L associated to sets $U^{(2)}, \ldots, U^{(k)}$. For sake of notation we let $U^{(1)} = \{a_1^{(1)}\}$ and $U^{(k+1)} = V^{(k+1)} \setminus \{a_1^{(k+1)}\}$ for any word in L.

Let us look at the computation of $\mathcal{A}$ on words from L. Since all words in L must be rejected, there must be some state q in the computation on the current word such that $\kappa(q)$ implies that there is no path from $a_1^{(1)}$ to $a_1^{(k+1)}$. We denote by q_i the first state that satisfies this property for the word $z^{(i)}$. Since q_s can not satisfy this property, there always exists a state r_i preceding q_i in the computation on $z^{(i)}$. We also call $U_i^{(h)}$, for $1 \le h \le k+1$, the sets of vertices associated to $z^{(i)}$, and we let $W_i^{(h)} = V_i^{(h)} \setminus U_i^{(h)}$.

We claim that, for each $z^{(i)}$, the state r_i must contain the information that, for any letter except the focus of r_i, there is no edge between U-vertices to the left and the W-vertices to the right of that symbol. Formally, the formula $\kappa(r_i)$ must contain variable $\neg e(b, c)$, for any $b \in U^{(\alpha)}, c \in W^{(\alpha+1)}$, where $1 \le \alpha \le k+1$ and $\alpha \ne \tau(r_i)$. In fact, if one such variable is missing, say $\neg e(\bar{b}, \bar{c})$, we consider the word $z^{(j)}$ having the same edges as $z^{(i)}$, plus the edge $(\bar{b}, \bar{c})$. Such a word clearly satisfies $\kappa(r_i)$ and, since the $\tau(r_i)$-th symbol is the same, we have $\delta(r_i, z_{\tau(r_i)}^{(i)}) = q_i$. However, in $z^{(j)}$ there exists a path from $a_1^{(1)}$ to $a_{k+1}^{(1)}$, therefore it does not satisfy $\kappa(q_i)$. This contradiction settles the claim.

We now show that for any $z^{(i)}, z^{(j)} \in L$ such that $i \ne j$, it holds that $r_i \ne r_j$. Since $z^{(i)} \ne z^{(j)}$, there exists a value $2 \le t \le k$ such that $U_i^{(t)} \ne U_j^{(t)}$. Let t be the highest value with such property: w.l.o.g., we assume there exists a vertex $v \in U_i^{(t)} \setminus U_j^{(t)}$ (otherwise swap i and j). If $\tau(r_i) \ne \tau(r_j)$, then we clearly have $r_i \ne r_j$, so we assume that $\tau(r_i) = \tau(r_j) = x$, i.e., they have the same focus x.

We first consider the case $x \neq t$. Since $v \in U_i^{(t)}$, for any vertex $c \in W_i^{(t+1)}$, the word $z^{(i)}$ does not contain the edge (v, c), so the above claim implies that $\neg e(v, c)$ must be part of $\kappa(r_i)$. However, because of how we chose t, we have that $U_i^{(t+1)} = U_j^{(t+1)}$, so the word $z^{(j)}$ contains the edge (v, c), therefore $\neg e(v, c)$ cannot be part of $\kappa(r_i)$, thus $r_i \neq r_j$.

Now we consider the case $x = t$. By going backwards from t, we find the first position where there is a vertex on the left side of the current letter which is a U-vertex for both words. More formally, we let $\ell = \max\{h \in \{1, 2, \ldots, t-1\} \mid \exists c \in V_i^{(h+1)}$ such that $c \in U_i^{(h+1)} \cap U_j^{(h+1)}\}$. Clearly such ℓ always exists since $a_1^{(1)} \in U_i^{(1)} \cap U_j^{(1)}$. Because of the maximality of ℓ, for $\ell < t - 1$ there exists a vertex b in the $\ell+1$-th layer such that, $b \in U_i^{(\ell+1)} \cap W_j^{(\ell+1)}$ or $b \in W_i^{(\ell+1)} \cap U_j^{(\ell+1)}$. We let, w.l.o.g., $b \in U_i^{(\ell+1)} \cap W_j^{(\ell+1)}$ (for the other case simply swap i and j), and we let $b = v$ in the case $\ell = t - 1$. Thus, we have that there is no edge between c and b in $z^{(i)}$, so because of the above claim $\neg e(c, b)$ must be in $\kappa(r_i)$. However, $z^{(j)}$ has the edge (c, b), therefore it must be $r_i \neq r_j$.

As there are $(2^n - 2)^{k-1}$ words in L, there must be at least $(2^n - 2)^{k-1}$ non accepting states, therefore the total number of states must be $\Omega(2^{(k-1)n})$. □

We now consider only atomic predicates representing reachability of a given vertex. When allowing only conjunctions as connectors, 2^n required for RAs for recognizing the n-th language from 1-LIV even in the case of only two-letter words [6]. Here we show that, by allowing any well-formed formula over such predicates, we can only improve the size by a square root factor.

Theorem 6. *Consider propositional variables $r(a)$ ($\neg r(a)$, resp.) with the interpretation that a is reachable (non reachable, resp.) from the left side, and let $\mathcal{F}$ be the set of all propositional formulæ on such variables. There exists a RA $\mathcal{A}$ over $\mathcal{F}$ accepting the n-th language from 1-LIV$_2$ by using $O(2^{\frac{n}{2}})$ states.*

Proof (Sketch). We call $b_1, \ldots, b_n$ the vertices on the middle side. We build $\mathcal{A}$ for 1-LIV$_2$ that works as follows: after reading the first symbol, $\mathcal{A}$ memorises in the state the set B_1 of reachable vertices in $S_1 = \{b_1, \ldots, b_{\lfloor \frac{n}{2} \rfloor}\}$. Then it reads the second symbol: if there are edges connecting any vertex of B_1 to the right side, $\mathcal{A}$ accepts the input word, otherwise it computes the set B_2 of vertices in $S_2 = \{b_{\lfloor \frac{n}{2} \rfloor + 1}, \ldots, b_n\}$ which have edges to the right side, and memorises in the state the information that the right side is reachable if and only if one of the vertices in B_2 is reachable. Finally, the automaton reads the first symbol again, and accepts the input word if and only if there exists a vertex in B_2 connected to the left side. □

We now focus on proving a matching lower bound for the upper bound stated in Theorem 6. For technical reasons, we need a normal form for RAs, where the automaton rejects (or accepts) as soon as possible, that is, whenever from the formula and the current symbol we can infer a definite answer. This form can be easily obtained by redirecting the transitions to the accepting/rejecting states as soon as enough information can be inferred from the configuration of the system.

Theorem 7. *Consider propositional variables $r(a)$ ($\neg r(a)$, resp.) with the interpretation that the vertex a is reachable (non reachable, resp.) from the left side, and let $\mathcal{F}$ be the set of all propositional formulæ on such variables. Then, any reasonable automaton over $\mathcal{F}$ solving the n-th language from $1\text{-}\mathrm{LIV}_2$ must have $\Omega(2^{\frac{n}{2}})$ states.*

Proof. Let us call the vertices on left side $a_1, \ldots, a_n$, the vertices in the middle $b_1, \ldots, b_n$ and the vertices on the right side $c_1, \ldots, c_n$. We show that this lower bound holds for a restricted version of the problem, in which edges to c_2, c_3, $\ldots$, c_n are not allowed. The bound can be then transfered to the general version of the problem – if there existed an automaton with less states in the general case, this automaton could be directly used for solving the restricted case as well, since the restriction is purely syntactical. Let us therefore consider only the restricted version of the problem.

We first notice that, in order to accept or reject the input word, any reasonable automaton must first read the entire input, therefore every computation has at least 3 states. We call *pre-halting* (*pre-pre-halting*, resp.) the state encountered one step (two steps, resp.) before the accepting or rejecting state.

Let $B = \{b_1, \ldots, b_n\}$. We consider a set L of $2^n - 2$ words from Σ_n^2 constructed as follows: for every set of vertices D in the middle column, such that $\emptyset \subset D \subset B$, we connect all vertices in D to the vertex a_1, and we connect all the vertices in the complement of D to the vertex c_1. Clearly, none of these words belongs to the n-th language from $1\text{-}\mathrm{LIV}_2$, and the restriction is valid in all of them. We also consider the set $L' = \{x_1 y_2 \in \Sigma_n^2 \mid x \in L \wedge y \in L\}$ of words obtained by taking pairs of words in L and swapping the first letter. Clearly, all words in L' belong to n-th language from $1\text{-}\mathrm{LIV}_2$.

By contradiction, suppose there exists a RA $\mathcal{A} = (\Sigma_n, Q, 2, q_s, Q_F, Q_R, \delta, \tau, \kappa)$ with $m < 2^{\frac{n}{2}}$ states solving the n-th language from $1\text{-}\mathrm{LIV}_2$. Since there are only $m^2 < 2^{\frac{n}{2}} - 2$ pairs of states of $\mathcal{A}$, there exist two input words x and y whose computation has the same pre-halting and pre-pre-halting state. Let us call those states s and t, respectively. Furthermore, let $D_x, D_y \subset B$ be the sets associated with the words x and y, respectively. Since we assumed $\mathcal{A}$ halts as soon as possible, $\kappa(t)$ together with the information carried by either $x_{\tau(t)}$ or $y_{\tau(t)}$ cannot imply either $r(c_1)$ nor $\neg r(c_1)$. Therefore, since t is encountered in the computations of both x and y, the formula $\kappa(t)$ must hold for all words with one of the following properties:

1. $\bigwedge_{b_i \in D_x} r(b_i) \wedge \bigwedge_{b_i \in B \setminus D_x} \neg r(b_i) \wedge r(c_1)$,
2. $\bigwedge_{b_i \in D_x} r(b_i) \wedge \bigwedge_{b_i \in B \setminus D_x} \neg r(b_i) \wedge \neg r(c_1)$,
3. $\bigwedge_{b_i \in D_y} r(b_i) \wedge \bigwedge_{b_i \in B \setminus D_y} \neg r(b_i) \wedge r(c_1)$,
4. $\bigwedge_{b_i \in D_y} r(b_i) \wedge \bigwedge_{b_i \in B \setminus D_y} \neg r(b_i) \wedge \neg r(c_1)$,

The word $x_1 y_2$ belongs to the first category. Therefore, by property 2 of κ in Definition 1, $\kappa(s)$ must be valid for $x_1 y_2$ as well, since $\delta(t, x_{\tau(t)}) = \delta(t, y_{\tau(t)}) = s$. Using property 2 once more, since s is a pre-halting state which leads to a rejecting state if the letter $x_{\tau(s)}$ or $y_{\tau(s)}$ is read, the word $x_1 y_2$ will be rejected, which contradicts the validity of the automaton $\mathcal{A}$. □

References

1. Berman, P.: A note on sweeping automata. In: de Bakker, J., van Leeuwen, J. (eds.) ICALP. LNCS, vol. 85, pp. 91–97. Springer, Heidelberg (1980)
2. Berman, P., Lingas, A.: On complexity of regular languages in terms of finite automata. Technical report, Institute of Computer Science, Polish Academy of Sciences, Warsaw (1977)
3. Chrobak, M.: Finite automata and unary languages. Theor. Comp. Sci. **47**(2), 149–158 (1986)
4. Geffert, V.: Magic numbers in the state hierarchy of finite automata. Information and Computation **205**(11), 1652–1670 (2007)
5. Geffert, V., Mereghetti, C., Pighizzini, G.: Converting two-way nondeterministic unary automata into simpler automata. Theor. Comp. Sci. **295**, 189–203 (2003)
6. Hromkovič, J., Královič, R., Královič, R., Štefanec, R.: Determinism vs. Nondeterminism for Two-Way Automata: Representing the Meaning of States by Logical Formulæ. Int. J. Found. Comput. Sci. **24**(7), 955–978 (2013)
7. Hromkovič, J., Schnitger, G.: Nondeterminism versus determinism for two-way finite automata: generalizations of Sipser's separation. In: Baeten, J.C.M., Lenstra, J.K., Parrow, J., Woeginger, G.J. (eds.) ICALP 2003. LNCS, vol. 2719, pp. 439–451. Springer, Heidelberg (2003)
8. Kapoutsis, C.A.: Removing bidirectionality from nondeterministic finite automata. In: Jedrzejowicz, J., Szepietowski, A. (eds.) MFCS 2005. LNCS, vol. 3618, pp. 544–555. Springer, Heidelberg (2005)
9. Kapoutsis, C.A.: Size complexity of two-way finite automata. In: Diekert, V., Nowotka, D. (eds.) DLT 2009. LNCS, vol. 5583, pp. 47–66. Springer, Heidelberg (2009)
10. Kapoutsis, C.A.: Nondeterminism is essential in small two-way finite automata with few reversals. Inf. Comput. **222**, 208–227 (2013)
11. Kapoutsis, C.A., Královič, R., Mömke, T.: Size complexity of rotating and sweeping automata. Journal of Computer and System Sciences **78**(2), 537–558 (2012)
12. Kolodin, A.N.: Two-way nondeterministic automata. Cybernetics and Systems Analysis **10**(5), 778–785 (1972)
13. Lupanov, O.: A comparison of two types of finite automata. Problemy Kibernet **9**, 321–326 (1993). (in Russian); German translation: Über den Vergleich zweier Typen endlicher Quellen. Probleme der Kybernetik **6**, 329–335 (1966)
14. Micali, S.: Two-way deterministic finite automata are exponentially more succinct than sweeping automata. Inf. Process. Lett. **12**, 103–105 (1981)
15. Moore, F.R.: On the bounds for state-set size in the proofs of equivalence between deterministic, nondeterministic, and two-way finite automata. IEEE Transactions on Computers **C–20**(10), 1211–1214 (1971)
16. Rabin, M.O., Scott, D.: Finite automata and their decision problems. IBM J. Res. Dev. **3**(2), 114–125 (1959)
17. Sakoda, W.J., Sipser, M.: Nondeterminism and the size of two way finite automata. In: Proc. of the 10th Annual ACM Symposium on Theory of Computing (STOC 1978), pp. 275–286. ACM Press, New York (1978)
18. Sipser, M.: Lower bounds on the size of sweeping automata. Journal of Computer and System Sciences **21**, 195–202 (1980)
19. Štefanec, R.: Semantically restricted two-way automata. PhD Thesis, Comenius University in Bratislava (2014)

Squareable Words

Francine Blanchet-Sadri[1]([✉]) and Abraham Rashin[2]

[1] Department of Computer Science, University of North Carolina, P.O. Box 26170,
Greensboro, NC 27402–6170, USA
blanchet@uncg.edu
[2] Department of Mathematics, Rutgers University, 110 Frelinghuysen Rd.,
Piscataway, NJ 08854–8019, USA

Abstract. For a word w and a partial word u of the same length, say
w derives u if w can be transformed into u by inserting holes, i.e., by
replacing letters with don't cares, with the restriction that no two holes
may be within distance two. We present and prove a necessary and sufficient condition for a word of even length (at least eight) to not derive
any squares (such word is called non-squareable). The condition can be
decided in $O(n)$ time, where n is the length of the word.

Keywords: Combinatorics on words · Partial words · Squareable word ·
Squares

1 Introduction

For an alphabet A, a partial word is a word over the extended alphabet $A \cup \{\diamond\}$,
where $\diamond$ denotes a so-called "hole" or "don't care". These holes are used to indicate positions which may contain letters from a given substitution alphabet. For
example, $\diamond ab\diamond abb$ is a partial word with two holes at positions 0 and 3 that can
be transformed into any of the four words $aabaabb, aabbabb, babaabb, babbabb$ by
using the substitution that maps $\diamond$ into $\{a, b\}$. In the context of string matching,
Fischer and Paterson [5] introduced partial words in 1974 as "strings with don't
cares". Berstel and Boasson [1] began the study of their combinatorics in 1999
(they also introduced the terminology "partial words"). Both algorithms and
combinatorics on partial words have been developing since (see, e.g., [2]).

The concepts of repetition-freeness such as square-freeness, cube-freeness,
and overlap-freeness in words were studied in the early papers of Thue [8,9].
These concepts were extended to partial words in some recent papers [4,6,7]
where the authors studied the problem of whether or not infinite repetition-
free total words (those without holes) exist with the property that an arbitrary
number of holes can be inserted while remaining repetition-free. Here, inserting
holes means replacing the letters at a number of positions of the word with
holes. This insertion of holes is subject to the restriction that no two holes may
be within distance two, i.e., any positions i, j satisfying $0 < |i - j| \le 2$ are not
both holes. Such insertion, called 2-restricted, avoids the introduction of trivial

© Springer International Publishing Switzerland 2015
I. Potapov (Ed.): DLT 2015, LNCS 9168, pp. 132–142, 2015.
DOI: 10.1007/978-3-319-21500-6_10

squares and cubes. This problem is equivalent to determining whether an infinite partial word u, constructed by a 2-restricted insertion of holes into a total word w, exists such that none of u's factors of length kp has period p, for any rational $k \geq 2$ and integer $p \geq 1$.

Blanchet-Sadri et al. [3] described an algorithm that given as input a finite total word w of length n, and integers $p \geq 1$ and $d \geq 2$, determines whether w can be transformed into a p-periodic partial word u by a d-restricted insertion of holes, i.e., a partial word with period p and with the property that no two holes are within distance d. Their algorithm outputs (if any exists) such a "maximal" partial word u in $O(nd)$ time. Consequently, they constructed an infinite overlap-free total word over a 5-letter alphabet that remains overlap-free by an arbitrary 2-restricted insertion of holes. It had been shown earlier that there exists an infinite overlap-free total word over a 6-letter alphabet that remains overlap-free by an arbitrary insertion of holes, and that none exists over a 4-letter alphabet [4].

In this paper, we only deal with finite (total or partial) words. We consider the question of determining whether or not a total word w of length 2ℓ, with $\ell \geq 4$, can be turned into a partial word u that is a square (i.e., $u_0 \cdots u_{\ell-1}$ and $u_\ell \cdots u_{2\ell-1}$ are compatible) by inserting holes into w, subject to a 2-restriction. If such u exists, we say that w derives the square u or that w is squareable. We give a complete characterization of when a total word is squareable and as a consequence, we describe an algorithm for determining whether a total word derives a square, whose runtime is linear in the length of the word. Using this algorithm, we can also determine whether a total word derives a non-square-free square, whose runtime is cubic in the length of the word. Our approach is graph-theoretical.

The contents of our paper are as follows: In Section 2, we recall some basic concepts of combinatorics on partial words. In Section 3, we consider the set of all partial words that are squares derivable from a fixed total word w. Since it is a finite set, it is non-empty if and only if it has a "maximal" element. We show that there is a one-to-one correspondence between the set of all "maximal" partial words that are squares derivable from w and the set of 2-colorings of some graph, associated with w, subject to three conditions. In Section 4 we show that, aside from the trivial case of having 3-cycles, the associated graph is acyclic with maximal degree two. The abovementioned one-to-one correspondence allows us to determine whether or not w is squareable based on the structure of this graph. Finally in Section 5, we conclude with some remarks.

2   Preliminaries

We denote by $[i..j)$ (respectively, $[i..j]$) the discrete set of integers $\{i, \ldots, j-1\}$ (respectively, $\{i, \ldots, j\}$), where $i \leq j$.

Let A be a non-empty finite alphabet. A *total word* over A is a sequence of characters from A. A *partial word* over A is a sequence of characters from $A_\diamond = A \cup \{\diamond\}$, the alphabet A being extended with the hole character $\diamond$ (a total word is a partial word that does not contain the $\diamond$ character). The *length* of a

$$
\begin{array}{cccccc}
0 & 1 & 2 & 3 & 4 & 5 \\
a & b & c & a & b & a \\
c & a & c & b & b & b \\
6 & 7 & 8 & 9 & 10 & 11
\end{array}
\qquad
\begin{array}{cccccc}
0 & 1 & 2 & 3 & 4 & 5 \\
\diamond & b & c & \diamond & b & a \\
c & \diamond & c & b & b & \diamond \\
6 & 7 & 8 & 9 & 10 & 11
\end{array}
$$

Fig. 1. The squareable total word $w = abcabacacbbb$ (positions aligned in two rows on the left) derives the partial word $u = \diamond bc\diamond bac\diamond cbb\diamond$ (positions aligned in two rows on the right)

partial word u is denoted by $|u|$ and represents the total number of characters in u (the *empty word* is the word of length zero). We sometimes abbreviate the character in position i of u by u_i (the labelling of positions starts at 0). We let $H(u)$ be the set of positions where a partial word u has holes, i.e., $H(u) = \{i \mid i \in [0..|u|) \text{ and } u_i = \diamond\}$. A partial word u is a *factor* of a partial word v if $v = xuy$ for some x, y.

If u and v are two partial words of equal length over A, then u is *contained* in v, denoted by $u \subset v$, if $u_i = v_i$ whenever $u_i \in A$. For example, $\diamond bc\diamond ba \subset abc\diamond ba$. Partial words u and v are *compatible*, denoted by $u \uparrow v$, if there exists a partial word w such that $u \subset w$ and $v \subset w$. For example, $\diamond bc\diamond ba \uparrow c\diamond c\diamond b\diamond$ and $\diamond bcaba \not\uparrow c\diamond c\diamond bb$.

A *period* of a partial word u over A is a positive integer p such that $u_i = u_j$ whenever $u_i, u_j \in A$ and $i \equiv j \bmod p$. We say that u is p-*periodic*. The partial word $u = abba\diamond bacba$ is not 3-periodic since $u_1 = b$ and $u_7 = c$.

3   Squareable Words and Colorings of Associated Graphs

We say that a fixed total word w of length 2ℓ, with $\ell \geq 4$, *derives* a partial word u of the same length (u is said to be *derivable* from w) if u can be constructed by inserting holes into w, subject to the condition that no two holes may be within distance two. We consider the problem of determining whether or not w can derive a partial word u that is a square, i.e., $u_0 \cdots u_{\ell-1} \uparrow u_\ell \cdots u_{2\ell-1}$. Any such w is called *squareable*. Fig. 1 gives an example of a squareable total word.

It turns out that this problem is easier to tackle if we restrict our attention to partial words u that are maximal under the $\subset$ relation, if any exist. Accordingly, we define the following.

Definition 1. *Let U be the (possibly empty) set of squares u derivable from w. Call a partial word u in U maximal in U if for all $v \in U$, $u \subset v$ implies that $u = v$. Let $U_{\max}$ be the set of partial words u that are maximal in U.*

Since U is a finite set, it is non-empty if and only if it has a maximal element. We show that $U_{\max}$ is in bijection with the 2-colorings of a certain graph, subject to three conditions. This allows us to determine whether or not $U_{\max}$, and hence U, is empty based on the structure of this graph.

Let $V = \{i \mid i \in [0..\ell) \text{ and } w_i \neq w_{i+\ell}\}$. The following lemma gives a characterization of $U_{\max}$ in terms of V.

Lemma 1. *Let u be a partial word of length 2ℓ such that w derives u. Then $u \in U_{\max}$ if and only if both of the following conditions hold:*

1. *for $i \in V$, exactly one of u_i or $u_{i+\ell}$ is a hole,*
2. *for $i \in [0..\ell) \setminus V$, neither u_i nor $u_{i+\ell}$ is a hole.*

Proof. First note that by definition, $u \in U$ if and only if u is derivable from w and $u_0 \cdots u_{\ell-1} \uparrow u_\ell \cdots u_{2\ell-1}$. This is true precisely when for every $i \in V$, at least one of positions i and $i + \ell$ has been turned into a hole. Now, suppose $u \in U$ and that there is some $i \in V$ for which $u_i = \diamond = u_{i+\ell}$, or some $i \in [0..\ell) \setminus V$ for which either $u_i = \diamond$ or $u_{i+\ell} = \diamond$. Then we can remove one of these holes (replace it with the corresponding character in w), and the two halves of the partial word are still compatible. (The resulting partial word is still in U.) Thus u is not maximal. Therefore Conditions 1 and 2 are necessary and sufficient for u to be in $U_{\max}$. ☐

Lemma 1 applies to partial words derived from the given total word w. But because of the restriction that no two holes can be too close together, not every insertion of holes into w corresponds to a derived word. This condition, however, corresponds to independence of the hole positions on a certain graph.

Definition 2. *For $n \in \mathbb{N}$, let G_n be the graph with vertex set $[0..n)$ and an edge between two vertices i and j if and only if $|i - j| \in \{1, 2\}$.*

Fig. 2 gives an example of such a graph and illustrates Lemma 2 below.

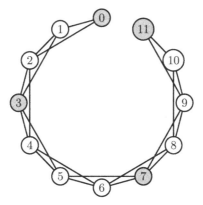

Fig. 2. $\{0, 3, 7, 11\}$ is an independent set of vertices on the graph G_{12}

Lemma 2. *Let u be a partial word of length 2ℓ. Then w derives u if and only if $u \subset w$ and $H(u)$ is an independent set of vertices on $G_{2\ell}$.*

Proof. This is precisely the condition that w can be transformed into u by insertion of holes, with no two holes within distance two (the definition of "w derives u"). ☐

Now let G be the subgraph of G_ℓ induced by V, i.e., the graph with vertex set V and an edge between i and j if and only if there is such an edge in G_ℓ. Fig. 3 gives an example and illustrates the following theorem.

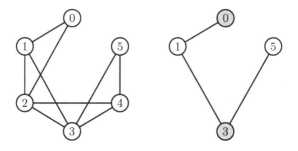

Fig. 3. A 2-coloring of the subgraph G (right) of G_6 (left) induced by $V = \{0, 1, 3, 5\}$

Theorem 1. *The set U is non-empty if and only if the graph G has a 2-coloring (say, grey and white) such that all three of the following conditions hold:*

1. *if $0, \ell - 1 \in V$ and 0 is white, then $\ell - 1$ is white,*
2. *if $0, \ell - 2 \in V$ and 0 is white, then $\ell - 2$ is white,*
3. *if $1, \ell - 1 \in V$ and 1 is white, then $\ell - 1$ is white.*

Proof. Recall that U is non-empty if and only if $U_{\max}$ is non-empty. Let C be the set of {grey, white}-colorings of G satisfying the above conditions. For every coloring $c \in C$, let $f(c)$ be the word derived from w whose set of holes is $\{i \mid i \in V \text{ and } c(i) = grey\} \cup \{i + \ell \mid i \in V \text{ and } c(i) = white\}$.

We claim that f is a bijection from C to $U_{\max}$. Let $c \in C$. Each of the sets $\{i \mid i \in V \text{ and } c(i) = grey\}$ and $\{i + \ell \mid i \in V \text{ and } c(i) = white\}$ is independent on $G_{2\ell}$ since no two grey vertices neighbor in G and no two white vertices neighbor in G. The union of these two sets is the hole set of $f(c)$, i.e., $H(f(c))$. Condition 1 prevents $\ell - 1$ and $\ell = 0 + \ell$ (which are within distance two) from both being holes in $f(c)$. Similarly, Condition 2 prevents $\ell - 2$ and ℓ from both being holes in $f(c)$ and Condition 3 prevents $\ell - 1$ and $\ell + 1$ from both being holes in $f(c)$. Together with the independence of each of the two sets that make up $H(f(c))$, this suffices to prove that $H(f(c))$ is independent on $G_{2\ell}$. Furthermore, since each vertex in G has exactly one color under c, Condition 1 of Lemma 1 is satisfied, and since G only has vertices in V, Condition 2 of Lemma 1 is also satisfied. Therefore, by Lemma 1, $f(c) \in U_{\max}$.

The set of grey-colored vertices in any $c \in C$ can be recovered from $f(c)$ as $\{i \mid i \in H(f(c)) \cap [0..\ell)\}$. Therefore f is injective.

0	1	2	3	4	5	6	7	8	9	10	11
a	b	c	a	b	a	a	b	c	a	b	a
c	a	c	b	b	b	c	c	c	b	b	a
12	13	14	15	16	17	18	19	20	21	22	23

Fig. 4. A non-squareable total word (positions aligned in two rows)

Let $u \in U_{\max}$. Then w derives u. By Lemma 2, $u \subset w$ and $H(u)$ is an independent set of vertices on $G_{2\ell}$ and by Lemma 1, exactly one of $i, i+\ell$ is in $H(u)$ if $i \in V$ and precisely zero if $i \in [0..\ell)\backslash V$. Let c be the coloring of G defined by $c(i) = grey$ if $i \in H(u)$, $c(i) = white$ if $i+\ell \in H(u)$. Each of these vertices is colored only by a single color, and the set of grey vertices and the set of white vertices are each independent on G. Therefore c is a 2-coloring of G. Furthermore, since $\ell - 1$ and ℓ (which are within distance two) are not both in $H(u)$, it is not the case that both 0 and $\ell - 1$ are in V and that $\ell - 1$ is colored grey while 0 is colored white. Therefore Condition 1 is satisfied. Similarly, Conditions 2 and 3 are satisfied. So $c \in C$. Now $f(c) = u$. Therefore f is surjective.

Therefore $U_{\max}$ (and hence U) is non-empty if and only if C is non-empty, i.e., there exists a coloring of G satisfying the three conditions above.   □

Fig. 4 gives an example of a non-squareable total word and Fig. 5 shows the associated induced subgraph G.

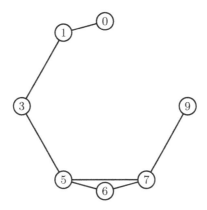

Fig. 5. The subgraph G of G_{12} induced by $V = \{0, 1, 3, 5, 6, 7, 9\}$; it has no 2-coloring

4 Linear Time Algorithm for Deciding Squareability

To assist in this graph-coloring problem, we first describe the structure of the graph G. It turns out that aside from a trivial case (having 3-cycles), it is always acyclic with maximal degree two. Therefore, unless V contains three consecutive integers (the only way to have a 3-cycle), G is always 2-colorable, and we need only determine whether one of these colorings satisfies the three conditions of Theorem 1.

Lemma 3. *Let $n \in \mathbb{N}$ and let $V' \subseteq [0..n)$. Let G' be the subgraph of G_n induced by V'. If V' does not contain any three consecutive integers, then the following hold:*

1. *G' is acyclic,*
2. *every vertex in G' has degree at most two.*

Proof. First, suppose for a later contradiction that there is a vertex $i \in V'$ of degree at least 3. Then three of $i - 2$, $i - 1$, $i + 1$ and $i + 2$ are in V', so it must be the case that either $i - 2, i - 1, i \in V'$ or $i, i + 1, i + 2 \in V'$. In either case, V' contains three consecutive integers, a contradiction.

Next, suppose for a later contradiction that G' has cycles, and let $i_1, \ldots, i_m, i_1$ be a minimal length cycle in G'. Now let $k \in [1..m]$ so that $i_k \leq i_l$ for all $l \in [1..m]$. We claim that $i_k + 1$ and $i_k + 2$ are in V'. To see this, consider the subgraph H of G' induced by $\{j \mid j \in V' \text{ and } j \geq i_k\}$. Since the above cycle has no vertex less than i_k, it is also a minimal cycle on H. Therefore the vertex i_k must have degree at least two in H, from which it follows that $i_k + 1$ and $i_k + 2$ are in V'. Therefore V' contains three consecutive integers, a contradiction. □

In other words, unless V contains three consecutive integers (checkable given w in ℓ-linear time), G simply consists of a union of disjoint paths. Such a graph admits 2^m possible 2-colorings, where m is the number of components.

Lemma 4. *Let $I = \min(V)$ and $J = \max(V)$. Then G is connected if and only if there is no $i \in \mathbb{N}$, $I < i < J$, for which $i \notin V$ and $i + 1 \notin V$.*

Proof. First, suppose that i is an integer between I and J for which $i, i + 1 \notin V$. Partition the vertex set V of G into $V_1 = \{j \mid j \in V \text{ and } j \leq i\}$ and $V_2 = \{j \mid j \in V \text{ and } j \geq i\}$. These sets are non-empty because $I \in V_1$ and $J \in V_2$. But since $i, i + 1 \notin V$, we have for all $j_1 \in V_1, j_2 \in V_2$ that $j_1 \leq i - 1$ and $j_2 \geq i + 2$, and hence that $|j_2 - j_1| = j_2 - j_1 > 2$. Therefore there are no edges between V_1 and V_2, so G is not connected.

Second, suppose that G is disconnected. Let $V_1, \ldots, V_r$ $(r \geq 2)$ be a partition of V into components, ranked in increasing order of maximal element. (Each V_k is connected, no V_k and V_l have an edge between them unless $k = l$, and for $k < l$, $\max(V_k) < \max(V_l)$.) We claim that $J - 1, J \notin V_1$. First, $J \in V_r$ since it is maximal in V. Now if $J - 1 \in V$, then there is an edge between J and $J - 1$, implying that $J - 1$ is in V_r, and hence not in V_1. If $J - 1 \notin V$, then we also have that $J - 1 \notin V_1 \subset V$. In either case, $J - 1$ and J are not in V_1.

Now let $J_1 = \max(V_1)$. By the above, $J_1 + 1 < J$. We also have that $J_1 \geq I$ (since $J_1 \in V$), so $J_1 + 1 > I$. Also, if $J_1 + 1$ or $J_1 + 2$ is in V, then there is an edge between J_1 and this greater vertex, violating the maximality of J_1 in the component V_1. Therefore, since $J_1 \in V$, we have that $I < J_1 + 1 < J$ and that $J_1 + 1, J_1 + 2 \notin V$. Hence $J_1 + 1$ satisfies the required properties. □

We can now prove our necessary and sufficient conditions for a total word of even length (at least 8) to be squareable.

Theorem 2. *The total word w derives a square (i.e., $U \neq \emptyset$) if and only if neither of the following two conditions holds:*

1. *V contains three consecutive integers,*
2. *$0, 1, \ell - 2, \ell - 1 \in V$, $[0..\ell] \setminus V$ does not contain any two consecutive integers, and V has an odd number of elements.*

Proof. Suppose that Condition 1 holds, i.e., V contains three consecutive integers. Then these integers form a 3-cycle in G, so G is not bipartite and hence not 2-colorable. By Theorem 1, w does not derive any squares.

So, suppose that Condition 1 does not hold. By Lemma 3, G consists of a union of (one or more) disjoint paths. Let m be the number of components, and let ℓ_k be the length of component k for $k \in [0..m)$. Also, for $k \in [0..m)$ and $l \in [0..\ell_k]$, let $i_{k,l}$ be element l of component k of G, with the ends chosen so that $i_{k,0} \leq i_{k,\ell_k}$. In this case, G is always 2-colorable. There are two ways to color each of the m components (each component must have the colors of vertices alternating along the path), making for 2^m possible 2-colorings. The question is whether any of these colorings satisfies the three additional conditions of Theorem 1 that none of the three pairs $(\ell - 2, 0), (\ell - 1, 0), (\ell - 1, 1)$ be colored (grey,white).

First, suppose that Condition 2 holds. Then by Lemma 4, G is connected since there are no two consecutive integers outside V between $0 = \min(V)$ and $\ell - 1 = \max(V)$. Thus $m = 1$, and G is a path $i_{0,0}, \ldots, i_{0,\ell_0}$ of length $\ell_0 = |V|$, an odd number by assumption. The first element $i_{0,0}$ must be 0, since if 0 has degree two in G, it has both 1 and 2 as neighbors, forming a 3-cycle (but we assume that Condition 1 does not hold). Similarly, i_{0,ℓ_0} must be $\ell - 1$. Also, since these vertices have unique neighbors, $i_{0,1} = 1$ and $i_{0,\ell_0-1} = \ell - 2$. There are only two ways to 2-color a path, since the colors must be alternating. In one case, $0, 1, \ell - 2$ and $\ell - 1$ are grey, white, white and grey, respectively, violating Condition 3 of Theorem 1, that $\ell - 1$ not be grey when 1 is white. In the other case, $0, 1, \ell - 2$ and $\ell - 1$ are white, grey, grey and white, respectively, violating Condition 2 of Theorem 1, that $\ell - 2$ not be grey when 0 is white. Thus, there are no appropriate 2-colorings of G (satisfying the conditions of Theorem 1), so by Theorem 1, w does not derive any squares.

Second, suppose that Condition 2 does not hold. We consider several cases and show that in each case, there is a 2-coloring of G that satisfies the three conditions of Theorem 1 (and thus w derives a square).

Assume that G is of even size, connected and includes the first two and last two elements of $[0..\ell)$. This is similar to that immediately above, except that the chain is of odd length. It is easy to verify that the alternating coloring with 0 colored grey satisfies the three conditions of Theorem 1, since it leaves $0, 1, \ell - 2$ and $\ell - 1$ colored grey, white, grey and white, respectively, with none of the three ordered pairs $(\ell - 2, 0), (\ell - 1, 0), (\ell - 1, 1)$ colored (grey,white). (They are colored (grey,grey), (white,grey), (white,white), instead. The opposite coloring, which is the only other coloring of this graph, has $(\ell - 1, 0)$ colored (grey,white), violating Condition 1 of Theorem 1.)

Now, assume that G includes the first two and last two elements of $[0..\ell)$ and is not connected. Then it follows from the reasoning of Lemma 4 that 0 and 1 lie in a component different from that in which $\ell - 2$ and $\ell - 1$ lie. These components can therefore be colored independently. Color the first component so that 0 is grey and 1 is white, and color the latter component so that $\ell - 2$ is grey and $\ell - 1$ is white. Color the other components (if any more exist) any other way. Now the pairs $(\ell - 2, 0), (\ell - 1, 0), (\ell - 1, 1)$ are colored (grey,grey), (white,grey), (white,white), so none of them are colored the forbidden (grey,white).

Consider the case when G lacks at least three of $0, 1, \ell - 1, \ell - 2$. Then no three of the conditions of Theorem 1 apply. Any coloring suffices.

Next, consider the case when G includes $0, \ell - 2$ but lacks at least one of $1, \ell - 1$. In this case, Condition 3 of Theorem 1 is already satisfied, since we do not have that both 1 and $\ell - 1$ are in V. Choose a coloring of G in which 0 is grey. Then neither $(\ell - 2, 0)$ nor $(\ell - 1, 0)$ is colored (grey,white).

Finally, consider the case when G includes $1, \ell - 1$ but lacks at least one of $0, \ell - 2$. In this case, Condition 2 of Theorem 1 is already satisfied, since we do not have that both 0 and $\ell - 2$ are in V. Choose a coloring of G in which $\ell - 1$ is white. Then neither $(\ell - 1, 0)$ nor $(\ell - 1, 1)$ is colored (grey,white).   □

Returning to Figs. 4 and 5, note that the set V contains three consecutive integers.

When V does not contain three consectutive integers, the proof of Theorem 2 shows that it is impossible to find a coloring satisfying the three conditions of Theorem 1 exactly when G is a single (connected) path of even length that includes vertices $0, 1, \ell - 2$ and $\ell - 1$. This provides a ℓ-linear time algorithm for determining whether w derives a square.

Algorithm 1. Deciding Squareability of a Total Word w of Length 2ℓ, $\ell \geq 4$

1: $V \leftarrow \{i \mid i \in [0..\ell)$ and $w_i \neq w_{i+\ell}\}$
2: **if** V contains three consective integers **then**
3:    **return** w is non-squareable
4: $G \leftarrow$ the subgraph of G_ℓ induced by V
5: **if** G is a single (connected) path of even length that includes vertices $0, 1, \ell - 2$ and $\ell - 1$ **then**
6:    **return** w is non-squareable
7: **return** w is squareable

Corollary 1. *Algorithm 1 determines whether a total word of length 2ℓ, where $\ell \geq 4$, derives a square. Its runtime is linear in the length of the word.*

Note that the cases for $\ell \in \{0, 1, 2, 3\}$ are achievable in constant time.

By cycling through possible starting positions and periods and using Algorithm 1, we obtain the following.

Corollary 2. *There is an algorithm for determining whether a total word derives a non-square-free word, whose runtime is cubic in the length of the word.*

5 Conclusion

In this paper, we considered the problem of deciding whether or not a given total word w of length 2ℓ, with $\ell \geq 4$, can be turned into a partial word that is a square by inserting holes into w, subject to the restriction that no two holes may be within distance two. We gave a complete characterization of such squareable words w and as a consequence, we described an algorithm for determining whether a total word derives a square, whose runtime is linear in the length of the word. Using this algorithm, we can also determine whether a total word derives a non-square-free square, whose runtime is cubic in the length of the word. Our approach was based on the existence of 2-colorings, satisfying some conditions, of a graph associated with the given total word. It turned out that, aside from the trivial case of having 3-cycles, this graph is acyclic with maximal degree two.

Acknowledgements. Project sponsored by the National Security Agency under Grant Number H98230-15-1-0232. The United States Government is authorized to reproduce and distribute reprints notwithstanding any copyright notation herein. This manuscript is submitted for publication with the understanding that the United States Government is authorized to reproduce and distribute reprints.

This material is based upon work supported by the National Science Foundation under Grant No. DMS–0754154. The Department of Defense is also gratefully acknowledged.

We thank the referees of a preliminary version of this paper for their very valuable comments and suggestions.

References

1. Berstel, J., Boasson, L.: Partial words and a theorem of Fine and Wilf. Theoretical Computer Science **218**, 135–141 (1999)
2. Blanchet-Sadri, F.: Algorithmic Combinatorics on Partial Words. Chapman & Hall/CRC Press, Boca Raton (2008)
3. Blanchet-Sadri, F., Mercaş, R., Rashin, A., Willett, E.: Periodicity algorithms and a conjecture on overlaps in partial words. Theoretical Computer Science **443**, 35–45 (2012)
4. Blanchet-Sadri, F., Mercaş, R., Scott, G.: A generalization of Thue freeness for partial words. Theoretical Computer Science **410**, 793–800 (2009)
5. Fischer, M., Paterson, M.: String matching and other products. In: Karp, R. (ed.) 7th SIAM-AMS Complexity of Computation, pp. 113–125 (1974)
6. Halava, V., Harju, T., Kärki, T., Séébold, P.: Overlap-freeness in infinite partial words. Theoretical Computer Science **410**, 943–948 (2009)
7. Manea, F., Mercaş, R.: Freeness of partial words. Theoretical Computer Science **389**, 265–277 (2007)

8. Thue, A.: Über unendliche Zeichenreihen. Norske Vid. Selsk. Skr. I, Mat. Nat. Kl. Christiana **7**, 1–22 (1906). (Reprinted in Selected Mathematical Papers of Axel Thue, T. Nagell, editor, Universitetsforlaget, Oslo, Norway (1977), pp. 139–158)

9. Thue, A.: Über die gegenseitige Lage gleicher Teile gewisser Zeichenreihen. Norske Vid. Selsk. Skr. I, Mat. Nat. Kl. Christiana **1**, 1–67 (1912). (Reprinted in Selected Mathematical Papers of Axel Thue, T. Nagell, editor, Universitetsforlaget, Oslo, Norway (1977), pp. 413–478)

Complexity Analysis: Transformation Monoids of Finite Automata

Christian Brandl$^{(\boxtimes)}$ and Hans Ulrich Simon

Department of Theoretical Computer Science, Faculty of Mathematics,
Ruhr-University Bochum, 44780 Bochum, Germany
{christian.brandl,hans.simon}@rub.de

Abstract. We examine the computational complexity of some problems from algebraic automata theory and from the field of communication complexity: testing Green's relations (relations that are fundamental in monoid theory), checking the property of a finite monoid to have only Abelian subgroups, and determining the deterministic communication complexity of a regular language. By well-known algebraizations, these problems are closely linked with each other. We show that all of them are PSPACE-complete.

Keywords: Green's relations · Finite monoids · Regular languages · Communication complexity · PSPACE-completeness

1 Introduction

The Green's relations $\mathcal{L}, \mathcal{R}, \mathcal{J}, \mathcal{H}$ are ubiquitous tools for studying the buildup of a (finite) monoid M. For example, the maximal subgroups of M can be characterized as $\mathcal{H}$-classes of M containing an idempotent element (e.g. [7]). Such $\mathcal{H}$-classes are called *regular*. As illustrated in [3], there are also important applications in automata theory: star-free languages, factorization forests, and automata over infinite words. The former application, due to Schützenberger, characterizes the star-free languages as regular languages with syntactic monoids having only trivial subgroups [8]. Finite monoids having only trivial subgroups are called *aperiodic* and form a variety denoted as **A**. In [2], Cho and Huynh prove Stern's conjecture from [9] that testing for aperiodicity (star-freedom, alternatively) is PSPACE-complete if the regular language is given as a minimum DFA. **A** is contained in the variety $\overline{\mathbf{Ab}}$ of monoids having only Abelian subgroups (regular $\mathcal{H}$-classes, alternatively). $\overline{\mathbf{Ab}}$ plays a decisive role for the communication complexity of a regular language [10]. Since its introduction by Yao [11], communication complexity has developed to one of the major complexity measures with plenty of applications (e.g., listed in [6]). In [10], Tesson and Thérien categorize the communication complexity of an arbitrary regular language L by some properties of its underlying syntactic monoid $M(L)$. This algebraic classification can be achieved in the following models of communication: deterministic, randomized (bounded error), simultaneous, and Mod_p-counting. For the deterministic

© Springer International Publishing Switzerland 2015
I. Potapov (Ed.): DLT 2015, LNCS 9168, pp. 143–154, 2015.
DOI: 10.1007/978-3-319-21500-6_11

model, the authors of [10] show that the communication complexity of L can only be constant, logarithmic, or linear (in the sense of Θ-notation). Thereby, the condition $M(L) \notin \overline{\mathbf{Ab}}$ is a sufficient (but not necessary) condition for the linear case.

In this paper, we focus on monoids which are generated by mappings with domain and range Q for some finite set Q (like the syntactic monoid where these mappings are viewed as state transformations). We primarily analyze the computational complexity of problems related to Green's relations, the monoid-variety $\overline{\mathbf{Ab}}$, and the deterministic communication complexity of regular languages. Our main contributions are summarized in the following theorem:

Theorem 1. *(Main results) Let M be a finite monoid given by the generators (=state transformations) $f_1, \ldots, f_l$. Let g, h be 2 elements of M. Let $\mathcal{G} \in \{\mathcal{L}, \mathcal{R}, \mathcal{J}, \mathcal{H}\}$ be one of Green's relations w.r.t. the monoid M. Let A be a minimum DFA with syntactic monoid $M(A)$. Let N be an NFA recognizing language L. Let $cc(L)$ be the deterministic communication complexity of L. With these notations, the following problems are PSPACE-complete:*

Problem	Input	Question
(i) $\mathcal{G}$-TEST	$g, h \in M$ and generators $f_1, \ldots, f_l$	$g \, \mathcal{G} \, h$?
(ii) DFA-$\overline{\mathbf{Ab}}$	minimum DFA A	$M(A) \in \overline{\mathbf{Ab}}$?
(iii) NFA-CC	NFA N	Is $cc(L)$ const., log., or lin.?

This paper is organized as follows. In Section 2, we introduce the necessary background and notations. In Section 3, we show the PSPACE-hardness of deciding Green's relations. In Section 4, we show the PSPACE-hardness of the problem DFA-$\overline{\mathbf{Ab}}$. In Section 5.1, we show that all problems listed in Theorem 1 are members of PSPACE even when the underlying monoid is a syntactic monoid of a regular language that is given by an NFA. In Section 5.2, we show the PSPACE-hardness of the problem NFA-CC. Moreover, this problem remains PSPACE-hard even when the language L is given by a regular expression β at the place of the NFA N.

2   Preliminaries

We assume that the reader is familiar with the basic concepts of computational complexity (e.g. [4]), communication complexity (e.g. [6]), regular languages (e.g. [4]), and algebraic automata theory (e.g. [7]). In the sequel, we will briefly recapitulate some definitions and facts from these fields and thereby fix some notation.

A Deterministic Finite Automaton (DFA) is formally given as a 5-tuple $A = (Q, \Sigma, \delta, s, F)$ where Q denotes the finite set of states, Σ the input alphabet, $\delta : Q \times \Sigma \to Q$ the transition function, $s \in Q$ the initial state and $F \subseteq Q$ the set of final (accepting) states. As usual, the mapping δ can be extended by morphism to $\delta^* : Q \times \Sigma^* \to Q$. Throughout the paper, we will make use of the notation $q \cdot w := \delta^*(q, w)$. Intuitively, $q \cdot w$ is the state that is reached after A, started in state q, has completely processed the string w. The change from state q to state $q \cdot w$ is sometimes called the *w-transition from q*. The language recognized by A is given by $L(A) = \{w \in \Sigma^* : s \cdot w \in F\}$. Languages recognizable by DFA are called *regular*. The DFA with the minimal number of states that recognizes a regular language L is called the *minimum DFA* for L. DFA-minimization can be executed efficiently.

Let $L \subseteq \Sigma^*$ be a formal language. We write $w_1 \sim_L w_2$ iff the equivalence $uw_1v \in L \Leftrightarrow uw_2v \in L$ holds for every choice of $u, v \in \Sigma^*$. $\sim_L$ defines a congruence relation on Σ^* named *syntactic congruence*. For every $w \in \Sigma^*$, $[w]_{\sim_L}$ denotes the equivalence class represented by w. The quotient monoid $\Sigma^* / \sim_L$, denoted as $M(L)$, is called the *syntactic monoid* of L. $M(L)$ is finite iff L is regular. Moreover, $M(L)$ coincides with the monoid consisting of the state transformations $f_w(q) := q \cdot w$ of a minimum DFA for L. Clearly, this monoid is generated by $\{f_a : a \in \Sigma\}$. If $L = L(A)$ for some DFA A, we often simply write $M(A)$ instead of $M(L(A))$. The analogous convention applies to NFA.

Let M be a monoid and $a, b \in M$ two arbitrary elements. $a \mathcal{J} b :\Leftrightarrow MaM = MbM$; $a \mathcal{L} b :\Leftrightarrow Ma = Mb$; $a \mathcal{R} b :\Leftrightarrow aM = bM$; $a \mathcal{H} b :\Leftrightarrow a \mathcal{L} b \wedge a \mathcal{R} b$ define four equivalence relations on M named *Green's relations*. An element $e \in M$ is called *idempotent* iff $e^2 = e$. A *subgroup* of M is a subsemigroup of M that is a group.

We denote by PSPACE the class of all problems that can be solved by a Deterministic Turing-Machine (DTM) in polynomial space. We use the symbol $\mathcal{M}$ for such a DTM. PSPACE is closed under complement. The following two decision problems FAI [5] and RENU [1] are known to be complete problems for PSPACE. FAI: Given DFAs $A_1, \ldots, A_k$ with a common input alphabet and unique final states, is there an input word accepted by all of $A_1, \ldots, A_k$? RENU: Given a regular expression β over Σ (i.e., an expression with operands from $\Sigma \cup \{\varepsilon\}$ and operations "+" (for union), "·" (for concatenation) and "∗" (for Kleene closure)), do we have $L(\beta) \neq \Sigma^*$?[1]

Let $L \subseteq \Sigma^*$ be a formal language. Let ε denote the empty string. In the so-called communication game, 2 parties, say X and Y, exchange bits in order to decide if the string $x_1 y_1 \ldots x_n y_n$ belongs to L. Thereby, X (resp. Y) only knows $(x_1, \ldots, x_n) \in (\Sigma \cup \{\varepsilon\})^n$ (resp. $(y_1, \ldots, y_n) \in (\Sigma \cup \{\varepsilon\})^n$). The communication may depend only on the input of the bit-sending party and the bits already exchanged. The minimal number of communication bits needed to decide membership is called *deterministic communication complexity* and is denoted as $cc(L)$. By [10], a regular language L has constant communication complexity iff $M(L)$ is commutative.

The definition of an NFA $N = (Q, \Sigma, \delta, s, F)$ is similar to the definition of a DFA with the notable exception that $\delta(q, w)$ is not an element but a subset of Q, i.e., an element of the powerset 2^Q. Again the mapping δ can be extended by morphism to a mapping $\delta : Q \times \Sigma^* \to 2^Q$ or even to a mapping $\delta^* : 2^Q \times \Sigma^* \to 2^Q$ by setting

$$\delta^*(R, w) = \cup_{z \in R} \delta^*(z, w) \ . \tag{1}$$

---

[1] The language $L(\beta)$ induced by a regular expression is defined in the obvious manner.

3   Testing Green's Relations

In this section, we prove the PSPACE-hardness of testing Green's relations (see Theorem 1,(i)). To this end, we design 2 logspace-reductions that start from the problem FAI. Recall that an instance of FAI is given by DFA $A_1, \ldots, A_k$ with a common input alphabet $\Sigma = \{a_1, \ldots, a_l\}$ and unique final states.
$\mathcal{L}$-,$\mathcal{H}$-TEST: The instance of FAI is transformed to the mappings

$$f_{a_1}, \ldots, f_{a_l}, h_0, g_+, h_+ : Z \uplus \{\sigma_0, \sigma_1, \sigma_2, \tau_0, \tau_1\} \rightarrow Z \uplus \{\sigma_0, \sigma_1, \sigma_2, \tau_0, \tau_1\} \ , \quad (2)$$

which we view as state transformations. Here, $Z = \uplus_{j=1}^{k} Z(A_j)$ denotes the disjoint union of the state sets of the DFAs, and $\sigma_0, \sigma_1, \sigma_2, \tau_0, \tau_1$ are five additional *special states*. In the sequel, the notion *state diagram* refers to the total diagram that is formed by the disjoint union of the state diagrams for all DFAs whereas the diagram for a particular DFA A_i is called *sub-diagram*. On the *ordinary states* (as opposed to the special states), the state transformations act as follows:

$$\forall a \in \Sigma : f_a(z) = z \cdot a \ \text{ and } \ h_0(z) = z_0, g_+(z) = h_+(z) = z_+ \ .$$

Here, z_0 denotes the unique initial state in the sub-diagram containing z. Likewise, z_+ denotes the unique accepting state in this sub-diagram. The special states $\sigma_0, \sigma_1, \sigma_2$ are transformed as follows:

$$\sigma_0 \xmapsto{g_+, h_+} \sigma_1 \xmapsto{g_+, h_+} \sigma_2 \qquad (3)$$

Moreover, σ_0, σ_1 are fix-points for $f_{a_1}, \ldots, f_{a_l}, h_0$, and σ_2 is a fix-point for every transformation. The analogous interpretation applies to

$$\tau_0 \xmapsto{h_0, h_+} \tau_1, \tau_1 \xmapsto{h_0, h_+} \tau_0 \qquad (4)$$

Concerning the $\mathcal{L}$-TEST (resp. the $\mathcal{H}$-TEST), we ask whether $g_+ \mathcal{L} h_+$ (resp. $g_+ \mathcal{H} h_+$) w.r.t. the monoid generated by the mappings in (2). We claim that the following equivalences are valid:

$$\bigcap_{j=1}^{k} L(A_j) \neq \emptyset \Leftrightarrow g_+ \mathcal{H} h_+ \Leftrightarrow g_+ \mathcal{L} h_+ \ . \qquad (5)$$

To prove this claim, we first suppose $\bigcap_{j=1}^{k} L(A_j) \neq \emptyset$. Now, pick a word w from $\bigcap_{j=1}^{k} L(A_j)$, and observe that the following holds:

$$h_+ = f_w \circ h_0 \circ g_+, \qquad\qquad g_+ = f_w \circ h_0 \circ h_+ \qquad (6)$$
$$h_+ = g_+ \circ h_0, \qquad\qquad g_+ = h_+ \circ h_0 \qquad (7)$$

Thus, we have $g_+ \mathcal{L} h_+$ by (6) and $g_+ \mathcal{R} h_+$ by (7). Hence, $g_+ \mathcal{H} h_+$, as required. The implication from $g_+ \mathcal{H} h_+$ to $g_+ \mathcal{L} h_+$ holds for trivial reasons. Now, suppose $g_+ \mathcal{L} h_+$. Certainly, this implies that h_+ can be written as $h_+ = P \circ g_+$ where

P is a product of generators, i.e., a composition of functions from (2). Since h_+ and g_+ are the only generators that do not leave states of type σ fixed and act on them according to (3), it follows that P neither contains h_+ nor g_+. For ease of later reference, we call this kind of reasoning the σ-*argument*. Since h_+, h_0 are the only generators that do not leave states of type τ fixed and act on them according to (4), the product P must contain h_0 an odd number of times. Focusing on the leftmost occurrence of h_0, P can be written as $P = P' \circ h_0 \circ P''$ where P' does not contain h_0, and P'' contains h_0 an even number of times. It is easily verified that $h_0 = h_0 \circ P''$, so $P = P' \circ h_0$ and $h_+ = P' \circ h_0 \circ g_+$ where product P' contains exclusively generators from $\{f_{a_1}, \ldots, f_{a_l}\}$. Thus, there exists a word $w \in \Sigma^*$ such that $P' = f_w$ and $h_+ = f_w \circ h_0 \circ g_+$. Now, we are done with the proof of (5) since this implies $w \in \bigcap_{j=1}^{k} L(A_j)$. (5) directly implies the desired hardness result for the $\mathcal{L}$- and the $\mathcal{H}$-TEST, respectively.

$\mathcal{R}$-,$\mathcal{J}$-TEST: This time, we map $A_1, \ldots, A_k$ to the following list of generators:

$$f_{a_1}, \ldots, f_{a_l}, f_0, f, g, g_+ : Z \uplus Z' \uplus \{\sigma_0, \sigma_1, \sigma_2\} \to Z \uplus Z' \uplus \{\sigma_0, \sigma_1, \sigma_2\} \qquad (8)$$

Here, Z is chosen as in (2), $Z' = \{z' : z \in Z\}$ contains a marked state z' for every ordinary state z, and $\sigma_0, \sigma_1, \sigma_2$ are special states (put into place to apply the σ-argument). The marked states are fix-points for every mapping. Mappings g, g_+ act on states of type σ according to (3) but now with g in the role of h_+. The remaining mappings leave states of type σ fixed. For every $a \in \Sigma$, $f_a(z) = z \cdot a$ is defined as in the previous logspace-reduction. Mappings f_0, f, g, g_+ act on ordinary states (with the same notational conventions as before) as follows:

$$f_0(z) = z_0, f(z) = g(z) = z', g_+(z) = z'_+$$

Concerning the $\mathcal{R}$-TEST (resp. the $\mathcal{J}$-TEST), we ask whether $g\mathcal{R}g_+$ (resp. $g\mathcal{J}g_+$) w.r.t. the monoid generated by the mappings in (8). We claim that the following equivalences are valid:

$$\bigcap_{j=1}^{k} L(A_j) \neq \emptyset \Leftrightarrow g\mathcal{R}g_+ \Leftrightarrow g\mathcal{J}g_+ \qquad (9)$$

To prove this claim, we first suppose $\bigcap_{j=1}^{k} L(A_j) \neq \emptyset$. Now, pick a word w from $\bigcap_{j=1}^{k} L(A_j)$, and observe that the following holds:

$$g = g_+ \circ f \;,\; g_+ = g \circ f_w \circ f_0$$

Thus, $g\mathcal{R}g_+$, as required. The implication from $g\mathcal{R}g_+$ to $g\mathcal{J}g_+$ holds for trivial reasons. Now, suppose $g\mathcal{J}g_+$. Certainly, this implies that g_+ can be written as $g_+ = P \circ g \circ Q = g \circ Q$ where P and Q are products of generators, respectively. The second equation is valid simply because g marks ordinary states and marked states are left fixed by all generators (so that P is redundant). It follows from the σ-argument that neither g nor g_+ can occur in Q (or P). We may furthermore assume that f does not occur in Q because a decomposition of $g \circ Q$ containing

f could be simplified according to $g \circ Q = g \circ Q' \circ f \circ Q'' = g \circ Q''$. The last equation holds because f (like g) marks ordinary states which are then kept fixed by all generators. We may conclude that $g_+ = g \circ Q$ for some product of Q that does not contain any factor from $\{g, g_+, f\}$. Because of the simplification $Q' \circ f_0 \circ Q'' = Q' \circ f_0$, we may furthermore assume that either Q does not contain f_0, or it contains f_0 as the rightmost factor only. Thus, there exists some word $w \in \Sigma^*$ such that either $g_+ = g \circ Q = g \circ f_w$ or $g_+ = g \circ Q = g \circ f_w \circ f_0$. In both cases, this implies that $w \in \bigcap_{j=1}^{k} L(A_j)$ so that the proof of (9) is now complete. (9) directly implies the desired hardness result for the $\mathcal{R}$- and the $\mathcal{J}$-TEST, respectively.   □

4   Finite Monoids: Testing for a Non-Abelian Subgroup

Recall from Section 1 that **A** denotes the variety of finite monoids with only trivial subgroups (the so-called aperiodic monoids). Let DFA-**A** be defined in analogy to the problem DFA-$\overline{\textbf{Ab}}$ from Theorem 1. In [2], Cho and Huynh show the PSPACE-hardness of DFA-**A** by means of a generic reduction that proceeds in two stages with the first one ending at a special version of FAI. We will briefly describe this reduction and, thereafter, we will modify it so as to obtain a generic reduction to the problem DFA-$\overline{\textbf{Ab}}$.

Let $\mathcal{M}$ be an arbitrary but fixed polynomially space-bounded DTM with input word x. In a first stage, Cho and Huynh efficiently transform $(\mathcal{M}, x)$ into a collection of prime p many minimum DFAs $A_1, \ldots, A_p$ with aperiodic syntactic monoids $M(A_i)$, initial states s_i, unique accepting states f_i, and unique (non-accepting) dead states such that $L(A_1) \cap \ldots \cap L(A_p)$ coincides with the strings that describe an accepting computation of $\mathcal{M}$ on x. Consequently, $L(A_1) \cap \ldots \cap L(A_p)$ is either empty or the singleton set that contains the (representation of the) unique accepting computation of $\mathcal{M}$ on x. In a second stage, Cho and Huynh connect $A_1, \ldots, A_p$ in a cyclic fashion by using a new symbol $\#$ that causes a state-transition from the accepting state f_i of A_i to the initial state s_{i+1} of A_{i+1} (or, if $i = p$, from f_p to s_1). This construction of a single DFA A (with $A_1, \ldots, A_p$ as sub-automata) is completed by amalgamating the p dead states, one for every sub-automaton, to a single dead state, and by declaring s_1 as the only initial state and f_1 as the only accepting state. (All $\#$-transitions different from the just described ones end up in the dead state.) By construction, A is a minimum DFA. Moreover, $M(A)$ is not aperiodic iff $\mathcal{M}$ accepts x. The latter result relies on the following general observation:

Lemma 1 ([2]). *Let B be a minimum DFA: $M(B)$ is not aperiodic iff there is a state q and an input word u such that u defines a non-trivial cycle starting at q, i.e., $q \cdot u \neq q$ and, for some positive integer r, $q \cdot u^r = q$.*

For ease of later reference, we insert the following notation here:

$$r(q, u) := \min \left(\{ r \in \mathbb{Z} : r \geq 1, q \cdot u^r = q \} \right)$$

with the convention that $\min(\emptyset) = \infty$.

Our modification of the reduction by Cho and Huynh is based on the following general observation:

Lemma 2. *Let B be a minimum DFA with state set Q and alphabet Σ: If $M(B)$ contains a non-Abelian subgroup G, then there exists a state q and a word u with $r(q, u) \geq 3$.*

Proof. Since every subgroup whose elements are of order at most 2 is Abelian, G contains an element $u \in \Sigma^+$ (identified with the element in $M(B)$ that it represents) of order r at least 3. Because u^r fixes the states from $Q' := Q \cdot u$, for every $q' \in Q'$, u defines a cycle starting at q'. Therefore, we obviously get $r = \mathrm{lcm}\{r(q', u) : q' \in Q'\}$. Because of $r \geq 3$, this directly implies the claim. □

We modify the first stage of the reduction by Cho and Huynh by introducing 2 new symbols $\vdash, \dashv$ (so-called *endmarkers*). Moreover, each sub-automaton A_i gets s_i' as its new initial state and f_i' as its new unique accepting state. We set $s_i' \cdot \vdash = s_i$ and $f_i \cdot \dashv = f_i'$. All other transitions involving s_i', f_i' or $\vdash, \dashv$ end into the dead state of A_i. It is obvious that A_i still satisfies the conditions that are valid for the construction by Cho and Huynh: it has a unique accepting state and a unique (non-accepting) dead state; it is a minimum DFA whose syntactic monoid, $M(A_i)$, is aperiodic so that, within a single sub-automaton A_i, a word can define a trivial cycle only. In an intermediate step, we perform a duplication and obtain $2p$ sub-automata, say $A_1', A_2', \ldots, A_{2p-1}', A_{2p}'$ such that A_{2i-1}' and A_{2i}' are state-disjoint duplicates of A_i.

In stage 2, we build a DFA A' by concatenating the sub-automata $A_1', A_2', \ldots, A_{2p-1}', A_{2p}'$ in a cyclic fashion in analogy to the original construction (using symbol #) but now with s_i', f_i' in the role of s_i, f_i. Again in analogy, we amalgamate the $2p$ dead states to a single dead state denoted REJ, and we declare s_1' as the initial state and f_{2p}' as the unique accepting state of A'. The most significant change to the original construction is the introduction of a new symbol swap that, as indicated by its naming, causes transitions from s_{2i-1}' to s_{2i}' and vice versa, and transforms any other state into the unique dead state. The following result is obvious:

Lemma 3. *A' is a minimum DFA.*

The following two results establish the hardness result from Theorem 1,(ii).

Lemma 4. *If $\mathcal{M}$ accepts x, then $M(A')$ contains a non-Abelian subgroup.*

Proof. Let α denote the string that describes the accepting computation of $\mathcal{M}$ on x. Then, for every $i = 1, \ldots, 2p - 1$ and for every state $q \notin \{s_1', \ldots, s_{2p}'\}$,

$$s_i' \cdot \vdash \alpha \dashv \# = s_{i+1}', \quad s_{2p}' \cdot \vdash \alpha \dashv \# = s_1', \quad q \cdot \vdash \alpha \dashv \# = \mathrm{REJ} .$$

Thus, string $\vdash \alpha \dashv \#$ represents the cyclic permutation $\langle s_1', s_2', \ldots, s_{2p-1}', s_{2p}' \rangle$ in $M(A')$. A similar argument shows that the letter swap represents the permutation $\langle s_1', s_2' \rangle \ldots \langle s_{2p-1}', s_{2p}' \rangle$ in $M(A')$. The strings $\vdash \alpha \dashv \#$ and swap generate a non-Abelian subgroup of $M(A')$. □

Lemma 5. *If $M(A')$ contains a non-Abelian subgroup, then $\mathcal{M}$ accepts x.*

Proof. According to Lemma 2, there exists a state q and a word u such that u defines a cycle C starting at q and $r := r(q, u) \geq 3$. Clearly, q must be different from the dead state. Let $S := \{s'_1, \ldots, s'_{2p}\}$. Let $C(q, u)$ be the set of states occurring in the computation that starts (and ends) at q and processes u^r letter by letter. $C(q, u) \cap S$ cannot be empty because, otherwise, the cycle C would be contained in a single sub-automaton A'_i which, however, is impossible because A'_i is aperiodic. By reasons of symmetry, we may assume that $s'_1 \in C(q, u)$. After applying an appropriate cyclic permutation to the letters of u, we may also assume that u defines a cycle C starting (and ending) at s'_1 and $r = r(s'_1, u) \geq 3$ (the same r as before). Since $C(q, u)$ does not contain the dead state, u must decompose into segments of two types:

Type 1: segments of the form $\vdash \alpha \dashv \#$ with no symbol from $\{\underline{\text{swap}}, \vdash, \dashv, \#\}$ between the endmarkers

Type 2: segments consisting of the single letter $\underline{\text{swap}}$

Since $r \geq 3$, there must be at least one segment of type 1. Applying again the argument with the cyclic permutation, we may assume that the first segment in u, denoted $\bar{u}_1$ in what follows, is of type 1. Every segment of type 1 transforms s'_i into s'_{i+1}.[2] Every segment of type 2 transforms s'_{2i-1} into s'_{2i} and vice versa. Now, consider the computation path, say P, that starts at s'_1 and processes u letter by letter. Let k be the number of segments of type 1 in u, let k' be the number of occurrences of $\underline{\text{swap}}$ in u that hit a state $s'_i \in P$ for an odd index i, and finally let k'' be the number of occurrences of $\underline{\text{swap}}$ in u that hit a state $s'_i \in P$ for an even index i. Thus, $s'_{2i-1} \cdot u = s'_{2i-1+k+k'-k''}$ and $s'_{2i} \cdot u = s'_{2i+k-k'+k''}$. Let $d := k + k' - k''$.

Case 1: d is even.

Note that $d \not\equiv 0 \pmod{2p}$ (because, otherwise, $s'_1 \cdot u = s'_1$ - a contradiction to $r \geq 3$). Since the sequence $s'_1, s'_1 \cdot u, s'_1 \cdot u^2, \ldots$ exclusively runs through states of odd index from S and there are p (prime number) many of them, the sequence runs through all states of odd index from S. It follows that at some point every sub-automaton A'_{2i-1} will process the first segment $\bar{u}_1 = \vdash \alpha \dashv \#$ of u (which is of type 1) and so it will reach state f'_{2i-1}. We conclude that $L(A'_1) \cap L(A'_3) \cap \ldots \cap L(A'_{2p-1})$ is not empty (as witnessed by $\bar{u}_1$). Thus, α represents an accepting computation of $\mathcal{M}$ on input x.

Case 2: d is odd.

Note that, for every $i = 1, \ldots, 2p$, $s'_i \cdot u^2 = s'_{i+2k}$. Thus, $2k \not\equiv 0 \pmod{2p}$ (because, otherwise, $s'_1 \cdot u^2 = s'_1$ - a contradiction to $r \geq 3$). Now, the sequence $s'_1, s'_1 \cdot u^2, s'_1 \cdot u^4, \ldots$ exclusively runs through states of odd index from S, and we may proceed as in Case 1.   □

---

[2] Throughout this proof, we identify an index of the form $2pm + i, 1 \leq i \leq 2p$, with the index i. For example, s'_{2p+1} is identified with s'_1.

5   Complexity of Communication Complexity

5.1   Space-Efficient Algorithms for Syntactic Monoids

Let $N = (Z, \Sigma, \delta, z_1, F)$ be an NFA with states $Z = \{z_1, \ldots, z_n\}$, alphabet Σ, initial state z_1, final states $F \subseteq Z$, transition function $\delta : Z \times \Sigma \to 2^Z$, and let $\delta^* : 2^Z \times \Sigma^* \to 2^Z$ be the extension of δ as defined in Section 2. Let $L = L(N)$ be the language recognized by N, and let $\mathcal{A}$ be the minimum DFA for L. It is well-known that the syntactic monoid $M := M(L)$ of L coincides with the transformation monoid of $\mathcal{A}$, and that $\mathcal{A}$ may have up to 2^n states. We aim at designing space-efficient algorithms that solve questions related to M. These algorithms will never store a complete description of $\mathcal{A}$ (not to speak of M). Instead, they will make use of the fact that *reachable sets* $R \subseteq Z$ represent states of $\mathcal{A}$ in the following sense:

- R is called *reachable (by w)* if there exists $w \in \Sigma^*$ such that $R = \delta^*(z_1, w)$.
- Two sets Q, R are called *equivalent*, denoted as $Q \equiv R$, if, for all $w \in \Sigma^*$, $\delta^*(Q, w) \cap F \neq \emptyset \Leftrightarrow \delta^*(R, w) \cap F \neq \emptyset$, which is an equivalence relation.
- For reachable $R \subseteq Z$, $[R]$ denotes the class of reachable sets Q such that $Q \equiv R$.

The following should be clear from the power-set construction combined with DFA-minimization (e.g. [4]): the states of $\mathcal{A}$ are in bijection with the equivalence classes $[R]$ induced by reachable sets. Moreover, the transition function $\delta_{\mathcal{A}}$ of $\mathcal{A}$ satisfies $\delta_{\mathcal{A}}([R], a) = [\delta^*(R, a)]$ for every $a \in \Sigma$ (and this is well-defined). The extension $\delta_{\mathcal{A}}^*$ is related to δ^* according to $\delta_{\mathcal{A}}^*([R], w) = [\delta^*(R, w)]$ for every $w \in \Sigma^*$.

We now move on and turn our attention to M. Since M coincides with the transformation monoid of $\mathcal{A}$, it precisely contains the mappings

$$T_w([R]) := \delta_{\mathcal{A}}^*([R], w) = [\delta^*(R, w)] \ , \ \text{reachable } R \subseteq Z \tag{10}$$

for $w \in \Sigma^*$. M is generated by $\{T_a | a \in \Sigma\}$. Because of (1) and (10), every transformation T_w is already determined by $A^w := (A_1^w, \ldots, A_n^w)$ where

$$A_i^w := \delta^*(z_i, w) \subseteq Z, i = 1, \ldots, n \ . \tag{11}$$

In particular, the following holds for $A := A^w$, $A_i := A_i^w$, and $T_A := T_w$:

$$T_A([R]) = \left[\bigcup_{i : z_i \in R} \delta^*(z_i, w) \right] = \left[\bigcup_{i : z_i \in R} A_i \right] \ , \ \text{reachable } R \subseteq Z \tag{12}$$

Thus, given a reachable R and $A = A^w$, one can time-efficiently calculate a representant of $T_A([R]) = T_w([R])$ without knowing w. In order to emphasize this, we prefer the notation T_A to T_w in what follows. We call A a *transformation-vector* for T_A.

The next lemma presents a list of problems some of which can be solved in polynomial time (p.t.), and all of which can be solved in polynomial space (p.s.):

Lemma 6. *The NFA N is part of the input of all problems in the following list.*

1. *Given $R \subseteq Z$, the reachability of R can be decided in p.s..*
2. *Given a reachable set $R \subseteq Z$ and a transformation-vector A (for an unknown T_w), a representant of $T_A([R]) = T_w([R])$ can be computed in p.t..*
3. *Given $a \in \Sigma$, a transformation-vector for T_a can be computed in p.t..*
4. *Given transformation-vectors A (for an unknown T_w), B (for an unknown $T_{w'}$), a transformation-vector for $T_B \circ T_A$, denoted as $B \circ A$, can be computed in p.t..*
5. *Given a transformation-vector A, its validity (i.e., does there exist $w \in \Sigma^*$ such that $A = A^w$) can be decided in p.s..*
6. *Given $Q, R \subseteq Z$, it can be decided in p.s. whether $Q \equiv R$.*
7. *Given valid transformation-vectors A, B, their equivalence (i.e., $T_A = T_B$) can be decided in p.s..*
8. *It can be decided in p.s. whether M is commutative.*
9. *Given a valid transformation-vector A, it can be decided in p.s. whether T_A is idempotent.*
10. *Given valid transformation-vectors A, B, it can be decided in p.s. whether $T_A \in M T_B M$ (similarly for $T_A \in M T_B$, or for $T_A \in T_B M$).*

Proof. By Savitch's Theorem, membership in PSPACE can be proved by means of non-deterministic procedure. We shall often make use of this option.

1. Initialize Q to $\{z_1\}$. While $Q \neq R$ do
   (a) Guess a letter $a \in \Sigma$.
   (b) Replace Q by $\delta^*(Q, a)$.
2. Apply formula (12).
3. Apply formula (11) for $i = 1, \ldots, n$ and $w = a$ (so that δ^* collapses to δ).
4. For $i = 1, \ldots, n$, apply the formula $C_i = \cup_{j : z_j \in A_i} B_j$. Then $T_C = T_{ww'}$.
5. Initialize B to $(\{z_1\}, \ldots, \{z_n\})$ which is a transformation-vector for T_ε. While $B \neq A$ do
   (a) Guess a letter $c \in \Sigma$. Compute the (canonical) transformation-vector C for T_c.
   (b) Replace B by the (canonical) transformation-vector for $T_C \circ T_B$.
6. It suffices to present a non-deterministic procedure that recognizes inequivalence: While $Q \cap F \neq \emptyset \Leftrightarrow R \cap F \neq \emptyset$ do
   (a) Guess a letter $a \in \Sigma$.
   (b) Replace Q by $\delta^*(Q, a)$ and R by $\delta^*(R, a)$, respectively.
7. It suffices to present a non-deterministic procedure that recognizes inequivalence:
   (a) Guess $Q \subseteq Z$ and verify that Q is reachable.
   (b) Compute a representant R of $T_A([Q])$.
   (c) Compute a representant S of $T_B([Q])$.
   (d) Verify that $R \not\equiv S$.
8. The syntactic monoid is commutative iff its generators commute. It suffices to present a non-deterministic procedure that recognizes non-commutativity:
   (a) Guess two letters $a, b \in \Sigma$.

(b) Compute transformation-vectors A for T_{ab} and B for T_{ba}.

(c) Given these transformation-vectors, verify their inequivalence.

9. Compute $A \circ A$ and decide whether A and $A \circ A$ are equivalent.

10. Guess two transformations-vectors C, D and verify their validity. Compute the transformation-vector $D \circ B \circ C$ and accept iff it is equivalent to A.

$\square$

Note that the 10th assertion of Lemma 6 is basically saying that Green's relations w.r.t. the syntactic monoid of $L(N)$ can be decided in polynomial space.

Corollary 1. *Given NFA N, the deterministic communication complexity $cc(L)$ of the language $L = L(N)$ can be determined in polynomial space. Moreover, the membership of the syntactic monoid $M(L)$ in $\overline{\mathbf{Ab}}$ can be decided in polynomial space.*

Proof. The following facts are known from [10]: $cc(L)$ is constant iff $M(L)$ is commutative. If $cc(L)$ is not constant, it is either logarithmic or linear. The linear case occurs iff there exist $a, b, c, d, e \in M(L)$ such that (i) $a\mathcal{H}b\mathcal{H}c, a^2 = a, bc \neq cb$ or (ii) $a\mathcal{J}b, a^2 = a, b^2 = b, (ab)^2 \neq ab \vee a \mathcal{J} ab$. Condition (i) is equivalent to the condition $M(L) \notin \overline{\mathbf{Ab}}$. The assertion of the corollary is now immediate from Lemma 6. $\square$

5.2   Hardness Result for Regular Expressions

Definition 1. *Let L be a formal language over an alphabet Σ. Let $w = a_1 \ldots a_m$ be an arbitrary Σ-word of length m. We say that L is invariant under permutation if*

$$w = a_1 \ldots a_m \in L \Longrightarrow \pi(w) := a_{\pi(1)} \ldots a_{\pi(m)} \in L$$

holds for every permutation π of $1, \ldots, m$.

The following result is folklore:

Lemma 7. *Let L be a formal language over an alphabet Σ. Then $M(L)$ is commutative iff L is invariant under permutation.*

We are now ready for the main result in this section:

Theorem 2. *For every $f(n) \in \{1, \log n, n\}$, the following problem is PSPACE-hard: given a regular expression β over an alphabet Σ, decide whether $L(\beta)$ has deterministic communication complexity $\Theta(f(n))$.*

Proof. We know from [10] (see Section 2) that a regular language has constant deterministic communication complexity iff its syntactic monoid is commutative. In [1], the authors show (by means of a generic reduction) that the problem of deciding whether $L(\beta) \neq \Sigma^*$ is PSPACE-hard, even if either $L(\beta) = \Sigma^*$ or $L(\beta) = \Sigma^* \setminus \{w\}$ for some word $w \in \Sigma^*$ that contains at least two distinct letters. Clearly, Σ^* is invariant under permutation whereas $\Sigma^* \setminus \{w\}$ is not. According to Lemma 7, the syntactic monoid of Σ^* is commutative whereas the syntactic monoid of $\Sigma^* \setminus \{w\}$ is not. It readily follows that deciding "$cc(L(\beta)) = O(1)$?" is PSPACE-hard. It is easy to show that the deterministic communication complexity of $\Sigma^* \setminus \{w\}$ is $\Theta(\log n)$. Thus, deciding "$cc(L(\beta)) = \Theta(\log n)$?" is PSPACE-hard, too. It is easy to modify the proof of [1] so as to obtain the PSPACE-hardness of the problem "$L(\beta) \neq \Sigma^*$?" even when either $L(\beta) = \Sigma^*$ or $L(\beta) = \Sigma^* \setminus w^*$ for some word w that contains at least two distinct letters. It is easy to show that the deterministic communication complexity of $\Sigma^* \setminus w^*$ is $\Theta(n)$. Thus, deciding "$cc(L(\beta)) = \Theta(n)$?" is PSPACE-hard, too.   □

As is well-known, a regular expression can be transformed into an equivalent NFA in polynomial time. Thus, the decision problems from Theorem 2 remain PSPACE-hard when the language L is given by an NFA.

References

1. Aho, A.V., Hopcroft, J.E., Ullman, J.D.: The Design and Analysis of Computer Algorithms. Addison-Wesley, Reading (1974)
2. Cho, S., Huynh, D.T.: Finite-Automaton Aperiodicity is PSPACE-Complete. Theor. Comput. Sci. **88**(1), 99–116 (1991)
3. Colcombet, T.: Green's relations and their use in automata theory. In: Dediu, A.-H., Inenaga, S., Martín-Vide, C. (eds.) LATA 2011. LNCS, vol. 6638, pp. 1–21. Springer, Heidelberg (2011)
4. Hopcroft, J.E., Ullman, J.D.: Introduction to Automata Theory, Languages and Computation. Addison-Wesley, Reading (1979)
5. Kozen, D.: Lower bounds for natural proof systems. In: 18th Annual Symposium on Foundations of Computer Science, pp. 254–266. IEEE Computer Society, Washington (1977)
6. Kushilevitz, E., Nisan, N.: Communication Complexity. Cambridge University Press, Cambridge (1997)
7. Pin, J.E.: Varieties of Formal Languages. Plenum Publishing, New York (1986)
8. Schützenberger, M.P.: On Finite Monoids Having Only Trivial Subgroups. Information and Control **8**(2), 190–194 (1965)
9. Stern, J.: Complexity of Some Problems from the Theory of Automata. Information and Control **66**(3), 163–176 (1985)
10. Tesson, P., Thérien, D.: Complete Classifications for the Communication Complexity of Regular Languages. Theory Comput. Syst. **38**(2), 135–159 (2005)
11. Yao, A.C.: Some complexity questions related to distributive computing. In: 11th Annual Symposium on Theory of Computing, pp. 209–213. ACM, New York (1979)

Palindromic Complexity of Trees

Srečko Brlek[1], Nadia Lafrenière[1], and Xavier Provençal[2(✉)]

[1] Université du Québec à Montréal, Montréal, QC, Canada
brlek.srecko@uqam.ca, lafreniere.nadia.2@courrier.uqam.ca
[2] Université de Savoie, Chambéry, France
xavier.provencal@univ-savoie.fr

Abstract. We consider finite trees with edges labeled by letters on a finite alphabet Σ. Each pair of nodes defines a unique labeled path whose trace is a word of the free monoid Σ^*. The set of all such words defines the language of the tree. In this paper, we investigate the palindromic complexity of trees and provide hints for an upper bound on the number of distinct palindromes in the language of a tree.

Keywords: Words · Trees · Language · Palindromic complexity · Sidon sets

1 Introduction

The palindromic language of a word has been extensively investigated recently, see for instance [1] and more recently [2,5]. In particular, Droubay, Justin and Pirillo [10] established the following property:

Theorem 1 (Proposition 2 [10]). *A word w contains at most $|w| + 1$ distinct palindromes.*

Several families of words have been studied for their total palindromic complexity, among which periodic words [4], fixed points of morphism [15] and Sturmian words [10].

Considering words as geometrical objects, we can extend some definitions. For example, the notion of palindrome appears in the study of multidimensional geometric structures, thus introducing a new characterization. Some known classes of words are often redefined as digital planes [3,16], and the adjacency graph of structures obtained by symmetries appeared more recently [9]. In the latter article, authors show that the obtained graph is a tree and its palindromes have been described by Domenjoud, Provençal and Vuillon [8]. The trees studied by Domenjoud and Vuillon [9] are obtained by iterated palindromic closure, just as Sturmian [7] and episturmian [10,13] words. It has also been shown [8] that the total number of distinct nonempty palindromes in these trees is equal to the number of edges in the trees. This property highlights the fact that these trees form a multidimensional generalization of Sturmian words.

© Springer International Publishing Switzerland 2015
I. Potapov (Ed.): DLT 2015, LNCS 9168, pp. 155–166, 2015.
DOI: 10.1007/978-3-319-21500-6_12

A finite word is identified with a tree made of only one branch. Therefore, (undirected) trees appear as generalizations of words and it is natural to look forward to count the patterns occurring in it. Recent work by Crochemore et al. [6] showed that the maximum number of squares in a tree of size n is in $\Theta(n^{4/3})$. This is asymptotically bigger than in the case of words, for which the number of squares is known to be in $\Theta(n)$ [12]. We discuss here the number of palindromes and show that, as for squares, the number of palindromes in trees is asymptotically bigger than in words. Figure 1, taken from [8], shows an example of a tree having more nonempty palindromes than edges, so that Theorem 1 does not apply to trees.

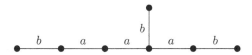

Fig. 1. A tree T with 6 edges and 7 nonempty palindromes, presented in [8]

Indeed, the number of nonempty factors in a tree is at most the ways of choosing a couple of edges (e_i, e_j), and these factors correspond to the unique shortest path from e_i to e_j. Therefore, the number of nonempty palindromes in a tree cannot exceed the square of its number of edges. In this article, we exhibit a family of trees with a number of palindromes substantially larger than the bound given by Theorem 1. We give a value, up to a constant, for the maximal number of palindromes in trees having a particular language, and we conjecture that this value holds for any tree.

2   Preliminaries

Let Σ be a finite alphabet, Σ^* be the set of finite words over Σ, $\varepsilon \in \Sigma^*$ be the empty word and $\Sigma^+ = \Sigma^* \setminus \{\varepsilon\}$ be the set of nonempty words over Σ. We define the *language* of a word w by $\mathcal{L}(w) = \{f \in \Sigma^* \mid w = pfs, \ p, s \in \Sigma^*\}$ and its elements are the *factors* of w. The *reverse* of w is defined by $\tilde{w} = w_{|w|}w_{|w|-1}\dots w_2w_1$, where w_i is the i-th letter of w and $|w|$, the length of the word. The number of occurrences of a given letter a in the word w is denoted $|w|_a$. A word w is a *palindrome* if $w = \tilde{w}$. The restriction of $\mathcal{L}(w)$ to its palindromes is denoted $\mathrm{Pal}(w) = \{u \in \mathcal{L}(w) \mid u = \tilde{u}\}$.

Some notions are issued from graph theory. We consider a *tree* to be an undirected, acyclic and connected graph. It is well known that the number of nodes in a tree is exactly one more than the number of edges. The *degree* of a node is given by the number of edges connected to it. A *leaf* is a node of degree 1. We consider a tree T whose edges are labeled by letters in Σ. Since in a tree there exists a unique simple path between any pair of nodes, the function $p(x, y)$ that returns the list of edges along the path from the node x to the node y is well defined, and so is the sequence $\pi(x, y)$ of its labels. The word $\pi(x, y)$ is called a *factor* of T and the set of all its factors, noted $\mathcal{L}(T) = \{\pi(x, y) \mid x, y \in \text{Nodes}(T)\}$, is called the *language* of T. As for words, we define the palindromic language of a tree T by $\text{Pal}(T) = \{w \in \mathcal{L}(T) \mid w = \widetilde{w}\}$. Even though the *size* of a tree T is usually defined by its nodes, we define it here to be the number of its edges and denote it by $|T|$. This emphasizes the analogy with words, where the length is defined by the number of letters. Observe that, since a nonempty path is determined by its first and last edges, the size of the language of T is bounded by:

$$\mathcal{L}(T) \le |T|^2 + 1. \tag{1}$$

Using the definitions above, we can associate a threadlike tree W to a pair of words $\{w, \widetilde{w}\}$. We may assume that x and y are its extremal nodes (the leaves). Then, $w = \pi(x, y)$ and $\widetilde{w} = \pi(y, x)$. The size of W is equal to $|w| = |\widetilde{w}|$. Analogously, $\text{Pal}(W) = \text{Pal}(w) = \text{Pal}(\widetilde{w})$. The language of W corresponds to the union of the languages of w and of $\widetilde{w}$. For example, Figure 2 shows the word $ababb$ as a threadlike tree. Any factor of the tree is either a factor of $\pi(x, y)$, if the edges are read from left to right, or a factor of $\pi(y, x)$, otherwise.

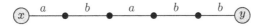

Fig. 2. A threadlike tree represents a pair formed by a word and its reverse

For a given word w, we denote by $\Delta(w)$ its run-length-encoding, that is the sequence of constant block lengths. For example, for the French word "appelle", $\Delta(\text{appelle}) = 12121$. As well, for the sequence of integers $w = 11112111211211$, $\Delta(w) = 4131212$. Indeed, each letter of $\Delta(w)$ represents the length of a block, while the length of $\Delta(w)$ can be associated with the number of blocks in w.

Given a fixed alphabet Σ, we define an infinite sequence of families of trees

$$\mathcal{T}_k = \{\text{tree } T \mid |\Delta(f)| \le k \text{ for all } f \in \mathcal{L}(T)\}.$$

For any positive integer k, we count the maximum number of palindromes of any tree of $\mathcal{T}_k$ according to its size. To do so, we define the function

$$\mathcal{P}_k(n) = \max_{T \in \mathcal{T}_k, |T| \le n} |\text{Pal}(T)|.$$

This value is at least equal to $n+1$. It is known [10] that each prefix p of a Sturmian word contains $|p|$ nonempty palindromes. This implies that $\mathcal{P}_\infty(n) \in \Omega(n)$. On the other hand, equation (1) provides a trivial upper bound on the growth rate of $\mathcal{P}_k(n)$ since it implies $\mathcal{P}_\infty(n) \in \mathcal{O}(n^2)$. We point out that $\mathcal{P}_k(n)$ is an increasing function with respect to k. In the following sections we provide the asymptotic growth, in Θ-notation, of $\mathcal{P}_k(n)$, for $k \le 4$. Although we have not been able to prove the asymptotic growth for $k \ge 5$, we explain why we conjecture that $\mathcal{P}_\infty(n) \in \Theta(\mathcal{P}_4(n))$ in section 5.

3   Trees of the Family $\mathcal{T}_2$

First recall that, by definition, every nonempty factor of a tree T in $\mathcal{T}_2$ has either one or two blocks of distinct letters. In other terms, up to a renaming of the letters, every factor in T is of the form a^*b^*. Therefore, any palindrome in T is on a single letter. From this, we can deduce a value for $\mathcal{P}_2(n)$:

Proposition 2. *The maximal number of palindromes for the family $\mathcal{T}_2$ is* $\mathcal{P}_2(n) = n + 1$.

Proof. The number of nonempty palindromes on a letter a is the length of the longest factor containing only a's. Thus, the total number of palindromes is at most the number of edges in T, plus one (for the empty word). This leads directly to $\mathcal{P}_2(n) \le n+1$. On the other hand, a word of length n on a single-letter alphabet contains $n + 1$ palindromes. This word is associated to threadlike tree in $\mathcal{T}_1$. Therefore, $\mathcal{P}_2(n) = n + 1$.   □

4   Trees of the Families $\mathcal{T}_3$ and $\mathcal{T}_4$

In this section, we show that $\{\mathcal{P}_3(n), \mathcal{P}_4(n)\} \subseteq \Theta(n^{\frac{3}{2}})$. To do so, we proceed in two steps. First, we present a construction that allows to build arbitrary large trees in $\mathcal{T}_3$ such that the number of palindromes in their languages is large enough to show that $\mathcal{P}_3(n) \in \Omega(n^{\frac{3}{2}})$. Then, we show that, up to a constant, this construction is optimal for all trees of $\mathcal{T}_3$ and $\mathcal{T}_4$.

4.1   A Lower Bound for $\mathcal{P}_3(n)$.

Some Elements from Additive Combinatorics. An integer sequence is a *Sidon set* if the sums (equivalently, the differences) of all distinct pairs of its elements are distinct. There exists infinitely many of these sequences. For example, the powers of 2 are an infinite Sidon set. The maximal size of a Sidon set $A \subseteq \{1, 2, \ldots, n\}$ is only known up to a constant [14]. This bound is easily obtained since A being Sidon set, there are exactly $\frac{|A|(|A|+1)}{2}$ sums of pairs of elements of A and all their sums are less or equal to $2n$. Thus,

$$\frac{|A|(|A| + 1)}{2} \le 2n$$

and $|A| \le 2\sqrt{n}$. Erdős and Turán [11] showed that for any prime number p, the sequence

$$A_p = (2pk + (k^2 \mod p))_{k=1,2,\ldots,p-1}, \tag{2}$$

is a Sidon set. The reader should notice that, since there exists arbitrarily large prime numbers, there is no maximal size for sequences constructed in this way.

Moreover, the sequence A_p is, up to a constant, the densest possible. Indeed, the maximum value of any element of A_p is less than $2p^2$ and $|A_p| = p - 1$. Since a Sidon set in $\{1, 2, \ldots, n\}$ is of size at most $2\sqrt{n}$, the density of A_p is $\sqrt{8}$ (around 2.83) times smaller, for any large p.

The Hair Comb Construction. Our goal is to describe a tree having a palindromic language of size substantially larger than the size of the tree. In this section, we build a tree $\mathcal{C}_p \in \mathcal{T}_3$ for any prime p containing a number of palindromes in $\Theta(|\mathcal{C}_p|^{\frac{3}{2}})$.

For each prime number p, let $B = (b_1, \ldots, b_{p-2})$ be the sequence defined by $b_i = a_{i+1} - a_i$, where the values a_i are taken in the sequence A_p presented above, equation (2), and let $\mathcal{C}_p$ be the tree constructed as follows :

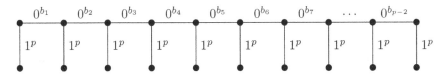

Proposition 3. *The sums of the terms in each contiguous subsequence of B are pairwise distinct.*

Proof. By contradiction, assume that there exists indexes k, l, m, n such that $\sum_{i=k}^{l} b_i = \sum_{j=m}^{n} b_j$. By definition of B,

$$\sum_{i=k}^{l} b_i = \sum_{i=k}^{l} (a_i - a_{i-1}) = a_l - a_{k-1} \text{ and } \sum_{j=m}^{n} b_j = a_n - a_{m-1}.$$

This implies that $a_l + a_{m-1} = a_n + a_{k-1}$, which is impossible. □

Lemma 4. *The number of palindromes in $\mathcal{C}_p$ is in $\Theta(p^3)$.*

Proof. The nonempty palindromes of $\mathcal{C}_p$ are of three different forms. Let c_0 be the number of palindromes of the form 0^+, c_1 be the number of palindromes of the form 1^+ and c_{101} be the number of palindromes of the form $1^+0^+1^+$. The number of palindromes of $\mathcal{C}_p$ is clearly $|\text{Pal}(\mathcal{C}_p)| = c_0 + c_1 + c_{101} + 1$, where one is added for the empty word.

$$c_0 = b_1 + b_2 + \cdots + b_{p-2} = a_{p-1} - a_1 = 2p^2 - 4p,$$
$$c_1 = p,$$
$$c_{101} = |\{1^x 0^y 1^x \in \text{Pal}(\mathcal{C}_p)\}|$$
$$= |\{x \mid 1 \le x \le p\}| \cdot |\{y \mid y = \sum_{i=k}^{l} b_i \text{ for } 1 \le k \le l \le p - 2\}|$$
$$= \tfrac{1}{2} p(p-1)(p-2).$$

The last equality comes from the fact that there are $(p-1)(p-2)/2$ possible choices of pairs (k, l) and proposition 3 guarantees that each choice sums up to a different value. The asymptotic behavior of the number of palindromes is determined by the leading term p^3. □

Lemma 5. *The number of edges in C_p is in $\Theta(p^2)$.*

Proof. The number of edges labeled by 0 is $b_1 + b_2 + \ldots + b_{p-2} = 2p^2 - 4p$. For those labeled with 1, there are exactly $p-1$ sequences of edges labeled with 1's and they all have length p. The total number of edges is thus $2p^2 - 4p + p(p-1) = 3p^2 - 5p$. □

Theorem 6. $\mathcal{P}_3(n) \in \Omega(n^{\frac{3}{2}})$.

Proof. Lemmas 4 and 5 implies that the number of palindromes in C_p is in $\Theta(|C_p|^{\frac{3}{2}})$. Since there are infinitely many trees of the form C_p and since their size is not bounded, these trees provide a lower bound on the growth rate of $\mathcal{P}_3(n)$. □

4.2 The Value of $\mathcal{P}_4(n)$ is in $\Theta(n^{\frac{3}{2}})$.

In this subsection, we show that the asymptotic value of $\mathcal{P}_3(n)$ is reached by the hair comb construction, given above, and that it is the same value for $\mathcal{P}_4(n)$.

Theorem 7. $\mathcal{P}_4(n) \in \Theta(n^{\frac{3}{2}})$.

Before giving a proof of this theorem, we need to explain some arguments. We first justify why we reduce any tree of $\mathcal{T}_4$ to a tree in $\mathcal{T}_3$. Then, we present some properties of the latter trees in order to establish an upper bound on $\mathcal{P}_4(n)$.

Lemma 8. *For any $T \in \mathcal{T}_4$, there exists a tree $S \in \mathcal{T}_3$ on a binary alphabet satisfying $|S| \leq |T|$, and with $\frac{1}{|\Sigma|^2}|\operatorname{Pal}(T)| - |T| \leq |\operatorname{Pal}(S)| \leq |\operatorname{Pal}(T)|$.*

Proof. If there is in T no factor with three blocks starting and ending with the same letter, this means that all the palindromes are repetitions of a single letter. We then denote by a the letter on which the longest palindrome is constructed. It might not be unique, but it does not matter. Let S be the longest path labeled only with a's. Then, $|\operatorname{Pal}(T)| \leq |\Sigma||\operatorname{Pal}(S)| \leq |\Sigma||\operatorname{Pal}(T)|$.

Otherwise, let a and b be letters of Σ and let (a, b) be a pair of letters for which $|\mathcal{L}(T) \cap \operatorname{Pal}(a^+b^+a^+)|$ is maximal. We define the set

$$E_S = \cup\left(p(u, v) \mid \pi(u, v) \in \operatorname{Pal}(a^+b^+a^+) \right)$$

and let S be the subgraph of T containing exactly the edges of E_S and the nodes connected to these edges. Then, there are three things to prove :

– S is a tree: Since S is a subgraph of T, it cannot contain any cycle. We however need to prove that S is connected. To do so, assume that S has two connected components named C_1 and C_2. Of course, $\mathcal{L}(C_1) \subseteq a^*b^*a^*$ and C_1 has at least one factor in $a^+b^+a^+$. The same holds for C_2. Since T is a tree, there is a unique path in $T\backslash S$ connecting C_1 and C_2. We call it q. There are paths in C_1 and in C_2 starting from an extremity of q and containing factors in b^+a^+. Thus, by stating that w is the trace of q, T has a factor $f \in a^+b^+a^*wa^*b^+a^+$. By hypothesis, $T \in \mathcal{T}_4$ so any factor of T contains at most four blocks. Then, f has to be in $a^+b^+wb^+a^+$, with $w \in b^*$ and so q is a path in S. A contradiction.

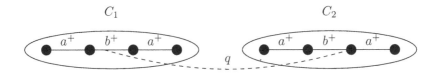

– $S \in \mathcal{T}_3$ is on a binary alphabet: By construction, S contains only edges labeled by a or b and has no leaf connected to an edge labeled by b. This implies that if S contains a factor $f \in a^+b^+a^+b^+$, f may be extended to $f' \in a^+b^+a^+b^+a^+$, which does not appear in T.
– $|\mathrm{Pal}(S)| \geq \frac{1}{|\Sigma|^2}|\mathrm{Pal}(T)| - |T|$: We chose (a,b) to be the pair of letters for which the number of palindromes on an alphabet of size at least 2 was maximal. The number of palindromes on a single letter is at most $|T|$. Thus,

$$\frac{1}{|\Sigma|^2}|\mathrm{Pal}(T)| - |T| \leq |\mathrm{Pal}(S)| \leq |\mathrm{Pal}(T)|.$$

$\square$

Lemma 9. *For any $T \in \mathcal{T}_3$, T cannot contain both factors of $0^+1^+0^+$ and of $1^+0^+1^+$.*

Proof. We proceed by contradiction. Assume that there exists in T four nodes u, v, x, y such that $\pi(u,v) \in 0^+1^+0^+$ and $\pi(x,y) \in 1^+0^+1^+$. Since T is a tree, there exists a unique path between two nodes. In particular, there is a path from $w \in \{u,v\}$ to $w' \in \{x,y\}$ containing a factor of the form $0^+1^+0^+\Sigma^*1^+$, which contradicts the hypothesis that $T \in \mathcal{T}_3$. $\square$

We now define the restriction $\mathcal{R}_a(T)$ of a tree T to the letter a by keeping from T only the edges labeled by a and the nodes connected to them.

Lemma 10. *Let T be in $\mathcal{T}_3$. There exists at least one letter $a \in \Sigma$ such that $\mathcal{R}_a(T)$ is connected.*

Proof. If T does not contain a factor on at least two letters that starts and ends with the same letter, that is of the form $b^+a^+b^+$, then $\mathcal{R}_a(T)$ is connected for any letter a.

Otherwise, assume that a factor $f \in b^+a^+b^+$ appears in T. Then, $\mathcal{R}_a(T)$ must be connected. By contradiction, suppose there exists an edge labeled with a that is connected to the sequence of a's in f, by a word w that contains another letter than a. Then, there exists a word of the form awa^+b^+ in $\mathcal{L}(T)$ and this contradicts the hypothesis that $T \in \mathcal{T}_3$. □

Given a node u in a tree, we say that u is a *splitting on the letter* a if $\deg(u) \geq 3$ and there is at least two edges labeled with a connected to u.

Lemma 11. *Let T be in $\mathcal{T}_3$. Then, there is a tree T' of size $|T|$ such that $\mathcal{L}(T) \subseteq \mathcal{L}(T')$ and there exists a letter $a \in \Sigma$ such that any splitting of T' is on the letter a.*

Proof. If T is in $\mathcal{T}_2$, we apply the upcoming transformation to every branches. Otherwise, assume that a factor of the form $b^+a^+c^+$ appears in T (note that b might be equal to c). We allow splittings only on the letter a. Let v be a node of T that is a splitting on $b \in \Sigma\backslash\{a\}$ (if it does not exist, then $T' = T$). By the hypothesis on T, this means that there exists, starting from v, at least two paths labeled only with b's leading to leaves x and y.

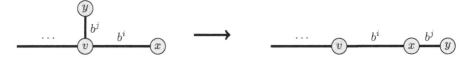

Fig. 3. The destruction of a splitting on the letter b

We assume that $|\pi(v, x)| \geq |\pi(v, y)|$. Then, the words having $\pi(v, y)$ as suffix are a subset of those for which $\pi(v, x)$ is suffix. Therefore, the only case where $\pi(v, y)$ may contribute to the language of T is when both the edges of $\pi(v, x)$ and $\pi(v, y)$ are used. The words of this form are composed only of b's and are of length at most $|\pi(v, x)| + |\pi(v, y)|$. Moving the edges between s and y to the other extremity of x, we construct a tree for which the language contains $\mathcal{L}(T)$ and having the same number of nodes. Finally, we can apply this procedure until the only remaining splittings are on the letter a. This leads to T'. □

We are now ready to prove the main theorem.

Proof. [Theorem 7: $\mathcal{P}_4(n) \in \Theta(n^{\frac{3}{2}})$.] Let T be in $\mathcal{T}_4$. By assumption, each factor of T contains at most four blocks of distinct letters.

1.   Let $S \in \mathcal{T}_3$ be such that $|S| \leq |T|$, $\mathcal{L}(S) \subseteq \{0,1\}^*$ and $\frac{|\mathrm{Pal}(T)|-|T|}{|\Sigma|^2} \leq |\mathrm{Pal}(S)| \leq |\mathrm{Pal}(T)|$. Using lemma 8, we know that this exists. We know by lemma 9 that S may contain factors in $1^+0^+1^+$, but not in $0^+1^+0^+$.

2.   By lemma 11, there exists a tree S' with $|S'| = |S|$, such that $\mathcal{L}(S) \subseteq \mathcal{L}(S')$, and with no splitting on the letter 1.

3.   Finally, we count the palindromes in S'. The form of these palindromes is either 0^+, 1^+ or $1^+0^+1^+$. For the palindromes on a one-letter alphabet, their number is bounded by n, where n is the size of S'. We now focus on the number of palindromes of the form $1^+0^+1^+$. Call c_{101} this number. We show that $c_{101} \leq 2n\sqrt{n}$.

Since S' does not admit any splitting on the letter 1, each connected component of $R_1(S')$ is a threadlike branch going from a leaf of S' to a node of $R_0(S')$. We name these connected components $b_1, \ldots, b_m$ and by lemma 10, we know that $R_0(S')$ is connected.

Let b_i and b_j be two distinct branches of S'. By abuse of notation, we note $\pi(b_i, b_j)$ the word defined by the unique path from b_i to b_j. Let l be such that $\pi(b_i, b_j) = 0^l$ and suppose that $|b_i| \leq |b_j|$. Then, for any node u in b_i, there exists a unique node v in b_j, such that the word $\pi(u, v) = 1^k 0^l 1^k$ is a palindrome. Moreover, if $|b_i| < |b_j|$, then there are nodes in b_j that cannot be paired to a node of b_i in order to form a palindrome. From this observation, a first upper bound is:

$$c_{101} \leq \sum_{1 \leq i < j \leq m} \min(|b_i|, |b_j|). \tag{3}$$

Another way to bound c_{101} is to count the palindromes of the form $1^+0^+1^+$ according to the length of the block of 0's. For each length l from 1 to n, there might be more than one pair $\{b_i, b_j\}$ that produces palindromes with central factor 0^l. This provides a second upper bound:

$$c_{101} \leq \sum_{l=1}^{n} \max_{\substack{1 \leq i < j \leq m \\ \pi(b_i, b_j) = 0^l}} (\min(|b_i|, |b_j|)) \tag{4}$$

In order to obtain the desired bound on c_{101} we combine these two bounds. Let $B' = \{i \mid |b_i| \geq \sqrt{n}\}$. Since n is the size of S', we have that $|B'| \leq \sqrt{n}$ and that the average size of the branches b_i is such that $i \in B'$ is bounded by $n/|B'|$. By applying the bound from (3) to the palindromes formed by two branches in B', we obtain that the number of such palindromes is:

$$\sum_{\substack{1 \leq i < j \leq m \\ \{i,j\} \subseteq B'}} \min(|b_i|, |b_j|) \leq \frac{|B'|(|B'|-1)}{2} \frac{n}{|B'|} \leq n\sqrt{n}. \tag{5}$$

Finally, it remains to count the number of palindromes that are defined by pairs of branches $\{b_i, b_j\}$ such that i or j is not in B'. In such case, we always find that $\min(|b_i|, |b_j|) < \sqrt{n}$. The number of such palindromes is:

$$\sum_{l=1}^{n} \max_{\substack{1 \le i < j \le m \\ \pi(b_i, b_j) = 0^l \\ \{i,j\} \not\subset B'}} (\min(|b_i|, |b_j|)) < n\sqrt{n}. \tag{6}$$

Since each palindrome in S' is counted by equation (5) or (6), we obtain, summing both, $c_{101} < 2n\sqrt{n} = 2|S'|^{\frac{3}{2}}$. We deduce that, for any tree T in $\mathcal{T}_4$, the number of palindromes is bounded by

$$|\operatorname{Pal}(T)| \le |\Sigma|^2 |\operatorname{Pal}(S)| + |T| < 2|\Sigma|^2 |S'|^{\frac{3}{2}} + |T| \le 2|\Sigma|^2 |T|^{\frac{3}{2}} + |T|.$$

Using the fact that the alphabet is fixed (so its size is given by a constant), it is enough to prove that $\mathcal{P}_4(n) \in \mathcal{O}(n^{\frac{3}{2}})$. Combining this result with the one given in section 4.1, one may assert that both $\mathcal{P}_3(n)$ and $\mathcal{P}_4(n)$ are in $\Theta(n^{\frac{3}{2}})$. $\square$

5   Hypotheses for the Construction of Trees with a Lot of Distinct Palindromes

Let T be a tree that maximizes the number of palindromes for its size. It is likely that T contains triples of nodes (u, v, w) such that $\pi(u, v)$, $\pi(u, w)$ and $\pi(v, w)$ are all palindromes. Suppose it is the case, and define T' as the restriction of T to the paths that join u, v and w. We have that either T' is a threadlike tree, or T' has three leaves and a unique node of degree 3. The first case is of no interest here since it is equivalent to words, while the latter case implies a restrictive structure on the factors $\pi(u, v)$, $\pi(u, w)$ and $\pi(v, w)$. We now focus on the second case and call x the unique node of T' with degree 3.

Let $U = \pi(u, x)$, $V = \pi(v, x)$, $W = \pi(w, x)$ and, without loss of generality, suppose that $|U| \le |V| \le |W|$. Then, as shown in Figure 4, $U\widetilde{V}$, $U\widetilde{W}$ and $V\widetilde{W}$ are all palindromes.

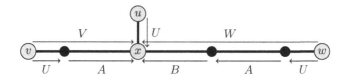

Fig. 4. The structure of the tree T'. The palindromicity of $U\widetilde{V}$, $U\widetilde{W}$ and $V\widetilde{W}$ forces that V starts with U while W starts with both factors U and V.

Let A be the suffix of length $|V| - |U|$ of V. Since, by hypothesis, $U\widetilde{V}$ is a palindrome, $V = UA$ and A is a palindrome. Similarly, let B be the suffix of length $|W| - |V|$ of W. This implies that $W = VB = UAB$ and both B and AB are palindromes. Using a well-known lemma from Lothaire [17], we prove that AB is periodic.

Lemma 12 (Proposition 1.3.2 in [17]). *Two words commute if and only if they are powers of the same word.*

The next proposition states that the word ABA is periodic and that its period is at most the gcd of the difference of length of the three paths between u, v and w. More formally, let

$$p = \gcd\left(|\pi(u, w)| - |\pi(u, v)|, |\pi(v, w)| - |\pi(u, v)|, |\pi(v, w)| - |\pi(u, w)|\right).$$

Proposition 13. *There exists a word S and two integers i, j such that $|S|$ divides p and $A = S^i$ and $B = S^j$.*

Proof. Since A, B and AB are palindromes, $AB = \widetilde{AB} = \widetilde{B}\widetilde{A} = BA$. Thus, by lemma 12, there exists a word S such that $A = S^i$ and $B = S^j$. This implies that $|S|$ divides $\gcd(|A|, |B|)$ and, by construction, $\gcd(|A|, |B|) = p$. □

From the above proposition, we deduce that a triple of nonaligned nodes with any path from a node to another being a palindrome forces a local structure isomorphic to that of the hair comb tree, as illustrated in Figure 5.

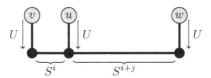

Fig. 5. A triple of nodes with palindromes between each pair of them is isomorphic to a part of a hair comb

In a more general way, suppose that a tree contains m leaves $(u_i)_{1 \leq i \leq m}$, and that each $\pi(u_i, u_j)$ is a palindrome. Let T' be the restriction of this tree to the paths that connect these leaves and, for each i, let v_i be the first node of degree higher than 2 accessible from the leaf u_i in T'. By applying the above proposition to each triplet (u_i, u_j, u_k), for all $i \neq j$, the word $\pi(u_i, u_j)$ is of the form

$$\pi(u_i, u_j) = US^+\widetilde{U},$$

where $|U| = \min_i(\pi(u_i, v_i))$ and $|S|$ divides $\gcd_{i \neq j, k \neq l}\left(\left|\,|\pi(u_i, u_j)| - |\pi(u_k, u_l)|\,\right|\right)$.

Moreover, in order to maximize the number of palindromes relatively to the size of the tree, we can choose S to be a single letter. This is indeed possible since the only condition on the length of S is that it divides all the differences of lengths between any palindromic path from a leaf to another.

This gives a tree analogous to those presented in section 4.1, $\mathcal{C}_p$, and for which we have established that $|\operatorname{Pal}(\mathcal{C}_p)| \in \Theta(|\mathcal{C}_p|^{\frac{3}{2}})$. Therefore, we conjecture that $\mathcal{P}_\infty(n) \in \Theta(n^{\frac{3}{2}})$.

References

1. Allouche, J.P., Baake, M., Cassaigne, J., Damanik, D.: Palindrome complexity. Theoretical Computer Science **292**(1), 9–31 (2003)
2. Balková, L., Pelantová, E., Starosta, S.: Proof of the Brlek-Reutenauer conjecture. Theoretical Computer Science **475**, 120–125 (2013)
3. Berthé, V., Vuillon, L.: Tilings and rotations on the torus: a two-dimensional generalization of Sturmian sequences. Discrete Mathematics **223**(1–3), 27–53 (2000)
4. Brlek, S., Hamel, S., Nivat, M., Reutenauer, C.: On the palindromic complexity of infinite words. International Journal on Foundation of Computer Science **15**(2), 293–306 (2004)
5. Brlek, S., Reutenauer, C.: Complexity and palindromic defect of infinite words. Theoretical Computer Science **412**(4–5), 493–497 (2011)
6. Crochemore, M., Iliopoulos, C.S., Kociumaka, T., Kubica, M., Radoszewski, J., Rytter, W., Tyczyński, W., Waleń, T.: The maximum number of squares in a tree. In: Kärkkäinen, J., Stoye, J. (eds.) CPM 2012. LNCS, vol. 7354, pp. 27–40. Springer, Heidelberg (2012)
7. de Luca, A.: Sturmian words: Structure, combinatorics, and their arithmetics. Theoretical Computer Science **183**(1), 45–82 (1997)
8. Domenjoud, E., Provençal, X., Vuillon, L.: Palindromic language of thin discrete planes (to appear)
9. Domenjoud, E., Vuillon, L.: Geometric palindromic closure. Uniform Distribution Theory **7**(2), 109–140 (2012)
10. Droubay, X., Justin, J., Pirillo, G.: Episturmian words and some constructions of de Luca and Rauzy. Theoretical Computer Science **255**(1–2), 539–553 (2001)
11. Erdös, P., Turán, P.: On a problem of Sidon in additive number theory, and on some related problems. Journal of the London Mathematical Society. Second Series **16**, 212–215 (1941)
12. Fraenkel, A.S., Simpson, J.: How many squares can a string contain? J. Combin. Theory Ser. A **82**(1), 112–120 (1998)
13. Glen, A., Justin, J.: Episturmian words: a survey. Theoretical Informatics and Applications. Informatique Théorique et Applications **43**(3), 403–442 (2009)
14. Gowers, T.: What are dense Sidon subsets of $\{1, 2, \ldots, n\}$ like? (2012). gowers.wordpress.com/2012/07/13/what-are-dense-sidon-subsets-of-12-n-like/
15. Hof, A., Knill, O., Simon, B.: Singular continuous spectrum for palindromic Schrödinger operators. Comm. in Mathematical Physics **174**(1), 149–159 (1995)
16. Labbé, S., Reutenauer, C.: A d-dimensional extension of Christoffel words. Discrete & Computational Geometry (2015). http://arxiv.org/abs/1404.4021
17. Lothaire, M.: Combinatorics on Words. Cambridge University Press, Cambridge (1997)

Deciding Proper Conjugacy of Classes of One-Sided Finite-Type-Dyck Shifts

Marie-Pierre Béal$^{(\boxtimes)}$ and Pavel Heller$^{(\boxtimes)}$

LIGM UMR 8049, Université Paris-Est,
77454 Marne-la-Vallée Cedex 2, France
{marie-pierre.beal,pavel.heller}@u-pem.fr

Abstract. One-sided sofic-Dyck shifts are sets of infinite sequences of symbols avoiding a visibly pushdown language of finite words. One-sided finite-type-Dyck shifts constitute a subclass of these sets of sequences. A (one-sided) finite-type-Dyck shift is defined as the set of infinite sequences avoiding both some finite set of words and some finite set of matching patterns. We prove that proper conjugacy is decidable for a large class of one-sided finite-type-Dyck shifts, the matched-return extensible shifts. This class contains many known non-sofic one-sided shifts like Dyck shifts and Motzkin shifts. It contains also strictly all extensible one-sided shifts of finite type. Our result is thus an extension of the decidability of conjugacy between one-sided shifts of finite type obtained by Williams.

1 Introduction

A shift of (infinite or bi-infinite) sequences may be defined as the set of sequences of symbols over a finite alphabet avoiding some set F of finite words. A sequence belongs to the shift if and only if no finite factor of the sequence belongs to F. Such a set F defining the shift is called a set of forbidden factors. Several sets of forbidden factors may define the same shift and their combinatorial nature or complexity induces some dynamic properties of the shift. For instance, when F can be chosen regular, the shift is called a sofic shift. When F can be chosen finite, it is called a shift of finite type. A shift of bi-infinite or left-infinite sequences is implicitly equipped with the operation of shifting one place to the left.

Sofic-Dyck shifts are generalizations of Dyck shifts introduced by Krieger in [11] and Markov-Dyck shifts studied by Inoue, Krieger and Matsumoto [12], [9]. They are exactly the sets of sequences avoiding a visibly pushdown language (or a regular language of nested-words) of finite words (see [6], [7]).

These sequences of symbols are defined over a tri-partitioned alphabet called a pushdown alphabet. Symbols may be either call, return or internal symbols.

M.-P. Béal and P. Heller—This work is supported by the French National Agency (ANR) through "Programme d'Investissements d'Avenir" (Project ACRONYME n°ANR-10-LABX-58), through the ANR EQINOCS, and by the region of Île-de-France through the DIM RDM-IdF.

© Springer International Publishing Switzerland 2015
I. Potapov (Ed.): DLT 2015, LNCS 9168, pp. 167–178, 2015.
DOI: 10.1007/978-3-319-21500-6_13

Call symbols correspond to open tags or parentheses and return symbols correspond to close tags or parentheses. Sofic-Dyck shifts are defined as sets of labels of infinite paths of a labeled directed graph where some symbols have to be matched with other symbols. Visibly pushdown languages over a tri-partitioned alphabet were introduced by Alur *et al.* [1],[2]. Note that the partitioning of the alphabet is fixed since languages may be visibly pushdown for some partitioning of their alphabet and no more visibly pushdown with another partitioning of the same alphabet. These languages are higher than regular languages in the Chomsky hierarchy but close to them. Indeed, they share many interesting properties with regular languages like stability by intersection and complementation. They are used as models for structured data files like XML files.

A (topological) conjugacy is a bijective continuous map between two shifts which commutes with the shift transformations. It is a bijective map which may be defined with a sliding window of bounded length, the inverse being also defined with a sliding window of bounded length. Conjugate shifts may be seen as "recoded" versions of a same shift sharing all its properties. The decidability of the conjugacy of shifts is thus an interesting problem which was solved only for one-sided shifts of finite type by Williams [14] (see also [10]). Williams' theorem was extended to tree-shifts of finite type in [3]. It is unknown whether conjugacy between two two-sided shifts of finite-type is decidable (see for instance [13]). It is also unknown whether conjugacy between two one-sided sofic shifts is decidable (see [8]).

A notion of conjugacy suitable for sofic-Dyck shifts is the notion of proper conjugacy. A proper conjugacy between a shift over a tri-partitioned alphabet A and a shift over a tri-partitioned alphabet B is a conjugacy which maps call (resp. return, internal) symbols of A to call (resp. return, internal) symbols of B. It is shown in [7] that the class of sofic-Dyck shifts is stable under proper conjugacy.

In this paper, we consider one-sided finite-type-Dyck shifts, a subclass of sofic-Dyck shifts introduced in [6] and [7]. A (one-sided) finite-type-Dyck shift is defined as the set of (left-infinite) sequences avoiding both some finite set of words and some finite set of matching patterns. This class contains strictly the (non-Dyck) shifts of finite type. We consider the subclass of matched-return extensible shifts for which any block may be extended by a matched-return block, *i.e.* a block where every return symbol is matched with a call symbol. This class contains for instance the Dyck shifts and Motzkin shifts. It contains also strictly the class of extensible (one-sided) shifts of finite type.

We prove that proper conjugacy between two one-sided finite-type-Dyck shifts is decidable when these shifts are matched-return extensible.

The proof of Williams for deciding conjugacy between one-sided shifts of finite type is based on a decomposition theorem which states that every conjugacy between the shifts can be decomposed into a finite sequence of splitting and merging maps. Splitting and merging are elementary operations performed on automata presenting the two shifts that induce bijections between left-infinite paths of the two underlying graphs. Since the shifts are one-sided, the splittings may be all input-state splittings and are therefore commuting. This commutation property guarantees the existence of a common in-amalgamation for two conjugate shifts. We extend this scheme to one-sided finite-type-Dyck shifts. The problem is first reduced to deciding proper conjugacy between one-sided edge-Dyck shifts. A decomposition theorem is obtained for these shifts using input state-splittings of the one-sided edge-Dyck shifts. State-splitting for two-sided finite-type-Dyck shifts was introduced in [5]. The matched-return extensible property can be seen as a property guaranteeing the unicity of a minimal amalgamated presentation of an edge-Dyck shift. In general, the unicity is not guaranteed, although it holds trivially for one-sided edge-shifts.

Our result is thus an extension of the decidability of conjugacy between one-sided shifts of finite type obtained by Williams. We prove the decidability of proper conjugacy for a class of shifts going beyond the regular sets of sequences, the (left) one-sided finite-type-Dyck shifts which are matched-return extensible.

The paper is organized as follows. Section 2 provides some background on finite-type-Dyck shifts and introduces the notion of edge-Dyck shifts. In Section 3 we define the notion of state-splitting of Dyck graphs. The decomposition of proper conjugacy is presented in Section 4 and the decidability of proper conjugacies is proved in Section 5. Due to space restrictions, some proofs are omitted in this short version of the paper.

2   One-Sided Finite-Type Dyck Shifts

2.1   Preliminaries

We consider an alphabet A which is a disjoint union of three finite sets of letters, the set A_c of *call letters*, the set A_r of *return letters*, and the set A_i of *internal letters*. The set $A = A_c \sqcup A_r \sqcup A_i$ is called a *pushdown alphabet* or a *tri-partitioned alphabet*.

We denote by $\mathrm{MR}(A)$ the set of all finite words over A where every return symbol is matched with a call symbol, *i.e.* $u \in \mathrm{MR}(A)$ if for every prefix u' of u, the number of call symbols of u' is greater than or equal to the number of return symbols of u'. These words are called *matched-return*. Similarly, $\mathrm{MC}(A)$ denotes the set of *matched-call* words where every call symbol is matched with a return symbol. We say that a word is a *Dyck word* if it belongs to the intersection of $\mathrm{MC}(A)$ and $\mathrm{MR}(A)$. Dyck words are well-parenthesized or well-formed words. Note that the empty word or all words over A_i are Dyck words. The set of Dyck words over A is denoted by $\mathrm{Dyck}(A)$.

2.2   Finite-Type Dyck Shifts

We consider in this paper *one-sided shifts* over A which are sets of left-infinite sequences in $A^{-\mathbb{N}}$ avoiding some set of finite words. The reason for considering left-infinite sequences instead of right-infinite ones is due to the fact that finite-type-Dyck shifts of left-infinite sequences have natural deterministic presentations.

A shift is implicitly equipped with the shift transformation, denoted by σ, which maps a left-infinite sequence $(x_i)_{i \in -\mathbb{N}}$ to $\sigma((x_i)_{i \in -\mathbb{N}}) = (x_{i-1})_{i \in -\mathbb{N}}$.

Let A be a tri-partitioned alphabet. If u and u' are two words over A, we note $u \preceq u'$ if u is a suffix of u'.

Let $F \subseteq A^*$ and $U \subseteq (A^* A_c \times A^* A_r)$. We say that a finite or infinite sequence x *avoids* F if, for each finite factor u of x, one has $u \notin F$. We say that a finite or infinite sequence x *avoids* U if for each finite factor $u = vawb$ of x with $a \in A_c, b \in A_r, w \in \mathrm{Dyck}(A)$, there is no pair $(u_1 a, v_1 b)$ in U such that $u_1 \preceq v$ and $v_1 \preceq vaw$. We denote by $\mathsf{X}_{F,U}$ the set of left-infinite sequences avoiding F and U.

A *(one-sided) finite-type-Dyck shift* over A is a subset X of $A^{-\mathbb{N}}$ for which there are two finite sets F, U such that $X = \mathsf{X}_{F,U}$.

Hence a subset X of $A^{-\mathbb{N}}$ is a finite-type Dyck shift if there is a nonnegative integer m and $F \subseteq A^{m+1}$, $U \subseteq A^m A_c \times A^m A_r$ with $X = \mathsf{X}_{F,U}$.

We also define a class of finite automata called Dyck automata accepting a class of shifts which is larger than the class of finite-type-Dyck shifts. A *(finite) Dyck automaton* $\mathcal{A}$ over A is a pair (G, M) of an automaton (or a directed labeled graph) $G = (Q, E, A)$ over A where Q is the finite set of states and $E \subseteq Q \times A \times Q$ is the set of edges, and a set M of pairs of edges $((p, a, q), (r, b, s))$ such that $a \in A_c$ and $b \in A_r$, referred to as *matched pairs of edges*.

A finite path (a finite sequence of consecutive edges) π of $\mathcal{A}$ is said to be an *admissible path* if for any factor $(p, a, q) \cdot \pi_1 \cdot (r, b, s)$ of π with $a \in A_c$, $b \in A_r$, and the label of π_1 being a Dyck word on A, $((p, a, q), (r, b, s))$ is a matched pair. Hence any path of length zero is admissible and factors of finite admissible paths are admissible. An infinite path is *admissible* if all its finite factors are admissible. The (left) one-sided shift *presented* by $\mathcal{A}$ is the set of labels of (left)-infinite admissible paths of $\mathcal{A}$ and $\mathcal{A}$ is called a *presentation* of the shift. Shifts presented by Dyck automata are called *sofic-Dyck shifts*. Finite-type-Dyck shifts are sofic-Dyck shifts (see [7]).

If x is a (finite or infinite) sequence of symbols over A, we denote by $x[i, j]$ its finite factor $x_i \cdots x_j$ for $i \leq j$. If X is a shift we denote by $\mathcal{B}(X)$ the set of finite factors of sequences in X. This set is called the set of *blocks* of X.

Example 1. The *(one-sided) Dyck shift* X over $A = A_c \sqcup A_r \sqcup A_i$ with $A_c = \{(, [\}$, $A_r = \{),]\}$ is the set of left-infinite sequences for which the round and square brackets are open and closed in the right order. For instance the sequences $\cdots (([ ] [ )$ or $\cdots (([[[[($ are legal while the sequence $\cdots (([ ]]] ($ does not belong to X. It is a one-sided finite-type-Dyck shift since $X = \mathsf{X}_{F,U}$ where F is the empty set and U contains the forbidden pairs $\{(,]\}^1$ and $\{[,)\}$.

Example 2. Consider the Motzkin shift X over $A = A_c \sqcup A_r \sqcup A_i$ with $A_c = \{(, [\}$, $A_r = \{),]\}$ and $A_i = \{j\}$. It is presented by the Dyck automaton on the left part of Fig. 1. The Motzkin shift over A is $X = \mathsf{X}_{F,U}$ where F is empty and U contains the forbidden pairs $\{(,]\}$ and $\{[,)\}$. The shift Y over $B = B_c \sqcup B_r \sqcup B_i$

---

[1] Here the curly brackets are used to denote a pair of words in order to avoid confusion with the symbols of the alphabet.

with $B_c = \{(, [\}, B_r = \{),]\}$ and $B_i = \{j, k\}$ presented by the Dyck automaton on the right part of Fig. 1 is a finite-type-Dyck shift too. It is the shift $X_{G,U}$ with $G = \{jj, kk, ja, ak \mid a \in B_c \sqcup B_r\}$ and U containing the forbidden pairs $\{(,]\}$ and $\{[,)\}$. For instance, the left-infinite sequences $\cdots ((([j\,k][]) \text{ and } \cdots)))))$ belong to Y while the sequences $\cdots (([][]] \text{ or } \cdots ([j][])$ do not.

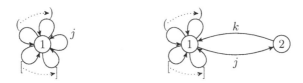

Fig. 1. A Motzkin shift X (on the left) over $A = A_c \sqcup A_r \sqcup A_i$ with $A_c = \{(, [\}, A_r = \{),]\}$ and $A_i = \{j\}$. A finite-type-Dyck shift Y (on the right) over $B = B_c \sqcup B_r \sqcup B_i$ with $B_c = \{(, [\}, B_r = \{),]\}$ and $B_i = \{j, k\}$. Matched pairs of edges are indicated with dotted arrows.

Let A and B be two tri-partitioned alphabets. Let $X \subseteq A^{-\mathbb{N}}$ be a shift and let m be a nonnegative integer. A map $\Phi : X \to B^{-\mathbb{N}}$ is called an $(m + 1)$-*block map with memory* m if there exists a function $\phi : \mathcal{B}_{m+1}(X) \to B$ such that, for all $x \in X$ and any $i \in -\mathbb{N}$, $\Phi(x)_i = \phi(x_{i-m} \cdots x_{i-1}x_i)$. The map ϕ extends naturally to a map from $\mathcal{B}_{m+k}(X) \to B^k$ for any positive integer k. A *block map* is a map which is an $(m + 1)$-block map for some nonnegative integer m.

We say that a block-map $\Phi : A^{\mathbb{Z}} \to B^{\mathbb{Z}}$ is *proper* if and only if $\Phi(x)_i \in A_c$ (resp. A_r, A_i) whenever $x_i \in A_c$ (resp. A_r, A_i). Note that the shift map σ is not proper.

A *conjugacy* is a bijective block map between two shifts.

2.3   (One-Sided) Edge-Dyck Shifts

In this section, we use the notation of [4] for graphs (meaning multigraphs) and graph morphisms. We extend the definitions to Dyck graphs.

A *graph* $G = (Q, E)$ is a pair composed of a finite set Q of *vertices* (or *states*) and a finite set E of *edges*. Every graph is equipped with two maps $i, t : E \to Q$ which associate to an edge e its *initial* and *terminal* vertex. We say that e *starts* in $i(e)$ and *ends* in $t(e)$. Sometimes, $i(e)$ is called the *source* and $t(e)$ is called the *target* of e.

We also say that e is an *incoming edge* for $t(e)$, and an *outgoing edge* for $i(e)$. Two edges $e, e' \in E$ are *consecutive* if $t(e) = i(e')$. A *path* is a sequence of consecutive edges.

A *Dyck graph* $\mathcal{G} = (G, M)$ consists of a graph $G = (Q, E)$ equipped with a set of pairs of edges M. The set of edges E of the graph is partitioned into three categories: the set of *call edges* (denoted E_c), the set of *return edges* (denoted E_r), and the set of *internal edges* (denoted E_i). Altogether we have $E = E_c \sqcup$

$E_r \sqcup E_i$. The set M is a set of selected pairs (e, f), with $e \in E_c$, $f \in E_r$ called the set of *matched edges*. Dyck graphs can be seen as Dyck automata where the labelings are the identity on the edges.

If a path of such graph belongs to $\mathrm{Dyck}(E)$, it is called a *Dyck path*. Less formally, in a Dyck path each call edge is paired against a return edge and vice versa.

A finite path $(e_i)_{k \le i \le k+n}$ of $\mathcal{G}$ is *admissible* if whenever $e_j \in E_c, e_{j+m} \in E_r$, with $k \le j, j + m \le k + n$, and $e[j + 1, j + m - 1]$ is path in $\mathrm{Dyck}(E)$, we have $(e_j, e_{j+m}) \in M$. An infinite path is *admissible* if all its finite factor are finite admissible paths.

A *(one-sided) edge-Dyck shift* is the set of admissible left-infinite paths of a Dyck graph. The edge-Dyck shift defined by the Dyck graph $\mathcal{G}$ is denoted by $X_{\mathcal{G}}$.

The following proposition shows that each finite-type-Dyck shift is properly conjugate to an edge-Dyck shift. It also holds for two-sided shifts (see [5]).

Proposition 1. *Each finite-type-Dyck shift is properly conjugate to an edge-Dyck shift.*

Proof. The proof is omitted.

3 State-Splitting of Dyck Graphs

We define below the notion of input state-splitting for one-sided edge-Dyck shifts and Dyck graphs. State splitting was introduced in [5] for two-sided finite-type-Dyck shifts and Dyck automata.

Let $G = (Q, E)$ and $H = (R, F)$ be two graphs. A pair (h, k) of surjective maps $k : R \rightarrow Q$ and $h : F \rightarrow E$ is called a *graph morphism* from H onto G if $i \circ h = k \circ i$ and $t \circ h = k \circ t$. We denote by $h_\infty : F^{-\mathbb{N}} \rightarrow E^{-\mathbb{N}}$ the map defined by $h_\infty((f_i)_{i \in -\mathbb{N}}) = (h(f_i))_{i \in -\mathbb{N}}$.

If $E = E_c \sqcup E_r \sqcup E_i$ and $F = F_c \sqcup F_r \sqcup F_i$, a graph morphism (h, k) from $H = (R, F)$ to $G = (Q, E)$ is said to be *proper* if h maps F_c to E_c (resp. F_r to E_r, F_i to E_i).

Let $\mathcal{G}$ and $\mathcal{H}$ be two Dyck graphs with $\mathcal{G} = (G, M)$ and $\mathcal{H} = (H, N)$. A pair (h, k) is called a *Dyck-graph morphism* from $\mathcal{H}$ onto $\mathcal{G}$ if it is a graph morphism from H to G and if $(e, f) \in N$ implies $(h(e), h(f)) \in M$. Hence a Dyck-graph morphism sends matched edges to matched edges.

For $p, q \in Q$, we denote by E_p^q the set of edges of a graph $G = (Q, E)$ starting in state p and ending in state q.

Let $\mathcal{G} = (G, M)$ and $\mathcal{H} = (H, N)$ be two Dyck graphs with $G = (Q, E)$, $H = (R, F)$. An *in-merge* from $\mathcal{H}$ onto $\mathcal{G}$ is a proper Dyck-graph morphism (h, k) from $\mathcal{H}$ onto $\mathcal{G}$ such that

- for each $p, q \in Q$ there is a partition[2] $(E_p^q(t))_{t \in k^{-1}(q)}$ of the set E_p^q such that for each $r \in k^{-1}(p)$ and $t \in k^{-1}(q)$, the map h is a bijection from F_r^t onto $E_p^q(t)$,
- for each $e, f \in E$, $(e, f) \in M$ if and only if $(h^{-1}(e), h^{-1}(f)) \in N$.

---

[2] A *partition* of a set X is a family $(X_i)_{i \in I}$ of pairwise disjoint, possibly empty subsets of X, indexed by a set I, such that X is the union of the sets X_i for $i \in I$.

If this holds, then $\mathcal{G}$ is called an *in-merge* of $\mathcal{H}$, and $\mathcal{H}$ is an *in-split* of $\mathcal{G}$. The map h_∞ from $X_\mathcal{H}$ to $X_\mathcal{G}$ is called an *edge in-merging map* and its inverse an *edge in-splitting map*. The edge-Dyck shift $X_\mathcal{G}$ is called an *in-merge* of $X_\mathcal{H}$ and $X_\mathcal{H}$ is called an *in-split* of $X_\mathcal{G}$.

Thus an in-split $\mathcal{H}$ is obtained from a Dyck graph $\mathcal{G}$ as follows: each state $q \in Q$ is split into copies which are the states of $\mathcal{H}$ in the set $k^{-1}(q)$. Each of these states t receives a copy of $E_p^q(t)$ starting in r and ending in t for each r in $k^{-1}(p)$. Each r in $k^{-1}(p)$ has the same number of edges going out of r and coming in s, for any $s \in R$. Moreover, for any $p, q \in Q$ and $e \in E_p^q$, all edges in $h^{-1}(e)$ have the same terminal vertex, namely the state t such that $e \in E_p^q(t)$.

Example 3. Let $\mathcal{G}$ and $\mathcal{H}$ be the Dyck graphs represented on Fig. 2. Here $G = (Q, E)$ with $Q = \{1, 2\}$, $E = E_c \sqcup E_r \sqcup E_i$ with $E_c = \{e_1, e_3\}$, $E_r = \{e_2, e_4\}$, and $E_i = \{e_5, e_6\}$. The matched pairs of edges of $\mathcal{G}$ are (e_1, e_2) and (e_3, e_4).

On the right is the Dyck graph $\mathcal{H}$ which is an in-split of $\mathcal{G}$. The state 1 of G is split into two states 3 and 4 of H. Roughly speaking, the edges coming into 1 in G are partitioned into two parts $\{e_1, e_2\}$ and $\{e_3, e_4, e_6\}$. The edges $\{e_1, e_2\}$ become the edges $\{e_1, e_2\}$ coming in 3 in H. The edges $\{e_3, e_4, e_6\}$ become the edges $\{e_3, e_4, e_6\}$ coming in 4 in H. Each edge e_1, e_2, e_3, e_4, e_5 going out of 1 in G is duplicated into two edges with the same target going out of 3 and 4 in H. Hence the edge e_1 is duplicated into e_1, f_1, the edge e_2 is duplicated into e_2, f_2, the edge e_3 is duplicated into e_3, f_3, the edge e_4 is duplicated into e_4, f_4 and the edge e_5 is duplicated into e_5, f_5.

More formally, an in-merge (h, k) from $X_\mathcal{H}$ onto $X_\mathcal{G}$ can be defined by $k(3) = k(4) = 1$ and $k(5) = 2$. The map h is associated to the following partition of edges of G: the edges from 1 to 1 are partitioned into two classes $E_1^1(3) = \{e_1, e_2\}$ and $E_1^1(4) = \{e_3, e_4\}$. The edges from 2 to 1 are partitioned into two classes $E_2^1(3) = \emptyset$ and $E_2^1(4) = \{e_6\}$. We have $h(e_i) = h(f_i) = e_i$ for $i = 1, .., 5$ and $h(e_6) = e_6$. The matched edges of $\mathcal{H}$ are (e_1, e_2), (e_1, f_2), (f_1, e_2), (f_1, f_2), (e_3, e_4), (e_3, f_4), (f_3, e_4), (f_3, f_4).

The Dyck graph $\mathcal{H}$ is therefore an in-split of the Dyck graph $\mathcal{G}$.

The following result is known for (non-Dyck) graphs (see [13, Theorem 2.4.1]). It shows that if a Dyck graph $\mathcal{H}$ is an in-split of a Dyck graph $\mathcal{G}$, then $X_\mathcal{G}$ and $X_\mathcal{H}$ are conjugate.

Proposition 2. *If (h, k) is an in-merge of a Dyck graph $\mathcal{H}$ onto a Dyck graph $\mathcal{G}$, then h_∞ is a 1-block proper conjugacy from $X_\mathcal{H}$ onto $X_\mathcal{G}$ and its inverse is 2-block.*

Proof. The proof is omitted.

The following result shows that the in-merge operations performed on a Dyck graph are confluent. Roughly speaking, whenever it is possible to in-merge two states p, q or two states r, s (with s possibly equal to q), one can perform first the in-merging of p, q and then the in-merging of r, s, or one can perform first the in-merging of r, s and then the in-merging of p, q getting the same final graph up to an isomorphism.

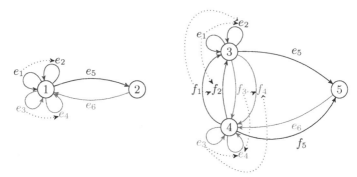

Fig. 2. Two Dyck graphs, $\mathcal{G}$ on the left and $\mathcal{H}$ on the right. The Dyck graph $\mathcal{H}$ is an in-split of $\mathcal{G}$. Matched edges are linked with a dotted arrow. The edges coming in 1 in $\mathcal{G}$ are partitioned into two parts $\{e_1, e_2\}$ (in blue) and $\{e_3, e_4, e_6\}$ (in red).

Proposition 3. *Let $\mathcal{G}_0$ be a Dyck graph and $\mathcal{G}_1$ and $\mathcal{G}_2$ be two in-merges of $\mathcal{G}_0$. Then there exists a Dyck graph $\mathcal{G}_3$ which is a common in-merge of $\mathcal{G}_1$ and $\mathcal{G}_2$.*

Proof. The proof is omitted.

We call *in-amalgamation* of a Dyck graph $\mathcal{G}$ a Dyck graph $\mathcal{H}$ for which there is a finite sequence of in-merges from $\mathcal{G}$ to $\mathcal{H}$. We call *minimal in-amalgamation* of a Dyck graph $\mathcal{G}$ an in-amalgamation $\mathcal{H}$ such that each in-merge of $\mathcal{H}$ is equal to $\mathcal{H}$ up to a Dyck-graph isomorphism. It is a consequence of Proposition 3 that any Dyck graph has a unique minimal in-amalgamation up to an isomorphism.

4   Decomposition of Proper Conjugacies

In this section, we show that whenever two (left)-one-sided edge-Dyck shifts $\mathsf{X}_\mathcal{G}$ and $\mathsf{X}_\mathcal{H}$ are conjugate through a proper conjugacy, there are edge-Dyck shifts $\mathsf{X}_\mathcal{K}$ and $\mathsf{X}_\mathcal{L}$ and sequences of in-merging maps from $\mathsf{X}_\mathcal{K}$ onto $\mathsf{X}_\mathcal{G}$ (resp. from $\mathsf{X}_\mathcal{H}$ onto $\mathsf{X}_\mathcal{L}$) such that $\mathsf{X}_\mathcal{K}$ and $\mathsf{X}_\mathcal{L}$ are conjugate through a 0-memory proper conjugacy whose inverse is also a 0-memory conjugacy. These results are known for (non-Dyck) edge shifts.

Proposition 4. *Let $\Phi : \mathsf{X}_\mathcal{G} \to \mathsf{X}_\mathcal{H}$ be a proper conjugacy between two edge-Dyck shifts with memory $m \geq 1$ and such that Φ^{-1} has memory m'. Then there is an in-split $\mathsf{X}_\mathcal{K}$ of $\mathsf{X}_\mathcal{G}$, and an in-merging map $h_\infty : \mathsf{X}_\mathcal{K} \to \mathsf{X}_\mathcal{G}$ such that $\alpha = \Phi \circ h_\infty$ is a proper conjugacy from $\mathsf{X}_\mathcal{K}$ onto $\mathsf{X}_\mathcal{H}$ with memory $m-1$ and α^{-1} has memory $m' + 1$.*

Proof. The proof is omitted.

Corollary 1. *Let $\Phi : \mathsf{X}_\mathcal{G} \to \mathsf{X}_\mathcal{H}$ be a proper conjugacy between two edge-Dyck shifts with memory m and such that Φ^{-1} has memory m'. Then there is an* edge-Dyck shift $\mathsf{X}_\mathcal{K}$ with a sequence of in-merging maps $\Psi : \mathsf{X}_\mathcal{K} \to \mathsf{X}_\mathcal{G}$ such that $\alpha : \mathsf{X}_\mathcal{K} \to \mathsf{X}_\mathcal{H}$ defined by $\alpha = \Phi \circ \Psi$ is a proper conjugacy with memory 0 and α^{-1} has memory $m' + m$.

Proof. Proposition 4 is applied m times. □

Proposition 5. *Let $\Phi : \mathsf{X}_\mathcal{G} \to \mathsf{X}_\mathcal{H}$ be a proper conjugacy between two edge-Dyck shifts with memory 0 and such that Φ^{-1} has memory $m' \geq 1$. Then there is an in-split $\mathsf{X}_\mathcal{K}$ of $\mathsf{X}_\mathcal{G}$, an in-merging map $h_\infty : \mathsf{X}_\mathcal{K} \to \mathsf{X}_\mathcal{G}$, an in-split $\mathsf{X}_\mathcal{L}$ of $\mathsf{X}_\mathcal{H}$, and an in-merging map $h'_\infty : \mathsf{X}_\mathcal{L} \to \mathsf{X}_\mathcal{H}$, such that $\alpha = h'^{-1}_\infty \circ \Phi \circ h_\infty$ is a proper conjugacy from $\mathsf{X}_\mathcal{K}$ onto $\mathsf{X}_\mathcal{L}$ with memory 0 and α^{-1} has memory $m' - 1$.*

Proof. The proof is omitted.

Corollary 2. *Let $\Phi : \mathsf{X}_\mathcal{G} \to \mathsf{X}_\mathcal{H}$ be a proper conjugacy between two edge-Dyck shifts with memory m and such that Φ^{-1} has memory m'. Then there is an edge-Dyck shift $\mathsf{X}_\mathcal{K}$ with a sequence of in-merging maps $\Psi : \mathsf{X}_\mathcal{K} \to \mathsf{X}_\mathcal{G}$, an edge-Dyck shift $\mathsf{X}_\mathcal{L}$ with a sequence of in-merging maps $\Theta : \mathsf{X}_\mathcal{L} \to \mathsf{X}_\mathcal{H}$, such that $\alpha : \mathsf{X}_\mathcal{K} \to \mathsf{X}_\mathcal{L}$ defined by $\alpha = \Theta^{-1} \circ \Phi \circ \Psi$ and α^{-1} are proper conjugacies with memory 0.*

Proof. Corollary 1 composed with $m' + m$ applications of Proposition 5. □

5   Decidability of Proper Conjugacy for a Class of Edge-Dyck Shifts

In this section we prove that proper conjugacy is decidable for the class of (left) one-sided matched-return extensible finite-type-Dyck shifts.

A shift X is said *(right) matched-return extensible* if for any block u of X there is a nonempty block w in $\mathrm{MR}(X)$ such that uw is a block of X.

A shift X is said *(right) Dyck-extensible* if for any block u of X there is a nonempty block w in $\mathrm{Dyck}(X)$ such that uw is a block of X.

Dyck-extensible shifts are also MR-extensible since $\mathrm{Dyck}(X) \subseteq \mathrm{MR}(X)$ for any shift X.

Example 4. The Motzkin shift X and the shift Y of Example 2 are both right (and left) Dyck extensible. Indeed, any block of the shift can be extended by the Dyck word j or k. The Dyck shifts are also matched-return extensible since any sequence is right-extensible by a call letter.

Finite-type-Dyck shifts may not have the MR-extensibility property but a large class of them do. This class contains also the class of (left)-one-sided shifts of finite type which are extensible. MR-extensible (or Dyck-extensible) shifts are stable by proper conjugacy as is shown in Proposition 6.

Proposition 6. *Let X, Y be two properly conjugate one-sided shifts. If X is MR-extensible (resp. Dyck-extensible), then Y also.*

Proof. The proof is omitted.

In order to obtain our decision result we need to work with trim presentations of the edge-Dyck shifts.

A Dyck graph $\mathcal{G}$ is said to be *trim* if it has no state without an incoming edge and for any pair (e, f) of matched edges there is an admissible Dyck path w of $\mathcal{G}$ such that ewf is a path of $\mathcal{G}$.

Thus any edge of a trim Dyck graph $\mathcal{G}$ is a block of $X_{\mathcal{G}}$. A Dyck graph can be made trim in an effective way.

The following proposition is a key point to ensure the decidability of proper conjugacy. It shows how the restriction to matched-return extensible shifts allows to obtain the result.

Proposition 7. *Let $\Phi : X_{\mathcal{G}} \to X_{\mathcal{H}}$ be a proper conjugacy between two edge-Dyck shifts which are matched-return extensible. If both Φ and Φ^{-1} have memory 0 and $\mathcal{G}$ and $\mathcal{H}$ are trim, then $\mathcal{G}$ and $\mathcal{H}$ are equal up to a Dyck-graph isomorphism.*

Proof. The proof is omitted.

Proposition 8. *Let $\mathcal{G}$ and $\mathcal{H}$ be two trim Dyck graphs such that $X_{\mathcal{G}}$ and $X_{\mathcal{H}}$ are MR-extensible. If $X_{\mathcal{G}}$ and $X_{\mathcal{H}}$ are properly conjugate, then $\mathcal{G}$ and $\mathcal{H}$ have a common in-amalgamation.*

Proof. Let $\Phi : X_{\mathcal{G}} \to X_{\mathcal{H}}$ be a proper conjugacy between two edge-Dyck shifts presented by Dyck graphs $\mathcal{G}$ and $\mathcal{H}$.

By Corollary 2 and Propositions 6 and 7 there is an edge-Dyck shift $X_{\mathcal{K}}$ with a sequence of in-merging maps $\Psi : X_{\mathcal{K}} \to X_{\mathcal{G}}$ and a sequence of in-merging maps $\Theta : X_{\mathcal{K}} \to X_{\mathcal{H}}$. Therefore $\mathcal{G}$ and $\mathcal{H}$ can be seen as two in-amalgamations of one Dyck graph, $\mathcal{K}$, and repeated application of proposition 3 (the confluency property) already gives us the result.

$\square$

For (left) one-sided shifts of finite type, being matched-return extensible is equivalent to being right extensible. Left one-sided shifts of finite type may not be right-extensible. Nevertheless, Proposition 8 is true for these shifts without this condition of extensibility.

Theorem 1. *It is decidable in an effective way whether two one-sided edge-Dyck shifts which are MR-extensible are properly conjugate.*

Proof. Let $\Phi : X_{\mathcal{G}} \to X_{\mathcal{H}}$ be a proper conjugacy between two edge-Dyck shifts presented by Dyck graphs $\mathcal{G}$ and $\mathcal{H}$. Without loss of generality, we may also assume that $\mathcal{G}$ and $\mathcal{H}$ are trim.

By Proposition 8 the Dyck graphs $\mathcal{G}$ and $\mathcal{H}$ have a common in-amalgamation. As a consequence $\mathcal{G}$ and $\mathcal{H}$ have the same minimal in-amalgamation. Conversely if $\mathcal{G}$ and $\mathcal{H}$ have the same minimal in-amalgamation the shifts $X_{\mathcal{G}}$ and $X_{\mathcal{H}}$ are properly conjugate.

Hence proper conjugacy can be decided as follows. One can show that the minimal in-amalgamations of the two Dyck graphs can be computed in polynomial time. One then checks the equality of these graphs up to a Dyck-graph isomorphism. It is however not known whether the last step can be performed in polynomial time.   □

Example 5. We consider the two trim Dyck graphs $\mathcal{G}$ and $\mathcal{H}$ of Fig. 3 where matched edges are indicated with dotted arrows. The edge-Dyck shifts $\mathsf{X}_\mathcal{G}$ and $\mathsf{X}_\mathcal{H}$ are Dyck-extensible. These two shifts are conjugate since the minimal amalgamations of $\mathcal{G}$ and $\mathcal{H}$ are equal to the Dyck graph $\mathcal{L}$ of Fig. 4 with the same representation of matched edges.

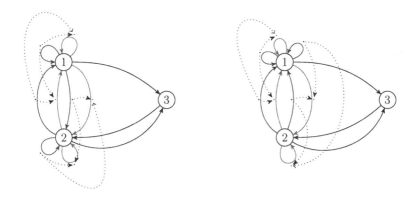

Fig. 3. The Dyck graphs $\mathcal{G}$ on the left and $\mathcal{H}$ on the right. Call edges are colored in blue, return edges in red and internal edges in black.

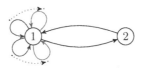

Fig. 4. The minimal amalgamation $\mathcal{L}$ of $\mathcal{G}$ and $\mathcal{H}$

Theorem 2. *It is decidable whether two one-sided finite-type-Dyck shifts which are MR-extensible are properly conjugate.*

Proof. By Propositions 1 and 6, we may assume that the two MR-extensible finite-type-Dyck shifts are edge-Dyck shifts. We then apply Theorem 1.   □

Proposition 8 is no more true in general when the edge-Dyck shifts are not MR-extensible. Indeed, there are examples of edge-Dyck graphs having disctint numbers of vertices while being minimal in-amalgamations of a same edge-Dyck shift The decidability of proper conjugacy of edge-Dyck shifts which are not matched-return extensible is hence open.

References

1. Alur, R., Madhusudan, P.: Visibly pushdown languages. In: Proceedings of the 36th Annual ACM Symposium on Theory of Computing, pp. 202–211 (electronic). ACM, New York (2004)
2. Alur, R., Madhusudan, P.: Adding nesting structure to words. J. ACM **56**(3) (2009)
3. Aubrun, N., Béal, M.-P.: Tree-shifts of finite type. Theoret. Comput. Sci. **459**, 16–25 (2012)
4. Béal, M.-P., Berstel, J., Eilers, S., Perrin, D.: Symbolic dynamics. to appear in Handbook of Automata (2010)
5. Béal, M.-P., Blockelet, M., Dima, C.: Finite-type-Dyck shift spaces. CoRR (2013). http://arxiv.org/1311.4223
6. Béal, M.-P., Blockelet, M., Dima, C.: Sofic-Dyck shifts. In: Csuhaj-Varjú, E., Dietzfelbinger, M., Ésik, Z. (eds.) MFCS 2014, Part I. LNCS, vol. 8634, pp. 63–74. Springer, Heidelberg (2014)
7. Béal, M.-P., Blockelet, M., Dima, C.: Sofic-Dyck shifts. CoRR (2014). http://arxiv.org/1305.7413
8. Fujiwara, M.: Conjugacy for one-sided sofic systems. In: Dynamical Systems and Singular Phenomena (Kyoto, 1986), World Sci. Adv. Ser. Dynam. Systems, vol. 2, pp. 189–202. World Sci. Publishing, Singapore (1987)
9. Inoue, K., Krieger, W.: Subshifts from sofic shifts and Dyck shifts, zeta functions and topological entropy. CoRR (2010). abs/1001.1839
10. Kitchens, B.P.: Symbolic Dynamics, One-sided, two-sided and countable state Markov shifts. Universitext. Springer, Berlin (1998)
11. Krieger, W.: On the uniqueness of the equilibrium state. Math. Systems Theory **8**(2), 97–104 (1974)
12. Krieger, W., Matsumoto, K.: Zeta functions and topological entropy of the Markov-Dyck shifts. Münster J. Math. **4**, 171–183 (2011)
13. Lind, D., Marcus, B.: An Introduction to Symbolic Dynamics and Coding. Cambridge University Press, Cambridge (1995)
14. Williams, R.F.: Classification of subshifts of finite type. In: Beck, A. (ed.) Recent Advances in Topological Dynamics. LNCS, vol. 318, pp. 281–285. Springer, Heidelberg (1973)

Transfinite Lyndon Words

Luc Boasson and Olivier Carton[(⊠)]

LIAFA, Université Paris Diderot, Paris, France
Olivier.Carton@liafa.univ-paris-diderot.fr

Abstract. In this paper, we extend the notion of Lyndon word to transfinite words. We prove two main results. We first show that, given a transfinite word, there exists a unique factorization in Lyndon words that are locally decreasing, a relaxation of the condition used in the case of finite words.

In a second part, we prove that the factorization of a rational word has a special form and that it can be computed in polynomial time from a rational expression describing the word.

1 Introduction

Lyndon words were introduced in [6,7] as *standard lexicographic sequences* in the study of the derived series of the free group over some alphabet A. These words can be used to construct a basis of the free Lie algebra over A and their enumeration yields the well-known Witt's formula for the dimension of the homogeneous component $\mathcal{L}_n(A)$ of this free Lie algebra.

They are several equivalent definitions of these words but they are usually defined as those words which are primitive and minimal for the lexicographic ordering in their conjugacy class. The nice properties they enjoy in linear algebra are actually closely related to their properties in the free monoid. Lyndon words provide a nice factorization of the free monoid.

Lyndon words can be studied with the tools of the combinatorics on words, leaving aside the algebraic origin of these words. It then can be proved directly that each word w of the free monoid A^* has a unique decomposition as a product $w = u_1 \cdots u_n$ of a non-increasing sequence of Lyndon words $u_1 \geqslant \cdots \geqslant u_n$ for the lexicographic ordering. This uniqueness of the decomposition of each word is indeed remarkable. It has lead Knuth to call Lyndon words *prime words* [4, p. 305] and we will use this terminology. As usual, such a result raises the two related following questions: first how to efficiently test whether a given word is prime and second, more ambitious, how to compute its prime factorization. It has been shown that this factorization can be computed in linear time in the size of the given word w [3].

Very often, in the field of combinatorics of words, classical results give rise to an attempt to some generalization. These can be achieved by adapting the results to trees or to infinite words. The notion of prime word does not make exception: the unique prime decomposition has already been extended to ω-words by Siromoney et al in [11] where it is shown that any ω-word x can be uniquely factorized either

© Springer International Publishing Switzerland 2015
I. Potapov (Ed.): DLT 2015, LNCS 9168, pp. 179–190, 2015.
DOI: 10.1007/978-3-319-21500-6_14

$x = u_0 u_1 u_2 \cdots$ where $(u_i)_{i \geq 0}$ is a non-increasing sequence of finite prime words or $x = u_0 u_1 \cdots u_n$ where $u_0, \ldots, u_{n-1}$ is a non-increasing sequence of finite prime words and u_n is a prime ω-word such that $u_{n-1} \geqslant u_n$. In [11], prime ω-words are defined as ω-words having infinitely many prime prefixes but Proposition 2.2 in the same paper states that an ω-word is prime if and only if it is lexicographically strictly smaller than all its suffixes. Prime ω-words are also considered in [8,9] where the prime factorization of some well-known ω-words like the Fibonacci word is also given. The property that a limit of prime words is still a prime word also holds in our context and it is stated in Proposition 4.

The goal of this paper is to extend further such results to transfinite words, that is, words indexed by ordinals. For simplicity, we restrict ourselves to countable ordinals. First we extend the factorization theorem to all words and second we provide an algorithm that computes this factorization for words that can be finitely described by a rational expression.

The first task is to find a suitable notion of transfinite prime words. This is not easy as the different equivalent definitions for finite prime words do not coincide any more on transfinite words. Since the factorization property is presumably their most remarkable one, it can be used as a gauge to measure the accuracy of a definition. If a definition allows us to prove that each transfinite word has a unique decomposition in prime words, it can be considered as the right one. The two main points are that the factorization should always exist and that it should be unique. Of course, the definition should also satisfy the following additional requirement: it has to be an extension of the classical one for finite words, meaning that it must coincide with the classical definition for finite words. We introduce such a definition. The existence and uniqueness of the factorization is obtained by relaxing slightly the property of being non-increasing. It is replaced by the property of being locally non-increasing (see Section 4 for the precise definition). As requested, the two properties coincide for finite sequences. Our results extend the ones of Siromoney et al. [11] as we get the same definition of prime words of length ω and the same decomposition for words of length ω.

The second task is to extend the algorithmic property of the decomposition of a word in prime words. It is, of course, not possible to compute the factorization of any transfinite word but we have focused on the so-called rational words, that is, words which can be described from the letters using product and ω-operations (possibly nested). We prove that the factorization of these rational words have a special form. It can be a transfinite sequence of primes, but only finitely many different ones occur in it. Furthermore, all the prime words occurring are also rational and the sequence is really non-increasing in that case. We give an algorithm which computes, in polynomial time, the factorization of a rational word given by an expression involving products and ω-operations.

The paper is organized as follows. Basic definitions of ordinals and transfinite words are recalled in Section 2. The definition of prime words and some comments are given in Section 3. The existence and uniqueness of the prime factorization is stated in Section 4. Section 5 is devoted to rational words and to the algorithm for computing their prime factorization.

2   Transfinite Words

We refer the reader to [10] for a complete introduction to the theory of ordinals. An ordinal is a class for isomorphism of well-founded linear orderings.

Let A be a finite set called the *alphabet* equipped with a linear ordering $<$. Its elements are called *letters*. In the examples, we often assume that $A = \{a, b\}$ with $a < b$. This ordering on A is necessary to define the lexicographic ordering on words (see below). For an ordinal α, an α-sequence of letters is also called a *word* of length α or an α-*word* over A. The sequence of length 0 which has no element is called the *empty word* and it is denoted by ε. The length of a word x is denoted by $|x|$. The set of all words of countable length over A is denoted by $A^\#$.

Let x be a word $(a_\beta)_{\beta < \alpha}$ of length α. For any $\gamma \leq \gamma' \leq \alpha$, we denote by $x[\gamma, \gamma')$ the word $(b_\beta)_{\beta < \gamma' - \gamma}$ of length $\gamma' - \gamma$ defined by $b_\beta = a_{\gamma + \beta}$ for any $0 \leq \beta < \gamma' - \gamma$. It is the empty word if $\gamma' = \gamma$ and it is a single letter if $\gamma' = \gamma + 1$. Such a word $x[\gamma, \gamma')$ is called a *factor* of x. A word of the form $x[0, \gamma)$ (resp. $x[\gamma, \alpha)$) for $0 \leq \gamma \leq \alpha$ is called a *prefix* (resp. *suffix*) of x. The prefix (resp. suffix) is called *proper* whenever $0 < \gamma < \alpha$. If x is the word $(ab)^\omega (bc)^\omega$ of length $\omega \cdot 2$, the prefix $x[0, \omega)$ is $(ab)^\omega$, the suffix $x[\omega, \omega \cdot 2)$ is $(bc)^\omega$ and the factor $x[5, \omega + 2)$ is the word $(ba)^\omega bc$. Remark that a proper suffix of a word x may be equal to x. For instance, the proper suffix $x[4, \omega \cdot 2)$ of the word $x = (ab)^\omega (bc)^\omega$ is equal to x. Remark however that a proper prefix y of a word x cannot be equal to x since it satisfies $|y| < |x|$.

Let $(x_\beta)_{\beta < \alpha}$ be an α-sequence of words. The word obtained by concatenating the words of the sequence $(x_\beta)_{\beta < \alpha}$ is denoted by $\prod_{\beta < \alpha} x_\beta$. Its length is the sum $\sum_{\beta < \alpha} |x_\beta|$. The product $\prod_{n < \omega} x$ for a given word x is denoted by x^ω. An α-*factorization* of a word x is a sequence $(x_\beta)_{\beta < \alpha}$ of words such that $x = \prod_{\beta < \alpha} x_\beta$.

We write $x \sqsubseteq x'$ whenever x is a prefix of x' and $x \sqsubset x'$ whenever x is a prefix of x' different from x'. The relation $\sqsubset$ is an ordering on $A^\#$. The ordering $\prec$ is defined by $x \prec x'$ if there exist two letters $a < b$ and three words y, z and z' such that $x = yaz$ and $x' = ybz'$. The *lexicographic ordering* $\leqslant$ is finally defined by $x \leqslant x'$ if $x \sqsubseteq x'$ or $x \prec x'$. We write $x < x'$ whenever $x \leqslant x'$ and $x \neq x'$. The relation $<$ is a linear ordering on $A^\#$.

We mostly use Greek letters $\alpha, \beta, \ldots$ to denote ordinals, letters $a, b, \ldots,$ to denote elements of the alphabet, letters $x, y, \ldots$ to denote transfinite words and letters $u, v, \ldots$ to denote prime transfinite words.

3   Prime Words

In this section, we introduce the crucial definition of a prime transfinite word.

Recall that a word x is *primitive* if it is not the power of another word, i.e., if the equality $x = y^\alpha$ for some ordinal α and some word y implies $\alpha = 1$ and $y = x$. Note that any word x is either primitive or the power y^α of some primitive word y for some ordinal $\alpha \geq 2$ [2].

A word w is *prime*, also called *Lyndon*, if w is primitive and any proper suffix x of w satisfies $w \leqslant x$. The terminology *prime* is borrowed from [4, p. 305].

It is justified by Theorem 5 which states that any word has a unique factorization in prime words which is almost non-increasing (see Section 4 for a precise statement).

Example 1. Both finite words a^2b and a^2bab are prime. Both finite words aba and $abab$ are not prime. Indeed, the suffix a of aba satisfies $a < aba$ and $abab$ is not primitive. The ω-words ab^ω and $abab^2ab^3ab^4 \cdots$ are prime. Both ω-words ba^ω and $(ab)^\omega$ are not prime. Indeed, the suffix a^ω of ba^ω satisfies $a^\omega < ba^\omega$ and the ω-word $(ab)^\omega$ is not primitive.

Let us make some comments about this definition. Note firstly that only proper suffixes are considered since the empty word ε is a suffix of any word w but does not satisfy $w \leqslant \varepsilon$ unless $w = \varepsilon$. Secondly, each suffix x of a prime word w must satisfy $w \leqslant x$, that is either $w \sqsubseteq x$ or $w \prec x$. Since the length of x is smaller than the length of w, the relation $w \sqsubseteq x$ is impossible since $w \sqsubseteq x$ would imply $|w| < |x|$. The relation $w \sqsubseteq x$ reduces then to $w = x$. Therefore, a word w is prime if it is primitive and any proper suffix x of w satisfies either $w = x$ or $w \prec x$.

Our definition of prime words coincides with the classical definition for finite words [5, Chap. 5]. A finite word is a prime word if it is minimal in its conjugacy class or, equivalently, if it is strictly smaller than any of its proper suffixes [5, Prop. 5.1.2]. A proper suffix of a finite word cannot be equal to the whole word and therefore, it does not matter whether it is required that any proper suffix is *strictly smaller* or just *smaller* than the whole word. For transfinite words, it does matter since some proper suffix might be equal to the whole word. Our definition allows indeed a suffix of a prime word to be equal to the whole word. The word $w = a^\omega b$ of length $\omega + 1$ is prime but some of its proper suffixes like $w[1, \omega + 2)$ or $w[2, \omega + 2)$ are equal to w.

Our definition also requires the word to be primitive. It is not needed for finite words since, in that case, being smaller than all its suffixes implies primitivity. Indeed, if the finite word x is equal to y^n for $n \geq 2$, then y is a proper suffix of x which is strictly smaller than x. Therefore, x cannot be prime. This argument does not hold anymore for transfinite words. The ω-word $x = a^\omega$ is, of course, not primitive but none of its proper suffixes is strictly smaller than itself. Each proper suffix of x is actually equal to x. The same property holds for each word of the form a^α where α is a power of ω, that is, $\alpha = \omega^\beta$ for some ordinal $\beta \geq 1$.

Our definition of prime words also coincides with the definition for ω-words given in [11] where an ω-word is called prime if it is the limit of finite prime words. It is also shown in [11, Prop. 2.2] that an ω-word is prime if and only if it is strictly smaller than any of its suffixes. Requiring that no suffix is equal to the whole ω-word prevents the ω-word from being periodic, that is, of the form x^ω for some finite word x. These last words are the only non-primitive ω-words. Let us now give a more involved example.

Example 2. Define by induction the sequence $(u_n)_{n<\omega}$ of words by $u_0 = a$ and $u_{n+1} = u_n^\omega b$. The first words of the sequence are $u_1 = a^\omega b$ and $u_2 = (a^\omega b)^\omega b$. It can be proved by induction on n that the length of u_n is $\omega^n + 1$ since

$(\omega^n + 1) \cdot \omega + 1 = \omega^{n+1} + 1$. Let u_ω be the word $u_0 u_1 u_2 \cdots$ of length ω^ω. Note that equality $u_n u_{n+1} = u_{n+1}$ holds for any $n \geq 0$ and that therefore equality $u_\omega = u_n u_{n+1} u_{n+2} \cdots$ also holds for any $n \geq 0$. The word u_ω is actually the limit of the sequence $(u_n)_{n<\omega}$. It can be shown that each word u_n is prime and that their limit u_ω is also prime.

In the case of finite words, the existence of the decreasing factorization follows for the following property: if the prime words u and v satisfy $u < v$, then the word uv is also prime [5, Prop 5.1.3]. Proposition 3 states a similar result for transfinite words. However, the existence of the factorization does not follow immediately from it. More involved arguments and Proposition 4 about limits are indeed needed.

Proposition 3. *Let u and v be two prime words such that $u < v$. Then $u^\alpha v$ is a prime word for any ordinal α. Furthermore, if $u^\alpha v < v$, then the word $u^\alpha v^\beta$ is prime for any ordinal $\beta \geq 1$.*

The second proposition states the property that Siromoney [11] uses as the definition of of prime words for ω-words. It allows us to get prime words of limit length.

Proposition 4. *Let $(u_n)_{n<\omega}$ be an ω-sequence of words such that the product $u_0 \cdots u_n$ is prime for each $n < \omega$. Then the ω-product $u_0 u_1 u_2 \cdots$ is also prime.*

4   Factorization in Prime Words

In this section, we present the first main result: that any word has a unique factorization into prime words which is almost non-increasing. The goal is to extend to transfinite word the classical result that any finite word is the product of a non-increasing sequence of prime words [5, Thm 5.1.5]. It turns out that this extension is not straightforward since some words are not equal to a product of a non-increasing sequence of prime words. Let us consider the ω-word $x = aba^2 ba^3 \cdots$ and the $(\omega + 1)$-word xb. The word x can be factorized $x = ab \cdot a^2 b \cdot a^3 b \cdots$ and the sequence $(a^n b)_{n<\omega}$ is indeed a non-increasing sequence of prime words. The word xb, however, cannot be factorized into a non-increasing sequence of prime words. A naive attempt could be $ab \cdot a^2 b \cdot a^3 b \cdots b$ but the sequence $(u_n)_{n \leq \omega}$ where $u_n = a^{n+1} b$ for $n < \omega$ and $u_\omega = b$ is not non-increasing since $u_n < u_\omega$ for each $n < \omega$. To cope with this difficulty, we have introduced a notion of a locally non-increasing sequence. This is a slightly weaker notion than the notion of a non-increasing sequence. A locally non-increasing sequence $(u_\beta)_{\beta<\alpha}$ may have some $\gamma < \gamma' < \alpha$ such that $u_\gamma < u_{\gamma'}$ but this may only happen if there exists a limit ordinal $\gamma < \gamma'' \leq \gamma'$ such that the sequence $(u_\beta)_{\beta<\alpha}$ is cofinally decreasing in γ''. Roughly speaking, an increase is allowed if it comes after an ω-sequence of strict decreases. The $(\omega + 1)$-sequence $(u_n)_{n \leq \omega}$ where $u_n = a^n b$ for $n < \omega$ and $u_\omega = b$ is locally non-increasing. One has indeed $u_n < u_\omega$ but also $u_n > u_{n+1}$ for each $n < \omega$.

We introduce now the formal definition of a locally non-increasing sequence. We will only use this notion for sequences of prime words lexicographically ordered but we give the definition for an arbitrary ordered set U. Let $(U, <)$ be a linear ordering and let $\bar{u} = (u_\beta)_{\beta < \alpha}$ be a sequence of elements of U. The sequence $\bar{u}$ is *constant* in the interval $[\gamma, \gamma')$ where $\gamma < \gamma' \leq \alpha$ if $u_\beta = u_\gamma$ holds for any $\gamma \leq \beta < \gamma'$. As usual, the sequence x is *non-increasing* if for any β and β', $\beta < \beta' < \alpha$ implies $u_\beta \geq u_{\beta'}$. It is *locally non-increasing* if for any interval $[\gamma, \gamma')$ where $\gamma < \gamma' \leq \alpha$, either it is constant in $[\gamma, \gamma')$ or there exist two ordinals $\gamma \leq \beta < \beta' < \gamma'$ such that $u_\beta > u_{\beta'}$.

It is clear that a non-increasing sequence is also locally non-increasing. The converse does not hold as it can be shown by the already considered $(\omega + 1)$-sequence $(u_\beta)_{\beta \leq \omega}$ defined by $u_n = a^n b$ for $n < \omega$ and $u_\omega = b$.

The following theorem is the main result of the paper. It extends the classical result which states that any finite word can be uniquely written as a non-increasing product of prime words [5, Thm 5.1.5].

Theorem 5. *For any word $x \in A^\#$, there exists a unique locally non-increasing sequence $(u_\beta)_{\beta < \alpha}$ of prime words such that $x = \prod_{\beta < \alpha} u_\beta$. This sequence is called the* prime factorization *of x.*

Example 6. The prime factorization of the finite words $aabab$ and $abaab$ are $aabab$ and $ab \cdot aab$ since ab, aab and $aabab$ are prime words. The prime factorization of the ω-words and $x_0 = aba^2 ba^3 b \cdots$ and $x_1 = abab^2 ab^3 \cdots$ are $x_0 = ab \cdot a^2 b \cdot a^3 b \cdots$ and $x_1 = abab^2 ab^3 \cdots$ since $ab, a^2 b, a^3 b, \ldots$ are prime words and $x_1 = abab^2 ab^3 \cdots$ is also prime by Proposition 4.

The prime factorization of the $(\omega + 1)$-word $x_2 = x_0 b$ is the $(\omega + 1)$-sequence $(u_\beta)_{\beta \leq \omega}$ given by $u_n = a^{n+1} b$ for $n < \omega$ and $u_\omega = b$. This factorization is not non-increasing since $u_0 = ab < b = u_\omega$ but it is, of course, locally non-increasing.

We discuss first the existence of the factorization and second its uniqueness. The existence can be proved in two ways. The first one is based on Zorn's lemma while the second one uses a transfinite induction on the length of words. Surprisingly, this latter one needs the uniqueness of the factorization which can be proved independently. The former proof is shorter but the latter one provides a much better insight. We only sketch here the former one.

In the case of finite words, the existence follows easily from the following property of finite prime words: if $u < v$ are two prime words, then uv is also prime. A similar result holds for transfinite words (see Proposition 3) but more involved arguments are needed to conclude.

Let x be a fixed word. Let X be the set of sequences $\bar{u} = (u_\beta)_{\beta < \alpha}$ of prime words such that $x = \prod_{\beta < \alpha} u_\beta$. Note that it is not assumed that the sequence $\bar{u}$ is locally non-increasing. We define an ordering $<$ on the sequences of words as follows. Two sequences $\bar{u} = (u_\beta)_{\beta < \alpha}$ and $\bar{u}' = (u'_\beta)_{\beta < \alpha}$ satisfy $\bar{u} < \bar{u}'$ if $\bar{u}$ refines $\bar{u}'$. This means that there exists a sequence $(\gamma_\beta)_{\beta < \alpha'}$ of ordinals such that $u'_\beta = \prod_{\gamma_\beta \leq \eta < \gamma_{\beta+1}} u_\eta$ for each $\beta < \alpha'$. For any totally ordered non-empty subset Y of X, there exists a least sequence $\bar{u}' = (u'_\beta)_{\beta < \alpha'}$ such that $\bar{u} < \bar{u}'$ for any sequence $\bar{u} = (u_\beta)_{\beta < \alpha} \in Y$. Each word u'_β can be shown to be prime and the sequence $\bar{u}' = (u'_\beta)_{\beta<\alpha'}$ belongs to X. This result follows from non obvious properties of transfinite prime words and in particular Proposition 3. It is then possible to apply Zorn's lemma: the set X has a maximal element $\bar{v} = (v_\beta)_{\beta<\alpha}$. It remains to show that this sequence $\bar{v}$ is indeed locally non-increasing. This again follows from the closure properties of prime words.

Now we turn to the uniqueness of the factorization. In the case of finite words, the last prime word of the prime factorization of a word x is the least suffix (for the lexicographic ordering) of x [5, Prop. 5.1.6]. A similar result does not hold for transfinite words and thus the uniqueness cannot be proved this way. Indeed, since the lexicographic ordering is not well founded, a word may not have a least suffix. Consider, for instance, the ω-word $x_0 = aba^2ba^3b\cdots$. It does not have a least suffix and its prime factorization $x_0 = ab \cdot a^2b \cdot a^3b\cdots$ does not have a last factor. Even when the prime factorization of x has a last prime factor, the word x may not have a least suffix. Consider the $(\omega + 1)$-word $x_2 = x_0b$. The prime factorization $x_2 = ab \cdot a^2b \cdot a^3b\cdots b$ has a last factor b but this word x_2 does not have a least suffix.

On the other hand, in the case of finite words, the first prime word of the prime factorization is exactly the longest prime prefix [5, Prop. 5.1.5]. This result, Proposition 7, can be extended to transfinite words. The uniqueness of the prime factorization follows then easily.

Proposition 7. *Let $\bar{u} = (u_\beta)_{\beta<\alpha}$ be a locally non-increasing sequence of prime words. The word u_0 is the longest prime prefix of the product $\prod_{\beta<\alpha} u_\beta$.*

5   Rational Words

The second half of this paper is devoted to prove that, for a special kind of transfinite words, the prime factorization can be effectively computed. The class of *rational words* is the smallest class of words which contains the empty word ε and the letters and which is closed under product and the iteration ω. This means that each letter a is a rational word and that if u and v are two rational words, then both words uv and u^ω are also rational. A rational word that can be described by a rational expression using only concatenation and the ω operator.

The following theorem states that the prime factorization of a rational word has a very special form, namely it has a finite range made of rational words.

Theorem 8. *For any rational word x, there exists a finite decreasing sequence of rational prime words $u_1 > \cdots > u_n$ and ordinals $\alpha_1, \ldots, \alpha_n$ less than ω^ω such that $x = u_1^{\alpha_1} \cdots u_n^{\alpha_n}$.*

Now we introduce special factorizations of x called main and secondary cuts. Main cuts occur between two different prime factors and secondary cuts occur between two occurrences of the same prime factor. More formerly, each factorization $x = (u_1^{\alpha_1} \cdots u_k^{\alpha_k})(u_{k+1}^{\alpha_{k+1}} \cdots u_n^{\alpha_n})$ for $1 \leq k \leq n - 1$ is called a *main cut* of x. Each factorization $x = (u_1^{\alpha_1} \cdots u_{k-1}^{\alpha_{k-1}} u_k^\beta)(u_k^\gamma u_{k+1}^{\alpha_{k+1}} \cdots u_n^{\alpha_n})$ with $\alpha_k = \beta + \gamma$ and $\beta, \gamma \neq 0$ is called a *secondary cut* of x.

5.1   Automata

Rational words are described by rational expressions using the product and the iteration ω. These expressions can be viewed as automata with special transitions for the ω-operation. Having a finite number of states, these automata are convenient to compute the prime factorization stated in Theorem 8. Besides, it turns out that these automata are a very special case of finite automata accepting transfinite words [1]: they just accepts a single word which is then rational.

The algorithm computing the prime factorization of rational word x, actually processes a finite automaton accepting x. As already mentioned, this automaton is made of two kinds of transitions corresponding respectively to letters and ω-powers.

The transitions associated with letters have the form $p \xrightarrow{a} q$ and are called *successor transitions*. Transition associated with ω-powers have the form $P \rightarrow q$ where P is a subset of states and are called *limit transitions*. The subset P is the set of states repeated in the corresponding ω-power.

Consider for instance the rational expression $(ba^\omega)^\omega a^\omega$. It is first flatten to give the word $(ba\omega)\omega a\omega$ over the extended alphabet $A \cup \{(,),\omega\}$. Let n be the number of letters in $A \cup \{\omega\}$ in this word. (parentheses are ignored). In our example, this number n is equal to 6. The integers from 0 to $n-1$ are then inserted before the letters in $A \cup \{\omega\}$ and the integer n is added at the end of the word to obtain the word $(0b1a2\omega3)\omega4a5\omega6$ over the alphabet $A \cup \{(,),\omega\} \cup \{0,1,\ldots,n\}$.

The automaton is then constructed as follows. For each $0 \leq i \leq n-1$, if i lies just before a letter $a \in A$, there is a successor transition $i \xrightarrow{a} (i+1)$. If i lies before a symbol ω, there are a successor transition from i and a limit transition defined as follows. Let $j-1$ be the integer just before the first letter a of the sub-expression under this ω. Note that j satisfies $j \leq i$. The successor transition is then the transition $i \xrightarrow{a} j$ and the limit transition is $\{j, j+1, \ldots, i\} \rightarrow (i+1)$. Note that both transitions $(j-1) \xrightarrow{a} j$ and $i \xrightarrow{a} j$ enter the same state j and have the same label.

The automaton obtained from the expression $(0b1a2\omega3)\omega4a5\omega6$ is pictured in Figure 1.

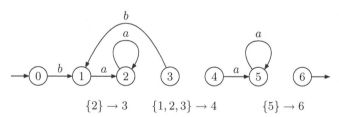

$$\{2\} \rightarrow 3 \qquad \{1,2,3\} \rightarrow 4 \qquad \{5\} \rightarrow 6$$

Fig. 1. Automaton for $(ba^\omega)^\omega a^\omega$

Given the automaton $\mathcal{A}$, we now introduce two families of automata called $_i\mathcal{A}_j$ and $_i\mathcal{A}_j^\#$ built from $\mathcal{A}$. Let $0 \leq i < j \leq n$ be two states such that there exists no backwards transition $k \xrightarrow{a} k'$ with $k' \leq i \leq k$. The automaton $_i\mathcal{A}_j$ is obtained by restricting the set of states to $[i, j]$, i being the initial state and j the final state. It accepts a unique transfinite word denoted $_ix_j$ which is a candidate for a prime factor.

The automaton $_i\mathcal{A}_j^{\#}$ is built by adding to $_i\mathcal{A}_j$ two transitions: a successor one and a limit one. The additional successor transition is $j \xrightarrow{a} (i + 1)$ where a is the label of the transition $i \xrightarrow{a} (i + 1)$. The additional limit transition is $\{i + 1, \ldots, j\} \to j$. This automaton accepts now the set $_ix_j^{\#} = \{_ix_j^{\alpha} | \alpha \text{ ordinal}\}$ of all powers of $_ix_j$.

5.2 Algorithm

We first define here a transformation τ on regular expressions. Given a regular expression e, $\tau(e)$ is another regular expression which defines the same word. This new expression permits the description of the prime factorization. The transformation τ is defined by induction on the expression as follows.

$$\tau(a) = a \qquad \tau(ee') = \tau(e)\tau(e') \qquad \tau(e^{\omega}) = \tau(e)\tau(e)^{\omega}$$

Example 9. If $e = (ba^{\omega})^{\omega}a^{\omega}$, then $\tau(e) = baa^{\omega}(baa^{\omega})^{\omega}aa^{\omega}$ and the corresponding automaton is pictured in Figure 2.

We denote by $|e|$ the *size* of a regular expression. This size is actually the number of letters in $A \cup \{\omega\}$ used in the expression. We also denote by $\mathrm{dp}(e)$ the *depth* that is the maximum number of nested ω in e. For any regular expression e, the relation $|\tau(e)| \leq 2^{\mathrm{dp}(e)}|e|$ holds.

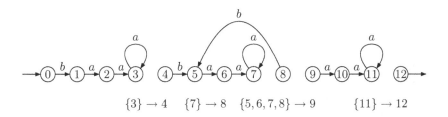

$$\{3\} \to 4 \qquad \{7\} \to 8 \qquad \{5, 6, 7, 8\} \to 9 \qquad \{11\} \to 12$$

Fig. 2. Automaton for $baa^{\omega}(baa^{\omega})^{\omega}aa^{\omega}$

Now we present the algorithm which computes the prime factorization of a rational word x. Such a word is given by a rational expression e. It turns out that, in general, e cannot be used to describe the prime factorization of x but the duplicated expression $\tau(e)$ can. The factorization is given by marking main and secondary cuts of x in $\tau(e)$ as illustrated in the following example. Consider the word x given by the expression $e = (ba^{\omega})^{\omega}a^{\omega}$. Then the duplicated expression $\tau(e)$ is $baa^{\omega}(baa^{\omega})^{\omega}aa^{\omega}$. The prime factorization of x is $b(a^{\omega}b)^{\omega}a^{\omega}$ that is $x = u_1u_2^{\omega}u_3^{\omega}$ where the three prime factors are $u_1 = b$, $u_2 = a^{\omega}b$ and $u_3 = a$. It can be given by inserting a marker | (respectively |) at main (respectively secondary) cuts in the expression $\tau(e)$ as $|b|aa^\omega(b|aa^\omega)^\omega|a|(a|)^\omega|$. Note that such a marking cannot be done in the expression e.

The algorithm given below works actually on the automaton $\mathcal{A}_{\tau(e)}$ associated to the expression $\tau(e)$. Rather that inserting markers in the expression, it marks states of $\mathcal{A}_{\tau(e)}$. Indeed it outputs two subsets Q_M and Q_S of distinguished states of the automaton $\mathcal{A}_{\tau(e)}$. As the states are in one-to-one correspondence with the positions in $\tau(e)$, distinguishing states is the same as inserting markers. In the above example, the algorithm produces the sets $Q_M = \{0, 1, 9, 12\}$ and $Q_S = \{5, 10, 11\}$.

Theorem 10. *Given a rational word x denoted by a regular expression e, there are two subsets Q_M and Q_S of states of $\mathcal{A}_{\tau(e)}$ such that the main and secondary cuts respectively correspond to states in Q_M and Q_S. Furthermore, these subsets can be computed in polynomial time in the size of $\tau(e)$.*

The algorithm is essentially inspired by its counterpart used in the classical case of finite words [3]. In this case, three variables i, j and k representing positions in the word are used. The variable i contains a position such that the prefix of the finite word ending at this position is already factorized. The variable j contains a position greater than i such that the factor between positions given by i and j is the possible next prime factor. The variable k is greater than j and contains the current position in the finite word. Moreover, to make the classical algorithm easier to understand, a fourth variable k' can be introduced. It contains a position in the possible next prime factor, this position ranges between the positions i and j. The classical algorithm is then directed by the comparison of the letters at positions k and k'. The algorithm for rational transfinite words proceeds similarly but uses states of automata rather positions in words. This allows the algorithm to run in a finite number of steps although the words are infinite. However, the algorithm must now cope with loops in the automata. This is solved, in our algorithm, by keeping track of already visited states. The algorithm produces two subsets of states of $\mathcal{A}$, the first one Q_M contains states corresponding to main cuts and the second one Q_S contains states corresponding to secondary cuts. Now we give a more detailed presentation of the algorithm.

Given a rational word x represented by an expression e, the algorithm, as already mentioned, works on the automaton $\mathcal{A}_{\tau(e)}$ denoted $\mathcal{A}$. The algorithm is described below. It uses also four variables i, j, k and k', but theses variables will not contain positions in the word x but rather states of $\mathcal{A}$ for the three first ones and a state of ${}_i\mathcal{A}_j^\#$ for the last one k'. Variable i contains the greatest state known to be in Q_M up to now. Variable j contains a state greater than i such that ${}_ix_j$ is a prime word which is a candidate for a prime factor. States k and k' are the current states of $\mathcal{A}$ and ${}_i\mathcal{A}_j^\#$ respectively. The algorithm simulates a run of the synchronized product $\mathcal{A} \times {}_i\mathcal{A}_j^\#$. It records the list of pairs of visited states in their appearance order. This list is called the *history* and its used to detect cycles. The last pair, called the *leading pair*, is the pair of states contained in variables k and k'. Let a_k and $a_{k'}$ be the unique letters labeling the successor transitions leaving k in $\mathcal{A}$ and k' in $_i\mathcal{A}_j^\#$. There are then three cases. The first one (Case 1) is $a_k = a_{k'}$ (Line 4 in the algorithm). The algorithm considers the pair $k \cdot a_k$ and $k' \cdot a_{k'}$ of states reached by the two successor transitions. If this pair is not yet in the history, it becomes the new leading pair and it is thus added to the history. If it is already in the history, both automata use limit transition giving raise to a new pair of states which becomes the new leading pair (Line 6). The second case (Case 2) is $a_k > a_{k'}$ (Line 14). The prime factor candidate $_i x_j$ has to be extended. Then, a new state is assigned to variable j. This may need the use of a limit transition in $\mathcal{A}$ (Lines 17 and 18). The set Q_S and the history are reset. The last case (Case 3) is $a_k < a_{k'}$ (Line 20 in the algorithm). The prime factor candidate $_i x_j$ is indeed a prime factor. Its last occurrence in x becomes the new main cut added to Q_M (Line 22).

1: Input: automaton $(\{0, \ldots, n\}, A, E, \{0\}, \{n\})$.
2: $i \leftarrow 0$, $j \leftarrow 1$, $k \leftarrow 1$, $k' \leftarrow 0$, $H = (i, j) = (0, 1)$, $Q_M \leftarrow \{0\}$, $Q_S \leftarrow \varnothing$
3: **while** $k < n$ **do**
4:       **if** $a_k = a_{k'}$ **then**
5:             $k \leftarrow k \cdot a_k$ in $\mathcal{A}$ and $k' \leftarrow k' \cdot a_{k'}$ in $_i\mathcal{A}_j^\#$
6:             **if** (k, k') occurs in H **then**
7:                   **if** j does not occur since the previous visit of (k, k') **then**
8:                         $k \leftarrow \max\{q \mid \exists q' \ (q, q') \text{ occurs in } H \text{ after } (k, k')\}$
9:                         $k' \leftarrow \max\{q' \mid \exists q \ (q, q') \text{ occurs in } H \text{ after } (k, k')\}$
10:                 **else**
11:                       $k \leftarrow \max\{q \mid \exists q' \ (q, q') \text{ occurs in } H \text{ after } (k, k')\}$
12:                       $k' \leftarrow j$
13:             $H \leftarrow H \cdot (k, k')$
14:       **else if** $a_k > a_{k'}$ **then**
15:             **if** $k \cdot a_k$ does not occur in H **then**
16:                   $j \leftarrow k \cdot a_k$
17:             **else**
18:                   $j \leftarrow \max\{q \mid \exists q' \ (q, q') \in H\} + 1$
19:             $k \leftarrow j$, $k' \leftarrow i$, $H \leftarrow (i, j)$
20:       **else**
21:             $Q_S \leftarrow Q_S \cup \{j\} \cup \{q \mid (q, j) \in H\}$.
22:             $i \leftarrow \max Q_S$, $Q_S \leftarrow Q_S \setminus \{i\}$, $Q_M \leftarrow Q_M \cup \{i\}$
23:             $j \leftarrow i \cdot a_i$, $k \leftarrow j$, $k' \leftarrow i$, $H \leftarrow (i, j)$
24: **Output** Q_M and Q_S

Algorithm 1. LYNDONFACTORIZE

A complete detailed description of the algorithm is given in the appendix. Its correctness follows from several invariants given in the appendix. Furthermore it can be shown that the number of steps of the algorithm is at most n^3 where n is the number of state of the automaton $\mathcal{A}_{\tau(e)}$.

6   Conclusion

To conclude, let us sketch a few problems that are raised by our work.

In order to obtain a prime factorization for each transfinite word, we have only required the sequence of prime words to be locally decreasing. It seems interesting to characterize those words which have a decreasing factorization. We have proved in Theorem 8 that rational words do have but they are not the only ones. The ω-word $x = aba^2ba^3b\cdots$ has also the decreasing factorization $x = ab \cdot a^2b \cdot a^3b\cdots$.

The algorithm given in Section 5.2 outputs the factorization of a rational word given by an expression e by inserting markers in the duplicated expression $\tau(e)$. Even if the complexity of this algorithm is polynomial in the size of $\tau(e)$, the algorithm is indeed exponential in the size of the expression e. This is due to the exponential blow up generated by the duplication. It could then be interesting to design a better algorithm: this new algorithm could determine which parts of the expression e have to be duplicated in order to get a better complexity. Along the same lines, it seems that it is possible to design an algorithm such that, given an expression e, it decides if the expression can be used to describe the factorization of the corresponding rational word without any duplication.

References

1. Büchi, J.R.: Transfinite automata recursions and weak second order theory of ordinals. In: Proc. Int. Congress Logic, Methodology, and Philosophy of Science, Jerusalem 1964, North Holland, pp. 2–23 (1965)
2. Carton, O., Choffrut, C.: Periodicity and roots of transfinite strings. Theoret. Informatics and Applications **35**(6), 525–533 (2001)
3. Duval, J.P.: Mots de Lyndon et périodicité. RAIRO Informat. Théor. **14**, 181–191 (1980)
4. Knuth, D.E.: Combinatorial Algorithms, The Art of Computer Programming. Addison-Wesley Professional, vol. 4A (2011)
5. Lothaire, M.: Combinatorics on Words, Encyclopedia of Mathematics and its Applications, vol. 17. Addison-Wesley, Reading (1983)
6. Lyndon, R.C.: On Burnside problem I. Trans. Am. Math. Soc. **77**, 202–215 (1954)
7. Lyndon, R.C.: On Burnside problem II. Trans. Am. Math. Soc. **78**, 329–332 (1954)
8. Melançon, G.: Lyndon factorization of infinite words. In: Puech, C., Reischuk, R. (eds.) STACS 96. LNCS, vol. 1046, pp. 147–154. Springer, Heidelberg (1996)
9. Melançon, G.: Viennot factorization of infinite words. Inf. Process. Lett. **60**(2), 53–57 (1996)
10. Rosenstein, J.G.: Linear Ordering. Academic Press, New York (1982)
11. Siromoney, R., Mathew, L., Dare, V.R., Subramanian, K.G.: Infinite Lyndon words. Inf. Process. Lett. **50**(2), 101–104 (1994)

Unary Patterns with Permutations

James Currie[1], Florin Manea[2], and Dirk Nowotka[2(✉)]

[1] Department of Mathematics and Statistics,
University of Winnipeg, Winnipeg, Canada
j.currie@uwinnipeg.ca
[2] Department of Computer Science, Kiel University, Kiel, Germany
{flm,dn}@informatik.uni-kiel.de

Abstract. Thue characterized completely the avoidability of unary patterns. Adding function variables gives a general setting capturing avoidance of powers, avoidance of patterns with palindromes, avoidance of powers under coding, and other questions of recent interest. Unary patterns with permutations have been previously analysed only for lengths up to 3. Consider a pattern $p = \pi_{i_1}(x) \ldots \pi_{i_r}(x)$, with $r \geq 4$, x a word variable over an alphabet Σ and π_{i_j} function variables, to be replaced by morphic or antimorphic permutations of Σ. If $|\Sigma| \geq 3$, we show the existence of an infinite word avoiding all pattern instances having $|x| \geq 2$. If $|\Sigma| = 3$ and all π_{i_j} are powers of a single π, the length restriction is removed. In general, the restriction on x cannot be removed, even for powers of permutations: for every positive integer n there exists N and a pattern $\pi^{i_1}(x) \ldots \pi^{i_n}(x)$ which is unavoidable over all Σ with $|\Sigma| \geq N$.

1 Introduction

The avoidability of patterns by infinite words is a core topic in combinatorics on words, going back to Thue [7,8]. Important results are surveyed in, e.g., [3,5].

Recently, a natural generalisation of classical patterns, in which functional dependencies between variables are allowed, has been considered [1,2,6]. More precisely, patterns consist of word variables, as usual, together with function variables (standing for either morphic or antimorphic extensions of permutations on the alphabet) which act on the words. For example, consider the pattern $x\pi(x)x\pi(x)$ whose instances are words $uvuv$ that consist of four parts of equal length, that is, $|u| = |v|$, where v is the image of u under some permutation of the alphabet. For example, $aab|bba|aab|bba$ (respectively, $aab|abb|aab|abb$) is an instance of $x\pi(x)x\pi(x)$ for the morphic (respectively, antimorphic) extension of permutation $a \mapsto b$ and $b \mapsto a$.

We note that, while patterns x^k describe all repetitions of some exponent k, patterns of the type $\pi^{i_1}(x) \ldots \pi^{i_k}(x)$ describe words that have an intrinsic repetitive structure, hidden by the application of the different iterations of the function π, which encode of the original root of the repetition.

Patterns with involutions were studied in [1,2]; motivation for considering involutions includes word reversal and DNA/RNA complementation. The main

I. Potapov (Ed.): DLT 2015, LNCS 9168, pp. 191–202, 2015.
DOI: 10.1007/978-3-319-21500-6_15

result obtained was that for each unary pattern with one variable involution, one can identify all alphabets over which it is avoidable. In the more general setting of patterns with permutations, the only results obtained so far (see [6]) regarded cube-like patterns under morphisms or antimorphisms (anti-/morphisms, for short) which are powers of a single (variable) permutation, i.e., patterns of the form $\pi^i(x)\pi^j(x)\pi^k(x)$, where $i, j, k \geq 0$. The avoidability of such patterns was completely characterised: for each $\pi^i(x)\pi^j(x)\pi^k(x)$ one can determine exactly the alphabets over which the pattern is avoidable. Contrary to both the classical and to the involution settings, where once a pattern is avoidable for some alphabet size it remains avoidable in larger alphabets, a cubic pattern with permutations may become unavoidable over a larger alphabet.

We extend the results of [6] as follows. First, we construct a ternary word that avoids all patterns $\pi_{i_1}(x)\ldots\pi_{i_r}(x)$ where $r \geq 4$, x a word variable over some alphabet Σ, with $|x| \geq 2$ and $|\Sigma| \geq 3$, and the π_{i_j} function variables that may be replaced by anti-/morphic permutations of Σ. This is the first result where the avoidability of patterns involving more functions, which are not powers of the same initial variable permutation, has been shown; even more, we do not restrict these functions so that all have the same type: we can mix both morphic and antimorphic permutations. On the down side, the result only works when we restrict the length of x to be at least 2. However, we also show that such a restriction is needed. Indeed, for each $n \geq 1$ there exists a unary pattern $\pi_1(x)\ldots\pi_n(x)$ where all functions are powers of the same anti-/morphic permutation π, i.e., $\pi_j = \pi^{i_j}$ with $1 \leq j \leq n$, and an integer N such that $\pi^{i_1}(x)\ldots\pi^{i_n}(x)$ has as instances all the words of length n over an alphabet of size at least N; in other words, $\pi^{i_1}(x)\ldots\pi^{i_n}(x)$ is unavoidable over all alphabets Σ with $|\Sigma| \geq N$. In between these two results, we show that for $n \geq 4$ all patterns $\pi^{i_1}(x)\ldots\pi^{i_n}(x)$ under anti-/morphic permutations are avoidable in Σ_3. So, just like in the case of cubes with permutations, there are patterns under anti-/morphic permutations (including the eventually unavoidable patterns we construct) which are avoidable in small alphabets (here, Σ_3) but become unavoidable in larger alphabets. On the other hand, unlike the case of cubes with permutations, where there exist patterns unavoidable in Σ_2 and Σ_3 (e.g., $x\pi(x)\pi^2(x)$, see [6]), all unary patterns of length at least 4 under anti-/morphic permutations are avoidable in both Σ_2 (see [1]) and Σ_3 (as shown here).

2   Definitions

We freely use the usual notations of combinatorics on words. (See [5], for example.) Define alphabets $\Sigma_k = \{0, \ldots, k-1\}$ and $\Sigma'_k = \{1, 2, \ldots, k\}$. We use w^R, to denote the reversal of word w.

A morphism f (respectively, antimorphism) of Σ_k^* is defined by its values on letters; $f(uv) = f(u)f(v)$ (respectively, $f(uv) = f(v)f(u)$) for all words $u, v \in \Sigma_k^*$. When we define an anti-/morphism it is enough to define $f(a)$, for all $a \in \Sigma_k$. If the restriction of f to Σ_k, is a permutation of Σ_k, we call f an anti-/morphic permutation. Denote by $\mathbf{ord}(f)$ the order of f, i.e., the minimum positive integer m such that f^m is the identity. If $\mathbf{ord}(f) = 2$, we call f an involution. If $a \in \Sigma_k$ is a letter, the order of a with respect to f, denoted $\mathbf{ord}_f(a)$, is the minimum number m such that $f^m(a) = a$.

A pattern which involves functional dependencies is a term over (word) variables and function variables (where concatenation is an implicit functional constant); a pattern with only one word variable is called unary. For example, $x\pi(x)\pi(\pi(x))x = x\pi(x)\pi^2(x)x$ is a unary pattern involving the variable x and the function variable π. An instance of a pattern p in Σ_k is the result of substituting every variable by a word in Σ_k^+ and every function variable by a function over Σ_k^*. A pattern is avoidable in Σ_k if there is an infinite word over Σ_k that does not contain any instance of the pattern.

In this paper, we consider patterns with morphic and antimorphic permutations, that is, all function variables are substituted by morphic or antimorphic permutations only.

Let $\mathbf{h}$ be the infinite word defined as $\mathbf{h} = \lim_{n\to\infty} \phi_h^n(0)$, where $\phi_h : \Sigma_3^* \to \Sigma_3^*$ is a morphism due to Hall [4], defined by $\phi_h(0) = 012$, $\phi_h(1) = 02$ and $\phi_h(2) = 1$. For the simplicity of the exposure, if $\mathbf{h} = \prod_{i=0}^{\infty} h_i$ with $h_i \in \Sigma_3$, we define the infinite word $\mathbf{v}$ over Σ_3' as $\mathbf{v} = \prod_{i=0}^{\infty} v_i$, with $v_i = h_i + 1$. The infinite word $\mathbf{v}$ avoids squares xx and does not contain the factors 121 and 323.

We investigate the factors of an infinite word $\mathbf{g}$ that have the form

$$\pi_{i_1}(x)\pi_{i_2}(x)\ldots\pi_{i_r}(x)$$

with x a non-empty word and each π_{i_j} a morphic or antimorphic permutation for $1 \le j \le r$. Replacing x by $\pi_{i_1}^{-1}(x)$ and $\pi_{i_j}(x)$ by $\pi_{i_j}(\pi_{i_1}^{-1}(x))$ for $1 \le j \le r$, this is equivalent to investigating factors of $\mathbf{g}$ of the form $x\pi_{j_1}(x)\ldots\pi_{j_{r-1}}(x)$ with x a non-empty word, and each π_{j_ℓ} a morphic or antimorphic permutation for all $1 \le \ell \le r - 1$.

3   A General Result

We use the word $\mathbf{v}$ defined above to define the word $\mathbf{u} \in \Sigma_3^\omega$ given by

$$\mathbf{u} = \prod_{i=0}^{\infty} (0^{v_{3i}} 1^{v_{3i+1}} 2^{v_{3i+2}}).$$

Theorem 1. *The word $\mathbf{u}$ has no factor of the form $x\pi_i(x)\pi_j(x)\pi_k(x)$ with $|x| \ge 2$ and π_i, π_j and π_k are each a morphic or antimorphic permutation.*

Proof. **(Morphic case)** Suppose, to the contrary, that $\mathbf{u}$ has a factor $w = x\pi_i(x)\pi_j(x)\pi_k(x)$ with $|x| \ge 2$, where each π_r is a morphic permutation. We consider the block structure of x; that is, we parse x as

$$x = a_1^{k_1} a_2^{k_2} \cdots a_{n-1}^{k_{n-1}} a_n^{k_n}$$

where the $a_i \in \Sigma_3$, with $a_\ell \neq a_{\ell+1}$, $k_\ell \geq 1$, $1 \leq \ell \leq n$. Certainly, $\pi_r(x)$ has the same block structure for each r:

$$\pi_r(x) = (\pi_r(a_1))^{k_1}(\pi_r(a_2))^{k_2} \cdots (\pi_r(a_{n-1}))^{k_{n-1}}(\pi_r(a_n))^{k_n}$$

and letters $\pi_r(a_\ell)$ and $\pi_r(a_{\ell+1})$ are distinct, since π_r is a permutation.

We consider several cases based on n, k_1 and k_n as follows:

Case 1: $n = 1$. This means that $w = a_1^{k_1}(\pi_i(a_1))^{k_1}(\pi_j(a_1))^{k_1}(\pi_k(a_1))^{k_1}$.

Since $|x| \geq 2$, we have $k_1 \geq 2$. If $a_1 = \pi_i(a_1)$, then w contains the factor $a_1^{k_1}(\pi_i(a_1))^{k_1} = a_1^{2k_1}$. Since $2k_1 \geq 4$, this is impossible; the block lengths in **u** are 1, 2 or 3. We conclude that $a_1 \neq \pi_i(a_1)$. Similarly, $\pi_i(a_1) \neq \pi_j(a_1)$, and $\pi_j(a_1) \neq \pi_k(a_1)$. It follows that, in the context of w, $(\pi_i(a_1))^{k_1}$ and $(\pi_j(a_1))^{k_1}$ are successive blocks of **u**; however, this implies that $k_1 k_1$ is a factor of **v**. Since **v** is square-free, this is impossible.

Case 2a: $n > 1$, and $k_1 = 3$ or $k_n = 3$ Suppose $k_1 = 3$. This implies that $a_n \neq \pi_i(a_1)$; otherwise w contains a block $a_n^{k_n}(\pi_i(a_1))^3 = a_n^{k_n+3}$, of length 4 or greater. Similarly, $\pi_i(a_n) \neq \pi_j(a_1)$ and $\pi_j(a_n) \neq \pi_k(a_1)$. Each $(\pi_i(a_\ell))^{k_\ell}$ and $(\pi_j(a_\ell))^{k_\ell}$ is thus a complete block of **u**, and **v** contains the factor $(k_1 k_2 \cdots k_n)^2$. This is impossible. Similarly, one argues that $k_n = 3$ gives a contradiction.

Case 2b: $n > 1$, and $k_1, k_n \leq 2$ If $a_n = \pi_i(a_1)$ and $\pi_i(a_n) = \pi_j(a_1)$, then **u** contains the factor

$$a_{n-1}a_n^{k_n+k_1}(\pi_i(a_2))^{k_2} \cdots (\pi_i(a_{n-1}))^{k_{n-1}}\pi_i(a_n)^{k_n+k_1}(\pi_j(a_2))^{k_2} \cdots (\pi_j(a_{n-1}))^{k_{n-1}}\pi_j(a_n),$$

and **v** contains the square factor $((k_n + k_1)k_2 k_3 \cdots k_{n-1})^2$, which is impossible. Similarly, if $a_n \neq \pi_i(a_1)$ and $\pi_i(a_n) \neq \pi_j(a_1)$, then **v** contains the factor

$$a_1 a_2^{k_2} \cdots a_n^{k_n}(\pi_i(a_1))^{k_1}(\pi_i(a_2))^{k_2} \cdots (\pi_i(a_n))^{k_n}(\pi_j(a_1))^{k_1}\pi_j(a_2),$$

and then v contains the factor $(k_2 k_3 \cdots k_1)^2$, which is again impossible. In conclusion, exactly one of the equations $a_n = \pi_i(a_1)$ and $\pi_i(a_n) = \pi_j(a_1)$ holds. Similarly, exactly one of $\pi_i(a_n) = \pi_j(a_1)$ and $\pi_j(a_n) = \pi_k(a_1)$ holds.

Case 2bi: $k_1, k_n \leq 2$, and $n \geq 3$. Suppose that $a_n = \pi_i(a_1)$ and $\pi_i(a_n) \neq \pi_j(a_1)$. (The other case is similar.)

Since $\pi_i(a_n) \neq \pi_j(a_1)$, but

$$\pi_i(a_{n-1})(\pi_i(a_n))^{k_n}(\pi_j(a_1))^{k_1}\pi_j(a_2)$$

is a factor of **u**, we see that $k_n k_1$ is a factor of **v**, whence $k_n \neq k_1$. Since we have already reasoned that $k_1, k_n \leq 2$, we see that $k_1 + k_n = 3$. Now $a_{n-2}(a_{n-1})^{k_{n-1}}a_n^3$ is a factor of **u**, so that $k_{n-1} \neq 3$. On the other hand, since

$$\pi_i(a_{n-2})(\pi_i(a_{n-1}))^{k_{n-1}}(\pi_i(a_n))^{k_n}\pi_j(a_1)$$

is a factor of **u**, and $\pi_i(a_n) \neq \pi_j(a_1)$, we conclude that $k_{n-1}k_n$ is a factor of **v**; therefore, $k_{n-1} \neq k_n$, and since $k_n, k_{n-1} < 3$, we have $k_{n-1} = 3 - k_n = k_1$.

Similar reasoning shows that $k_2 = k_n$. But then

$$\pi_i(a_{n-2})(\pi_i(a_{n-1}))^{k_{n-1}}(\pi_i(a_n))^{k_n}(\pi_j(a_1))^{k_1}(\pi_j(a_2))^{k_2}\pi_j(a_3)$$

is a factor of $\mathbf{u}$, so that $k_{n-1}k_nk_1k_2 = (k_1k_n)^2$ is a factor of $\mathbf{v}$. This is impossible.
Case 2bii: $k_1, k_n \leq 2$, **and** $n = 2$. We make four subcases, depending on whether $(k_1, k_2) = (1, 2)$ or $(2, 1)$, and on whether $a_2 = \pi_i(a_1)$, $\pi_i(a_2) \neq \pi_j(a_1)$ and $\pi_j(a_2) = \pi_k(a_1)$, or alternatively, $a_2 \neq \pi_i(a_1)$, $\pi_i(a_2) = \pi_j(a_1)$ and $\pi_j(a_2) \neq \pi_k(a_1)$.

1. $(k_1, k_2) = (1, 2)$, $a_2 = \pi_i(a_1)$, $\pi_i(a_2) \neq \pi_j(a_1)$, $\pi_j(a_2) = \pi_k(a_1)$:
   In this case, $\mathbf{u}$ contains the word

   $$a_1 a_2^2 \pi_i(a_1)(\pi_i(a_2))^2 \pi_j(a_1)(\pi_j(a_2))^2 \pi_k(a_1)(\pi_k(a_2))^2$$

   $$= a_1 a_2^3 (\pi_i(a_2))^2 \pi_j(a_1)(\pi_j(a_2))^3 (\pi_k(a_2))^2$$

   so that $\mathbf{v}$ contains a word $\alpha 3213\beta$, $\alpha, \beta \in \{1, 2, 3\}$, $\beta \geq 2$. In fact, if $\beta = 3$, then $\mathbf{v}$ contains the square 3^2. Assume then that $\beta = 2$. Thus 32132 is a factor of $\mathbf{v}$; however, 32132 has no right extension in $\mathbf{v}$, since 321321 and 321322 end in squares, while 321323 ends in 323. This is impossible.
2. $(k_1, k_2) = (2, 1)$, $a_2 = \pi_i(a_1)$, $\pi_i(a_2) \neq \pi_j(a_1)$, $\pi_j(a_2) = \pi_k(a_1)$:
   In this case, $\mathbf{u}$ contains the word

   $$a_1^2 a_2 (\pi_i(a_1))^2 \pi_i(a_2)(\pi_j(a_1))^2 \pi_j(a_2)(\pi_k(a_1))^2 \pi_k(a_2)$$

   $$= a_1^2 a_2^3 \pi_i(a_2)(\pi_j(a_1))^2 (\pi_j(a_2))^3 \pi_k(a_2)$$

   so that $\mathbf{v}$ contains a word $\alpha 3123\beta$, $\alpha, \beta \in \{1, 2, 3\}$, $\alpha \geq 2$. In fact, if $\alpha = 3$, then $\mathbf{v}$ contains 3^2. Assume then that $\alpha = 2$, so that $\mathbf{v}$ contains 23123. Since $\mathbf{v}$ is recurrent, 23123 must have a left extension in $\mathbf{v}$; however, none of 123123, 223123 and 323123 is a possible factor of $\mathbf{v}$.
3. $(k_1, k_2) = (1, 2)$, $a_2 \neq \pi_i(a_1)$, $\pi_i(a_2) = \pi_j(a_1)$, $\pi_j(a_2) \neq \pi_k(a_1)$:
   In this case, w contains the word

   $$a_1 a_2^2 \pi_i(a_1)(\pi_i(a_2))^2 \pi_j(a_1)(\pi_j(a_2))^2 \pi_k(a_1)(\pi_k(a_2))^2$$

   $$= a_1 a_2^2 \pi_i(a_1)(\pi_i(a_2))^3 (\pi_j(a_2))^2 \pi_k(a_1)(\pi_k(a_2))^2$$

   so that v contains a word $\alpha 21321\beta$. No left extension of this word is a factor of $\mathbf{v}$.
4. $(k_1, k_2) = (2, 1)$, $a_2 \neq \pi_i(a_1)$, $\pi_i(a_2) = \pi_j(a_1)$, $\pi_j(a_2) \neq \pi_k(a_1)$: In this case, w contains the word

   $$a_1^2 a_2 (\pi_i(a_1))^2 \pi_i(a_2)(\pi_j(a_1))^2 \pi_j(a_2)(\pi_k(a_1))^2 \pi_k(a_2)$$

   $$= a_1^2 a_2 (\pi_i(a_1))^2 (\pi_i(a_2))^3 \pi_j(a_2)(\pi_k(a_1))^2 \pi_k(a_2)$$

   so that v contains a word $\alpha 12312\beta$. No right extension of this word is a factor of $\mathbf{v}$.

We see that w contains no instance $x\pi_i(x)\pi_j(x)\pi_k(x)$ with $|x| \geq 2$ where each π_r is a morphic permutation.

(Antimorphic case). Suppose, for the sake of getting a contradiction, that **u** has a factor $w = x\pi_i(x)\pi_j(x)\pi_k(x)$ with $|x| \geq 2$, where one of the π_r is an antimorphic permutation.

For notational simplicity, we will suppose that π_i is antimorphic; the other cases are similar.

We consider the block structure of x:

$$x = a_1^{k_1} a_2^{k_2} \cdots a_{n-1}^{k_{n-1}} a_n^{k_n}$$

where the $a_i \in \Sigma_3$, with $a_\ell \neq a_{\ell+1}$, $k_\ell \geq 1, 1 \leq \ell \leq n$. Since π_i is antimorphic,

$$\pi_i(x) = (\pi_i(a_n))^{k_n} (\pi_i(a_{n-1}))^{k_{n-1}} \cdots (\pi_i(a_2))^{k_2} (\pi_i(a_1))^{k_1}.$$

If $k_n = 3$, then **u** has the factor $a_n^{k_n} \pi_i(a_n)^{k_n}$, and either **u** has a block of length 6 (if $a_n = \pi_i(a_n)$), or **v** has a factor 33; both cases are impossible.

If $k_n = 2$, we make cases based on n: If $n = 1$, then $w = a_1^2 \pi_i(a_1^2)\pi_j(a_1^2)\pi_k(a_1^2)$, and the factor $\pi_i(a_1^2)\pi_j(a_1^2)$ of w implies that either **u** has a block of length 4, or **v** has a factor 22; both cases are impossible.

If $n > 1$, then factor $a_{n-1}^{k_{n-1}} a_n^{k_n} \pi_i(a_n^{k_n})\pi_i(a_{n-1}^{k_{n-1}})$ gives the same contradiction. We conclude that $k_n = 1$. Since $|x| \geq 2$, we have $n \geq 2$. If $a_n \neq \pi_i(a_n)$, then the factor $a_{n-1}^{k_{n-1}} a_n^1 \pi_i(a_n^1)\pi_i(a_{n-1}^{k_{n-1}})$ of w implies that 11 is a factor of **v**, which is impossible. We conclude that $a_n = \pi_i(a_n)$.

Suppose $|x| \geq 3$. If $k_{n-1} = 1$, then w contains $a_{n-2}^{k_{n-2}} a_{n-1}^1 a_n^2 \pi_i(a_{n-1}^1)\pi_i(a_{n-2})$, so that **v** has the factor 121. This is impossible. Thus $k_{n-1} > 1$. This forces **u** to contain a block $a_{n-1}^y a_n^2 \pi_i(a_{n-1}^z)$, where $y, z \geq 2$ and $y2z$ is a factor of **v**. However, then **v** has 22 or 323 as a factor, which is impossible. We conclude that $|x| = 2$. It follows that $n = 2$, $k_1 = k_2 = 1$.

Write $w = a_1 a_2 b_1 b_2 c_1 c_2 d_1 d_2$, each $a_i, b_i, c_i, d_i \in \Sigma_3$, and $a_1 \neq a_2$, $b_1 \neq b_2$, $c_1 \neq c_2$, $d_1 \neq d_2$. We have arrived at this case by considering the word $x\pi_i(x)$, assuming that π_i is antimorphic. If, instead, π_i is morphic and π_j is antimorphic, (resp., π_i and π_j are morphic, π_k is antimorphic) the same analysis goes through considering the word $\pi_i(x)\pi_j(x)$ (resp., $\pi_j(x)\pi_k(x)$).

We must have $a_2 = b_1$, or **v** contains the square 11. Similarly, $b_2 = c_1$. Now, however, **v** contains the square 22. This is a contradiction.   □

Consequently, **u** has no factor of the form $\pi_\ell(x)\pi_i(x)\pi_j(x)\pi_k(x)$ with $|x| \geq 2$ and π_i, π_j and π_k are each a morphic or antimorphic permutation. This means that **u** has not factor that contains, at its turn, an instance of a pattern $\pi_\ell(x)\pi_i(x)\pi_j(x)\pi_k(x)$ with $|x| \geq 2$ and π_i, π_j and π_k are each a morphic or antimorphic permutation. So, the following general theorem follows.

Theorem 2. *The word **u** has no factor of the form $\pi_{i_1}(x)\pi_{i_2}(x)\ldots\pi_{i_r}(x)$ with $|x| \geq 2$, $r \geq 4$, and π_{i_j} is a morphic or antimorphic permutation for $1 \leq j \leq r$.*

4   Avoidability for Ternary Alphabets

We now show that all patterns of length at least 4 under anti-/morphisms which are powers of the same permutation are avoidable in Σ_3. More precisely, we show that for each pattern $\mathcal{P} = x\pi^i(x)\pi^j(x)\pi^k(x)$ there exists an infinite word (that depends on $\mathcal{P}$) that does not contain any instance of $\mathcal{P}$ with π an anti-/morphic permutation of Σ_3.

Let us note from the beginning that the permutations of Σ_3 are either cycles (i.e., $\mathbf{ord}(a) = 3$ for all $a \in \Sigma_3$) or involutions (i.e., $\mathbf{ord}(a) \leq 2$ for all $a \in \Sigma_3$).

We use as a basic lemma the following corollary of Theorem 1.

Corollary 1. *The word* $\mathbf{u}$ *has no factor of the form* $x\pi^i(x)\pi^j(x)x$, *where* π *is an anti-/morphic permutation of* Σ_3.

Proof. By Theorem 1 we know that $\mathbf{u}$ has no factor of the form $x\pi^i(x)\pi^j(x)x$ with $|x| \geq 2$, by just taking in the statement of the theorem $\pi_i = \pi^i$, $\pi_j = \pi^j$, and π_k the identity on Σ_3. We assume, for the sake of a contradiction, that $\mathbf{u}$ has a factor $x\pi^i(x)\pi^j(x)x$ with $|x| = 1$. Say $x = a \in \Sigma_3$. Due to the form of $\mathbf{u}$ we get that between the two occurrences of the letter $x = a$ we must find the two other letters of Σ_3 (that is, both letters from $\Sigma_3 \setminus \{a\} = \{b, c\}$). Indeed $\mathbf{u}$ does not contain a block of 4 consecutive identical letters, so the two occurrences of the letter $x = a$ belong to separate maximal blocks made of letters $x = a$ of the word $\mathbf{u}$, and between two such blocks the other two letters of Σ_3 must occur. But this would mean that $\mathbf{u}$ contains the factor $abca$, so $\mathbf{h}$ should contain the factor 11, a contradiction.   □

The following lemma is immediate, as $\mathbf{v}$ avoids squares.

Lemma 1. *The word* $\mathbf{v}$ *has no factor of the form* $xx\pi^j(x)\pi^k(x)$, $x\pi^i(x)\pi^i(x)x$, $x\pi^i(x)\pi^j(x)\pi^j(x)$ *where* π *is an anti-/morphic permutation and* i, j, k *are non-negative integers.*

In [6] the following was shown.

Lemma 2. *There exists an infinite word* $\mathbf{v}_m$ *(respectively,* $\mathbf{v}_a$*) over* Σ_3 *that has no factor of the form* $x\pi(x)x$, *when* π *is replaced by a morphic (respectively, antimorphic) permutation.*

The final result we need is from [1].

Lemma 3. *For each pattern* $\mathcal{P} = x\pi^i(x)\pi^j(x)\pi^k(x)$, *where* i, j, k *are non-negative integers, there exists an infinite word* $\mathbf{u}_{\mathcal{P}} \in \Sigma_2^\omega$ *(respectively,* $\mathbf{u}'_{\mathcal{P}} \in \Sigma_2^\omega$*) that does not contain an instance of* $\mathcal{P}$ *when* π *is replaced by a morphic (respectively, antimorphic) involution of* Σ_2.

We can now show the two main results of this section.

Lemma 4. *For each pattern* $\mathcal{P} = x\pi^i(x)\pi^j(x)\pi^k(x)$, *where* i, j, k *are non-negative integers, there exists an infinite ternary word* $\mathbf{w}_{\mathcal{P}}$ *that does not contain any instance of this pattern with* π *a morphic permutation of* Σ_3.

Proof. Clearly, each morphic permutation π of Σ_3 is either a cycle or an involution. In all cases, π^6 is the identity morphism on Σ_3^*. Consequently, we can replace the exponents i, j, k by their values modulo 6.

By Corollary 1 and Lemmas 1 and 2, all the patterns $x\pi^i(x)\pi^j(x)\pi^k(x)$ with one of i, j, k equal to 0 and π replaced by a morphic permutation are avoidable, either by $\mathbf{v}$ (when $i = 0$), either by $\mathbf{v}_m$ (when $j = 0$), or by $\mathbf{u}$ (when $k = 0$). Similarly, the patterns $x\pi^i(x)\pi^j(x)\pi^k(x)$ with $i = k$ are avoided by $\mathbf{v}_m$, since this word does not contain instances of any pattern $\pi^i(x)\pi^j(x)\pi^i(x)$, while those with $i = j$ or $j = k$ are avoided by $\mathbf{v}$.

Consequently, we only have to consider the case when $0, i, j, k$ are pairwise distinct, and each is at most 5 in the following.

We look at the reminders of i, j and k modulo 3.

If $\{i(\mathrm{mod}\ 3), j(\mathrm{mod}\ 3), k(\mathrm{mod}\ 3)\} = \{1, 2\}$, we get that when replacing π with a cycle of Σ_3 (e.g., $\pi(0) = 1, \pi(1) = 2, \pi(2) = 3$), the instance of $\mathcal{P} = x\pi^i(x)\pi^j(x)\pi^k(x)$ will contain all the three letters 0, 1, and 2. It follows that $\mathbf{u}_\mathcal{P}$ (from Lemma 3) avoids p. Indeed, when π is replaced by an involution of Σ_2 the result follows from the definition of $\mathbf{u}_\mathcal{P}$, while when π is replaced by any other permutation of Σ_3, its instances will contain the letter 2, so $\mathbf{u}_\mathcal{P}$ canonically avoids all of them.

So, the only case left to consider is when $\{i(\mathrm{mod}\ 3), j(\mathrm{mod}\ 3), k(\mathrm{mod}\ 3)\}$ is either $\{0, 1\}$ or $\{0, 2\}$. If i, j, k are all equal modulo 3 it follows that at least two of them are actually equal, a contradiction to our earlier assumption that each two of the exponents are different. So, one of i, j, and k should be 3.

It is not hard to see that $x\pi^i(x)\pi^3(x)\pi^k(x)$ is avoided by $\mathbf{v}$. Indeed, an instance of this pattern will always contain a square. In the case when π is a cycle of Σ_3 we can only obtain words which have the form $xf(x)xf(x)$ for some morphic permutation f of Σ_3, while for π an involution the words we obtain definitely contain an instance of either xx or $\pi(x)\pi(x)$. So, in all cases, the instances of $x\pi^i(x)\pi^3(x)\pi^k(x)$ contain squares. By a similar argument, every pattern $x\pi^i(x)\pi^j(x)\pi^3(x)$ or $x\pi^3(x)\pi^j(x)\pi^k(x)$ is avoided by $\mathbf{v}$, as each instance of such a pattern contains a square. $\square$

Lemma 5. *For each pattern* $\mathcal{P} = x\pi^i(x)\pi^j(x)\pi^k(x)$ *, where* i, j, k *are nonnegative integers, there exists an infinite ternary word* $\mathbf{w}_\mathcal{P}$ *that does not contain any instance of this pattern with* π *an antimorphic permutation of* Σ_3.

Proof. Just like in the previous proof, for each antimorphic permutation π of Σ_3, we have that π^6 is the identical morphism on Σ_3^*. Consequently, we can replace the exponents i, j, k by their values modulo 6.

Using Lemma 2, we get that the patterns $x\pi^i(x)\pi^j(x)\pi^k(x)$ with one of i, j, k equal to 0 and π replaced by a antimorphic permutation are avoidable, either by $\mathbf{v}$ (when $i = 0$), either by $\mathbf{v}_a$ (when $j = 0$), or by $\mathbf{u}$ (when $k = 0$). The patterns $x\pi^i(x)\pi^j(x)\pi^k(x)$ with $i = k$ are avoided by $\mathbf{v}_a$, while those with $i = j$ or $j = k$ contain squares, so are avoided by $\mathbf{v}$.

So, just like before, we only have to consider in the following the case when each two of $0, i, j, k$ are distinct and each is at most 5. And, again, if we have that $\{i(mod\ 3), j(mod\ 3), k(mod\ 3)\} = \{1, 2\}$, we get that when replacing π with a cycle of Σ_3 the instance of $\mathcal{P} = x\pi^i(x)\pi^j(x)\pi^k(x)$ will contain all the three letters $0, 1$, and 2. So, by Lemma 3, it follows that $\mathbf{u}'_{\mathcal{P}}$ avoids $\mathcal{P}$.

Hence, the only case left to consider is when $\{i(mod\ 3), j(mod\ 3), k(mod\ 3)\}$ is either $\{0, 1\}$ or $\{0, 2\}$. The simple case is again when i, j, k are all equal modulo 3, as it follows that at least two of them are actually equal, which is a contradiction to our assumption that each two of the exponents are different. So, one of i, j, and k should be 3. A more detailed analysis is needed now.

Let us first look at patterns $x\pi^3(x)\pi^j(x)\pi^k(x)$. Obviously, j and k are not of the same parity; actually, the pair (j, k) is one of the pairs $(1, 4), (4, 1), (2, 5), (5, 2)$. Generally, when substituting π with a cycle of Σ_3, the pattern $x\pi^3(x)\pi^j(x)\pi^k(x)$ equals $xx^R\pi^j(x)\pi^k(x)$. But the instances of $xx^R\pi^j(x)\pi^k(x)$ always contain a square: the last letter of x equals the first, where i, j, k are non-negative integers, letter of x^R. When substituting π with an involution of Σ_3, the pattern $x\pi^3(x)\pi^j(x)\pi^k(x)$ either equals $x\pi(x)x\pi(x)$ if j is even and k is odd, or $x\pi(x)\pi(x)x$ if j is odd and k is even. Also in these cases the instances of the pattern contain squares. So, every instance of the pattern $x\pi^3(x)\pi^j(x)\pi^k(x)$ contains a square. This means that $\mathbf{v}$ avoids $x\pi^3(x)\pi^j(x)\pi^k(x)$.

Next we consider the patterns $x\pi^i(x)\pi^j(x)\pi^3(x)$. Like before, i and j do not have the same parity as (i, j) must be one of the pairs $(1, 4), (4, 1), (2, 5), (5, 2)$. Let us assume that i is even and j is odd. If π is a cycle, we have that a factor of the form $x\pi^i(x)\pi^j(x)\pi^3(x)$ has the form $xf(x)f(x^R)x^R$ for some morphic permutation f, which contains the square formed by the last letter of $f(x)$ and the first letter of $f(x^R)$. If π is an involution then each factor of the form $x\pi^i(x)\pi^j(x)\pi^3(x)$ starts with xx. Therefore, $\mathbf{v}$ avoids $x\pi^i(x)\pi^j(x)\pi^3(x)$ with i even and j odd. Further, we assume that i is odd and j is even. If π is a cycle, we have that a factor of the form $x\pi^i(x)\pi^j(x)\pi^3(x)$ has the form $xf(x^R)f(x)x^R$ for some morphic permutation f, which contains the square formed by the last letter of $f(x^R)$ and the first letter of $f(x)$. If π is an involution then each factor of the form $x\pi^i(x)\pi^j(x)\pi^3(x)$ is, in fact, $x\pi(x)x\pi(x)$. So, $\mathbf{v}$ avoids $x\pi^i(x)\pi^j(x)\pi^3(x)$ also for i odd and j even.

Finally, we consider the patterns $x\pi^i(x)\pi^3(x)\pi^k(x)$. Let us assume first that i is odd; consequently, k is even (the pair (i, k) is either $(1, 4)$ or $(2, 5)$). By Theorem 1, the word $\mathbf{u}$ contains no instances of $x\pi^i(x)\pi^3(x)\pi^k(x)$ with $|x| \geq 2$. We show that $\mathbf{u}$ does not contain instances of this pattern with $|x| = 1$. Assume that $x = a \in \Sigma_3$. If π is a cycle then the factors $x\pi^i(x)\pi^3(x)\pi^k(x)$ are, in fact, $abab$ with $b \in \Sigma_3$ such that $\pi^i(x) = b$. If π is an involution then the factors $x\pi^i(x)\pi^3(x)\pi^k(x)$ are $abba$ with $b \in \Sigma^3$ such that $\pi(x) = b$. By the structure of $\mathbf{u}$ (which has the form $(0^+1^+2^+)^\omega$), we get that it cannot contain such factors. So $\mathbf{u}$ avoids such patterns.

We now consider the case when i is even and k is odd. Let us write the Thue-Morse word as $\mathbf{t} = \prod_{i=0}^{\infty} t_i$ with $t_i \in \{0, 1\}$. Consider the word $\mathbf{t}' \in \{0, 1\}^\omega$ (also used in [1]) given by $\mathbf{t}' = \prod_{i=0}^{\infty} 01^{t_i+2}$.

We show now that $\mathbf{t}'$ avoids $x\pi^i(x)\pi^3(x)\pi^k(x)$ with i even and k odd. If π is a cycle then $x\pi^i(x)\pi^3(x)\pi^k(x)$ equals $xf(x)x^Rf(x^R)$ for some morphic permutation f (which is also a cycle). If x starts with 0, then $f(x)$ starts with 1, x^R ends with 0, and $f(x^R)$ ends with 1. But 0101 is not a factor $\mathbf{t}'$ (there are always at least 2 symbols 1 in a block). Thus, if $\mathbf{t}'$ contains an instance of $xf(x)x^Rf(x^R)$ with x starting with 0, then $|x| \geq 2$. Now, 0 is always followed by an 1, so x should start with 01. This means that $f(x)$ starts with 10, x^R ends with 10, and $f(x^R)$ ends with 01. Clearly, 01100110 is not a factor of $\mathbf{t}'$ (as this infinite word does not contain consecutive 0 letters), so if $\mathbf{t}'$ contains an instance of our pattern, then $|x| \geq 3$. Now, as $f(x^R)$ ends with 01 and there are no two consecutive 0's in $\mathbf{t}'$ we get that $f(x^R)$ should end with 101. This means that x should start with 010, a contradiction, as 1 letters always occur in blocks of length at least 2. In conclusion $\mathbf{t}'$ contains no instance of $x\pi^i(x)\pi^3(x)\pi^k(x)$ with x starting with 0 and π an antimorphic cycle; analogously, one can show that $\mathbf{t}'$ contains no instance of $x\pi^i(x)\pi^3(x)\pi^k(x)$ with x starting with 1 and π an antimorphic cycle. Moreover, by a very similar analysis one can show that $\mathbf{t}'$ does not contain any instance of $x\pi^i(x)\pi^3(x)\pi^k(x)$ with π an antimorphic involution. We have, thus, shown that $\mathbf{t}'$ avoids the pattern $x\pi^i(x)\pi^3(x)\pi^k(x)$ with i even and k odd.

This concludes the proof of this lemma.   □

By Lemmas 4 and 5 we obtain the following theorem.

Theorem 3. *All patterns $x\pi^i(x)\pi^j(x)\pi^k(x)$, where i, j, k are non-negative integers, and π is substituted by an anti-/morphic permutation, are avoidable over Σ_3.*

We conclude this section with the following result, which follows from the previous theorem by the arguments already presented in the end of Section 2.

Theorem 4. *All patterns $\pi^{i_1}(x)\pi^{i_2}(x)\ldots\pi^{i_r}(x)$ with $r \geq 4$, the i_j non-negative integers, and π an anti-/morphic permutation, are avoidable over Σ_3.*

5   Eventually Unavoidable Patterns

Let n be a positive integer, and let i_j be non-negative integers, $0 \leq j \leq n - 1$. Consider the unary pattern of length n given by

$$\mathcal{P} = \pi^{i_1}(x)\pi^{i_2}(x)\cdots\pi^{i_{n-1}}(x)\pi^{i_n}(x).$$

We say that $\mathcal{P}$ is **eventually unavoidable** if there exists an integer N such that, whenever Σ is an alphabet with $|\Sigma| \geq N$, and $w \in \Sigma^n$, there is a permutation π of Σ and a letter $a \in \Sigma$, such that

$$w = \pi^{i_1}(a)\pi^{i_2}(a)\cdots\pi^{i_{n-1}}(a)\pi^{i_n}(a).$$

Theorem 5. *Let n be a non-negative integer. There is an eventually unavoidable pattern of length n.*

Proof. Consider all partitions of $\{1, 2, \ldots, n\}$ into non-empty subsets; there are B_n of these, where B_n is the nth Bell number. Let the rth such partition be

$$P_r = \langle A_{1,r}, A_{2,r}, \ldots, A_{j_r,r} \rangle,$$

where $\{1, 2, \ldots, n\} = A_{1,r} \dot\cup A_{2,r} \dot\cup \cdots \dot\cup A_{j_r,r}$. We may assume that the sets $A_{\ell,r}$ of the partition are ordered in increasing order of their least element.

For $1 \le k \le n$, $1 \le r \le B_n$, let $q_{k,r}$ be the integer such that $k \in A_{q_{k,r}+1,r}$. In other words, k is in the $(q_{k,r}+1)$st piece of the rth partition. Let p_m denote the mth prime number.

For $1 \le k \le n$, consider the system of congruences

$$i_k \equiv q_{k,r} (mod\ p_r), 1 \le r \le B_n.$$

By the Chinese Remainder Theorem, choose i_k satisfying these congruences. (We remark in passing that $i_0 = 0$ is always possible, since 1 is always in the first piece of each partition, by our notational choice.) Let

$$\mathcal{P} = \pi^{i_1}(x)\pi^{i_2}(x) \cdots \pi^{i_{n-1}}(x)\pi^{i_n}(x).$$

Let $N = p_{B_n}$. Suppose $|\Sigma| \ge N$, and $w \in \Sigma^n$.

Suppose w contains exactly m distinct letters; say $w \in T^n$, where $|T| = m$. Let $f : T \to \{1, 2, \ldots, m\}$ be given by $f(x) = \ell$ if x first occurs in the length ℓ prefix of w. Thus, for example, the first letter of $f(w)$ is always 1. We will show that there is a permutation $\pi \in S_N$, such that

$$f(w) = \pi^{i_1}(1)\pi^{i_2}(1) \cdots \pi^{i_{n-1}}(1)\pi^{i_n}(1).$$

The desired result follows, replacing π by $f^{-1}\pi f$.

To find permutation π, let $P = \langle A_1, A_2, \ldots A_m \rangle$, where $\ell \in A_j$ if and only if the ℓth letter of $f(w)$ is j. For some r, $P = P_r$. Since m, $p_r \le p_{B_n} = N$, S_N will contain a p_r-cycle, π such that $\pi = (1, 2, \ldots, m, \ldots)$. Here the elements other than the first m can be arbitrary distinct elements of $\{m+1, m+2, \ldots, N\}$.

Now $\pi^j(1) = j + 1$, $j = 0, 1, \ldots, m - 1$. Since π is a p_r-cycle, if $i_k \equiv q_{k,r}(\text{modulo } p_r)$, then $\pi^{i_k}(1) = \pi^{q_{k,r}}(1) = q_{k,r} + 1$.

The kth letter of $\pi^{i_1}(1)\pi^{i_2}(1) \cdots \pi^{i_{n-1}}(1)\pi^{i_n}(1)$ is $\pi^{i_k}(1)$, which is $q_{k,r} + 1$. However, by definition of the $q_{k,r}$, this means that k is in the $(q_{k,r}+1)$st piece of P. By the definition of P, the kth letter of $f(w)$ is $q_{k,r} + 1$. Since k was arbitrary, we conclude that $\pi^{i_1}(1)\pi^{i_2}(1) \cdots \pi^{i_{n-1}}(1)\pi^{i_n}(1) = f(w)$, as claimed. □

Example 1. Let $n = 3$. The partitions of $\{1, 2, 3\}$ are

$$P_1 = \langle\{1, 2, 3\}\rangle, P_2 = \langle\{1\}, \{2, 3\}\rangle, P_3 = \langle\{1, 2\}, \{3\}\rangle,$$
$$P_4 = \langle\{1, 3\}, \{2\}\rangle, P_5 = \langle\{1\}, \{2\}, \{3\}\rangle.$$

This gives

$$q_{1,1} = 0, q_{2,1} = 0, q_{3,1} = 0, q_{1,2} = 0, q_{2,2} = 1, q_{3,2} = 1,$$
$$q_{1,3} = 0, q_{2,3} = 0, q_{3,3} = 1, q_{1,4} = 0, q_{2,4} = 1, q_{3,4} = 0,$$
$$q_{1,5} = 0, q_{2,5} = 1, q_{3,5} = 2.$$

As mentioned, we can always choose $i_1 = 0$. For i_2, we get these congruences:

$$i_2 \equiv 0 \pmod 2, i_2 \equiv 1 \pmod 3, i_2 \equiv 0 \pmod 5,$$
$$i_2 \equiv 1 \pmod 7, i_2 \equiv 1 \pmod{11}.$$

and $i_2 = 2080$ is the smallest integer solution.

For i_3, we get these congruences:

$$i_3 \equiv 0 \pmod 2, i_3 \equiv 1 \pmod 3, i_3 \equiv 1 \pmod 5,$$
$$i_3 \equiv 0 \pmod 7, i_3 \equiv 2 \pmod{11}.$$

and $i_3 = 1036$ is the smallest integer solution.

We conclude that $x\pi^{2080}(x)\pi^{1036}(x)$ is eventually unavoidable. As soon as $|\Sigma| \geq 11$, any length 3 word encounters this pattern. For example, to see that $w = aba$ encounters the pattern, we look at $f(w) = 121$, and the partition $P = \langle\{1,3\},\{2\}\rangle = P_4$. We thus let π be a 7-cycle $\pi = (1, 2, \ldots)$. Now

$$\pi^{2080}(1) = \pi^{2080(mod\ 7)}(1) = \pi(1) = 2, \quad \text{while } \pi^{1036}(1) = \pi^{1036(mod\ 7)}(1) = \pi^0(1) = 1,$$

so that $\pi^0(1)\pi^{2080}(1)\pi^{1036}(1) = 121 \sim aba$.   □

By Theorem 5 the following result is immediate.

Theorem 6. *Let n be a non-negative integer. There exists a pattern $\mathcal{P}$ of length n and an integer N such that for all alphabets Σ with $|\Sigma| \geq N$, the pattern $\mathcal{P}$ is unavoidable over Σ.*

This last theorem highlights the main open problem of this work. Each pattern under anti-/morphic permutations is avoidable in Σ_3, but some patterns become unavoidable for larger alphabets. Is there a way to determine exactly, for a given pattern $\mathcal{P}$ which are the alphabets Σ_k in which it is avoidable? Note that such a result was obtained in [6] for cubic patterns.

References

1. Bischoff, B., Currie, J., Nowotka, D.: Unary patterns with involution. Int. J. Found. Comput. Sci. **23**(8), 1641–1652 (2012)
2. Chiniforooshan, E., Kari, L., Xu, Z.: Pseudopower avoidance. Fundam. Inform. **114**(1), 55–72 (2012)
3. Currie, J.: Pattern avoidance: themes and variations. Theoret. Comput. Sci. **339**(1), 7–18 (2005)
4. Hall, M.: Lectures on Modern Mathematics, vol. 2, chap. Generators and relations in groups - The Burnside problem, pp. 42–92. Wiley, New York (1964)
5. Lothaire, M.: Combinatorics on Words. Cambridge University Press (1997)
6. Manea, F., Müller, M., Nowotka, D.: The avoidability of cubes under permutations. In: Yen, H.-C., Ibarra, O.H. (eds.) DLT 2012. LNCS, vol. 7410, pp. 416–427. Springer, Heidelberg (2012)
7. Thue, A.: Über unendliche Zeichenreihen. Norske Vid. Skrifter I. Mat.-Nat. Kl., Christiania **7**, 1–22 (1906)
8. Thue, A.: Über die gegenseitige Lage gleicher Teile gewisser Zeichenreihen. Norske Vid. Skrifter I. Mat.-Nat. Kl., Christiania **1**, 1–67 (1912)

Finite Automata Over Infinite Alphabets: Two Models with Transitions for Local Change

Christopher Czyba, Christopher Spinrath$^{(\boxtimes)}$, and Wolfgang Thomas

RWTH Aachen University, Aachen, Germany
{christopher.czyba,christopher.spinrath}@rwth-aachen.de,
thomas@informatik.rwth-aachen.de

Abstract. Two models of automata over infinite alphabets are presented, mainly with a focus on the alphabet $\mathbb{N}$. In the first model, transitions can refer to logic formulas that connect properties of successive letters. In the second, the letters are considered as columns of a labeled grid which an automaton traverses column by column. Thus, both models focus on the comparison of successive letters, i.e. "local changes". We prove closure (and non-closure) properties, show the decidability of the respective non-emptiness problems, prove limits on decidability results for extended models, and discuss open issues in the development of a generalized theory.

1 Introduction

Automata over infinite alphabets have been studied since the 1990's, starting with the work of Kaminski and Francez [8] on "register automata". Further motivation was added by application areas, e.g. in verification or data base theory. The latter led to the theory of "data words" and "data automata" [4]. Another approach was pursued by Bès [1]. In view of algorithmic applications, a common interest in these theories is to devise models (in automata theory or in logic) where satisfiability problems are decidable, in particular the non-emptiness problem for automata over infinite alphabets.

A fundamental question in this context is the mechanism by which "memory of past inputs" is realized (over an infinite alphabet of possible inputs). The models mentioned above differ in the way this memory is implemented. The technical details in the definitions are not always simple; for example, the access to the memory of register automata is subject to some constraints which are subtle and to some extent artificial. The subject as a whole seems far from finished.

The purpose of this paper is to present two models of automata where the non-emptiness problem is decidable. Both models follow the intuition that the relation between *successive* letters is relevant and has to be controlled by automaton transitions. The results given in this paper are initial; we comment at the end of the paper on further research questions that are motivated by the present work.

© Springer International Publishing Switzerland 2015
I. Potapov (Ed.): DLT 2015, LNCS 9168, pp. 203–214, 2015.
DOI: 10.1007/978-3-319-21500-6_16

The first approach is an extension of the model introduced by A. Bès in [1]. In that paper, the alphabet letters are elements of a relational structure $\mathcal{M}$ over an infinite domain M, and a logic $\mathcal{L}$ allows to express conditions on alphabet letters by formulas $\varphi(y)$. The transition relation of the automaton is specified by such formulas $\varphi_{pq}(y)$ where p, q are from the finite set Q of states. The automaton can proceed from state p via letter $a \in M$ to state q if $\mathcal{M} \models \varphi_{pq}[a]$, i.e. a satisfies $\varphi_{pq}(y)$ in $\mathcal{M}$. A central weakness of this model is the inability to connect successive letters, for example to check the property that two identical successive letters exist (which gives the language $\bigcup_{a \in M} M^* aa M^*$).

Continuing work of Spelten [12], we study an extension of this model where this defect is repaired to some extent, due to a description of the relation between successive letters: We use now formulas $\varphi_{pq}(x, y)$ which allow to move from state p to state q via letter y if state p was reached via x (of course, for the initial transition of a run one uses a transition formula $\varphi_{q_0 q}(y)$). We call them "two-letter transitions". Since successive letters are subject to conditions expressed in the considered logic $\mathcal{L}$, we speak of "local change". In the paradigmatic case of the alphabet $\mathbb{N}$ of the natural numbers, one may express, for example, that in the step from one letter to the next the value can increase or decrease by 1 (or by at most by some constant k). But a transition could also specify that odd and even numbers have to alternate.

The main contribution of the first part of the paper is a proof showing that the resulting automata have a decidable non-emptiness problem (thereby repairing an unclear point in [12]), provided that the MSO-theory of the alphabet structure $\mathcal{M}$ is decidable. It is also shown that the result fails for two natural extensions of the model, namely for the case that alphabet letters are elements of M^2, i.e. pairs of M-elements, and for the case that three successive letters are related in automaton transitions (by formulas $\varphi(x, y, z)$).

The second approach is presented here for the case of the alphabet $\mathbb{N}$ rather than an arbitrary structure, in order to allow for a concise presentation (see the final section for a more general framework). A word $n_1 n_2 \dots n_\ell$ is represented by a $\{1, \bot\}$-labeled grid: we imagine a sequence of grid columns where the i-th column represents number n_i; this column starts from below with n_i nodes

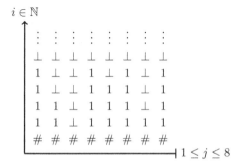

Fig. 1. Grid representing the word $4\ \ 2\ \ 0\ \ 4\ \ 3\ \ 4\ \ 1\ \ 4 \in \mathbb{N}^*$

colored 1, followed by nodes colored $\perp$. For technical reasons we add a #-labeled boundary at the bottom. For instance, Figure 1 shows the grid for the word 4 2 0 4 3 4 1 4. This grid is traversed by a finite automaton in a three-way mode (hence a special type of the grid-walking automata of Blum, Hewitt and Rosenfeld [2],[10]). Each column can be traversed in a two-way mode up and down (which means that an input letter is analyzed), and after a move to the right via a horizontal edge no return to a previous column is allowed. The move to the right enables the automaton to check, in the example of the alphabet $\mathbb{N}$ as in the Figure, whether the move to a next letter gives the same value as before, respectively a larger or smaller value. Thus, again a natural test on "local change" is possible. But also non-local properties of words can be checked, for example (in the non-deterministic version) the existence of two equal letters from $\mathbb{N}$ in the input word.

In the second part of the paper we study this model regarding closure properties, show that the non-emptiness problem is decidable but that the universality problem is undecidable. The decidability of the emptiness problem resolves a question of [7], where a weaker model is studied, processing finite grids labeled with a single letter alphabet.

The paper is structured as follows. The subsequent section is devoted to automata with two-letter transitions. In order to have a shorter name for them, we call them "strong automata", following [12]. After this, we present the mentioned three-way automata, called "progressive grid automata". In the final section we relate these automata, discuss connections to known models of automata on infinite alphabets, and address a number of open issues.

Throughout we assume that the reader is acquainted with the basics of automata and logic, in particular first-order logic FO and monadic second-order logic MSO; see e.g. [14]. Due to lack of space, proofs are outlined; more details can be found in [5].

2   Automata with Two-Letter Transitions

2.1   Definitions

We allow elements of an arbitrary structure $\mathcal{M}$ as alphabet letters. Central examples are $(\mathbb{N}, +1)$ and $(\mathbb{N}, +, 0)$. If $\mathcal{M}$ is a structure with domain M and $\mathcal{L}$ a logic[1] (used with the appropriate signature), we call $(\mathcal{M}, \mathcal{L})$ an "alphabet frame". We only use logics in which the Boolean connectives are available and where first-order variables $x, y, \ldots$ can be employed to express properties of (tuples of) M-elements by formulas $\varphi(x), \varphi(x, y)$, etc. We denote the set of formulas with k free first-order variables by Ψ_k (note that nevertheless such a formula may contain quantified second-order variables).

Let $(\mathcal{M}, \mathcal{L})$ be an alphabet frame where M is the domain of $\mathcal{M}$. A *strong automaton* over $(\mathcal{M}, \mathcal{L})$ has the form $\mathfrak{A} = (Q, M, q_0, \Delta, F)$ where Q is a finite set of states, the input alphabet is M (the domain of $\mathcal{M}$), $q_0 \in Q$ is the initial

---

[1] The term "logic" is used here as in abstract model theory, see [6] Chapter 13.1.

state, $\Delta \subseteq Q \times (\Psi_1 \cup \Psi_2) \times Q$ is the finite transition relation, and lastly $F \subseteq Q$ is the set of accepting states.

A transition for the initial state has the form $(q_0, \varphi(y), q)$, other transitions have the form $(p, \varphi(x, y), q)$. A run ρ of $\mathfrak{A}$ on the word $w = a_1...a_m \in M^*$ is a finite sequence $\rho = \rho(0)...\rho(m)$ where

1. $\rho(0) = q_0$,
2. $\rho(1) = q$ with $(q_0, \varphi(x), q) \in \Delta$ such that $\mathcal{M} \models \varphi[a_1]$,
3. $\rho(i) = q$ where for $p = \rho(i-1)$, $(p, \varphi(x, y), q) \in \Delta$ such that $\mathcal{M} \models \varphi[a_{i-1}, a_i]$ for $1 < i \leq m$.

Note that several transitions from p to q may be condensed into one, by taking the disjunction of the respective transition formulas. Then one can indicate a transition just by writing $\varphi_{pq}(x, y)$ as done in the Introduction.

A run ρ of $\mathfrak{A}$ on the word w is successful if $\rho(m) \in F$. We say $\mathfrak{A}$ accepts w if there is a successful run of $\mathfrak{A}$ on w. Furthermore, $L(\mathfrak{A}) := \{w \in M^* \mid \mathfrak{A}$ accepts $w\}$.

Call a strong automaton deterministic if Δ is functional, i.e. for all $p \in Q$ and all $a, b \in M$ there is exactly one state q reachable by a corresponding transition.

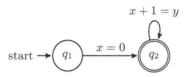

Fig. 2. Strong automaton recognizing $\{01 \cdots n \mid n \in \mathbb{N}\}$

A very simple example language recognized by a strong automaton is $L = \{01 \cdots n \mid n \in \mathbb{N}\}$ (for the alphabet frame $((\mathbb{N}, +1), FO)$), as depicted by the automaton in Figure 2.

The next two subsections lift results of Bès [1] from automata with single-letter transitions to strong automata (with two-letter transitions).

2.2 Boolean Closure Properties

The languages (over a given alphabet frame) that are recognized by strong automata form a Boolean algebra. The key for this claim is the following lemma on determinization, proved in analogy to [1].

Lemma 1. *Given a strong automaton $\mathfrak{A}$, one can construct a deterministic strong automaton $\mathfrak{A}'$ such that $L(\mathfrak{A}) = L(\mathfrak{A}')$.*

Proof. The idea is to take all the (finitely many) formulas $\varphi_i(x, y)$ $(i = 1, \ldots, m)$ that occur in a given automaton $\mathfrak{A} = (Q, M, q_1, \Delta, F)$ over the alphabet frame $(\mathcal{M}, \mathcal{L})$, and to form the Boolean min-terms of the relations $R_1, \ldots, R_m$ defined by the φ_i. Each of these min-terms is defined by a conjunction of formulas φ_i and their negations, and each φ_i is equivalent to a disjunction of these min-terms. The transitions of the given automaton can now be partitioned into sets of transitions labeled by disjoint properties of letter pairs, each of them given by a min-term formula (for the initial transitions with formulas $\varphi(x)$ one proceeds analogously). Now the standard subset construction is applied to the automaton with these new transitions; whence a deterministic strong automaton as desired is obtained. Moreover, this construction shows that $\mathfrak{A}'$ can be obtained such that its size is at most exponential in the size of $\mathfrak{A}$.

For deterministic strong automata, complementation is shown by exchanging final with non-final states. Closure of strong automata under union is trivial by taking the union of given automata and a new initial state. Thus, we obtain the claim:

Proposition 1. *The languages recognized by strong automata (over a given alphabet frame) form an effective Boolean algebra (where the Boolean operations are realized by effective constructions of strong automata).*

2.3   The Non-Emptiness Problem

Theorem 1. *If the MSO-theory of $\mathcal{M}$ is decidable, then the non-emptiness problem for strong automata over the alphabet frame $(\mathcal{M},\ MSO)$ is decidable.*

Proof. The main point in this result is to exploit MSO-logic for expressing a reachability condition. Given a strong automaton, say $\mathfrak{A} = (\{0,\ldots,n\}, M, 0, \Delta, F)$ over the alphabet frame $(\mathcal{M},\ MSO)$, we have to express that a successful run of $\mathfrak{A}$ exists. Then the existence of a successful run is captured by a path through the domain $M \times \{0,\ldots,n\}$,

1. starting with an arbitrary element $(m, 0)$ succeeded by an element (m', j') such that the triple $(0, m', j')$ is an admissible initial transition,
2. continuing with steps from a pair (m, j) to (m', j') if (m, j, m', j') is an admissible transition,
3. ending with a pair (m, j) with $j \in F$.

This existence claim can be expressed in MSO-logic over $M \times \{0,\ldots,n\}$ by saying

> each set X containing the pairs satisfying condition 1, and closed under the steps according to condition 2, contains an element according to condition 3.

It is easy to express all three parts of this statement in MSO, using the fact that the transition formulas of $\mathfrak{A}$ are MSO-formulae themselves.

Finally, one employs the assumption that the MSO-theory of $\mathcal{M}$ is decidable. It is well known that this implies that also the MSO-theory of $\mathcal{M} \times \{0,\ldots,n\}$ is decidable (see e.g. [3], Prop. 3.12). Alternatively, one can show the claim directly by working over the domain M and using a tuple of universal set quantifiers over variables $X_0,\ldots,X_n$, rather than using "$\forall X$" over $M \times \{0,\ldots,n\}$.

Due to the effective closure of the class of languages recognized by strong automata over an alphabet frame $(\mathcal{M}, MSO)$ under Boolean operations, we conclude the following:

Corollary 1. *Over an alphabet frame $(\mathcal{M}, MSO)$ where the MSO-theory of $\mathcal{M}$ is decidable, the inclusion problem, the equivalence problem, and the universality problem for strong automata are decidable.*

2.4   Extensions and Undecidability Results

There are two natural extensions of the model of strong automaton: First we can proceed from input letters of M to input letters of M^2 when M is the domain of an alphabet structure, and secondly we can introduce transitions that connect more than two successive input letters. In both cases, the transition formulae have more free variables. We show that both extensions cause the non-emptiness problem to be undecidable, even for the basic alphabet frame $((\mathbb{N}, +1), FO)$.

In these results we use – without going into all details – the obvious extension of the framework of strong automata to "two-dimensional" input letters. The transition formulas can now be written as $\varphi_{pq}(x_1, x_2, y_1, y_2)$, connecting two successive input letters (x_1, x_2), (y_1, y_2) from $M \times M$, where M is the alphabet.

Theorem 2. *The non-emptiness problem for strong automata over $\mathbb{N}^2$ is undecidable when the alphabet frame is $((\mathbb{N}, +1), FO)$.*

We use the well-known result that for 2-register machines the reachability problem is undecidable (cf. [9]). The two registers are x_1, x_2, and the operations are $\mathtt{INC}_j$ (increase the value of x_j by one), $\mathtt{DEC}_j$, $\mathtt{IF}\ x_j = 0\ \mathtt{GOTO}\ m$, and $\mathtt{GOTO}\ m$ with their usual semantics. The last instruction is always the $\mathtt{HALT}$ instruction. A configuration of such a machine R (with k instructions) is a tuple $(i, r_1, r_2) \in \{1, \ldots, k\} \times \mathbb{N}^2$ where i is the number of an instruction and r_j is the value stored in x_j. The reachability problem for 2-register machines with k instructions asks if a configuration (k, r_1, r_2) is reachable from $(1, 0, 0)$.

Proof. Let k be the number of instructions of the given 2-register machine R. One constructs a strong automaton $\mathfrak{A}_R = (Q, \mathbb{N}^2, q_0, \Delta, \{q_k\})$ whose language is non-empty iff R terminates from configuration $(1, 0, 0)$. Indeed, it is easy to build $\mathfrak{A}_R$ such that for a sequence of register contents $(0, 0)(m_2, n_2) \cdots (m_r, n_r)$ of a *halting* computation this word of $(\mathbb{N}^2)^*$ is the only word accepted by $\mathfrak{A}_R$, while the lack of such a halting computation causes $L(\mathfrak{A}_R)$ to be empty.

Let us turn to strong automata with higher "history extension", taking the view that the strong automata defined above have a "1-history" (they are allowed to look back one letter). Now we use transitions of the form $(p, \varphi(x_0, x_1, x_2), q)$, allowing to move from p to q via m_2 if the previous *two* letters were m_0, m_1, in this order, and $\mathcal{M} \models \varphi[m_0, m_1, m_2]$. We call these automata "strong automata with 2-history".

Theorem 3. *The non-emptiness problem for a strong automaton with 2-history is undecidable over the alphabet frame* $((\mathbb{N}, +1), FO)$.

Proof. From a 2-register machine R we construct a strong automaton $\mathfrak{A}_R$ with 2-history such that R has a terminating computation from $(1, 0, 0)$ iff $L(\mathfrak{A}_R) \neq \emptyset$. In order to simulate a computation of R, the strong automaton will in the even steps take care of simulation for the first register while in the odd steps will do the simulation for the second register. Since it is possible to look back 2 letters, an update of the two register contents can be done in two successive steps.

3   Three-Way-Grid Traversal Automata

3.1   Definitions

In this section, we consider automata that accept words whose letters are again words (over some finite alphabet Σ). We may write a sequence $w_1 w_2 \ldots w_n$ of words w_i as a sequence of entries in a grid structure, where each w_i is noted in a column, starting from the bottom line of the grid upwards and using the symbol $\perp$ after the last letter of w_i is written. In special cases, we use words w_i as unary representations of natural numbers; here each w_i is a word on $\Sigma = \{1\}$ (cf. Figure 1).

Let Σ be a finite alphabet and $n \in \mathbb{N}$. Formally, a *grid word* of length n is given by a function that maps grid positions to letters, i.e. $w : \mathbb{N} \times \{1, \ldots, n\} \to \Sigma \dot{\cup} \{\#, \perp\}$ such that for all $j \in \{1, \ldots, n\}$:

1. $w(i, j) = \# \Leftrightarrow i = 0$,
2. $w(i, j) \in \Sigma \Rightarrow w(k, j) \in \Sigma$ for all $1 \leq k \leq i$,
3. $w(i, j) = \perp$ for some $i > 0$

With $\mathcal{G}(\Sigma) := \{w : \mathbb{N} \times \{1, \ldots, n\} \to \Sigma \dot{\cup} \{\#, \perp\} \mid n \in \mathbb{N}, w \text{ is a grid word}\}$ we denote the set of all grid words on alphabet Σ. Given a subset $L \subseteq \mathcal{G}(\Sigma)$ the complement of L is defined as $\overline{L} = \mathcal{G}(\Sigma) \setminus L$. Finally, as mentioned above, note that $(\mathcal{G}(\{1\}), \cdot)$ is isomorphic to $(\mathbb{N}^*, \cdot)$ where "·" is the usual word concatenation in both cases. For instance, the word $2\,1 \in \mathbb{N}^*$ corresponds to the grid word which has two 1's in the first column and a single 1 in the second column.

A *progressive grid automaton* is a tuple $\mathcal{A} = (Q, \Sigma, q_0, \Delta, F)$ where Q is a finite set of states, Σ is the finite grid label alphabet, q_0 is the initial state, $\Delta \subseteq Q \times \Sigma \dot{\cup} \{\#, \perp\} \times Q \times \{\uparrow, \downarrow, \to\}$ is the transition relation (with $\Delta \cap Q \times \{\#\} \times Q \times \{\downarrow\} = \emptyset$), and $F \subseteq Q$ is the set of accepting states.

We call $\mathcal{A}$ a deterministic progressive grid automaton if Δ is functional. A *configuration* of $\mathcal{A}$ is an element in $Q \times \mathbb{N} \times \mathbb{N}$, consisting of the "current" state of $\mathcal{A}$ and its position in a grid word.

Let $w \in \mathcal{G}(\Sigma)$. Then a run of $\mathcal{A}$ on w is a finite sequence $\pi = c_0 \ldots c_m$ of configurations that satisfies the following conditions (which define the three-way movement through the grid):

1. $c_0 = (q_0, 1, 1)$,
2. for all $0 \leq i < m$ it holds that if $c_i = (q_i, h_i, \ell_i)$ then
   - $c_{i+1} = (q_{i+1}, h_i, \ell_i + 1)$ and $(q_i, w(h_i, \ell_i), q_{i+1}, \rightarrow) \in \Delta$, or
   - $c_{i+1} = (q_{i+1}, h_i + 1, \ell_i)$ and $(q_i, w(h_i, \ell_i), q_{i+1}, \uparrow) \in \Delta$, or
   - $c_{i+1} = (q_{i+1}, h_i - 1, \ell_i)$ and $(q_i, w(h_i, \ell_i), q_{i+1}, \downarrow) \in \Delta$,
3. for all $0 \leq i < m : c_i \in Q \times \mathbb{N} \times \{0, \ldots, |w|\}$,
4. and $c_m = (q_m, h_m, |w| + 1)$ for some $q_m \in Q$ and $h_m \in \mathbb{N}$.

If $c_m \in F \times \mathbb{N} \times \mathbb{N}$ then π is called an accepting run. A language $L \subseteq \mathcal{G}(\Sigma)$ is called grid recognizable, if L is recognized by a progressive grid automaton $\mathcal{A}$. We will mainly consider words and languages over $\mathbb{N}$, where the automaton operates on the corresponding grid words in $\mathcal{G}(\{1\})$.

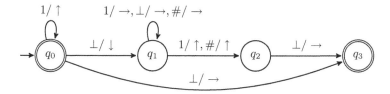

Fig. 3. A progressive grid automaton recognizing $L := \{n_0 \ldots n_p \in \mathbb{N}^* \mid n_0 = n_p\}$

Example 1. The progressive grid automaton pictured in Figure 3 recognizes the language $L := \{n_0 \ldots n_p \in \mathbb{N}^* \mid n_0 = n_p\}$. It operates as follows on a grid word in $\mathcal{G}(\{1\})$. In the beginning, the automaton moves to the "highest" 1 in the first column (or # if there is none). Afterwards the internal state is q_1. Then it moves non-deterministically to the last column and verifies that it equals the first one. Indeed, if this is the case, then the current position is labeled 1 (or #) and the position above is labeled with $\perp$. Otherwise, the last letter would be smaller or greater, respectively. This property is checked in states q_2 and q_3.

3.2   Closure Properties

Proposition 2. *The class of deterministically grid recognizable languages is effectively closed under complement.*

This result can be proven similarly to the complementation proof given in [7]. The basic idea is the introduction of several finite counters to check whether the automaton halts (i.e. moves to the right infinitely often).

Theorem 4. *Neither the class of non-deterministically grid recognizable nor the class of deterministically grid recognizable languages is closed under intersection (even over $\mathcal{G}(\{1\})$).*

Intuitively, theorem 4 can be justified as follows: A progressive grid automaton (deterministic or not) can only remember one letter (i.e. column) at a time it has seen before. This can be done by using the vertical position of the automaton upon a move to the right. Moreover, while remembering some letter the movement of the automaton is restricted.

Proof. We consider the following two languages:

$$L_1 := \{n_0 \dots n_k \mid k \in \mathbb{N}, \forall i \in \{2j \mid 0 \le j \le \lfloor \frac{k}{2} \rfloor\} : n_i = 0\}$$

$$L_2 := \{n_0 \dots n_k \mid k \in \mathbb{N}, \forall i_0, i_1 \in \{2j + 1 \mid 0 \le j \le \lfloor \frac{k}{2} \rfloor\} : n_{i_0} = n_{i_1}\}$$

Both are clearly deterministically grid recognizable. It suffices to show that $L := L_1 \cap L_2$ is not grid recognizable. Let $L' := \{(0m)^n \mid n, m \in \mathbb{N}, n > 0\} \subseteq L$. Using a pumping argument it can be shown that every progressive grid automaton recognizing a superset of L' accepts a word not in L. Informally, for sufficient large m, n the automaton has to forget the value of m to check that every second column is labeled only with $\bot$.

Corollary 2.   *1. The class of grid recognizable languages is not closed under complement.*

  2. The class of deterministically grid recognizable languages is not closed under union.

  3. The class of deterministically grid recognizable languages is strictly included in the class of (non-deterministically) grid recognizable languages.

Proof. For claim 1, observe that the class of grid recognizable languages is naturally closed under union (using non-determinism). Suppose it is closed under complement. Then closure under intersection follows immediately. However, this contradicts Theorem 4. Claim 2 follows immediately from Theorem 4. Thus, we also obtain claim 3.

3.3   Decidability of the Non-emptiness Problem

Theorem 5. *The non-emptiness problem for progressive grid automata is decidable.*

To prove this result, we will use the MSO-theory of $(\mathbb{N}, +1)$ (known to be decidable). We describe a run of a progressive grid automaton $\mathcal{A}$ on some grid word w in MSO. W.l.o.g. we can assume $\mathcal{A} = (\{1, \dots, n\}, \Sigma, \Delta, 1, F)$ for some $n \in \mathbb{N}$. At first, a single column of a grid word can be described in MSO-logic with the help of second order variables. The existence of an accepting run is certified by a sequence over $\{1, \dots, n\} \times \mathbb{N}$ that

1. starts with $(1,1)$,
2. contains $(p,k)(q,\ell)$ if and only if there is a well-labeled column (i.e. the labeling is in $\#1^*\bot^\omega$) and a tuple (p',ℓ), such that there is an applicable transition $(p',a,q,\rightarrow)$ and (p',ℓ) is reachable from (p,k) in the considered column,
3. and ends with a pair in $F \times \mathbb{N}$.

Note that the third component of a configuration, which is the index of the column, is dropped here. The reachability within a single fixed column used in the second condition, i.e. if there is a path from a configuration (p,k,i) to (p',ℓ,i), can be described by the following least fixed point:

– (p,k) is in the fixed point, and
– if (q,m) is in the fixed point, the m-th row is labeled a, and $(q,a,q',\uparrow)$ or $(q,a,q',\downarrow)$ is a transition, then $(q',m+1)$ resp. $(q',m-1)$ is in the fixed point.

Then (p',ℓ,i) is reachable iff (p',ℓ) is in the fixed point. This fixed point can be expressed in MSO using universal set quantification as in Theorem 1. Finally, the conditions 1 to 3 can be formalized in MSO analogously.

3.4   The Non-universality Problem

Since the non-emptiness problem is decidable and deterministic progressive grid automata are closed under complement, we can immediately conclude that the non-universality problem is decidable. We show that the situation changes for progressive grid automata in general.

Theorem 6. *The non-universality problem for non-deterministic progressive grid automata is undecidable.*

Proof. We fix $\Sigma = \{1\}$ and consider a 2-register machine R. A computation of R can be encoded in a grid word u. For the i-th computation step the unary encodings of the first and second register correspond to the $(2i-1)$-th and $2i$-th column of u, respectively. The claim follows by constructing a progressive grid automaton $\mathcal{B}$ with

$$L(\mathcal{B}) := \mathcal{G}(\{1\}) \setminus \{uv \mid u \text{ encodes a terminating computation of } R\}.$$

That is, $L(\mathcal{B}) = \mathcal{G}(\{1\})$ iff there is no terminating computation of $\mathcal{M}$. To achieve this property, $\mathcal{B}$ needs to accept all grid words that do not have a prefix encoding a valid computation of R starting in $(1,0,0)$. The idea is that $\mathcal{B}$ remembers the current instruction in the state space and guesses if the encoding is wrong at some point.

4   Discussion, Related Work, Perspectives

4.1   Comparison Between the Two Models

The language classes given by strong automata and progressive grid automata cannot be compared directly since the underlying alphabet structures differ (alphabet frames involving a logic vs. input letters as sequences ("columns")). On the other hand, we can provide a comparison for the fixed alphabet frame $((\mathbb{N}, +1), MSO)$ on one side and the alphabet $\mathbb{N}$ on the other: *Each language $L \subseteq \mathbb{N}^*$ recognized by a strong automaton over $((\mathbb{N}, +1), MSO)$ is non-deterministically grid recognizable.*[2] *The converse fails in general (e.g., using the language L presented in Example 1).*

4.2   Comparison with Other Models

Let us compare our models with the register automata introduced by Kaminski and Francez [8] which recognize the "quasi-regular" languages over $\mathbb{N}$. The language of words $0123 \ldots \ell$ is not quasi-regular, but recognizable both by strong automata and progressive grid automata. On the other hand, the language $L = \{(0m)^n \mid m, n \in \mathbb{N}, n > 0\}$ is quasi-regular, but neither recognizable by a strong automaton nor by a progressive grid automaton. Also, the "data automata" of [4] only refer to the equality (and non-equality) of input letters from an abstract data domain and do not incorporate a possibility of comparison (as with $<$ over $\mathbb{N}$).

4.3   Extension to Infinite Words

It is not difficult to generalize the main results of this paper to the case of infinite words. There are technical details that require some care (e.g. in the proofs for the decidability of the emptiness problem and the complementation of Büchi automata). We do not present these details here, but refer the reader to [5],[11].

4.4   General Framework of Progressive Grid Automata

In this paper, we considered progressive grid automata over the infinite alphabets Σ^* and $\mathbb{N}$. It should be mentioned that the model is easily extensible to all alphabets whose letters are presentable as labellings of a fixed infinite graph (i.e. with fixed edge relation). In the cases of the alphabets Σ^* and $\mathbb{N}$, this graph is the successor structure of the natural numbers (into which elements of Σ^* or $\mathbb{N}$ were encoded by vertex labellings). Using the same underlying graph, one can also handle the alphabet Σ^ω. Using the binary infinite tree instead, different (finite) labellings code different binary terms. The "columns" of the present paper are then replaced by different labellings of the infinite binary tree, and a progressive grid automaton would work through such a labeled tree as a

---

[2] A proof may be found in [5].

tree-walking automaton before jumping (on a certain tree node position) to the next tree. In this way a theory of recognizable languages over the alphabet of (binary) terms can be developed. Similarly, one can work with any fixed infinite graph and its possible labellings as the letters of an infinite alphabet.

Pursuing a different direction (which is subject of current work) we can lift the idea of progressive grid *word* automata to progressive grid *tree* automata, e.g. working on ℕ-labeled trees.

Acknowledgments. We thank the reviewers of DLT 2015 for their remarks which led to improvements of the presentation.

References

1. Bès, A.: An application of the feferman-vaught theorem to automata and logics for words over an infinite alphabet. Logical Methods in Computer Science **4**(1) (2008)
2. Blum, M., Hewitt, C.: Automata on a 2-dimensional tape. In: IEEE Conference Record of the Eighth Annual Symposium on Switching and Automata Theory, SWAT 1967, pp. 155–160. IEEE (1967)
3. Blumensath, A., Colcombet, T., Löding, C.: Logical theories and compatible operations. In: Flum, J., et al. (eds.) Logic and Automata: History and Perspectives, pp. 73–106. Amsterdam Univ. Press (2008)
4. Bojańczyk, M., David, C., Muscholl, A., Schwentick, T., Segoufin, L.: Two-variable logic on data words. ACM Transactions on Computational Logic (TOCL) **12**(4), 27 (2011)
5. Czyba, C., Spinrath, C., Thomas, W.: Finite Automata Over Infinite Alphabets: Two Models with Transitions for Local Change (Full Version). RWTH Aachen University (2015). https://www.lii.rwth-aachen.de/en/86-finite-automata-over-infinite-alphabets.html
6. Ebbinghaus, H.-D., Flum, J., Thomas, W.: Mathematical Logic, 2nd edn. Springer Undergraduate Texts in Mathematics and Technology. Springer (1996)
7. Inoue, K., Takanami, I.: A note on decision problems for three-way two-dimensional finite automata. Information Processing Letters **10**(4), 245–248 (1980)
8. Kaminski, M., Francez, N.: Finite-memory automata. Theoretical Computer Science **134**(2), 329–363 (1994)
9. Minsky, M.L.: Computation: finite and infinite machines. Prentice-Hall Inc, Upper Saddle River (1967)
10. Rosenfeld, A.: Picture languages. Academic Press (1979)
11. Spelten, A., Thomas, W., Winter, S.: Trees over infinite structures and path logics with synchronization. In: Yu, F., Wang, C. (eds.) Proceedings 13th International Workshop on Verification of Infinite-State Systems, INFINITY 2011, vol. 73, pp. 20–34. EPTCS, Taipei (2011)
12. Spelten, A.: Paths in Infinite Trees: Logics and Automata. PhD thesis, RWTH-Aachen University (2013)
13. Thomas, W.: On the bounded monadic theory of well-ordered structures. The Journal of Symbolic Logic **45**(02), 334–338 (1980)
14. Thomas, W.: Languages, automata, and logic. In: Rozenberg, G., Salomaa, A. (eds.) Handbook of Formal Languages, pp. 389–455. Springer, Heidelberg (1997)

Enumeration Formulæ in Neutral Sets

Francesco Dolce$^{(\boxtimes)}$ and Dominique Perrin

LIGM, Université Paris Est, Champs-sur-marne, France
`ceskino@gmail.com`

Abstract. We present several enumeration results holding in sets of words called neutral and which satisfy restrictive conditions on the set of possible extensions of nonempty words. These formulae concern return words and bifix codes. They generalize formulae previously known for Sturmian sets or more generally for tree sets. We also give a geometric example of this class of sets, namely the natural coding of some interval exchange transformations.

Keywords: Neutral sets · Bifix codes · Interval exchanges

1 Introduction

Sets of words of linear complexity play an important role in combinatorics on words and symbolic dynamics. This family of sets includes Sturmian sets, interval exchange sets and primitive morphic sets, that is, sets of factors of fixed points of primitive morphisms.

We study here a family of sets of linear complexity, called neutral sets. They are defined by a property of a graph $E(x)$ associated to each word x, called its extension graph and which expresses the possible extensions of x on both sides by a letter of the alphabet A. A set S is neutral if the Euler characteristic of the graph of any nonempty word is equal to 1, as for a tree. The Euler characteristic of the graph $E(\varepsilon)$ is called the characteristic of S and is denoted $\chi(S)$. These sets were first considered in [1] and in [5]. The factor complexity of a neutral set S on k letters is for $n \neq 1$

$$p_n = n(k - \chi(S)) + \chi(S). \tag{1}$$

We prove here several results concerning neutral sets. The first one (Theorem 2) is a formula giving the cardinality of a finite S-maximal bifix code X of S-degree n in a recurrent neutral set S on k letters as

$$\mathrm{Card}(X) = n(k - \chi(S)) + \chi(S). \tag{2}$$

The remarkable feature is that, for fixed S, the cardinality of X depends only on its S-degree. In the particular case where X is the set of all words of S of length n, we recover Equation (1). Formula (2) generalizes the formula proved in [2] for Sturmian sets and in [7] for neutral sets of characteristic 1.

© Springer International Publishing Switzerland 2015
I. Potapov (Ed.): DLT 2015, LNCS 9168, pp. 215–227, 2015.
DOI: 10.1007/978-3-319-21500-6_17

The second one concerns return words. The set of right first return words to a word x in a factorial set S, denoted $\mathcal{R}_S(x)$, is an important notion. It is the set of words u such that xu is in S and ends with x for the first time. In several families of sets of linear complexity, the set of first return words to x is known to be of fixed cardinality independent of x. This was proved for Sturmian words in [13], for interval exchange sets in [16] (see also [11]) and for neutral sets of characteristic zero in [1].

We first prove here (Theorem 3) that the set $\mathcal{CR}_S(X)$ of complete first return words to a bifix code X in a uniformly recurrent neutral set S on k letters satisfies $\mathrm{Card}(\mathcal{CR}_S(X)) = \mathrm{Card}(X) + k - \chi(S)$. The remarkable feature here is that, for fixed S, the cardinality of $\mathcal{CR}_S(X)$ depends only on $\mathrm{Card}(X)$. When X is reduced to one element x, we have $\mathcal{CR}_S(x) = x\mathcal{R}_S(x)$ and we recover the result of [1]. When $X = S \cap A^n$, then $\mathcal{CR}_S(X) = S \cap A^{n+1}$. This implies $p_{n+1} = p_n + k - \chi(S)$ and also gives Equation (1) by induction on n. The proofs of these formulæ use a probability distribution naturally defined on a neutral set.

A third result concerns the decoding of a neutral set by a bifix code. We prove that the decoding of any recurrent neutral set S by an S-maximal bifix code is a neutral set. This property is proved for uniformly recurrent tree sets in [8].

We finally prove a result which allows one to obtain a large family of neutral sets of geometric origin, namely using interval exchange transformations. More precisely, we prove that the natural coding of an interval exchange transformation without connections of length ≥ 1 is a neutral set. This extends a result in [6] concerning interval exchange without connections as well as a result of [9] concerning linear involutions without connection.

2   Extension Graphs

Let A be a finite alphabet. We denote by A^* the set of all words on A. We denote by ε or 1 the empty word. A *factor* of a word x is a word v such that $x = uvw$. If both u and w are nonempty, we say that x is an *internal factor*. A set of words on the alphabet A is said to be *factorial* if it contains the factors of its elements as well as the alphabet A.

Let S be a factorial set on the alphabet A. For $w \in S$, we denote $L_S(w) = \{a \in A \mid aw \in S\}$, $R_S(w) = \{a \in A \mid wa \in S\}$, $E_S(w) = \{(a,b) \in A \times A \mid awb \in S\}$, and further $\ell_S(w) = \mathrm{Card}(L_S(w))$, $r_S(w) = \mathrm{Card}(R_S(w))$, $e_S(w) = \mathrm{Card}(E_S(w))$.

We omit the subscript S when it is clear from the context. A word w is *right-extendable* if $r(w) > 0$, *left-extendable* if $\ell(w) > 0$ and *biextendable* if $e(w) > 0$. A factorial set S is called *right-extendable* (resp. *left-extendable*, resp. *biextendable*) if every word in S is right-extendable (resp. left-extendable, resp. biextendable).

A word w is called *right-special* if $r(w) \geq 2$. It is called *left-special* if $\ell(w) \geq 2$. It is called *bispecial* if it is both left-special and right-special. For $w \in S$, we denote

$$m_S(w) = e_S(w) - \ell_S(w) - r_S(w) + 1.$$

A word w is called *neutral* if $m_S(w) = 0$. We say that a set S is *neutral* if it is factorial and every nonempty word $w \in S$ is neutral. The *characteristic* of S is the integer $\chi(S) = 1 - m_S(\varepsilon)$.

Thus, a neutral set of characteristic 1 is such that all words (including the empty word) are neutral. This is what is called a neutral set in [5].

The following example of a neutral set is from [5].

Example 1. Let $A = \{a, b, c, d\}$ and let σ be the morphism from A^* into itself defined by $\sigma : a \mapsto ab$, $b \mapsto cda$, $c \mapsto cd$, $d \mapsto abc$. Let S be the set of factors of the infinite word $x = \sigma^\omega(a)$. One has $S \cap A^2 = \{ab, ac, bc, ca, cd, da\}$ and thus $m(\varepsilon) = -1$. It is shown in [5] that every nonempty word is neutral. Thus S is neutral of characteristic 2.

A set of words $S \neq \{\varepsilon\}$ is *recurrent* if it is factorial and for any $u, w \in S$, there is a $v \in S$ such that $uvw \in S$. An infinite factorial set is said to be *uniformly recurrent* if for any word $u \in S$ there is an integer $n \geq 1$ such that u is a factor of any word of S of length n. A uniformly recurrent set is recurrent.

The *factor complexity* of a factorial set S of words on an alphabet A is the sequence $p_n = \mathrm{Card}(S \cap A^n)$. Let $s_n = p_{n+1} - p_n$ and $b_n = s_{n+1} - s_n$ be respectively the first and second order differences sequences of the sequence p_n.

The following result is [12, Proposition3.5] (see also [10, Theorem4.5.4]).

Proposition 1. *Let S be a factorial set on the alphabet A. One has $b_n = \sum_{w \in S \cap A^n} m(w)$ and $s_n = \sum_{w \in S \cap A^n} (r(w) - 1)$ for all $n \geq 0$.*

One deduces easily from Proposition 1 the following result which shows that a neutral set has linear complexity.

Proposition 2. *The factor complexity of a neutral set on k letters is given by $p_0 = 1$ and $p_n = n(k - \chi(S)) + \chi(S)$ for every $n \geq 1$.*

Let S be a biextendable set of words. For $w \in S$, we consider the set $E(w)$ as an undirected graph on the set of vertices which is the disjoint union of $L(w)$ and $R(w)$ with edges the pairs $(a, b) \in E(w)$. This graph is called the *extension graph* of w. We sometimes denote $1 \otimes L(w)$ and $R(w) \otimes 1$ the copies of $L(w)$ and $R(w)$ used to define the set of vertices of $E(w)$. We note that since $E(w)$ has $\ell(w) + r(w)$ vertices and $e(w)$ edges, the number $1 - m_S(w)$ is the Euler characteristic of the graph $E(w)$.

A biextendable set S is called a *tree set* of characteristic c if for any nonempty $w \in S$, the graph $E(w)$ is a tree and if $E(\varepsilon)$ is a union of c trees (the definition of tree set in [5] corresponds to a tree set of characteristic 1). Note that a tree set of characteristic c is a neutral set of characteristic c.

Example 2. Let S be the neutral set of Example 1. The graph $E(\varepsilon)$ is represented in Figure 1. It is acyclic with two connected components. It is shown in [5] that the extension graph of any nonempty word is a tree. Thus S is a tree set of characteristic 2.

Fig. 1. The two trees forming the graph $E(\varepsilon)$. Vertices correspond to letters, while edges correspond to words of length 2 in S.

Let S be a factorial set. For $x \in S$, we define

$$\rho_S(x) = e_S(x) - \ell_S(x), \quad \lambda_S(x) = e_S(x) - r_S(x).$$

Thus, when x is neutral, $\rho_S(x) = r_S(x) - 1$ and $\lambda_S(x) = \ell_S(x) - 1$. The following result shows that in a biextendable neutral set, ρ_S is a left probability distribution on S (and λ_S is a right probability), except for the value on ε which is $\rho(\varepsilon) = e(\varepsilon) - \ell(\varepsilon) = m(\varepsilon) + r(\varepsilon) - 1 = \mathrm{Card}(A) - \chi(S)$ and can be different of 1 (see [2] for the definition of a right or left probability distribution). We omit the subscript S when it is clear from the context.

Proposition 3. *Let S be a biextendable neutral set. Then for any $x \in S$, one has $\lambda_S(x), \rho_S(x) \geq 0$ and*

$$\sum_{a \in L(x)} \rho_S(ax) = \rho_S(x), \quad \sum_{a \in R(x)} \lambda_S(xa) = \lambda_S(x).$$

Proof. Since S is biextendable, we have $\ell(x), r(x) \leq e(x)$. Thus $\lambda(x), \rho(x) \geq 0$. Next, $\sum_{a \in L(x)} \rho(ax) = \sum_{a \in L(x)}(r(ax) - 1) = e(x) - \ell(x) = \rho(x)$. The proof for λ is symmetric.

If $\rho(\varepsilon) = 0$, then $\rho(x) = 0$ for all $x \in S$. Otherwise, $\rho'(x) = \rho(x)/\rho(\varepsilon)$ is a left probability distribution. A symmetric result holds for λ.

3 Bifix Codes

A prefix code is a set of nonempty words which does not contain any proper prefix of its elements. A suffix code is defined symmetrically. A *bifix code* is a set which is both a prefix code and a suffix code (see [3] for a more detailed introduction). Let S be a recurrent set. A prefix (resp. bifix) code $X \subset S$ is S-maximal if it is not properly contained in a prefix (resp. bifix) code $Y \subset S$. Since S is recurrent, a finite S-maximal bifix code is also an S-maximal prefix code (see [2, Theorem 4.2.2]). For example, for any $n \geq 1$, the set $X = S \cap A^n$ is an S-maximal bifix code.

Given a set X, we denote $\rho(X) = \sum_{x \in X} \rho(x)$. We prove the following result. It accounts for the fact that, in a Sturmian set S, any finite S-maximal suffix code contains exactly one right-special word [2, Proposition 5.1.5].

Proposition 4. *Let S be a neutral set and let X be a finite S-maximal suffix code. Then $\rho(X) = \mathrm{Card}(A) - \chi(S)$.*

Proof. If $\rho(\varepsilon) = 0$, then $\chi(S) = \mathrm{Card}(A)$ and thus the formula holds. Otherwise, ρ' is a left probability distribution (as seen at the end of Section 2), and the formula holds by a well-known property of suffix codes (see [2, Proposition 3.3.4]).

Example 3. Let S be the neutral set of characteristic 2 of Example 1. The set $X = \{a, ac, b, bc, d\}$ is an S-maximal suffix code (its reversal is the $\tilde{S}$-maximal prefix code $\tilde{X} = \{a, b, ca, cb, d\}$). The values of ρ on X are represented in Figure 2 on the left. One has $\rho(X) = \rho(a) + \rho(bc) = 2$, in agreement with Proposition 4.

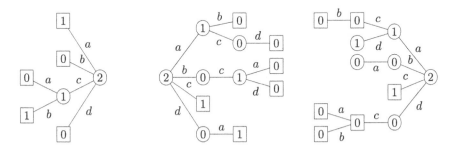

Fig. 2. An S-maximal suffix code (left) and an S-maximal bifix code represented as a prefix code (center) and as a suffix code (right)

Let X be a bifix code. Let Q be the set of words without any suffix in X and let P be the set of words without any prefix in X. A *parse* of a word w with respect to a bifix code X is a triple $(q, x, p) \in Q \times X^* \times P$ such that $w = qxp$. We denote by $d_X(w)$ the number of parses of a word w with respect to X. The S-degree of X, denoted $d_X(S)$ is the maximal number of parses with respect to X of a word of S. For example, the set $X = S \cap A^n$ has S-degree n.

Example 4. Let S be the neutral set of characteristic 2 of Example 1. The set $X = \{ab, acd, bca, bcd, c, da\}$ is an S-maximal bifix code of S-degree 2 (see Figure 2 on the center and the right).

Let S be a recurrent set and let X be a finite bifix code. By [2, Theorem 4.2.8], X is S-maximal if and only if its S-degree is finite. Moreover, in this case, a word $w \in S$ is such that $d_X(w) < d_X(S)$ if and only if it is an internal factor of a word of X. The following is [2, Theorem 4.3.7].

Theorem 1. *Let S be a recurrent set and let X be a finite S-maximal bifix code of S-degree n. The set of nonempty proper prefixes of X is a disjoint union of $n - 1$ S-maximal suffix codes.*

Example 5. Let S and X be as in Example 4. The set of nonempty proper prefixes of X is the S-maximal suffix code represented on the left of Figure 2.

The following statement is closely related with a similar statement concerning the average length of a bifix code, but which requires an invariant probability distribution (see [2, Corollary 4.3.8]).

Proposition 5. *Let S be a recurrent neutral set and let X be a finite S-maximal bifix code of S-degree n. The set P of proper prefixes of X satisfies $\rho_S(P) = n(\mathrm{Card}(A) - \chi(S))$.*

Proof. By Theorem 1, we have $P \setminus \{\varepsilon\} = \cup_{i=1}^{n-1} Y_i$, where the Y_i are S-maximal suffix codes. By Proposition 4, we have $\rho(Y_i) = \mathrm{Card}(A) - \chi(S)$ and thus $\rho(P) = \rho(\varepsilon) + (n-1)(\mathrm{Card}(A) - \chi(S)) = n(\mathrm{Card}(A) - \chi(S))$.

4   Cardinality Theorem for Bifix Codes

The following theorem is a generalization of [7, Theorem 3.6] where it is proved for a neutral set of characteristic 1. We consider a recurrent set S, and we implicitly assume that all words of S are on the alphabet A.

Theorem 2. *Let S be a neutral recurrent set. For any finite S-maximal bifix code X of S-degree n, one has*

$$\mathrm{Card}(X) = n(\mathrm{Card}(A) - \chi(S)) + \chi(S).$$

Note that we recover, as a particular case of Theorem 2 applied to the set X of words of length n in S, the fact that for a set S satisfying the hypotheses of the theorem, the factor complexity is $p_0 = 1$ and $p_n = n(\mathrm{Card}(A) - \chi(S)) + \chi(S)$. Note that Theorem 2 has a converse (see [4]).

Proof (of Theorem 2). Since X is a finite S-maximal bifix code, it is an S-maximal prefix code (see Section 3). By a well-known property of trees, this implies that $\mathrm{Card}(X) = 1 + \sum_{p \in P}(r(p) - 1)$ where P is the set of proper prefixes of X. Since $\rho(p) = r(p) - 1$ for p non empty and $\rho(\varepsilon) = m(\varepsilon) + r(\varepsilon) - 1$, we have

$$\mathrm{Card}(X) = 1 + \sum_{p \in P}(r(p) - 1) = 1 + \sum_{p \in P} \rho(p) - m(\varepsilon)$$
$$= \rho(P) + \chi(S) = n(\mathrm{Card}(A) - \chi(S)) + \chi(S)$$

since $\rho(P) = n(\mathrm{Card}(A) - \chi(S))$ by Proposition 5.

Example 6. Let S be the neutral set of Example 1 and let X be the S-maximal bifix code of Example 4. We have $\mathrm{Card}(X) = 2(4 - 2) + 2 = 6$ according to Theorem 2.

5   Cardinality Theorem for Return Words

Let S be a factorial set of words. For a set $X \subset S$ of nonempty words, a *complete first return word* to X is a word of S which has a proper prefix in X, a proper suffix in X and no internal factor in X. We denote by $\mathcal{CR}_S(X)$ the set of complete first return words to X. The set $\mathcal{CR}_S(X)$ is a bifix code. If S is uniformly recurrent, $\mathcal{CR}_S(X)$ is finite for any finite set X. For $x \in S$, we denote $\mathcal{CR}_S(x)$ instead of $\mathcal{CR}_S(\{x\})$.

Theorem 3. *Let S be a uniformly recurrent neutral set. For any bifix code $X \subset S$, we have*
$$\mathrm{Card}(\mathcal{CR}_S(X)) = \mathrm{Card}(X) + \mathrm{Card}(A) - \chi(S).$$

Proof. Let P be the set of proper prefixes of $\mathcal{CR}_S(X)$. For $q \in P$, we denote $\alpha(q) = \mathrm{Card}\{a \in A \mid qa \in P \cup \mathcal{CR}_S(X)\} - 1$ and $\alpha(P) = \sum_{q \in P} \alpha(p)$.

Since $\mathcal{CR}_S(X)$ is a finite nonempty prefix code, we have, by a well-known property of trees, $\mathrm{Card}(\mathcal{CR}_S(X)) = 1 + \alpha(P)$.

Let P' be the set of words in P which are proper prefixes of X and let $Y = P \setminus P'$. Since P' is the set of proper prefixes of X, we have $\alpha(P) = \mathrm{Card}(X) - 1$.

Since S is recurrent, any word of S with a prefix in X is comparable for the prefix order with a word of $\mathcal{CR}_S(X)$. This implies that for any $q \in Y$ and any $b \in R_S(q)$, one has $qb \in P \cup \mathcal{CR}_S(X)$. Consequently, we have $\alpha(q) = \rho_S(q)$ for any $q \in Y$. Thus we have shown that

$$\mathrm{Card}(\mathcal{CR}_S(X)) = 1 + \alpha(P') + \rho(Y) = \mathrm{Card}(X) + \rho(Y).$$

Let us show that Y is an S-maximal suffix code. This will imply our conclusion by Proposition 4. Suppose that $q, uq \in Y$ with u nonempty. Since q is in Y, it has a proper prefix in X. But this implies that uq has an internal factor in X, a contradiction. Thus Y is a suffix code. Consider $w \in S$. Since S is recurrent, there is some u and $x \in X$ such that $xuw \in S$. Let y be the shortest suffix of xuw which has a proper prefix in X. Then $y \in Y$. This shows that Y is an S-maximal suffix code.

Let S be a factorial set. A *right first return word* to x in S is a word w such that xw is a word of S which ends with x and has no internal factor equal to x (thus xw is a complete first return word to x). We denote by $\mathcal{R}_S(x)$ the set of right first return words to x in S. Since $\mathcal{CR}_S(x) = x\mathcal{R}_S(x)$, the sets $\mathcal{CR}_S(x)$ and $\mathcal{R}_S(x)$ have the same number of elements. Thus we have the following consequence of Theorem 3.

Corollary 1. *Let S be a uniformly recurrent neutral set. For any $x \in S$, the set $\mathcal{R}_S(x)$ has $\mathrm{Card}(A) - \chi(S) + 1$ elements.*

Example 7. Consider again the neutral set S of Example 1. We have $\mathcal{R}_S(a) = \{bca, bcda, cad\}$.

6   Bifix Decoding

Let S be a factorial set and let X be a finite S-maximal bifix code. A *coding morphism* for X is a morphism $f : B^* \to A^*$ which maps bijectively an alphabet B onto X. The set $f^{-1}(S)$ is called a *maximal bifix decoding* of S.

Theorem 4. *Any maximal bifix decoding of a recurrent neutral set is a neutral set with the same characteristic.*

Let S be a factorial set. For two sets of words X, Y and a word $w \in S$, we denote $L_S^X(w) = \{x \in X \mid xw \in S\}$, $R_S^Y(w) = \{y \in Y \mid wy \in S\}$, $E_S^{X,Y}(w) = \{(x, y) \in X \times Y \mid xwy \in S\}$, and further

$$e_S^{X,Y}(w) = \mathrm{Card}(E_S^{X,Y}(w)), \; \ell_S^X(w) = \mathrm{Card}(L_S^X(w)), \; r_S^Y(w) = \mathrm{Card}(R_S^Y(w)).$$

Finally, for a word w, we denote $m_S^{X,Y}(w) = e_S^{X,Y}(w) - \ell_S^X(w) - r_S^Y(w) + 1$. Note that $E_S^{A,A}(w) = E_S(w)$, $m_S^{A,A}(w) = m_S(w)$, and so on.

Proposition 6. *Let S be a neutral set, let X be a finite S-maximal suffix code and let Y be a finite S-maximal prefix code. Then $m_S^{X,Y}(w) = m_S(w)$ for every $w \in S$.*

Proof. We use an induction on the sum of the lengths of the words in X and in Y.

If X, Y contain only words of length 1, since X (resp. Y) is an S-maximal suffix (resp. prefix) code, we have $X = Y = A$ and there is nothing to prove.

Assume next that one of them, say Y, contains words of length at least 2. Let p be a nonempty proper prefix of Y of maximal length. Set $Y' = (Y \setminus pA) \cup p$. If $wp \notin S$, then $m^{X,Y}(w) = m^{X,Y'}(w)$ and the conclusion follows by induction hypothesis. Thus we may assume that $wp \in S$. Then

$$m^{X,Y}(w) - m^{X,Y'}(w) = e^{X,A}(wp) - \ell^X(wp) - r^A(wp) + 1 = m^{X,A}(wp).$$

By induction hypothesis, we have $m^{X,Y'}(w) = m(w)$ and $m^{X,A}(wp) = 0$, whence the conclusion.

Proof (of Theorem 4). Let S be a recurrent neutral set and let $f : B^* \to A^*$ be a coding morphism for a finite S-maximal bifix code X. Set $U = f^{-1}(S)$. Let $v \in U \setminus \{\varepsilon\}$ and let $w = f(v)$. Then $m_U(v) = m_S^{X,X}(w)$. Since S is recurrent, X is an S-maximal suffix code and prefix code. Thus, by Proposition 6, $m_U(v) = m_S(w)$, which implies our conclusion.

The following example shows that the maximal decoding of a uniformly recurrent neutral set need not be recurrent.

Example 8. Let S be the set of factors of the infinite word $(ab)^\omega$. The set $X = \{ab, ba\}$ is a bifix code of S-degree 2. Let $f : u \mapsto ab, v \mapsto ba$. The set $f^{-1}(S)$ is the set of factors of $u^\omega \cup v^\omega$ and it is not recurrent.

7   Neutral Sets and Interval Exchanges

Let $I =]\ell, r[$ be a nonempty open interval of the real line and A a finite ordered alphabet. For two intervals Δ, Γ, we denote $\Delta < \Gamma$ if $x < y$ for any $x \in \Delta$ and $y \in \Gamma$. A partition $(I_a)_{a \in A}$ of I (minus $\mathrm{Card}(A) - 1$ points) in open intervals is *ordered* if $a < b$ implies $I_a < I_b$.

We consider now two total orders $<_1$ and $<_2$ on A and two partitions $(I_a)_{a \in A}$ and $(J_a)_{a \in A}$ of I in open intervals ordered respectively by $<_1$ and $<_2$ and such that for every a, I_a and J_a have the same length λ_a. Let $\gamma_a = \sum_{b <_1 a} \lambda_b$ and $\delta_a = \sum_{b <_2 a} \lambda_a$.

An *interval exchange transformation* (with flips) relative to $(I_a)_{a \in A}$ and $(J_a)_{a \in A}$ is a map $T : I \to I$ such that for every $a \in A$, its restriction to I_a is either a translation or a symmetry from I_a to J_a (see for example [6] and [15] for interval exchanges with flips).

Observe that γ_a is the left boundary of I_a and that δ_a is the left boundary of J_a. If $\mathrm{Card}(A) = s$, we say that T is an *s-interval exchange transformation*.

Example 9. Let $A = \{a, b, c\}$. Consider the rotation of angle α with α irrational as a 3-transformation relative to the partition $(I_a)_{a \in A}$ of the interval $]0, 1[$, where $I_a =]0, 1-2\alpha[$, $I_b =]1-2\alpha, 1-\alpha[$ and $I_c =]1-\alpha, 1[$, while $J_c =]0, \alpha[$, $J_a =]\alpha, 1-\alpha[$ and $J_b =]1 - \alpha, 1[$ (see Figure 3). Then, for each letter a, the restriction to I_a is a translation to J_a. Note that one has $a <_1 b <_1 c$ and $c <_2 a <_2 b$.

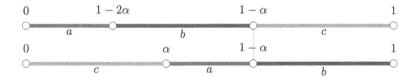

Fig. 3. A 3-interval exchange transformation

For a word $w = b_0 b_1 \cdots b_m$ let I_w be the set

$$I_w = I_{b_0} \cap T^{-1}(I_{b_1}) \cap \cdots \cap T^{-m}(I_{b_m}).$$

Set $J_w = T^{|w|}(I_w)$. We set by convention $I_\varepsilon = J_\varepsilon =]\ell, r[$. Note that each I_w is an open interval and so is each J_w (see [6]).

Let T be an interval exchange transformation on $I =]\ell, r[$. For a given $z \in I$, the *natural coding* of T relative to z is the infinite word $\Sigma_T(z) = a_0 a_1 \cdots$ on the alphabet A defined by $a_n = a$ if $T^n(z) \in I_a$. We denote by $\mathcal{L}(T)$ the set of factors of the natural codings of T. We also say that $\mathcal{L}(T)$ is the *natural coding* of T. Note that, for every $w \in \mathcal{L}(T)$, the interval I_w is the set of points z such that $\Sigma_T(z)$ starts with w, while the interval J_w is the set of points z such that $\Sigma_T(T^{-|w|}(z))$ starts with w. Moreover, it is easy to prove that a word u is in $\mathcal{L}(T)$ if and only if $I_u \neq \emptyset$ (and thus if and only if $J_u \neq \emptyset$).

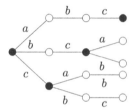

Fig. 4. The words of length ≤ 3 of $\mathcal{L}(T)$

Example 10. Let T be the interval exchange transformation of Example 9. The first element of $\mathcal{L}(T)$ are represented in Figure 4 (right-special words are colored).

A *connection* of an interval exchange transformation T is a triple (x, y, n) where x is a singularity of T^{-1}, y is a singularity of T, $n \geq 0$ and $T^n(x) = y$. We also say that (x, y, n) is a connection of length n ending in y. When $n = 0$, we say that $x = y$ is a connection.

Interval exchange transformations without connections, also called *regular* interval exchange transformations, are well studied (see, for example, [14] and [6]). The natural coding of a linear involutions without connection (see [9]) is essentially the coding of an interval exchange transformation with exactly one connection of length 0 ending in the midpoint of the interval.

Example 11. Let T be the transformation of Example 9. The point γ_c is a connection of length 0. This connection is represented with a dotted line in Figure 3.

Let T be an interval exchange transformation with exactly c connections all of length 0. Denote $\gamma_{k_0} = \ell$ and $\gamma_{k_1}, \ldots, \gamma_{k_c}$ the c connections of T. For every $0 \leq i < c$ the interval $]\gamma_{k_i}, \gamma_{k_{i+1}}[$ is called a *component* of I.

Example 12. Consider again the transformation T of Example 9. The two components of $]0, 1[$ are the two intervals $]0, 1 - \alpha[$ and $]1 - \alpha, 1[$.

In the next statement we generalize a result of [5] and show that the natural coding of an interval exchange is acyclic.

Theorem 5. *Let T be an interval exchange transformation with exactly c connections, all of length 0. Then $\mathcal{L}(T)$ is neutral of characteristic $c + 1$.*

Lemma 1. *Let T be an interval exchange transformation. For every nonempty word w and letter $a \in A$, one has*

(i) $a \in L(w) \Longleftrightarrow I_w \cap J_a \neq \emptyset$,
(ii) $a \in R(w) \Longleftrightarrow I_a \cap J_w \neq \emptyset$.

Proof. A letter a is in the set $L(w)$ if and only if $aw \in \mathcal{L}(T)$. As we have seen before, this is equivalent to $J_{aw} \neq \emptyset$. One has $J_{aw} = T(I_{aw}) = T(I_a) \cap I_w = J_a \cap I_w$, whence point (i). Point (ii) is proved symmetrically.

We say that a path in a graph is *reduced* if it does not use twice consecutively the same edge.

Lemma 2. *Let T be an interval exchange transformation over I without connection of length ≥ 1. Let $w \in \mathcal{L}(T)$ and $a, b \in L(w)$ (resp. $a, b \in R(w)$). Then $1 \otimes a, 1 \otimes b$ (resp. $a \otimes 1, b \otimes 1$) are in the same connected component of $E(w)$ if and only if J_a, J_b (resp. I_a, I_b) are in the same component of I.*

Proof. Let $a \in L(w)$. Since the set $\mathcal{L}(T)$ is biextendable, there exists a letter c such that $(1 \otimes a, c \otimes 1) \in E(w)$. Using the same reasoning as that in Lemma 1, one has $J_a \cap I_{wc} \neq \emptyset$. Since $I_{wc} \subset I_w$, one has in particular $J_a \cap I_w \neq \emptyset$. This proves that J_a and I_w belong to the same component of I for every $a \in L(w)$.

Conversely, suppose that $a, b \in L(w)$ are such that J_a and J_b belong to the same component of I. We may assume that $a <_2 b$. Then, there is a reduced path $(1 \otimes a_1, b_1 \otimes 1, \ldots, b_{n-1} \otimes 1, 1 \otimes a_n)$ in $E(w)$ (see Figure 5) with $a = a_1$, $b = a_n$, $a_1 <_2 \cdots <_2 a_n$ and $wb_1 <_1 \cdots <_1 wb_{n_1}$. Indeed, by hypothesis, we have no connection of length ≥ 1. Thus, for every $1 \leq i < n$, one has $J_{a_i} \cap I_{wb_i} \neq \emptyset$ and $J_{a_{i+1}} \cap I_{wb_i} \neq \emptyset$. Therefore, a and b are in the same connected component of $E(w)$.

The symmetrical statement is proved similarly.

We can now prove the main result of this section.

Proof (of Theorem 5). Let us first prove that for any $w \in \mathcal{L}(T)$, the graph $E(w)$ is acyclic. Assume that $(1 \otimes a_1, b_1 \otimes 1, \ldots, 1 \otimes a_n, b_n \otimes 1)$ is a reduced path in $E(w)$ with $a_1, \ldots, a_n \in L(w)$ and $b_1, \ldots, b_n \in R(w)$. Suppose that $n \geq 2$ and that $a_1 <_2 a_2$. Then one has $a_1 <_2 \cdots <_2 a_n$ and $wb_1 <_1 \cdots <_1 wb_n$ (see Figure 5). Thus one cannot have an edge (a_1, b_n) in the graph $E(w)$.

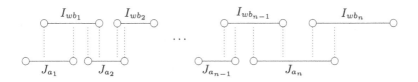

Fig. 5. A path from a_1 to a_n in $E(w)$

Let us now prove that the extension graph of the empty word is a union of $c + 1$ trees. Let $a, b \in A$. If J_a and J_b are in the same component of I, then $1 \otimes a, 1 \otimes b$ are in the same connected component of $E(\varepsilon)$ by Lemma 2. Thus $E(\varepsilon)$ is a union of $c + 1$ trees.

Finally, if $w \in \mathcal{L}(T)$ is a nonempty word and $a, b \in L(w)$, then J_a and J_b are in the same component of I, by Lemma 1, and thus a and b are in the same connected component of $E(w)$ by Lemma 2. Thus $E(w)$ is a tree.

The previous proof shows actually a stronger result: the set $\mathcal{L}(T)$ is a tree set of characteristic $c + 1$. This result generalizes the corresponding result for regular interval exchange in [5].

Example 13. Let T be the interval exchange transformation of Example 9. In Figure 6 are represented the extension graphs of the empty word (left) and of the letters a (center) and b (right).

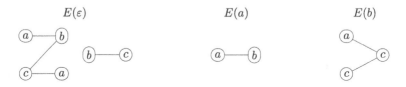

Fig. 6. Some extension graphs

Acknowledgments. This work was supported by grants from Région Île-de-France and ANR project Eqinocs.

References

1. Balková, Ĺ., Pelantová, E., Steiner, W.: Sequences with constant number of return words. Monatsh. Math. **155**(3–4), 251–263 (2008)
2. Berstel, J., De Felice, C., Perrin, D., Reutenauer, C., Rindone, G.: Bifix codes and Sturmian words. J. Algebra **369**, 146–202 (2012)
3. Berstel, J., Perrin, D., Reutenauer, C.: Codes and Automata. Cambridge University Press (2009)
4. Berthé, V., De Felice, C., Delecroix, V., Dolce, F., Leroy, J., Perrin, D., Reutenauer, C., Rindone, G.: Specular sets. In: Preparation (2015). (http://arxiv.org/abs/1505.00707)
5. Berthé, V., De Felice, C., Dolce, F., Leroy, J., Perrin, D., Reutenauer, C., Rindone, G.: Acyclic, connected and tree sets. Monatsh. Math. **176**(4), 521–550 (2015)
6. Berthé, V., De Felice, C., Dolce, F., Leroy, J., Perrin, D., Reutenauer, C., Rindone, G.: Bifix codes and interval exchanges. J. Pure Appl. Algebra **219**(7), 2781–2798 (2015)
7. Berthé, V., De Felice, C., Dolce, F., Leroy, J., Perrin, D., Reutenauer, C., Rindone, G.: The finite index basis property. J. Pure Appl. Algebra **219**, 2521–2537 (2015)
8. Berthé, V., De Felice, C., Dolce, F., Leroy, J., Perrin, D., Reutenauer, C., Rindone, G.: Maximal bifix decoding. Discrete Math (2015)
9. Berthé, V., Delecroix, V., Dolce, F., Perrin, D., Reutenauer, C., Rindone, G.: Return words of linear involutions and fundamental groups (2015). (http://arxiv.org/abs/1405.3529)
10. Berthé, V., Rigo, M.: Combinatorics, automata and number theory. Encyclopedia Math. Appl., vol. 135. Cambridge Univ. Press, Cambridge (2010)

11. Massé, A.B., Brlek, S., Labbé, S., Vuillon, L.: Palindromic complexity of codings of rotations. Theoret. Comput. Sci. **412**(46), 6455–6463 (2011)
12. Cassaigne, J.: Complexité et facteurs spéciaux. Bull. Belg. Math. Soc. Simon Stevin **4**(1), 67–88 (1997). Journées Montoises (Mons, 1994)
13. Justin, J., Vuillon, L.: Return words in Sturmian and episturmian words. Theor. Inform. Appl. **34**(5), 343–356 (2000)
14. Keane, M.: Interval exchange transformations. Math. Z. **141**, 25–31 (1975)
15. Nogueira, A., Pires, B., Troubetzkoy, S.: Orbit structure of interval exchange transformations with flip. Nonlinearity **26**(2), 525–537 (2013)
16. Vuillon, L.: On the number of return words in infinite words constructed by interval exchange transformations. Pure Math. Appl. (PU.M.A.) **18**(3–4), 345–355 (2007)

On the Density of Context-Free and Counter Languages

Joey Eremondi[1], Oscar H. Ibarra[2], and Ian McQuillan[3]([✉])

[1] Department of Information and Computing Sciences, Utrecht University,
P.O. Box 80.089, 3508 TB Utrecht, The Netherlands
j.s.eremondi@students.uu.nl
[2] Department of Computer Science, University of California,
Santa Barbara, CA 93106, USA
ibarra@cs.ucsb.edu
[3] Department of Computer Science, University of Saskatchewan,
Saskatoon, SK S7N 5A9, Canada
mcquillan@cs.usask.ca

Abstract. A language L is said to be *dense* if every word in the universe
is an infix of some word in L. This notion has been generalized from the
infix operation to arbitrary word operations ϱ in place of the infix operation (ϱ-dense, with infix-dense being the standard notion of dense). It
is shown here that it is decidable, for a language L accepted by a one-
way nondeterministic reversal-bounded pushdown automaton, whether
L is infix-dense. However, it becomes undecidable for both deterministic
pushdown automata (with no reversal-bound), and for nondeterministic
one-counter automata. When examining suffix-density, it is undecidable
for more restricted families such as deterministic one-counter automata
that make three reversals on the counter, but it is decidable with less
reversals. Other decidability results are also presented on dense languages, and contrasted with a marked version called ϱ-marked-density.
Also, new languages are demonstrated to be outside various deterministic language families after applying different deletion operations
from smaller families. Lastly, bounded-dense languages are defined and
examined.

1 Introduction

A language $L \subseteq \Sigma^*$ is dense if the set of all infixes of L is equal to Σ^* [3].
This notion is relevant to the theory of codes. Indeed, a language being dense
is connected with the notions of independent sets, maximal independent sets,
codes [11], and disjunctive languages [10,13].

Dense languages have been studied in [10,13] and generalized from density to
ϱ-density [11], where ϱ is an arbitrary word operation used in place of the infix-
operation in the definition. Some common examples are prefix-dense (coinciding

The research of O.H. Ibarra was supported, in part, by NSF Grant CCF-1117708.
The research of I. McQuillan was supported, in part, by the Natural Sciences and
Engineering Research Council of Canada.

I. Potapov (Ed.): DLT 2015, LNCS 9168, pp. 228–239, 2015.
DOI: 10.1007/978-3-319-21500-6_18

with left dense in [10]), suffix dense (coinciding with right dense in [10]), infix dense (usual notion of density), outfix dense, embedding dense, and others from [11]. Each type connects with a generalized notion of independent sets and codes.

It has long been known that universality of a language L (is $L = \Sigma^*$?) is undecidable for L accepted by a one-way nondeterministic one-counter automaton whose counter makes only one reversal, i.e., in an accepting computation, after decreasing the counter, it can no longer increase again [2]. This shows immediately that with the identity operation, it is undecidable if L in this family is identity-dense. In contrast, the universality problem is known to be decidable for one-way deterministic reversal-bounded multicounter languages [9], but these languages are not closed under taking suffix, infix or outfix [5]. However, to decide the property of infix-density, in this paper we can show contrasting results.

1. Infix-density is decidable for L accepted by a nondeterministic pushdown automaton where the pushdown is reversal-bounded (there is at most a fixed number of switches between increasing and decreasing the size of the pushdown).
2. Infix-density is undecidable for L accepted by a nondeterministic one-counter automaton (with no reversal-bound).
3. Infix-density is undecidable for L accepted by a deterministic pushdown automaton (with no reversal-bound).

Thus, it is surprisingly possible to decide if the set of all infixes of a nondeterministic reversal-bounded pushdown automaton gives universality, when it is undecidable with the identity operator for much smaller families.

Furthermore, if the question is altered to change the type of density from infix-density to either suffix-density or prefix-density, then it is undecidable even for nondeterministic one-counter automata that makes one counter reversal (coinciding with the result for identity-density). Suffix-density is decidable however for deterministic one-counter automata that makes one counter reversal, but is undecidable when there is either two more reversals, or two counters that both make one reversal. Thus suffix-density is often impossible to decide when infix-density is decidable. Prefix density is decidable for all deterministic reversal-bounded multicounter languages.

Contrasts are made between deciding if applying an operation ϱ to a language gives Σ^* and deciding if $\$\Sigma^*\$$ (with $\$ \notin \Sigma$) is a subset of ϱ applied to $L \subseteq (\Sigma \cup \{\$\})^*$. If this is true, the language is said to be ϱ-marked-dense. In contrast to infix-density, infix-marked-density is undecidable with only one-way deterministic one-counter 3-reversal-bounded languages, and for the outfix operation with many families as well.

In addition, new languages are established that can be accepted by a number of automata classes (deterministic one-counter machines that are 3-reversal-bounded, deterministic 2-counter machines that are 1-reversal-bounded, nondeterministic one-counter one-reversal-bounded machines), but taking any of the set of infixes, suffixes or outfixes of L produces languages that cannot be accepted by deterministic machines with an unrestricted pushdown and a fixed number of reversal-bounded counters. Hence, these deletion operations can create some very complex languages. It has been previously shown in [5] though, that the set of all infixes or suffixes of all deterministic one-counter one-reversal-bounded languages only produce deterministic reversal-bounded multicounter languages. Finally, the notion of ϱ-bounded-dense languages is defined and examined.

2 Definitions

In this section, some preliminary definitions are provided.

The set of non-negative integers is represented by $\mathbb{N}_0$. For $c \in \mathbb{N}_0$, let $\pi(c)$ be 0 if $c = 0$, and 1 otherwise.

We use standard notations for formal languages, referring the reader to [7]. The empty word is denoted by λ. We use Σ and Γ to represent finite alphabets, with Σ^* as the set of all words over Σ and $\Sigma^+ = \Sigma^* - \{\lambda\}$. For a word $w \in \Sigma^*$, if $w = a_1 \cdots a_n$ where $a_i \in \Sigma$, $1 \le i \le n$, the length of w is denoted by $|w| = n$, and the reversal of w is denoted by $w^R = a_n \cdots a_1$. Given a language $L \subseteq \Sigma^*$, the complement of L over Σ^*, $\Sigma^* - L$ is denoted by $\overline{L}$.

The definitions of deterministic and nondeterministic finite automata, deterministic and nondeterministic pushdown automata, deterministic Turing Machines, and instantaneous descriptions will be used from [7].

Notation for variations of word operations which we will use throughout the paper are presented next.

Definition 1. *For a language $L \subseteq \Sigma^*$, the prefix, suffix, infix, and outfix operations, respectively, are defined as follows:*

$$\text{pref}(L) = \{w \mid wx \in L, x \in \Sigma^*\}, \quad \text{suff}(L) = \{w \mid xw \in L, x \in \Sigma^*\},$$
$$\text{inf}(L) = \{w \mid xwy \in L, x, y \in \Sigma^*\}, \quad \text{outf}(L) = \{xy \mid xwy \in L, w \in \Sigma^*\}.$$

Different types of density are now given.

Definition 2. *Let Σ be an alphabet, and ϱ an operation from Σ^* to Σ^*. Then $L \subseteq \Sigma^*$ is ϱ-dense if $\varrho(L) = \Sigma^*$.*

The reader is referred to [9] and [2] for a comprehensive introduction to counter machines. A *nondeterministic multicounter machine* is an automaton which, in addition to having a finite set of states, has a fixed number of counters. At any point, the counters may be incremented, decremented, or queried for equality to zero. For our purposes, it will accept a word by final state.

Formally, a *one-way k-counter machine* is a tuple $M = (k, Q, \Sigma, \lhd, \delta, q_0, F)$, where $Q, \Sigma, \lhd, q_0, F$ are respectively the set of states, input alphabet, right input end-marker, initial state (in Q) and accepting states (a subset of Q). The transition function δ (defined as in [4]) is a mapping from $Q \times (\Sigma \cup \{\lhd\}) \times \{0, 1\}^k$ into $Q \times \{S, R\} \times \{-1, 0, +1\}^k$, such that if $\delta(q, a, c_1, \ldots, c_k)$ contains $(p, d, d_1, \ldots, d_k)$ and $c_i = 0$ for some i, then $d_i \ge 0$ (to prevent negative values in any counter). The symbols S and R indicate the direction that the input tape head moves, either *stay* or *right*. Further, M is *deterministic* if δ is a function. A configuration of M is a $k + 2$-tuple $(q, w\lhd, c_1, \ldots, c_k)$ representing that M is in state

q, with $w \in \Sigma^*$ still to read as input, and $c_1, \ldots, c_k \in \mathbb{N}_0$ are the contents of the k counters. The derivation relation $\vdash_M$ is defined between configurations, where $(q, aw, c_1, \ldots, c_k) \vdash_M (p, w', c_1 + d_1, \ldots, c_k + d_k)$, if $(p, d, d_1, \ldots, d_k) \in \delta(q, a, \pi(c_1), \ldots, \pi(c_k))$ where $d \in \{S, R\}$ and $w' = aw$ if $d = S$, and $w' = w$ if $d = R$. Let $\vdash_M^*$ be the reflexive, transitive closure of $\vdash_M$. A word $w \in \Sigma^*$ is accepted by M if $(q_0, w \triangleleft, 0, \ldots, 0) \vdash_M^* (q, \triangleleft, c_1, \ldots, c_k)$, for some $q \in F$, and $c_1, \ldots, c_k \in \mathbb{N}_0$. The language accepted by M, denoted by $L(M)$, is the set of all words accepted by M. Furthermore, M is l-reversal-bounded if it operates in such a way that in every accepting computation, the count on each counter alternates between increasing and decreasing at most l times.

We will denote the following families of languages (and classes of one-way machines) by: $\mathsf{NCM}(k, l)$ for nondeterministic l-reversal-bounded k-counter languages, $\mathsf{NCM} = \bigcup_{k, l \geq 0} \mathsf{NCM}(k, l)$, NCA for nondeterministic 1-counter languages (no reversal bound), NPDA for nondeterministic pushdown languages, $\mathsf{NPDA}(l)$ for nondeterministic l-reversal-bounded pushdown languages, and NPCM for languages accepted by nondeterministic machines with one unrestricted pushdown and a fixed number of reversal-bounded counters. For each, replacing N with D gives the deterministic variant.

It is known that for each $k \geq 1$, $\mathsf{DCM}(1, k) \subseteq \mathsf{DCM}(\lceil \frac{k+1}{2} \rceil, 1)$ [4] and thus $\mathsf{DCM}(1, 3) \subseteq \mathsf{DCM}(2, 1)$.

3 Deciding Types of Density

In addition to examining decidability of ϱ-density, a variant is defined called ϱ-marked-density that differs from ϱ-density only by an end-marker.

Definition 3. *Let Σ be an alphabet, $\$ \notin \Sigma$, $L \subseteq (\Sigma \cup \{\$\})^*$, and ϱ be an operation from $(\Sigma \cup \{\$\})^*$ to itself. Then L is ϱ-marked-dense if $\$\Sigma^*\$ \subseteq \varrho(L)$.*

It is only the marker $\$$ that differs from the usual ϱ-dense (i.e., $\Sigma^* \subseteq \varrho(L)$ if and only if $\Sigma^* = \varrho(L)$ for $L \subseteq \Sigma^*$). Yet we will see differences, as there are cases when the marked version is undecidable when the unmarked version is decidable.

First, deciding if languages are prefix-dense will be examined. It was recently shown in [5] that DCM languages are closed under prefix. The following is a result in that paper.

Proposition 1. *For $L \in \mathsf{DCM}$, $\mathrm{pref}(L) \in \mathsf{DCM}$.*

A main result in that paper was in fact far more general, showing that DCM is closed (with an effective construction) under right quotient with NPCM languages. Combining this with the known decidability of the inclusion problem for DCM [9], the following two corollaries are obtained, by testing if $\Sigma^* \subseteq \mathrm{pref}(L)$:

Corollary 1. *For $L_1, L_2 \in \mathsf{DCM}$, it is decidable whether $\mathrm{pref}(L_1) \subseteq L_2$ and whether $L_1 \subseteq \mathrm{pref}(L_2)$.*

Corollary 2. *It is decidable whether a given DCM language is prefix-dense, and prefix-marked-dense.*

Next, it is shown in [5] that the set of suffixes and infixes of a $\mathsf{DCM}(1,1)$ language is always in DCM (by sometimes increasing the number of counters). From this, the following is obtained:

Proposition 2. *For $L_1, L_2 \in \mathsf{DCM}(1,1)$, it is decidable whether $\inf(L_1) \subseteq L_2$ and whether $L_1 \subseteq \inf(L_2)$. It is also decidable whether $\mathrm{suff}(L_1) \subseteq L_2$ and whether $L_1 \subseteq \mathrm{suff}(L_2)$.*

Corollary 3. *It is decidable whether a $\mathsf{DCM}(1,1)$ language is infix-dense, suffix-dense, infix-marked-dense and suffix-marked-dense.*

This result will be improved shortly using a more general machine class for infix-density, but not for suffix-density, suffix-marked-density, or infix-marked-density.

Most undecidability proofs in this section use the halting problem for Turing machines. Let $U \subseteq \{a\}^*$ be a unary recursively enumerable language that is not recursive, i.e., not decidable (such a U exists [12]), and let Z be a deterministic Turing machine accepting U. Assume that Z accepts if and only if Z halts.

Let Q and Γ be the state set and worktape alphabet of Z, and $q_0 \in Q$ be the initial state of Z. Note that a is in Γ. Let $\Sigma = Q \cup \Gamma \cup \{\#\}$. Assume without loss of generality that if Z halts, it does so in a unique final state $q_f \neq q_0$, and a unique configuration, and that the initial state q_0 is never re-entered after the initial configuration, and that the length of every halting computation is even.

The halting computation of Z on the input a^d (if it accepts) can be represented by the string $x_d = ID_1 \# ID_2^R \# \cdots \# ID_{k-1} \# ID_k^R$ for some $k \geq 2$, where $ID_1 = q_0 a^d$ and ID_k are the initial and unique halting configurations of Z, and $(ID_1, ID_2, \cdots, ID_k)$ is a valid sequence of instantaneous descriptions (IDs, defined in [7]) of Z on input a^d, i.e., configuration ID_{i+1} is a valid successor of ID_i, and k is even.

Let $d \geq 0$. Let T be all strings w of the form $ID_1 \# ID_2^R \# \cdots \# ID_{k-1} \# ID_k^R$, where $k \geq 2$, $ID_1 = q_0 a^d$, and ID_k is the halting configuration of Z, and ID_i is any ID of the Turing machine, $1 < i < k$. Then T is a regular language, and thus a DFA M_T can be built accepting T, and also one can be built accepting $\overline{T}$. Let L_{na} be all strings $w \in T$ of the form $ID_1 \# ID_2^R \# \cdots \# ID_{k-1} \# ID_k^R$, where there is an i such that ID_{i+1} is not a valid successor of ID_i. Indeed, if ID_{i+1} is not a valid successor of ID_i, then this is detectable by scanning the state of ID_i, the letter after the state (symbol under the read/write head), and from these, the transition of Z applied to get the valid successor of ID_i can be calculated, as with whether the ID representing the valid successor to ID_i should be shorter or longer by one symbol. Then, there is some position j of ID_i such that examining positions $j - 2, j - 1, j, j + 1, j + 2$ of ID_i and ID_{i+1}, and the state of ID_{i+1} is enough to imply that ID_{i+1} is not a valid successor. Hence, let $L_{na}(p)$ be the set of words $w \in T$ of the form $w = ID_1 \# ID_2^R \# \cdots \# ID_{k-1} \# ID_k^R$, where the pth character of w is within the string ID_i for some i at position j of ID_i and examining characters $j - 2, j - 1, j, j + 1, j + 2$ of ID_i and ID_{i+1} (if they exist), plus the states of both, and the letter after the state, implies that ID_{i+1} is not a valid successor of ID_i. Thus, $\bigcup_{p \geq 0} L_{na}(p) = L_{na}$.

Let $L_d = L_{na} \cup \overline{T}$. Two lemmas are required for undecidability results.

Lemma 1. $L_d = \Sigma^*$ *if and only if* $T \subseteq L_{na}$ *if and only if* Z *does not halt on* a^d.

Proof. If $L_d = \Sigma^*$, then $T \subseteq L_{na}$, and if $T \subseteq L_{na}$ then $T \cup \overline{T} = \Sigma^* \subseteq L_d$. Thus the first two are equivalent.

Assume $L_d = \Sigma^*$. Thus, every sequence of IDs in T is in L_{na}, thus there is no sequence of IDs that halts on a^d.

Assume that Z does not halt on a^d. Let $w \in \Sigma^*$. If $w \notin T$, then $w \in L_d$. If $w \in T$, then w does not represent an accepting computation, thus, $w \in L_d$. $\square$

Let $\%$ be a new symbol not in Σ, and let $\Sigma_\% = \Sigma \cup \{\%\}$.

Lemma 2. $\bigcup_{p \geq 0} \%^p L_{na}(p)$ *and* $\bigcup_{p \geq 0} \%^p \$L_{na}(p)\$$ *are both in* $\mathsf{DCM}(1,3)$ *and* $\mathsf{DCM}(2,1)$. *Furthermore,* $L_{na}, \$L_{na}\$ \in \mathsf{NCM}(1,1)$.

Proof. We can construct a $\mathsf{DCM}(1,3)$ machine M_{na} to accept the strings of $\bigcup_{p \geq 0} \%^p L_{na}(p)$ as follows: when given $\%^p w$, it reads $\%^p$ and increments the counter by p. It then decrements the counter and verifies that when the counter becomes zero, the input head is within some ID_i (or ID_i^R if i is even). If i is odd, M_{na} then moves the input head incrementing the counter until it reaches the $\#$ to the right of ID_i. Let j be the value of the counter. M_{na} then decrements the counter while moving right on ID_{i+1}^R and after reaching zero, verifying that ID_{i+1} is not a valid successor of ID_i (this is possible as ID_{i+1}^R is reversed). Similarly when i is even. In the same way, we can construct a $\mathsf{DCM}(1,3)$ machine to accept $\bigcup_{p \geq 0} \%^p \$L_{na}(p)\$$. Both languages are in $\mathsf{DCM}(2,1)$ as $\mathsf{DCM}(1,3) \subseteq \mathsf{DCM}(2,1)$.

For L_{na} (and $\$L_{na}\$$), it is possible to nondeterministically guess the position p, and then when within ID_i, verify using the counter once that ID_{i+1} is not a valid successor to ID_i. $\square$

This is similar to the technique from [2] to show undecidability of universality for $\mathsf{NCM}(1,1)$.

Most of the undecidability results in this section build off of the above two lemmas, the input a^d, the languages T, L_{na}, etc.

Proposition 3. *Let* Σ *be an alphabet.*

1. *It is undecidable to determine, given* $L \in \mathsf{NCM}(1,1)$, *whether* L *is* ϱ-marked-dense, for $\varrho \in \{\mathrm{suff}, \mathrm{inf}, \mathrm{pref}\}$.
2. *It is undecidable to determine, given* $L \in \mathsf{DCM}(1,3)$, *whether* L *is* ϱ-marked-dense, for $\varrho \in \{\mathrm{suff}, \mathrm{inf}\}$.
3. *It is undecidable to determine, given* $L \in \mathsf{DCM}(2,1)$, *whether* L *is* ϱ-marked-dense, for $\varrho \in \{\mathrm{suff}, \mathrm{inf}\}$.

Proof. For part 1, we can accept $L' = \$L_{na}\$ \cup \$\overline{T}\$ \subseteq (\Sigma \cup \{\$\})^*$ in $\mathsf{NCM}(1,1)$ since $\$\overline{T}\$$ is a regular language (the complement is over Σ^*).

Then $\$\Sigma^*\$ \subseteq \mathrm{inf}(L')$ (resp., $\$\Sigma^*\$ \subseteq \mathrm{suff}(L')$, $\$\Sigma\$ \subseteq \mathrm{pref}(L')$) if and only if $\$\Sigma^*\$ = L'$ if and only if $L_d = \Sigma^*$, which we already know is true if and only if Z does not halt on a^d by Lemma 1, which is undecidable.

For parts 2 and 3, we instead use $L' = \bigcup_{p \geq 0} \%^p \$ L_{na}(p)\$ \cup \$\overline{T}\$$, the complement $\overline{T}$ is over $\Sigma_\%^* = (\Sigma \cup \{\%\})^*$ here, so it will also contain any word with $\%$ in it to allow for marked-density to be with $L' \subseteq (\Sigma_\% \cup \{\$\})^*$ where the goal is to decide whether $\$ \Sigma_\%^* \$ \subseteq \inf(L')$. Then L' is in $\mathsf{DCM}(1,3) \cap \mathsf{DCM}(2,1)$ by Lemma 2 and since $\mathsf{DCM}(k,l)$ is closed under union with regular languages, for every k, l. And $\$ \Sigma_\%^* \$ \subseteq \inf(L')$ if and only if $\$ \Sigma^* \$ \subseteq \inf(L')$ (since $\overline{T}$ contains all words with at least one $\%$) if and only if $L_d = \Sigma^*$. The proof in the case of the suffix operation is similar.   □

The proof for the outfix operation is similar.

Proposition 4. *It is undecidable, given $L \in \mathsf{NCM}(1,1)$, whether L is outf-marked-dense. Similarly with $L \in \mathsf{DCM}(2,1)$, and $L \in \mathsf{DCM}(1,3)$.*

Proof. For $L \in \mathsf{NCM}(1,1)$, we modify the language L' in the proof of Part 1 of Proposition 3. So $L' = \% \$ L_{na}\$ \cup \% \$\overline{T}\$$ ($\overline{T}$ over $\Sigma_\%^*$). For the other classes, L' in the proofs of parts 2, 3 also work for outf.   □

It follows from Propositions 3 and 4 that $\mathsf{DPDA}(3)$ has an undecidable ϱ-marked-density problem for suffix, infix, and outfix. The following shows that they are also undecidable for $\mathsf{DPDA}(1)$.

Proposition 5. *For $\varrho \in \{\mathrm{suff}, \mathrm{inf}, \mathrm{outf}\}$, it is undecidable given $L \in \mathsf{DPDA}(1)$, whether L is ϱ-marked-dense.*

Proof. The problem of whether the intersection of two $\mathsf{DPDA}(1)$ languages is empty is undecidable [2]. Let $L_1, L_2 \in \mathsf{DPDA}(1)$. Then $L_1 \cap L_2 = \emptyset$ if and only if $\overline{L_1 \cap L_2} = \Sigma^*$ if and only if $\overline{L_1} \cup \overline{L_2} = \Sigma^*$ if and only if $\$ \Sigma^* \$ \subseteq \$\overline{L_1}\$ \cup \$\overline{L_2}\$$.

Let $L' = \% \$\overline{L_1}\$ \cup \$\overline{L_2}\$ \cup \$\Sigma_\%^* \% \Sigma_\%^* \$$ (here, the complements are over Σ^*). Note that $L' \subseteq (\Sigma_\% \cup \{\$\})^* = (\Sigma \cup \{\%, \$\})^*$. L' is in $\mathsf{DPDA}(1)$ since $\mathsf{DPDA}(1)$ is closed under complement, the union of the first two sets is a $\mathsf{DPDA}(1)$ language (if $\%$ is the first letter then simulate the first set, otherwise simulate the second), and the third one is regular and $\mathsf{DPDA}(1)$ is closed under union with regular sets.

Then $\$ \Sigma_\%^* \$ \subseteq \inf(L')$ if and only if $\$ \Sigma_\%^* \$ \subseteq \$\overline{L_1}\$ \cup \$\overline{L_2}\$ \cup \$\Sigma_\%^* \% \Sigma_\%^* \$$ if and only if $\$ \Sigma^* \$ \subseteq \$\overline{L_1}\$ \cup \$\overline{L_2}\$$, which we know is undecidable. The proof is identical for suffix, as with outfix after preceding each word in L' by an additional $\%$.   □

Next, ϱ-density instead of ϱ-marked-density will be considered; specifically, the question of whether it is decidable to determine if a language L is ϱ-dense ($\varrho(L) = \Sigma^*$) for various operations and languages. For suffix-density, undecidability occurs for the same families as for marked-suffix-density. The proofs will again build on the Turing Machine Z, input a^d, and languages L_{na}, T, etc.

Proposition 6. *Let $L \in \mathsf{DCM}(1,3)$. It is undecidable to determine if L is suffix-dense. Similarly for $L \in \mathsf{DCM}(2,1)$ and $L \in \mathsf{NCM}(1,1)$.*

Proof. Let $L_1' = \{\%^p ux \mid u \in \Sigma \Sigma_\%^*, p = |u| + p', x \in L_{na}(p')\}$, $L_2' = \overline{\Sigma_\%^* T}$ (over $\Sigma_\%^*$), and $L' = L_1' \cup L_2'$. Then $L' \in \mathsf{DCM}(1,3)$ as one can build $M' \in \mathsf{DCM}(1,3)$

by adding p to the counter until hitting a letter that is not %. Then as M' reads the remaining input in $\Sigma\Sigma_\%^*$, for every character read, it decreases the counter, and each time M' hits state q_0 (which could be the beginning of a word in T), it runs M_T (the DFA accepting T) in parallel to check if the suffix starting at this position is in T. However, it is only required that a suffix of the input is in T. If the counter empties while M_T is running in parallel, then let ux be the input, where u is the input before reaching q_0 in the current run of M_T, and x be the input from q_0 to the end. Then M' tries to verify that $x \in L_{na}(p')$, where $p = |u| + p'$. When the counter reaches 0, M' has subtracted 1 from the counter the length of u plus $p - |u| = p'$ times. Thus, M' can continue the simulation of M_{na} from Lemma 2 from when the counter reaches 0, thereby verifying that $x \in L_{na}(p')$ (and $x \in T$). Then also L' must be in $\mathsf{DCM}(2,1)$.

It will be shown that $\mathrm{suff}(L') = \Sigma_\%^*$ if and only if $T \subseteq L_{na}$, which is enough by Lemma 1.

"$\Leftarrow$" Assume $T \subseteq L_{na}$. Let $w \in \Sigma_\%^*$.

Assume that there exists a (potentially not proper) suffix of w in T. Then $w = ux, x \in T, u \in \Sigma_\%^*$. Then $x \in L_{na}$, by assumption. Then there exists p such that $\%^p x \in L_{na}(p), x \in T$ and so $\%^{p'} aux \in L_1', au \in \Sigma\Sigma_\%^*$, where $p' = p + |au|$. Thus $ux = w \in \mathrm{suff}(L_1')$.

Assume that there does not exist a suffix of w in T. Then $w \in L_2'$, and $w \in \mathrm{suff}(L')$.

"$\Rightarrow$" Assume $\mathrm{suff}(L') = \Sigma_\%^*$. Let $w \in T$. Then $w \in \mathrm{suff}(L')$. Then there exists $\%^p uw \in L_1'$. This implies there exists p' such that $w \in L_{na}(p') \subseteq L_{na}$.

The case for $\mathsf{NCM}(1,1)$ is similar except using $L_1' = \{ux \mid x \in L_{na}, u \in \Sigma^*\}$ and $L_2' = \overline{\Sigma^* T}$, and $L' = L_1' \cup L_2' \subseteq \Sigma^*$, as u can be nondeterministically guessed without using the counter.   □

Corollary 4. *For $L \in \mathsf{NCM}(1,1)$, the question of whether L is prefix-dense is undecidable.*

Proof. It is known that $\mathsf{NCM}(k,l)$ is closed under reversal for each k, l. Also, $\mathrm{pref}(L^R) = \Sigma^*$ if and only if $\mathrm{suff}(L) = \Sigma^*$.   □

We are able to extend the undecidability results to infix-density, but only by using one unrestricted counter and with nondeterminism.

Proposition 7. *Let $L \in \mathsf{NCA}$. The question of whether L is infix-dense is undecidable.*

Proof. Let $L' = \overline{(\Sigma^* T \Sigma^*)}(L_{na}\overline{(\Sigma^* T \Sigma^*)})^* \subseteq \Sigma^*$. It is clear that $L' \in \mathsf{NCA}$. We will show that $T \subseteq L_{na}$ if and only if $\inf(L') = \Sigma^*$.

"$\Rightarrow$" Assume $T \subseteq L_{na}$. Let $w \in \Sigma^*$. If $w \in \overline{(\Sigma^* T \Sigma^*)}$, then $w \in L' \subseteq \inf(L')$. Assume $w \notin \overline{(\Sigma^* T \Sigma^*)}$. Then $w \in \Sigma^* T \Sigma^*$. Then $w = u_0 v_1 u_1 \cdots u_{n-1} v_n u_n$, where $n \geq 1, v_1, \ldots, v_n \in T$, and $u_0, \ldots, u_n \notin \Sigma^* T \Sigma^*$, and so $u_0, \ldots, u_n \in \overline{(\Sigma^* T \Sigma^*)}$. Also, $T \subseteq L_{na}$, and therefore $v_1, \ldots, v_n \in L_{na}$ and $w \in L' \subseteq \inf(L')$.

"$\Leftarrow$" Assume $\inf(L') = \Sigma^*$. Let $w \in T$. Then $w \in \inf(L')$. Since $w \in \inf(L') \cap T$, then $x = uwv \in L'$. Then $x = u_0 v_1 u_1 \cdots u_{n-1} v_n u_n$, where $n \geq 1, v_1, \ldots, v_n \in$

L_{na}, and $u_0, \ldots, u_n \in \overline{\Sigma^* T \Sigma^*}$. If w is an infix of u_i, for some i, then $u_i \in \Sigma^* T \Sigma^*$, a contradiction. If w overlaps with v_i for some i, then it must be exactly one v_i by the structure of T (initial and final states are only used once at beginning and end of words in T). Then $w \in L_{na}$.   □

The same undecidability is obtained with determinism, but an unrestricted pushdown automaton is used.

Proposition 8. *Let $L \in$ DPDA. The question of whether L is infix-dense is undecidable.*

Proof. Let $\Sigma_1 = \Sigma \cup \{\%, e, ¢\}$. Let

$$L' = \{ r_m r_{m-1} \cdots r_1 ¢ u_0 y_1 u_1 \cdots y_m u_m \mid u_i \in \overline{\Sigma_1^* T \Sigma_1^*}, 0 \le i \le m,$$
$$y_j \in T, r_j = \%^{p_j} e^{q_j}, q_j = |u_{j-1}|, y_j \in L_{na}(p_j) \text{ for } 1 \le j \le m \}.$$

(In the above set, the complementation is over Σ_1^*.) First, L' can be accepted by a DPDA as follows: create M' that reads $r_m \cdots r_1$ and pushes each symbol onto the pushdown, which is now (with bottom of pushdown marker Z_0)

$$Z_0 \%^{p_m} e^{q_m} \cdots \%^{p_1} e^{q_1}.$$

Then for each $\%^{p_j} e^{q_j}$ on the pushdown from 1 to m, M' reads one symbol at a time from the input while popping one e, while in parallel verifying $u_{j-1} \in \overline{\Sigma_1^* T \Sigma_1^*}$. Then M' verifies that $y_j \in L_{na}(p_j)$ as in Lemma 2 (by popping $\%^{p_j}$ one symbol at a time until zero and then pushing on the pushdown simulating the counter). Finally M' verifies $u_m \in \overline{\Sigma_1^* T \Sigma_1^*}$.

We claim that $\inf(L') = \Sigma_1^*$ if and only if $T \subseteq L_{na}$.

Assume $T \subseteq L_{na}$. Let $w \in \Sigma_1^*$. We will show $w \in \inf(L')$. Let $w = u_0 y_1 u_1 \cdots u_{m-1} y_m u_m$, where $m \ge 0, y_1, \ldots, y_m \in T, u_0, \ldots, u_m \in \overline{\Sigma_1^* T \Sigma_1^*}$. Then for each $y_j, 1 \le j \le m, y_j \in L_{na}(p_j)$, for some p_j, and thus there exists q_j such that $q_j = |u_{j-1}|$. Thus, $\%^{p_m} e^{q_m} \cdots \%^{p_1} e^{q_1} ¢ w \in L'$, and $w \in \inf(L')$.

Assume $\inf(L') = \Sigma_1^*$. Let $w \in T$. Then there must exist x, y such that $z = xwy \in L'$. Then $z = u_0 y_1 u_1 \cdots u_{m-1} y_m u_m$, where $y_1, \ldots, y_m \in T, u_0, \ldots, u_m \in \overline{\Sigma_1^* T \Sigma_1^*}$. Necessarily, one of $y_1, \ldots, y_m, y_i$ say, must be w. This implies $w = y_i \in L_{na}(p_i)$, for some p_i. Hence, $w \in L_{na}$.   □

In contrast to the undecidability of marked-infix-density and suffix-density for DCM(1, 3) and NCM(1, 1), for infix-density on reversal-bounded nondeterministic pushdown automata, it is decidable. The proof is quite lengthy and is omitted for space reasons. The main tool of the proof is the known fact that the language of all words over the pushdown alphabet that can appear on the pushdown in an accepting computation is a regular language [1].

Proposition 9. *It is decidable, given L accepted by a one-way reversal-bounded NPDA, whether L is infix-dense.*

Next, we briefly examine the reverse containments when testing if $\Sigma^* \subseteq \varrho(L)$ and $\$ \Sigma^* \$ \subseteq \varrho(L)$ for density and marked-density. Here, it is checked whether it is decidable to test $\varrho(L) \subseteq R$ for regular languages R. In fact, we will show a stronger result (the proof is omitted).

Proposition 10. *It is decidable, given $L_1 \in$ NPCM and $L_2 \in$ DCM, whether $\varrho(L_1) \subseteq L_2$, where $\varrho \in \{$suff, inf, pref, outf$\}$.*

The languages used in the proofs of Lemmas 1 and 2 are used next to show that $\varrho(L)$ does not belong in the same family as L, in general.

Proposition 11. *There is a language $L \in$ DCM(1, 3) (resp., NCM(1, 1), DCM(2, 1)) such that $\varrho(L)$ is not in DPCM, where $\varrho \in \{$suff, inf, outf$\}$.*

Proof. We first give a proof for DCM(1, 3). Consider $L' = \bigcup_{p \geq 0} \%^p\$L_{na}(p)\$ \in$ DCM(1, 3) by Lemma 2. For $\varrho \in \{$suff, inf, outf$\}$, we claim that $\varrho(L')$ cannot be accepted by any DPCM. We know $\$L_{na}\$ \subseteq \varrho(L')$. For suppose $\varrho(L')$ can be accepted by a DPCM M_1. Then, since the family of languages accepted by DPCMs is closed under complementation [8], we can construct a DPCM M_2 accepting $\overline{L(M_1)}$. Now using M_2, an algorithm can be constructed to determine whether $\$T\$ \not\subseteq \$L_{na}\$$, which we know is a subset of $\varrho(L')$.

1. Consider T, which can be accepted by a DFA M_3.
2. Construct a DPCM M_4 accepting $L(M_2) \cap L(M_3)$.
3. Check if the language accepted by M_4 is empty. This is possible since the emptiness problem for NPCMs (hence also for DPCMs) is decidable [9].

By Lemma 1, $a^d \in L(Z)$ if and only if $\$T\$ \not\subseteq \$L_{na}\$$ if and only if the language accepted by $L(M_4)$ is not empty. It follows that $\varrho(L) \notin$ DPCM.

Similarly with $\$L_{na}\$$ for NCM(1, 1) for suffix and infix, and $\%\$L_{na}\$$ for outfix.   □

We note that the proof above also shows that if $L \in$ NCM(1, 1), then pref(L) need not be in DPCM.

4   Bounded-Dense Languages

Let ϱ be an operation from Σ^* to Σ^*. Then a language L is ϱ-bounded-dense over given words $w_1, \ldots, w_k$ if $\varrho(L) = w_1^* \cdots w_k^*$. We will show below that determining bounded-denseness is decidable for NPCM languages.

The following lemma (whose proof is omitted) is a generalization of a similar result for NPDAs in [6]:

Lemma 3. *It is decidable, given two NPCMs M_1 and M_2, one of which accepts a bounded language that is a subset of $w_1^* \cdots w_k^*$ (for given non-null words $w_1, \ldots, w_k$), whether $L(M_1) \subseteq L(M_2)$.*

Corollary 5. *It is decidable, given two NPCMs M_1, M_2 accepting bounded languages $L(M_1), L(M_2) \subseteq w_1^* \cdots w_k^*$, whether $L(M_1) \subseteq L(M_2)$ (resp., $L(M_1) = L(M_2)$).*

Let $\varrho \in \{$suff, inf, pref, outf$\}$. Clearly, if M is an NPCM accepting a language $L(M) \subseteq w_1^* \cdots w_k^*$, we can construct an NPCM M' such that $L(M') = \varrho(L(M))$, and $L(M')$ is bounded, but over $v_1^* \cdots v_l^*$, which are effectively constructable from $w_1, \ldots, w_k$. From Lemma 3, by testing $w_1^* \cdots w_k^* \subseteq L(M')$ we have:

Proposition 12. *Let $\varrho \in \{\text{pref}, \text{inf}, \text{suff}, \text{outf}\}$. It is decidable, given an NPCM M accepting a language $L(M) \subseteq w_1^* \cdots w_k^*$ (for given $w_1, \ldots, w_k$), whether $L(M)$ is ϱ-bounded-dense.*

In the proposition above, the words $w_1, \ldots, w_k$ were given in advance. It was shown in [6] that it is decidable, given an NPDA (or, equivalently, a context-free grammar), whether the language L it accepts (generates) is bounded, and if so, to effectively find k and $w_1, \ldots, w_k$ such that $L \subseteq w_1^* \cdots w_k^*$. We believe that this also holds for NPCM, but have no proof at this time. However, we can show that it holds for a special case.

A language L is letter-bounded if there is a k such that $L \subseteq a_1^* \cdots a_k^*$ for some $a_1, \ldots, a_k$, where $a_1, \ldots, a_k$ are *not-necessarily distinct* symbols. So, e.g., $\{a^{n_1} b^{n_1} a^{n_2} b^{n_2} a^{n_3} b^{n_3} \mid n_1, n_2, n_3 > 0\}$ is letter-bounded (where $k = 6$), $\{ab\}^*$ is not letter-bounded but bounded, and $\{a^n b^n \mid n > 0\}^*$ is not bounded.

Proposition 13. *It is decidable, given an NPCM M, whether $L(M)$ is letter-bounded. If it is letter-bounded, we can effectively find a k and not-necessarily distinct symbols $a_1, \ldots, a_k$ such that $L(M) \subseteq a_1^* \cdots a_k^*$.*

Proof. We construct from M an NPCM M' accepting a unary language that is a subset of 1^*. M' has one more counter, C, than M. Then M', on input $1^k, k \geq 1$, simulates the computation of M on some input w by guessing the symbols of w symbol-by-symbol. During the simulation, M' increments the counter C if the next input symbol it guesses is different from the previous symbol it guessed. When M accepts, M' checks and accepts if the value of the counter C is at least k. Clearly, M is not letter-bounded if and only if $L(M')$ is infinite, which is decidable since the infiniteness problem for NPCMs is decidable [9].

Part 2 follows from part 1 and Lemma 3 by exhaustive search. □

Let $m \geq 1$. A language L is m-bounded if there exist k and not-necessarily distinct words $w_1, \ldots, w_k$ each of length m such that $L(M) \subseteq w_1^* \cdots w_k^*$. Note that 1-bounded is the same as letter-bounded. For example, $\{ab\}^*$ is 2-bounded, but not letter-bounded. We can generalize Proposition 13 as follows:

Proposition 14. *Let $m \geq 1$. It is decidable, given an NPCM M, whether $L(M)$ is m-bounded. If it is, we can effectively find k and (not-necessarily distinct) words $w_1, \ldots, w_k$ each of length m such that $L(M) \subseteq w_1^* \cdots w_k^*$.*

It is an interesting question whether Proposition 14 holds if we do not require that the (not-necessarily distinct) words $w_1, \ldots, w_k$ of length at most m are of the same length.

5   Conclusions and Open Questions

This paper studies decidability problems involving testing whether a language L is ϱ-dense and ϱ-marked-dense, depending on the language family of L. For the prefix operation, all are decidable for DCM, but undecidable for NCM(1, 1), and thus the problem has been completely characterized in terms of restrictions on reversal-bounded multicounter machines. For suffix, both density and marked-density are decidable for DCM(1, 1), but not for DCM(1, 3) and NCM(1, 1), and therefore this has also been completely characterized. For infix, marked-density is decidable for DCM(1, 1), but not for DCM(1, 3) and NCM(1, 1). For infix-density however, it is decidable for nondeterministic reversal-bounded pushdown automata, but undecidable for deterministic pushdown automata and nondeterministic one-counter automata. It remains open for DCM and NCM when there are at least two counters, and also for deterministic one-counter automata. For outfix, marked-density is undecidable for DCM(1, 3), DCM(2, 1) and NCM(1, 1) but is open for DCM(1, 1) as with all variants for outfix-density.

In Section 4, results on bounded-dense languages are presented where the words $w_1, \ldots, w_k$ are given. In particular, for each of prefix, infix, suffix and out-fix, it is decidable for NPCM languages that accept bounded languages, whether they are ϱ-bounded-dense. But it is still an open problem as to whether or not it is decidable, given an NPCM (resp., NCM) M, to determine if $L(M)$ is bounded, and if so, to effectively find k and $w_1, \ldots, w_k$ such that $L(M) \subseteq w_1^* \cdots w_k^*$.

References

1. Autebert, J., Berstel, J., Boasson, L.: Handbook of Formal Languages, vol. 1, chap. Context-Free Languages and Pushdown Automata. Springer-Verlag, Berlin (1997)
2. Baker, B.S., Book, R.V.: Reversal-bounded multipushdown machines. Journal of Computer and System Sciences **8**(3), 315–332 (1974)
3. Berstel, J., Perrin, D.: Theory of Codes. Academic Press, Orlando (1985)
4. Chiniforooshan, E., Daley, M., Ibarra, O.H., Kari, L., Seki, S.: One-reversal counter machines and multihead automata: Revisited. Theoretical Computer Science **454**, 81–87 (2012)
5. Eremondi, J., Ibarra, O.H., McQuillan, I.: Deletion operations on deterministic families of automata. In: Jain, R., Jain, S., Stephan, F. (eds.) TAMC 2015. LNCS, vol. 9076, pp. 388–399. Springer, Heidelberg (2015)
6. Ginsburg, S.: The Mathematical Theory of Context-Free Languages. McGraw-Hill Inc., New York (1966)
7. Hopcroft, J.E., Ullman, J.D.: Introduction to Automata Theory, Languages, and Computation. Addison-Wesley, Reading (1979)
8. Ibarra, O., Yen, H.: On the containment and equivalence problems for two-way transducers. Theoretical Computer Science **429**, 155–163 (2012)
9. Ibarra, O.H.: Reversal-bounded multicounter machines and their decision problems. Journal of the ACM **25**(1), 116–133 (1978)
10. Ito, M.: Dense and disjunctive properties of languages. In: Ésik, Z. (ed.) FCT 1993. Lecture Notes in Computer Science, vol. 710, pp. 31–49. Springer, Berlin Heidelberg (1993)
11. Jürgensen, H., Kari, L., Thierrin, G.: Morphisms preserving densities. International Journal of Computer Mathematics **78**, 165–189 (2001)
12. Minsky, M.L.: Recursive unsolvability of Post's problem of "tag" and other topics in theory of Turing Machines. Annals of Mathematics **74**(3), 437–455 (1961)
13. Shyr, H.J.: Free Monoids and Languages, 3rd edn. Hon Min Book Company, Taichung (2001)

*-Continuous Kleene ω-Algebras

Zoltán Ésik[1], Uli Fahrenberg[2]([✉]), and Axel Legay[2]

[1] University of Szeged, Szeged, Hungary
[2] Irisa / Inria Rennes, Rennes, France
ulrich.fahrenberg@irisa.fr

Abstract. We define and study basic properties of *-continuous Kleene ω-algebras that involve a *-continuous Kleene algebra with a *-continuous action on a semimodule and an infinite product operation that is also *-continuous. We show that *-continuous Kleene ω-algebras give rise to iteration semiring-semimodule pairs, and that for Büchi automata over *-continuous Kleene ω-algebras, one can compute the associated infinitary power series.

1 Introduction

A continuous (or complete) Kleene algebra is a Kleene algebra in which all suprema exist and are preserved by products. These have nice algebraic properties, but not all Kleene algebras are continuous, for example the semiring of regular languages over some alphabet. Hence a theory of *-continuous Kleene algebras has been developed to cover this and other interesting cases.

For infinite behaviors, complete semiring-semimodule pairs involving an infinite product operation have been developed. Motivated by some examples of structures which are not complete in this sense, cf. the energy functions of [5], we generalize here the notion of *-continuous Kleene algebra to one of *-continuous Kleene ω-algebra. These are idempotent semiring-semimodule pairs which are not necessarily complete, but have enough suprema in order to develop a fixed-point theory and solve weighted Büchi automata (*i.e.,* to compute infinitary power series).

We will define both a finitary and a non-finitary version of *-continuous Kleene ω-algebras. We then establish several properties of *-continuous Kleene ω-algebras, including the existence of the suprema of certain subsets related to regular ω-languages. Then we will use these results in our characterization of the free finitary *-continuous Kleene ω-algebras. We also show that each *-continuous Kleene ω-algebra gives rise to an iteration semiring-semimodule pair and that Büchi automata over *-continuous Kleene ω-algebras can be solved algebraically.

The work of the first author was supported by the National Foundation of Hungary for Scientific Research, Grant no. K 108448. The work of the second and third authors was supported by ANR MALTHY, grant no. ANR-13-INSE-0003 from the Frenc0h National Research Foundation.

I. Potapov (Ed.): DLT 2015, LNCS 9168, pp. 240–251, 2015.
DOI: 10.1007/978-3-319-21500-6_19

For proofs of the results in this paper, and also for further motivation and results related to energy functions, we refer to [4].

A Kleene algebra [12] is an idempotent semiring $S = (S, \vee, \cdot, \bot, 1)$ equipped with a star operation $^* : S \to S$ such that for all $x, y \in S$, yx^* is the least solution of the fixed point equation $z = zx \vee y$ and x^*y is the least solution of the fixed point equation $z = xz \vee y$ with respect to the natural order.

Examples of Kleene algebras include the language semiring $P(A^*)$ over an alphabet A, whose elements are the subsets of the set A^* of all finite words over A, and whose operations are set union and concatenation, with the languages $\emptyset$ and $\{\varepsilon\}$ serving as $\bot$ and 1. Here, ε denotes the empty word. The star operation is the usual Kleene star: $X^* = \bigcup_{n \geq 0} X^n = \{u_1 \ldots u_n : u_1, \ldots, u_n \in X, n \geq 0\}$.

Another example is the Kleene algebra $P(A \times A)$ of binary relations over any set A, whose operations are union, relational composition (written in diagrammatic order), and where the empty relation $\emptyset$ and the identity relation id serve as the constants $\bot$ and 1. The star operation is the formation of the reflexive-transitive closure, so that $R^* = \bigcup_{n \geq 0} R^n$ for all $R \in P(A \times A)$.

The above examples are in fact *continuous Kleene algebras*, i.e., idempotent semirings S such that equipped with the natural order, they are all complete lattices (hence all suprema exist), and the product operation preserves arbitrary suprema in either argument:

$$y(\bigvee X) = \bigvee yX \quad \text{and} \quad (\bigvee X)y = \bigvee Xy$$

for all $X \subseteq S$ and $y \in S$. The star operation is given by $x^* = \bigvee_{n \geq 0} x^n$, so that x^* is the supremum of the set $\{x^n : n \geq 0\}$ of all powers of x. It is well-known that the language semirings $P(A^*)$ may be identified as the *free* continuous Kleene algebras (in a suitable category of continuous Kleene algebras).

A larger class of models is given by the **-continuous Kleene algebras* [12]. By the definition of a *-continuous Kleene algebra $S = (S, \vee, \cdot, \bot, 1)$, only suprema of sets of the form $\{x^n : n \geq 0\}$ need to exist, where x is any element of S, and x^* is given by this supremum. Moreover, product preserves such suprema in both of their arguments:

$$y(\bigvee_{n \geq 0} x^n) = \bigvee_{n \geq 0} yx^n \quad \text{and} \quad (\bigvee_{n \geq 0} x^n)y = \bigvee_{n \geq 0} x^n y.$$

For any alphabet A, the collection $R(A^*)$ of all regular languages over A is an example of a *-continuous Kleene algebra which is not a continuous Kleene algebra. The Kleene algebra $R(A^*)$ may be identified as the free *-continuous Kleene algebra on A. It is also the free Kleene algebra on A, cf. [11]. There are several other characterizations of $R(A^*)$, the most general of which identifies $R(A^*)$ as the free iteration semiring on A satisfying the identity $1^* = 1$, cf. [1,13].

For non-idempotent extensions of the notions of continuous Kleene algebras, *-continuous Kleene algebras and Kleene algebras, we refer to [6,7].

When A is an alphabet, let A^ω denote the set of all ω-words (infinite sequences) over A. An *ω-language* over A is a subset of A^ω. It is natural to consider the set $P(A^\omega)$ of all languages of ω-words over A, equipped with the operation of set union as $\vee$ and the empty language $\emptyset$ as $\bot$, and the left action of $P(A^*)$ on $P(A^\omega)$ defined by $XV = \{xv : x \in X,\ v \in V\}$ for all $X \subseteq A^*$ and $V \subseteq A^\omega$. Then that $(P(A^\omega), \vee, \bot)$ is a $P(A^*)$-semimodule and thus $(P(A^*), P(A^\omega))$ is a semiring-semimodule pair. We may also equip $(P(A^*), P(A^\omega))$ with an infinite product operation mapping an ω-sequence $(X_0, X_1, \ldots)$ over $P(A^*)$ to the ω-language $\prod_{n\geq 0} X_n = \{x_0 x_1 \ldots \in A^\omega : x_n \in X_n\}$. (Thus, an infinite number of the x_n must be different from ε. Note that $1^\omega = \bot$ holds.) The semiring-semimodule pair so obtained is an example of a continuous Kleene ω-algebra.

More generally, we call a semiring-semimodule pair (S, V) a *continuous Kleene ω-algebra* if S is a continuous Kleene algebra (hence S and V are idempotent), V is a complete lattice with the natural order, and the action preserves all suprema in either argument. Moreover, there is an infinite product operation which is compatible with the action and associative in the sense that the following hold:

1. For all $x_0, x_1, \ldots \in S$, $\prod_{n\geq 0} x_n = x_0 \prod_{n\geq 0} x_{n+1}$.
2. Let $x_0, x_1, \ldots \in S$ and $0 = n_0 \leq n_1 \cdots$ be a sequence which increases without a bound. Let $y_k = x_{n_k} \cdots x_{n_{k+1}-1}$ for all $k \geq 0$. Then $\prod_{n\geq 0} x_n = \prod_{k\geq 0} y_k$.

Moreover, the infinite product operation preserves all suprema:

3. $\prod_{n\geq 0}(\bigvee X_n) = \bigvee\{\prod_{n\geq 0} x_n : x_n \in X_n,\ n \geq 0\}$, for all $X_0, X_1, \ldots \subseteq S$.

The above notion of continuous Kleene ω-algebra may be seen as a special case of the not necessarily idempotent complete semiring-semimodule pairs of [9].

Our aim in this paper is to provide an extension of the notion of continuous Kleene ω-algebras to *-*continuous Kleene ω-algebras*, which are semiring-semimodule pairs (S, V) consisting of a *-continuous Kleene algebra S acting on a necessarily idempotent semimodule V, such that the action preserves certain suprema in its first argument, and which are equipped with an infinite product operation satisfying the above compatibility and associativity conditions and some weaker forms of the last axiom.

2   Free Continuous Kleene ω-Algebras

We have defined continuous Kleene ω-algebras in the introduction as idempotent semiring-semimodule pairs (S, V) such that $S = (S, \vee, \cdot, \bot, 1)$ is a continuous Kleene algebra and $V = (V, \vee, \bot)$ is a continuous S-semimodule. Thus, equipped with the natural order $\leq$ given by $x \leq y$ iff $x \vee y = y$, S and V are complete lattices and the product and the action preserve all suprema in either argument. Moreover, there is an infinite product operation, satisfying the compatibility and associativity conditions, which preserves all suprema.

In this section, we offer descriptions of the free continuous Kleene ω-algebras and the free continuous Kleene ω-algebras satisfying the identity $1^\omega = \bot$.

A homomorphism between continuous Kleene algebras preserves all operations. A homomorphism is continuous if it preserves all suprema. We recall the following folklore result.

Theorem 1. *For each set A, the language semiring $(P(A^*), \vee, \cdot, \bot, 1)$ is the free continuous Kleene algebra on A.*

Equivalently, if S is a continuous Kleene algebra and $h : A \to S$ is any function, then there is a unique continuous homomorphism $h^\sharp : P(A^*) \to S$ extending h.

In view of Theorem 1, it is not surprising that the free continuous Kleene ω-algebras can be described using languages of finite and ω-words. Suppose that A is a set. Let A^ω denote the set of all ω-words over A and $A^\infty = A^* \cup A^\omega$. Let $P(A^*)$ denote the language semiring over A and $P(A^\infty)$ the semimodule of all subsets of A^∞ equipped with the action of $P(A^*)$ defined by $XY = \{xy : x \in X, y \in Y\}$ for all $X \subseteq A^*$ and $Y \subseteq A^\infty$. We also define an infinite product by $\prod_{n \geq 0} X_n = \{u_0 u_1 \ldots : u_n \in X_n\}$.

Homomorphisms between continuous Kleene ω-algebras (S, V) and (S', V') consist of two functions $h_S : S \to S'$, $h_V : V \to V'$ which preserve all operations. A homomorphism (h_S, h_V) is continuous if h_S and h_V preserve all suprema.

Theorem 2. $(P(A^*), P(A^\infty))$ *is the free continuous Kleene ω-algebra on A.*

Consider now $(P(A^*), P(A^\omega))$ with infinite product defined, as a restriction of the above infinite product, by $\prod_{n \geq 0} X_n = \{u_0 u_1 \ldots \in A^\omega : u_n \in X_n, \ n \geq 0\}$. It is also a continuous Kleene ω-algebra. Moreover, it satisfies $1^\omega = \bot$.

Theorem 3. *For each set A, $(P(A^*), P(A^\omega))$ is the free continuous Kleene ω-algebra satisfying $1^\omega = \bot$ on A.*

3   *-Continuous Kleene ω-Algebras

In this section, we define *-*continuous Kleene ω-algebras* and *finitary *-continuous Kleene ω-algebras* as an extension of the *-continuous Kleene algebras of [11]. We establish several basic properties of these structures, including the existence of the suprema of certain subsets corresponding to regular ω-languages.

We define a *-continuous Kleene ω-algebra (S, V) as a *-continuous Kleene algebra $(S, \vee, \cdot, \bot, 1, ^*)$ acting on a (necessarily idempotent) semimodule $V = (V, \vee, \bot)$ subject to the usual laws of unitary action as well as the following axiom

- Ax0: For all $x, y \in S$ and $v \in V$, $xy^*v = \bigvee_{n \geq 0} xy^n v$.

Moreover, there is an infinite product operation mapping an ω-word $x_0 x_1 \ldots$ over S to an element $\prod_{n \geq 0} x_n$ of V. Thus, infinite product is a function $S^\omega \to V$, where S^ω denotes the set of all ω-words over S.

The infinite product is subject to the following axioms relating it to the other operations of Kleene ω-algebras and operations on ω-words. The first two axioms are the same as for continuous Kleene ω-algebras. The last two are weaker forms of the complete preservation of suprema of continuous Kleene ω-algebras.

Ax1: For all $x_0, x_1, \ldots \in S$, $\prod_{n \geq 0} x_n = x_0 \prod_{n \geq 0} x_{n+1}$.

Ax2: Let $x_0, x_1, \ldots \in S$ and $0 = n_0 \leq n_1 \cdots$ be a sequence which increases without a bound. Let $y_k = x_{n_k} \cdots x_{n_{k+1}-1}$ for all $k \geq 0$. Then $\prod_{n\geq 0} x_n = \prod_{k\geq 0} y_k$.

Ax3: For all $x_0, x_1, \ldots$ and y, z in S, $\prod_{n\geq 0}(x_n(y \vee z)) = \bigvee_{x'_n \in \{y,z\}} \prod_{n\geq 0} x_n x'_n$.

Ax4: For all $x, y_0, y_1, \ldots \in S$, $\prod_{n\geq 0} x^* y_n = \bigvee_{k_n \geq 0} \prod_{n\geq 0} x^{k_n} y_n$.

It is clear that every continuous Kleene ω-algebra is *-continuous.

Some of our results will also hold for weaker structures. We define a *finitary* *-continuous Kleene ω-algebra as a structure (S, V) as above, equipped with a star operation and an infinite product $\prod_{n\geq 0} x_n$ restricted to *finitary ω-words* over S, i.e., to sequences $x_0, x_1, \ldots$ such that there is a finite subset F of S such that each x_n is a finite product of elements of F. (Note that F is not fixed and may depend on the sequence $x_0, x_1, \ldots$) It is required that the axioms hold whenever the involved ω-words are finitary.

Finally, a *generalized* *-continuous Kleene algebra (S, V) is defined as a *-continuous Kleene ω-algebra, but without the infinite product (and without Ax1–Ax4). However, it is assumed that Ax0 holds.

The above axioms have a number of consequences. For example, if $x_0, x_1, \ldots \in S$ and $x_i = \bot$ for some i, then $\prod_{n\geq 0} x_n = \bot$. Indeed, if $x_i = \bot$, then $\prod_{n\geq 0} x_n = x_0 \cdots x_i \prod_{n\geq i+1} x_n = \bot \prod_{n\geq i+1} x_n = \bot$. By Ax1 and Ax2, each *-continuous Kleene ω-algebra is an ω-semigroup.

Suppose that (S, V) is a *-continuous Kleene ω-algebra. For each word $w \in S^*$ there is a corresponding element $\overline{w}$ of S which is the product of the letters of w in the semiring S. Similarly, when $w \in S^*V$, there is an element $\overline{w}$ of V corresponding to w, and when $X \subseteq S^*$ or $X \subseteq S^*V$, then we can associate with X the set $\overline{X} = \{\overline{w} : w \in X\}$, which is a subset of S or V. Below we will denote $\overline{w}$ and $\overline{X}$ by just w and X, respectively, The following two lemmas are well-known (and follow from the fact that the semirings of regular languages are the free *-continuous Kleene algebras [11] and the free Kleene algebras [12]).

Lemma 4. *Suppose that S is a *-continuous Kleene algebra. If $R \subseteq S^*$ is regular, then $\bigvee R$ exists. Moreover, for all $x, y \in S$, $x(\bigvee R)y = \bigvee xRy$.*

Lemma 5. *Let S be a *-continuous Kleene algebra. Suppose that R, R_1 and R_2 are regular subsets of S^*. Then*

$$\bigvee(R_1 \cup R_2) = \bigvee R_1 \vee \bigvee R_2$$
$$\bigvee(R_1 R_2) = (\bigvee R_1)(\bigvee R_2)$$
$$\bigvee(R^*) = (\bigvee R)^*.$$

In a similar way, we can prove:

Lemma 6. *Let (S, V) be a generalized *-continuous Kleene algebra. If $R \subseteq S^*$ is regular, $x \in S$ and $v \in V$, then $x(\bigvee R)v = \bigvee xRv$.*

Lemma 7. *Let (S, V) be a *-continuous Kleene ω-algebra. Suppose that the languages $R_0, R_1, \ldots \subseteq S^*$ are regular and that the set $\{R_0, R_1, \ldots\}$ is finite. Moreover, let $x_0, x_1, \ldots \in S$. Then*

$$\prod_{n\geq 0} x_n\Big(\bigvee R_n\Big) = \bigvee \prod_{n\geq 0} x_n R_n.$$

Lemma 8. *Let (S,V) be a finitary *-continuous Kleene ω-algebra. Suppose that the languages $R_0, R_1, \ldots \subseteq S^*$ are regular and that the set $\{R_0, R_1, \ldots\}$ is finite. Moreover, let $x_0, x_1, \ldots$ be a finitary sequence of elements of S. Then*

$$\prod_{n\geq 0} x_n\Big(\bigvee R_n\Big) = \bigvee \prod_{n\geq 0} x_n R_n.$$

Note that each sequence $x_0, y_0, x_1, y_1, \ldots$ with $y_n \in R_n$ is finitary.

Corollary 9. *Let (S,V) be a finitary *-continuous Kleene ω-algebra. Suppose that $R_0, R_1, \ldots \subseteq S^*$ are regular and that the set $\{R_0, R_1, \ldots\}$ is finite. Then $\bigvee \prod_{n\geq 0} R_n$ exists and is equal to $\prod_{n\geq 0} \bigvee R_n$.*

When $v = x_0 x_1 \ldots \in S^\omega$ is an ω-word over S, it naturally determines the element $\prod_{n\geq 0} x_n$ of V. Thus, any subset X of S^ω determines a subset of V. Using this convention, Lemma 7 may be rephrased as follows. For any *-continuous Kleene ω-algebra (S,V), $x_0, x_1, \ldots \in S$ and regular sets $R_0, R_1, \ldots \subseteq S^*$ for which the set $\{R_0, R_1, \ldots\}$ is finite, it holds that $\prod_{n\geq 0} x_n(\bigvee R_n) = \bigvee X$ where $X \subseteq S^\omega$ is the set of all ω-words $x_0 y_0 x_1 y_1 \ldots$ with $y_i \in R_i$ for all $i \geq 0$, i.e., $X = x_0 R_0 x_1 R_1 \ldots$. Similarly, Corollary 9 asserts that if a subset of V corresponds to an infinite product over a finite collection of ordinary regular languages in S^*, then the supremum of this set exists.

In any (finitary or non-finitary) *-continuous Kleene ω-algebra (S,V), we define an ω-power operation $S \to V$ by $x^\omega = \prod_{n\geq 0} x$ for all $x \in S$. From the axioms we immediately have:

Corollary 10. *Suppose that (S,V) is a *-continuous Kleene ω-algebra or a finitary *-continuous Kleene ω-algebra. Then the following hold for all $x, y \in S$:*

$$x^\omega = x x^\omega$$
$$(xy)^\omega = x(yx)^\omega$$
$$x^\omega = (x^n)^\omega, \quad n \geq 2.$$

Thus, each *-continuous Kleene ω-algebra gives rise to a Wilke algebra [14].

Lemma 11. *Let (S,V) be a (finitary or non-finitary) *-continuous Kleene ω-algebra. Suppose that $R \subseteq S^\omega$ is ω-regular. Then $\bigvee R$ exists in V.*

Lemma 12. *Let (S,V) be a (finitary or non-finitary) *-continuous Kleene ω-algebra. For all ω-regular sets $R_1, R_2 \subseteq S^\omega$ and regular sets $R \subseteq S^*$ it holds that*

$$\bigvee(R_1 \cup R_2) = \bigvee R_1 \vee \bigvee R_2$$
$$\bigvee(RR_1) = (\bigvee R)(\bigvee R_1).$$

And if R does not contain the empty word, then

$$\bigvee R^\omega = (\bigvee R)^\omega.$$

4   Free Finitary *-Continuous Kleene ω-Algebras

Recall that for a set A, $R(A^*)$ denotes the collection of all regular languages in A^*. It is well-known that $R(A^*)$, equipped with the usual operations, is a *-continuous Kleene algebra on A. Actually, $R(A^*)$ is characterized up to isomorphism by the following universal property.

Call a function $f : S \to S'$ between *-continuous Kleene algebras a *-continuous homomorphism if it preserves all operations including star, so that it preserves the suprema of subsets of S of the form $\{x^n : n \geq 0\}$, where $x \in S$.

Theorem 13 ([12]). *For each set A, $R(A^*)$ is the free *-continuous Kleene algebra on A.*

Thus, if S is any *-continuous Kleene algebra and $h : A \to S$ is any mapping from any set A into S, then h has a unique extension to a *-continuous Kleene algebra homomorphism $h^\sharp : R(A^*) \to S$.

Now let $R'(A^\infty)$ denote the collection of all subsets of A^∞ which are finite unions of finitary infinite products of regular languages, that is, finite unions of sets of the form $\prod_{n \geq 0} R_n$, where each $R_n \subseteq A^*$ is regular, and the set $\{R_0, R_1, \ldots\}$ is finite. Note that $R'(A^\infty)$ contains the empty set and is closed under finite unions. Moreover, when $Y \in R'(A^\infty)$ and $u = a_0 a_1 \ldots \in Y \cap A^\omega$, then the alphabet of u is finite, i.e., the set $\{a_n : n \geq 0\}$ is finite. Also, $R'(A^\infty)$ is closed under the action of $R(A^*)$ inherited from $(P(A^*), P(A^\infty))$. The infinite product of a sequence of regular languages in $R(A^*)$ is not necessarily contained in $R'(A^\infty)$, but by definition $R'(A^\infty)$ contains all infinite products of finitary sequences over $R(A^*)$.

Example 14. Let $A = \{a, b\}$ and consider the set $X = \{aba^2 b \ldots a^n b \ldots\} \in P(A^\infty)$ containing a single ω-word. X can be written as an infinite product of subsets of A^*, but it cannot be written as an infinite product $R_0 R_1 \ldots$ of regular languages in A^* such that the set $\{R_0, R_1, \ldots\}$ is finite. Hence $X \notin R'(A^\infty)$.

Theorem 15. *For each set A, $(R(A^*), R'(A^\infty))$ is the free finitary *-continuous Kleene ω-algebra on A.*

Consider now $(R(A^*), R'(A^\omega))$ equipped with the infinite product operation $\prod_{n \geq 0} X_n = \{u_0 u_1 \in A^\omega : u_n \in X_n, \ n \geq 0\}$, defined on finitary sequences $X_0, X_1, \ldots$ of languages in $R(A^*)$.

Theorem 16. *For each set A, $(R(A^*), R'(A^\omega))$ is the free finitary *-continuous Kleene ω-algebra satisfying $1^\omega = \bot$ on A.*

5   *-Continuous Kleene ω-Algebras and Iteration Semiring-Semimodule Pairs

In this section, we will show that every (finitary or non-finitary) *-continuous Kleene ω-algebra is an iteration semiring-semimodule pair.

Some definitions are in order. Suppose that $S = (S, +, \cdot, 0, 1)$ is a semiring. Following [1], we call S a *Conway semiring* if S is equipped with a star operation $* : S \to S$ satisfying, for all $x, y \in S$,

$$(x + y)^* = (x^* y)^* x^*$$
$$(xy)^* = 1 + x(yx)^* y.$$

It is known [1] that if S is a Conway semiring, then for each $n \geq 1$, so is the semiring $S^{n \times n}$ of all $n \times n$-matrices over S with the usual sum and product operations and the star operation defined by induction on n so that if $n > 1$ and $M = \left(\begin{smallmatrix} a & b \\ c & d \end{smallmatrix} \right)$, where a and d are square matrices of dimension $< n$, then

$$M^* = \begin{pmatrix} (a + bd^*c)^* & (a + bd^*c)^* bd^* \\ (d + ca^*b)^* ca^* & (d + ca^*b)^* \end{pmatrix}.$$

The above definition does not depend on how M is split into submatrices.

Suppose that S is a Conway semiring and $G = \{g_1, \ldots, g_n\}$ is a finite group of order n. For each $x_{g_1}, \ldots, x_{g_n} \in S$, consider the $n \times n$ matrix $M_G = M_G(x_{g_1}, \ldots, x_{g_n})$ whose ith row is $(x_{g_i^{-1} g_1}, \ldots, x_{g_i^{-1} g_n})$, for $i = 1, \ldots, n$, so that each row (and column) is a permutation of the first row. We say that the group identity [2] associated with G holds in S if for each $x_{g_1}, \ldots, x_{g_n}$, the first (and then any) row sum of M_G^* is $(x_{g_1} + \cdots + x_{g_n})^*$. Finally, we call S an *iteration semiring* [1,3] if the group identities hold in S for all finite groups of order n.

Classes of examples of (idempotent) iteration semirings are given by the continuous and the *-continuous Kleene algebras defined in the introduction. As mentioned above, the language semirings $P(A^*)$ and the semirings $P(A \times A)$ of binary relations are continuous and hence also *-continuous Kleene algebras, and the semirings $R(A^*)$ of regular languages are *-continuous Kleene algebras.

When S is a *-continuous Kleene algebra and n is a nonnegative integer, then the matrix semiring $S^{n \times n}$ is also a *-continuous Kleene algebra and hence an iteration semiring, cf. [11]. The star operation is defined by

$$M_{i,j}^* = \bigvee_{m \geq 0, \; 1 \leq k_1, \ldots, k_m \leq n} M_{i,k_1} M_{k_1,k_2} \cdots M_{k_m,j},$$

for all $M \in S^{n \times n}$ and $1 \leq i, j \leq n$. It is not trivial to prove that the above supremum exists. The fact that M^* is well-defined can be established by induction on n together with the well-known matrix star formula mentioned above.

A semiring-semimodule pair (S, V) is a *Conway semiring-semimodule pair* if it is equipped with a star operation $* : S \to S$ and an omega operation $^\omega : S \to V$ such that S is a Conway semiring acting on the semimodule $V = (V, +, 0)$ and the following hold for all $x, y \in S$:

$$(x + y)^\omega = (x^* y)^* x^\omega + (x^* y)^\omega$$
$$(xy)^\omega = x(yx)^\omega.$$

It is known [1] that when (S, V) is a Conway semiring-semimodule pair, then so is $(S^{n \times n}, V^n)$ for each n, where V^n denotes the $S^{n \times n}$-semimodule of all

n-dimensional (column) vectors over V with the action of $S^{n \times n}$ defined similarly to matrix-vector product, and where the omega operation is defined by induction so that when $n > 1$ and $M = \left(\begin{smallmatrix} a & b \\ c & d \end{smallmatrix} \right)$, where a and d are square matrices of dimension $< n$, then

$$M^\omega = \begin{pmatrix} (a + bd^*c)^\omega + (a + bd^*c)^*bd^\omega \\ (d + ca^*b)^\omega + (d + ca^*b)^*ca^\omega \end{pmatrix}.$$

We also define *iteration semiring-semimodule pairs* [1,9] as those Conway semiring-semimodule pairs such that S is an iteration semiring and the omega operation satisfies the following condition: let $M_G = M_G(x_{g_1}, \ldots, x_{g_n})$ like above, with $x_{g_1}, \ldots, x_{g_n} \in S$ for a finite group $G = \{g_1, \ldots, g_n\}$ of order n, then the first (and hence any) entry of M_G^ω is equal to $(x_{g_1} + \cdots + x_{g_n})^\omega$.

Examples of (idempotent) iteration semiring-semimodule pairs include the semiring-semimodule pairs $(P(A^*), P(A^\omega))$ of languages and ω-languages over an alphabet A, mentioned in the introduction. The omega operation is defined by $X^\omega = \prod_{n \geq 0} X$. More generally, it is known that every continuous Kleene ω-algebra gives rise to an iteration semiring-semimodule pair. The omega operation is defined as for languages: $x^\omega = \prod_{n \geq 0} x_n$ with $x_n = x$ for all $n \geq 0$.

Other not necessarily idempotent examples include the *complete* and the *(symmetric) bi-inductive semiring-semimodule pairs* of [8,9].

Suppose now that (S, V) is a *-continuous Kleene ω-algebra. Then for each $n \geq 1$, $(S^{n \times n}, V^n)$ is a semiring-semimodule pair. The action of $S^{n \times n}$ on V^n is defined similarly to matrix-vector product (viewing the elements of V^n as column vectors). It is easy to see that $(S^{n \times n}, V^n)$ is a generalized *-continuous Kleene algebra for each $n \geq 1$.

Suppose that $n \geq 2$. We would like to define an infinite product operation $(S^{n \times n})^\omega \to V^n$ on matrices in $S^{n \times n}$ by

$$\left(\prod_{m \geq 0} M_m \right)_i = \bigvee_{1 \leq i_1, i_2, \ldots \leq n} (M_0)_{i,i_1} (M_1)_{i_1,i_2} \cdots$$

for all $1 \leq i \leq n$. However, unlike in the case of complete semiring-semimodule pairs [9], the supremum on the right-hand side may not exist. Nevertheless it is possible to define an omega operation $S^{n \times n} \to V^n$ and to turn $(S^{n \times n}, V^n)$ into an iteration semiring-semimodule pair.

Lemma 17. *Let (S, V) be a (finitary or non-finitary) *-continuous Kleene ω-algebra. Suppose that $M \in S^{n \times n}$, where $n \geq 2$. Then for every $1 \leq i \leq n$,*

$$\left(\prod_{m \geq 0} M \right)_i = \bigvee_{1 \leq i_1, i_2, \ldots \leq n} M_{i,i_1} M_{i_1,i_2} \cdots$$

exists, so that we define M^ω by the above equality. Moreover, when $M = \left(\begin{smallmatrix} a & b \\ c & d \end{smallmatrix} \right)$, where a and d are square matrices of dimension $< n$, then

$$M^\omega = \begin{pmatrix} (a \vee bd^*c)^\omega \vee (a \vee bd^*c)^*bd^\omega \\ (d \vee ca^*b)^\omega \vee (d \vee ca^*b)^*ca^\omega \end{pmatrix}. \tag{1}$$

Theorem 18. *Every (finitary or non-finitary) *-continuous Kleene ω-algebra is an iteration semiring-semimodule pair.*

Relation to Bi-Inductive Semiring-Semimodule Pairs. Recall that when P is a partially ordered set and f is a function $P \to P$, then a *pre-fixed point* of f is an element x of P with $f(x) \leq x$. Similarly, $x \in P$ is a *post-fixed point* of f if $x \leq f(x)$. Suppose that f is monotone and has x as its least pre-fixed point. Then x is a *fixed point*, i.e., $f(x) = x$, and thus the least fixed point of f. Similarly, when f is monotone, then the greatest post-fixed point of f, whenever it exists, is the greatest fixed point of f.

When S is a *-continuous Kleene algebra, then S is a Kleene algebra [11]. Thus, for all $x, y \in S$, x^*y is the least pre-fixed point (and thus the least fixed point) of the function $S \to S$ defined by $z \mapsto xz \vee y$ for all $z \in S$. Moreover, yx^* is the least pre-fixed point and the least fixed point of the function $S \to S$ defined by $z \mapsto zx \vee y$, for all $z \in S$. Similarly, when (S, V) is a generalized *-continuous Kleene algebra, then for all $x \in S$ and $v \in V$, x^*v is the least pre-fixed point and the least fixed point of the function $V \to V$ defined by $z \mapsto xz \vee v$.

A *bi-inductive semiring-semimodule pair* is defined as a semiring-semimodule pair (S, V) for which both S and V are partially ordered by the natural order relation $\leq$ such that the semiring and semimodule operations and the action are monotone, and which is equipped with a star operation $^* : S \to S$ and an omega operation $^\omega : S \to V$ such that the following hold for all $x, y \in S$ and $v \in V$:

- x^*y is the least pre-fixed point of the function $S \to S$ defined by $z \mapsto xz + y$,
- x^*v is the least pre-fixed point of the function $V \to V$ defined by $z \mapsto xz + v$,
- $x^\omega + x^*v$ is the greatest post-fixed point of the function $V \to V$ defined by $z \mapsto xz + v$.

A bi-inductive semiring-semimodule pair is said to be symmetric if for all $x, y \in S$, yx^* is the least pre-fixed point of the functions $S \to S$ defined by $z \mapsto zx + y$. It is known that every bi-inductive semiring-semimodule pair is an iteration semiring-semimodule pair, see [9]. By the above remarks we have:

Proposition 19. *Suppose that (S, V) is a finitary *-continuous Kleene ω-algebra. When for all $x \in S$ and $v \in V$, $x^\omega \vee x^*v$ is the greatest post-fixed point of the function $V \to V$ defined by $z \mapsto xz \vee v$, then (S, V) is a symmetric bi-inductive semiring-semimodule pair.*

6   Büchi Automata in *-Continuous Kleene ω-Algebras

A generic definition of Büchi automata in Conway semiring-semimodule pairs was given in [1,8]. In this section, we recall this general definition and apply it to *-continuous Kleene ω-algebras. We give two different definitions of the behavior of a Büchi automaton, an algebraic and a combinatorial, and show that these two definitions are equivalent.

Suppose that (S, V) is a Conway semiring-semimodule pair, S_0 is a subsemiring of S closed under star, and A is a subset of S. We write $S_0\langle A \rangle$ for the set of all finite sums $s_1a_1 + \cdots + s_ma_m$ with $s_i \in S_0$ and $a_i \in A$, for each $i = 1, \ldots, m$.

We define a (weighted) Büchi automaton over (S_0, A) of dimension $n \geq 1$ in (S, V) as a system $\mathbf{A} = (\alpha, M, k)$ where $\alpha \in S_0^{1 \times n}$ is the initial vector,

$M \in S_0 \langle A \rangle^{n \times n}$ is the transition matrix, and k is an integer $0 \leq k \leq n$. In order to define the behavior $|\mathbf{A}|$ of $\mathbf{A}$, let us split M into 4 parts as above, $M = \begin{pmatrix} a & b \\ c & d \end{pmatrix}$, with $a \in S_0 \langle A \rangle^{k \times k}$ the top-left k-by-k submatrix. Then we define

$$|\mathbf{A}| = \alpha \begin{pmatrix} (a + bd^*c)^\omega \\ d^*c(a + bd^*c)^\omega \end{pmatrix}.$$

We give another more combinatorial definition. Let (S, V) be a *-continuous Kleene ω-algebra. A Büchi automaton $\mathbf{A} = (\alpha, M, k)$ over (S_0, A) of dimension n may be represented as a transition system whose set of states is $\{1, \ldots, n\}$. For any pair of states i, j, the transitions from i to j are determined by the entry $M_{i,j}$ of the transition matrix. Let $M_{i,j} = s_1 a_1 \vee \cdots \vee s_m a_m$, say, then there are m transitions from i to j, respectively labeled $s_1 a_1, \ldots, s_n a_n$. An accepting run of the Büchi automaton from state i is an infinite path starting in state i which infinitely often visits one of the first k states, and the weight of such a run is the infinite product of the path labels. The behavior of the automaton in state i is the supremum of the weights of all accepting runs starting in state i. Finally, the behavior of the automaton is $\alpha_1 w_1 \vee \cdots \vee \alpha_n w_n$, where for each i, α_i is the ith component of α and w_i is the behavior in state i. Let $|\mathbf{A}|'$ denote the behavior of $\mathbf{A}$ according to this second definition.

Theorem 20. *For every Büchi automaton $\mathbf{A}$ over (S_0, A) in a (finitary or non-finitary) *-continuous Kleene ω-algebra, it holds that $|\mathbf{A}| = |\mathbf{A}|'$.*

For completeness we also mention a Kleene theorem for the Büchi automata introduced above, which is a direct consequence of the Kleene theorem for Conway semiring-semimodule pairs, cf. [8,10].

Theorem 21. *An element of V is the behavior of a Büchi automaton over (S_0, A) iff it is regular (or rational) over (S_0, A), i.e., when it can be generated from the elements of $S_0 \cup A$ by the semiring and semimodule operations, the action, and the star and omega operations.*

It is a routine matter to show that an element of V is rational over (S_0, A) iff it can be written as $\bigvee_{i=1}^n x_i y_i^\omega$, where each x_i and y_i can be generated from $S_0 \cup A$ by $\vee$, $\cdot$ and $*$.

7   Conclusion

We have introduced continuous and (finitary and non-finitary) *-continuous Kleene ω-algebras and exposed some of their basic properties. Continuous Kleene ω-algebras are idempotent complete semiring-semimodule pairs, and conceptually, *-continuous Kleene ω-algebras are a generalization of continuous Kleene ω-algebras in much the same way as *-continuous Kleene algebras are of continuous Kleene algebras: In *-continuous Kleene algebras, suprema of finite sets and of sets of powers are required to exist and to be preserved by the product; in

*-continuous Kleene ω-algebras these suprema are also required to be preserved by the infinite product.

It is known that in a Kleene algebra, *-continuity is precisely what is required to be able to compute the reachability value of a weighted automaton (or its power series) using the matrix star operation. Similarly, we have shown that the Büchi values of automata over *-continuous ω-algebras can be computed using the matrix omega operation.

We have seen that the sets of finite and infinite languages over an alphabet are the free continuous Kleene ω-algebras, and that the free finitary *-continuous Kleene ω-algebras are given by the sets of regular languages and of finite unions of finitary infinite products of regular languages. A characterization of the free (non-finitary) *-continuous Kleene ω-algebras (and whether they even exist) is left open.

We have seen that every *-continuous Kleene ω-algebra is an iteration semiring-semimodule pair, which permits to compute the behavior of Büchi automata with weights in a *-continuous Kleene ω-algebra using ω-powers of matrices.

References

1. Bloom, S.L., Ésik, Z.: Iteration Theories: The Equational Logic of Iterative Processes. EATCS monographs on theoretical computer science. Springer (1993)
2. Conway, J.H.: Regular Algebra and Finite Machines. Chapman and Hall (1971)
3. Ésik, Z.: Iteration semirings. In: Ito, M., Toyama, M. (eds.) DLT 2008. LNCS, vol. 5257, pp. 1–20. Springer, Heidelberg (2008)
4. Ésik, Z., Fahrenberg, U., Legay, A.: *-continuous Kleene ω-algebras. CoRR (2015). http://arxiv.org/abs/1501.01118
5. Ésik, Z., Fahrenberg, U., Legay, A., Quaas, K.: Kleene algebras and semimodules for energy problems. In: Van Hung, D., Ogawa, M. (eds.) ATVA 2013. LNCS, vol. 8172, pp. 102–117. Springer, Heidelberg (2013)
6. Ésik, Z., Kuich, W.: Rationally additive semirings. J. Univ. Comput. Sci. **8**(2), 173–183 (2002)
7. Ésik, Z., Kuich, W.: Inductive star-semirings. TCS **324**(1), 3–33 (2004)
8. Ésik, Z., Kuich, W.: A semiring-semimodule generalization of ω-regular languages, Parts 1 and 2. J. Aut. Lang. Comb. **10**, 203–264 (2005)
9. Ésik, Z., Kuich, W.: On iteration semiring-semimodule pairs. Semigroup Forum **75**, 129–159 (2007)
10. Ésik, Z., Kuich, W.: Finite automata. In: Handbook of Weighted Automata. Springer (2009)
11. Kozen, D.: On kleene algebras and closed semirings. In: Rovan, B. (ed.) MFCS 1990. LNCS, vol. 452, pp. 26–47. Springer, Heidelberg (1990)
12. Kozen, D.: A completeness theorem for Kleene algebras and the algebra of regular events. Inf. Comput. **110**(2), 366–390 (1994)
13. Krob, D.: Complete systems of B-rational identities. TCS **89**(2), 207–343 (1991)
14. Wike, T.: An eilenberg theorem for ∞-languages. In: Albert, J.L., Monien, B., Artalejo, M.R. (eds.) ICALP. LNCS, vol. 510, pp. 588–599. Springer, Heidelberg (1991)

Unary Probabilistic and Quantum Automata on Promise Problems

Aida Gainutdinova[1]($\boxtimes$) and Abuzer Yakaryılmaz[2]

[1] Kazan Federal University, Kazan, Russia
aida.ksu@gmail.com
[2] National Laboratory for Scientific Computing, Petrópolis, RJ 25651-075, Brazil
abuzer@lncc.br

Abstract. We continue the systematic investigation of probabilistic and quantum finite automata (PFAs and QFAs) on promise problems by focusing on unary languages. We show that bounded-error QFAs are more powerful than PFAs. But, in contrary to the binary problems, the computational powers of Las-Vegas QFAs and bounded-error PFAs are equivalent to deterministic finite automata (DFAs). Lastly, we present a new family of unary promise problems with two parameters such that when fixing one parameter QFAs can be exponentially more succinct than PFAs and when fixing the other parameter PFAs can be exponentially more succinct than DFAs.

1 Introduction

Promise problems are generalizations of language recognition. The aim is, instead of separating one language from its complement, to separate any two disjoint languages, i.e. the input is promised to be from the union of these two languages. Promise problems have served some important roles in the computational complexity. For example, it is not known whether the class BPP (BQP), bounded error probabilistic (quantum) polynomial time, has a complete problem, but, the class PromiseBPP (PromiseBQP), defined on promise problems, has some complete problems (see the surveys by Goldreich [12] and Watrous [27]).

In automata theory, the promise problems has also appeared in many different forms. For example, in 1989 Condon and Lipton [9] defined a promised version of emptiness problem for probabilistic finite automata (PFAs), and showed its undecidability by using a promised version of equality language ($\texttt{EQ} = \{a^n b^n | n > 0\}$), solved by two-way bounded-error PFAs, which was also used to show that there is a weak constant-space interactive proof system for any recursive enumerable language.

On the other hand, up to our knowledge, some systematic works on promise problems in automata theory have been started only recently. An initial result

---

The extended version of the paper that contains the missing proofs is [10].

Abuzer Yakaryılmaz— Yakaryılmaz was partially supported by CAPES with grant 88881.030338/2013-01.

I. Potapov (Ed.): DLT 2015, LNCS 9168, pp. 252–263, 2015.
DOI: 10.1007/978-3-319-21500-6_20

was given to compare exact quantum and deterministic pushdown automata [20], the former one was shown to be more powerful (see also [21] and [22] for the results in this direction). Then, the result given by Ambainis and Yakaryılmaz [6], the state advantages of exact quantum finite automata (QFAs) over deterministic finite automata (DFAs) cannot be bounded in the case of unary promise problems, has stimulated the topic and a series of papers appeared on the succinctness of QFAs and other models [1,8,13,14,30,31]. In parallel, the new results were given on classical and quantum automata models [11,24]:

- There is a promise problem solved by exact two-way QFAs but not by any sublogarithmic probabilistic Turing machine (PTM).
- There is a promise problem solved by an exact two-way QFA in quadratic expected time, but not by any bounded-error $o(\log \log n)$-space PTMs in polynomial expected time.
- There is a promise problem solvable by a Las Vegas realtime QFA, but not by any bounded-error PFA.
- The computational power of deterministic, nondeterministic, alternating, and Las Vegas PFAs are the same and two-wayness does not help.
- On the contrary to tight quadratic gap in the case of language recognition, Las-Vegas PFAs can be exponentially more state efficient than DFAs.
- The state advantages of bound-error unary PFAs over DFAs cannot be bounded.
- There is a binary promise problem solved by bounded-error PFAs but not by any DFA.

In this paper, we provide some new results regarding probabilistic and quantum automata on unary promise problems. We show that bounded-error QFAs are more powerful than PFAs. But, on contrary to the binary problems, the computational power of Las-Vegas QFAs and bounded-error PFAs are equivalent to DFAs. Lastly, we present a new family of unary promise problems with two parameters such that when fixing one parameter QFAs can be exponentially more succinct than PFAs and when fixing the other parameter PFAs can be exponentially more succinct than DFAs.

2 Preliminaries

In this section, we provide the necessary background to follow the remaining part. We start with the definitions of models and the notion of promise problems. Then, we give the basics of Markov chain which will be used in some proofs.

2.1 Definitions

A PFA $\mathcal{P}$ is a 5-tuple $\mathcal{P} = (Q, \Sigma, \{A_\sigma \mid \sigma \in \Sigma\}, v_0, Q_a)$, where

- Q is the set of (classical) states,
- Σ is the input alphabet,

- v_0 is a $|Q|$-dimensional stochastic initial column vector representing the initial probability distribution of the states at the beginning of the computation,
- A_σ is a (left) stochastic transition matrix for symbol $\sigma \in \Sigma$ where $A_\sigma(j, i)$ represents the probability of going from the ith state to the jth state after reading σ, and
- Q_a is the set of the accepting states.

The computation of $\mathcal{P}$ on the input $w \in \Sigma^*$ can be traced by a stochastic column vector, i.e.

$$v_j = A_{w_j} v_{j-1},$$

where $1 \leq j \leq |w|$. After reading the whole input, the final probabilistic state is $v_{|w|}$. Based on this, we can calculate the accepting probability of w by $\mathcal{P}$, denoted $f_{\mathcal{P}}(w)$, as follows:

$$f_{\mathcal{P}}(w) = \sum_{q_j \in Q_a} v_{|w|}(j).$$

If all stochastic elements of a PFA are restricted to have only 0s and 1s, then we obtain a DFA that starts in a certain state and switches to only one state in each step, and so the computation ends in only a single state. An input is accepted by the DFA if the final state is an accepting state.

There are different kinds of quantum finite automata (QFAs) models in the literature. The general ones (e.g. [3,15,29]) can exactly simulate PFAs (see [26] for a pedagogical proof). In this paper, we present our results based on the known simplest QFA model, called Moore-Crutcfield QFA [19]. Therefore, we only provide its definition. We assume the reader knows the basics of quantum computation (see [26] for a quick review and [23] for a complete reference).

A MCQFA $\mathcal{M}$ is 5-tuple $\mathcal{M} = (Q, \Sigma, \{U_\sigma \mid \sigma \in \Sigma\}, |v_0\rangle, Q_a)$ where, different from a PFA,

- $|v_0\rangle$ is a norm-1 complex-valued column initial vector that can be a superposition of classical states and represents the initial quantum state of $\mathcal{M}$ at the beginning of the computation, and,
- U_σ is a unitary transition matrix for symbol $\sigma \in \Sigma$ where $U_\sigma(j, i)$ represents the amplitude of going from the ith state to the jth state after reading σ.

Traditionally, vectors are represented with "ket" notation ($|\cdot\rangle$) in quantum mechanics and computations. The computation of $\mathcal{M}$ on the input $w \in \Sigma^*$ can be traced by a norm-1 complex-valued column vector, i.e.

$$|v_j\rangle = U_{w_j} |v_{j-1}\rangle,$$

where $1 \leq j \leq |w|$. After reading the whole input, the final quantum state is $|v_{|w|}\rangle$. Based on this, a measurement operator is applied to see whether the automaton is in an accepting or non-accepting state. The accepting probability of w by $\mathcal{M}$ is calculated as:

$$f_{\mathcal{M}}(w) = \sum_{q_j \in Q_a} ||v_{|w|}\rangle(j)|^2.$$

A Las Vegas PFA (or QFA) never gives a wrong decision, instead giving the decision of "don't know". Formally, its set of states is divided into three disjoint sets, the set of accepting states (Q_a), the set of neutral states (Q_n), and the set of rejecting states $(Q_r = Q \setminus Q_a \cup Q_n)$. At the end of the computation, the decision of "don't know" is given if the automaton ends with an neutral state. The probability of giving the decision of "don't know" (rejection) is calculated similar to the accepting probability by using Q_n (Q_r) instead of Q_a.

A promise problem $P \subset \Sigma^*$ is composed by two disjoint languages P_{yes} and P_{no}, where the former one is called the set of yes-instances and the latter one is called the set of no-instances.

A promise problem is said to be solved by a DFA if any yes-instance is accepted and any no-instance is rejected. A promise problem is said to be solved by a PFA or QFA with error bound $\epsilon < \frac{1}{2}$ if any yes-instance is accepted with probability at least $1 - \epsilon$ and any no-instance is rejected with probability at least $1 - \epsilon$. If all yes-instances are accepted exactly, then it is said the promise problem is solved with one-sided bounded error. In this case, the error bound can be greater than $\frac{1}{2}$ but it must be less than 1, i.e. $\epsilon < 1$. Lastly, a promise problem is said to be solved by a Las Vegas PFA or QFA with success probability $p > 0$,

- if any yes-instance is accepted with probability at least p and it is rejected with probability 0, and,
- if any no-instance is rejected with probability at least p and it is accepted with probability 0.

In the case of promise problems, we do not care about the decisions on the strings from $\Sigma^* \setminus P$.

2.2   The Theory of Markov Chains

The computation of a unary PFA can be described by a Markov chain. Here we present some basic facts and results from the theory of Markov chains that will be used in some proofs. We refer the reader to [16] for more details and [5] and [18] for some similar applications.

The states of a Markov chain are divided into ergodic and transient states. An *ergodic set of states* is a set which a process cannot leave once it has entered, a *transient set of states* is a set which a process can leave, but cannot return once it has left. An arbitrary Markov chain has at least one ergodic set. If a Markov chain C has more than one ergodic set, then there is absolutely no interaction between these sets. Hence we have two or more unrelated Markov chains lumped together and can be studied separately. If a Markov chain consists of a single ergodic set, then the chain is called an *ergodic chain*. According to the classification mentioned above, every ergodic chain is either regular or cyclic (see below).

If an ergodic chain is regular, then for sufficiently high powers of the state transition matrix, M has only positive elements. Thus, no matter where the process starts, after a sufficiently large number of steps it can be in any state. Moreover, there is a limiting vector of probabilities of being in the states of the chain, that does not depend on the initial state.

If a Markov chain is cyclic, then the chain has a period t and all of its states are subdivided into t cyclic subsets $(t > 1)$. For a given starting state a process moves through the cyclic subsets in a definite order, returning to the subset with the starting state after every t steps. It is known that after sufficient time has elapsed, the process can be in any state of the cyclic subset appropriate for the moment. Hence, for each of t cyclic subsets the t-th power of the state transition matrix M^t describes a regular Markov chain. Moreover, if an ergodic chain is a cyclic chain with the period t, it has at least t states.

Let $C_1, \ldots, C_l$ be cyclic subsets of states of Markov chain with periods $t_1, \ldots, t_l$, respectively, and D be the least common multiple of $t_1, \ldots, t_l$. For each cyclic subset C after every D steps, the process can be in any state of C and the Dth power of M describes a regular Markov chain for this subset. From the theory of Markov chains it is known that there exists an α_{acc} such that $\lim_{r \to \infty} \alpha_{acc}^{r \cdot D} = \alpha_{acc}$, where α_{acc}^i represents the probability of process being in accepting state(s) after the ith step. Hence, for any $\delta > 0$, there exists an $r_0 > 0$ such that
$$|\alpha_{acc}^{r \cdot D} - \alpha_{acc}^{r' \cdot D}| < \delta$$
for any $r, r' > r_0$.

Moreover, since $\alpha_{acc}^{r \cdot D}$ has a limit point α_{acc}, each $\alpha_{acc}^{r \cdot D + j}$ has also a limit point, say $\alpha_{acc(j)}$ for any $j \in \{1, \ldots, D - 1\}$.

3 The Computational Power of Unary PFAs and QFAs

First we show that any unary promise problem solved by a QFA exactly (without error) can also be solved by DFAs.

Theorem 1. *If a unary promise problem* $\mathsf{P} = (\mathsf{P_{yes}}, \mathsf{P_{no}})$ *is solved by a QFA exactly, then it is also solved by a DFA.*

Proof. Let $\mathcal{M}$ be a QFA solving P exactly. The automaton $\mathcal{M}$ also defines a language with cutpoint 0, say L, i.e. any string accepted with a non-zero (zero) probability is a member (non-member). Then, we can easily obtain the following two facts:

- Since each yes-instance of P is accepted with probability 1, it is also a member of L. Thus, $\mathsf{P_{yes}}$ is a subset of L.
- Since each no-instance of P is accepted with probability 0, it is also a member of $\bar{\mathsf{L}}$. Thus, $\mathsf{P_{no}}$ is a subset of $\bar{\mathsf{L}}$.

Any unary language defined by a QFA with cutpoint 0 (or equivalently recognized by a nondeterministic QFA [28]) is a unary exclusive language and it is known that any such language is regular (Page 89 of [25]). Thus, L is a unary regular language and there is a DFA, say $\mathcal{D}$, recognizing L. So, $\mathcal{D}$ can also solve promise problem P: $\mathcal{D}$ accepts all members of L including all $\mathsf{P_{yes}}$ and it rejects all members of $\bar{\mathsf{L}}$ including all $\mathsf{P_{no}}$.   □

We can extend this result also for Las Vegas QFAs.

Theorem 2. *If a unary promise problem* $P = (P_{yes}, P_{no})$ *is solvable by a Las Vegas QFA with a success probability* $p > 0$, *then it is also solvable by a DFA.*

Proof. Let $\mathcal{M}$ be our Las Vegas QFA solving P with success probability $p > 0$. We can obtain a new QFA $\mathcal{M}'$ by modifying $\mathcal{M}$ as follows: $\mathcal{M}'$ rejects the input when entering a neutral state at the end of the computation. Then, any member of P_{yes} is accepted by $\mathcal{M}$ with probability at least p and any member of P_{no} is accepted by $\mathcal{M}'$ with probability 0. After this, we can consider $\mathcal{M}'$ as a nondeterministic QFA and follow the same reasoning given in the previous proof. □

Since Las Vegas QFAs and DFAs define the same class of unary promise problems, one may ask how much state efficient QFAs can be over DFAs. Due to the result of Ambainis and Yakaryılmaz [6], we know that the gap (on unary promise problems) cannot be bounded. (Note that, in the case of language recognition, there is no gap between exact QFA and DFA [17] and the gap can be at most exponential between bounded-error QFAs and DFAs (see e.g. [2]).) On the other hand, as mentioned before, over binary promise problems, Las Vegas QFAs are known to be more powerful than bounded-error PFAs [24]. An open question here is whether exact QFAs can solve a binary promise problem that is beyond the capabilities of DFAs.

Las Vegas PFAs and DFAs have the same computational power even on binary promise problems and the tight gap on the number of states is exponential [11]. Currently we do not know whether this bound can be improved on unary case and we leave it as a future work. Here we show that making two-sided errors does not help to solve a unary promise problem that is beyond of the capability of DFAs. However, remark that, the state efficiency of bounded-error unary PFAs over unary DFAs also cannot be bounded [11].

Theorem 3. *If a unary promise problem* $P = (P_{yes}, P_{no})$ *is solved by a PFA, say* $\mathcal{P}$, *with error bound* $\epsilon < \frac{1}{2}$, *then it is also solvable by a DFA.*

Proof. The computation of $\mathcal{P}$ can be modelled as a Markov chain. Let $\mathcal{P}$ has n states and D be the least common multiple of periods of cycles of Markov chain (see the Section 2.2). So, $\mathcal{P}$ has D limiting accepting probabilities as described in Section 2.2, say

$$\alpha_{acc(0)}, \alpha_{acc(1)}, \ldots, \alpha_{acc(D-1)}.$$

For any small $\delta > 0$, there is an integer r_0 such that, for each $j \in \{0, \ldots, D-1\}$, we have the inequality $|f_{\mathcal{M}}(a^{r \cdot D + j}) - \alpha_{acc(j)}| < \delta$ for all $r \geq r_0$. Let's pick a $\delta' > 0$ such that, for any index $i \in \{0, \ldots, D-1\}$, the interval $|\alpha_{acc(i)} - \delta'|$ does contain at most one of the points $1 - \epsilon$ and ϵ, which is always possible since the gap between these two points $(1 - 2\epsilon)$ is non-zero. For this δ', we also have a r_0' such that, for any $j \in \{0, \ldots, D-1\}$, $f_{\mathcal{M}}(a^{r \cdot D + j})$ is in the interval $|\alpha_{acc(j)} - \delta'|$ for all $r \geq r_0'$.

We can classify $\alpha_{acc}(j)$ as follows:

– It is at least $\frac{1}{2}$. Then, $f_{\mathcal{M}}(a^{r \cdot D + j})$ cannot be ϵ or less than ϵ for any $r \geq r_0$.
– It is less than $\frac{1}{2}$. Then, $f_{\mathcal{M}}(a^{r \cdot D + j})$ cannot be $1 - \epsilon$ or greater than $1 - \epsilon$ for any $r \geq r_0$.

Thus, a D-state cyclic DFA with the following state transitions

$$q_0 \to q_1 \to \cdots \to q_j \to \cdots \to q_{D-1} \to q_0$$

can easily follow the periodicity of $\mathcal{P}$. Moreover, if $\alpha_{acc}(j)$ belongs the first (second) class of the above, then q_j is an accepting (a non-accepting) state. Thus, our cyclic DFA can give the same decisions of $\mathcal{P}$ on the promised strings with length at least $r_0 \cdot D$. The remaining (and shorter) promised strings form a finite set and a DFA with $(r_0' \cdot D - 1)$ states can give appropriate decisions on them. Therefore, by combining two DFAs, we can get a DFA with $r_0' \cdot D + D$ states that solves the promise problem P. □

Now we show that unary QFAs can define more promise problems than PFAs when the machines can err. We present our quantum result by a 2-state MCQFA. Then, we give our impossibility result for unary PFAs.

Let φ be a rotation angle which is an irrational fraction of 2π. For any $\theta \in (0, \frac{\pi}{4})$, we define a unary promise problem $L^\theta = \{L^\theta_{yes}, L^\theta_{no}\}$ as

– $L^\theta_{yes} = \{a^k \mid k\varphi \in [l\pi - \theta, l\pi + \theta] \text{ for some } l \geq 0\}$,
– $L^\theta_{no} = \{a^k \mid k\varphi \in [l\pi + \frac{\pi}{2} - \theta, l\pi + \frac{\pi}{2} + \theta] \text{ for some } l \geq 0\}$.

Theorem 4. *There is a 2-state MCQFA $\mathcal{M}$ solving the promise problem L^θ with error bound $\sin^2 \theta < \frac{1}{2}$. Moreover, $\mathcal{M}$ is defined only with real number transitions.*

Proof. Let $\{q_1, q_2\}$ be the set of states of $\mathcal{M}$ and q_1 be the initial and the only accepting state. The unitary operation is a rotation on $|q_1\rangle - |q_2\rangle$ plane with the angle φ. (Note that, there are infinitely many φ whose rotation matrices contain only rational numbers, e.g. $\arcsin \frac{3}{5}, \arcsin \frac{5}{13}, \arcsin \frac{7}{25}$, etc.). It is straightforward that, after reading a^k, the final quantum state becomes

$$|v_k\rangle = \cos(k\varphi)|q_1\rangle + \sin(k\varphi)|q_2\rangle,$$

and so a^k is accepted by $\mathcal{M}$ with probability $\cos^2(k\varphi)$. It is clear that $\mathcal{M}$ takes a^k and leaves it as $|v_k\rangle$ before the measurement, which can be seen as a map from an angle to a point on the unit circle. Therefore, the bounds on $k\varphi$ give similar bounds on $|v_k\rangle$, that allows $\mathcal{M}$ to solve the problem with bounded error. Now, we show that $\sin^2(\theta) < \frac{1}{2}$ can be a bound on the error.

If a^k is a yes-instance, we have $\cos\theta \leq |\cos(k\varphi)| \leq 1$. Then, the accepting probability can be bounded as $\cos^2 \theta \leq \cos^2(k\varphi) \leq 1$. That is, any yes-instance is accepted with probability at least $\cos^2(\theta)$, which is equal to $1 - \sin^2(\theta)$. In other word, the error for yes-instances can be at most $\sin^2 \theta$.

If a^k is a no-instance, $0 \leq |\cos(k\varphi)| \leq \sin\theta$. Then, the accepting probability can be bounded as $0 \leq \cos^2(k\varphi) \leq \sin^2 \theta$. That is, any no-instance is accepted with probability at most $\sin^2 \theta$, i.e. the error can be at most $\sin^2 \theta$ for any no-instance. □

Theorem 5. *There exists no PFA solving the promise problem* L^θ *for any error bound* $\epsilon < \frac{1}{2}$, *where* $\theta \in (0, \frac{\pi}{4})$.

Proof. We prove by contradiction. Let $\mathcal{P} = (Q, \{a\}, M, v_0, Q_a)$ be a PFA solving L^θ with error bound $\epsilon < \frac{1}{2}$. The computation of $\mathcal{P}$ can be described by a Markov chain and the states of $\mathcal{P}$ can be classified as described in Section 2.2. Let $C_1, \ldots, C_l$ be cyclic subsets of states of Markov chain with periods $t_1, \ldots, t_l$, respectively, and D be the least common multiple of $t_1, \ldots, t_l$.

We pick a yes-instance $a^n \in L^\theta_{\text{yes}}$ and define the set $A^n = \{a^{n+kD} \mid k \in \mathbb{Z}^+\}$. Now, we show that A^n contains some no-instances, i.e. $A^n \cap L^\theta_{no} \neq \emptyset$.

Claim. $A^n \cap L^\theta_{no} \neq \emptyset$.

Proof of the Claim. As verified from the definition of L^θ, each string can be associated to a point on the unit circle. Let γ_n be the angle of the point corresponding to our yes-instance a^n. So we have that $\gamma_n \in [-\theta, \theta] \cup [\pi - \theta, \pi + \theta]$. From now on, we consider all angles up to 2π and omit the period 2π from the value of angles. An input a^j is a no-instance ($a^j \in L^\theta_{no}$) if and only if $\gamma_j \in [\frac{\pi}{2} - \theta, \frac{\pi}{2} + \theta] \cup [3\frac{\pi}{2} - \theta, 3\frac{\pi}{2} + \theta]$. We need to show that there is an $l \in \mathbb{Z}^+$ such that $a^{n+lD} \in L^\theta_{no}$, that means $\gamma_{n+lD} \in [\frac{\pi}{2} - \theta, \frac{\pi}{2} + \theta] \cup [3\frac{\pi}{2} - \theta, 3\frac{\pi}{2} + \theta]$.

Reading D letters of the input corresponds to a rotation on the circle by the angle $D\varphi$. Let $\beta = D\varphi - 2\pi m$ for some $m \in \mathbb{N}$ satisfying $\beta \in (0, 2\pi)$. Since φ is an irrational multiple of π, β is also an irrational multiple of π. It is a well known fact that rotations with an angle of irrational multiple of π is dense on the unit circle. So the points corresponding to $\{a^{Dk} \mid k \in \mathbb{Z}^+\}$ are dense on the unit circle (and none of two strings from this set corresponds to the same point on the unit circle).

So for each point $\gamma_n \in [-\theta, \theta]$ (or for each point $\gamma_n \in [\pi - \theta, \pi + \theta]$), there exists an $l \in \mathbb{Z}^+$ such that $\gamma_{n+lD} \in [\frac{\pi}{2} - \theta, \frac{\pi}{2} + \theta] \cup [3\frac{\pi}{2} - \theta, 3\frac{\pi}{2} + \theta]$. Therefore, the set $A^n = \{a^{n+kD} \mid k \in \mathbb{Z}^+\}$ contains some no-instances. This completes the proof of the claim. ◁

After reading a^n, the final state is $v_n = M^n v_0$. Since there is no assumption on the length of a^n, it can be arbitrarily long. Assume that n is sufficiently big providing that

$$|f_P(a^{n+rD}) - f_P(a^{n+r'D})| < \frac{1}{2} - \epsilon$$

for any r, r'. Remember from Section 2.2 that this assumption follows from Markov chain theory and the bound approaches to 0 when $n \to \infty$. If a promise problem is solvable with an error bound ϵ, then the difference between the accepting probabilities of a yes-instance and a no-instance is at least $1 - 2\epsilon$. The set A^n has at least one no-instance whose accepting probability cannot be less than $\frac{1}{2}$, since (i) the minimal accepting probability for a member is $1 - \epsilon$ and (ii) we can obtain at least $\frac{1}{2}$ if we go away from $1 - \epsilon$ with $\frac{1}{2} - \epsilon$. However, this no-instance must be accepted with a probability at most $\epsilon < \frac{1}{2}$. Therefore, the PFA $\mathcal{P}$ cannot solve the promise problem L^θ with an error bound $\epsilon < \frac{1}{2}$. □

4   Succinctness

For each $n \in \mathbb{Z}^+$, we define a family of unary promise problems $F_n = \{\mathrm{L}^{k,n} \mid k \in \mathbb{Z}^+\}$ as follows. Let p_j be the j-th prime, $P_{k,n} = \{p_n, p_{n+1}, \ldots, p_{n+k-1}\}$ be the set of primes from n-th to $(n + k - 1)$-th one, and $N = p_n \cdot p_{n+1} \cdots p_{n+k-1}$.

The promise problem $\mathrm{L}^{k,n} = \{\mathrm{L}^{k,n}_{\text{yes}}, \mathrm{L}^{k,n}_{\text{no}}\}$ is defined as

- $\mathrm{L}^{k,n}_{\text{yes}} = \{a^m \mid m \equiv 0 \mod N\}$ and
- $\mathrm{L}^{k,n}_{\text{no}} = \{a^m \mid m \mod p_j \in \left[\frac{p_j}{8}, \frac{3p_j}{8}\right] \cup \left[\frac{5p_j}{8}, \frac{7p_j}{8}\right]$ for at least $\frac{2k}{3}$ different p_j from the set $P_{k,n}\}$.

Here we can use *Chinese remainder theorem* to show that the number of no-instances is infinitely many.

Lemma 1. *There are infinitely many strings in* $\mathrm{L}^{k,n}_{\text{no}}$.

Proof. If positive integers $p_1, p_2, \ldots, p_n$ are pairwise coprime, then for any integers $r_1, r_2, \ldots, r_n$ satisfying $0 \le r_i < p_i$ ($i \in \{1, 2, \ldots, n\}$), there exists a number K, such that $K = r_i \pmod{p_i}$ for each $i \in \{1, 2, \ldots, n\}$. Moreover, any such K is congruent modulo the product, $N = p_1 \cdots p_n$. That is all numbers of the form $K + N \cdot m$ will satisfy this condition, where $m \in \mathbb{Z}^+$. $\square$

Theorem 6. *For any* $n \in \mathbb{Z}^+$, *the promise problem* $\mathrm{L}^{k,n}$ *can be solvable by a* $2k$-*state MCQFA, say* $\mathcal{M}_{k,n}$, *such that yes-instances are accepted exactly and no-instance are rejected with probability at least* $\frac{1}{3}$.

Proof. We use the technique given in [4,5]. The set of states of automaton $\mathcal{M}_{k,n}$ is $\{q_1^0, q_1^1, \ldots, q_k^0, q_k^1\}$ and the ones with superscript "0" are the accepting states. The initial quantum state is

$$|v_0\rangle = \frac{1}{\sqrt{k}}|q_1^0\rangle + \frac{1}{\sqrt{k}}|q_2^0\rangle + \cdots + \frac{1}{\sqrt{k}}|q_k^0\rangle.$$

During reading the input, the states $|q_j^0\rangle$ and $|q_j^1\rangle$ form a small MCQFA isolated from the others, where $1 \le j \le k$. For each letter a, a rotation with the angle $\frac{2\pi}{p_j}$ is applied on $\{|q_j^0\rangle, |q_j^1\rangle\}$:

$$U_j = \begin{pmatrix} \cos(2\pi/p_j) & \sin(2\pi/p_j) \\ -\sin(2\pi/p_j) & \cos(2\pi/p_j) \end{pmatrix}.$$

Then, the overall transition matrix is

$$U = \begin{pmatrix} U_1 & \mathbf{0} & \cdots & \mathbf{0} \\ \mathbf{0} & U_2 & \cdots & \mathbf{0} \\ \vdots & \vdots & \ddots & \vdots \\ \mathbf{0} & \mathbf{0} & \cdots & U_k \end{pmatrix}$$

where $\mathbf{0}$ denotes 2×2 zero matrix.

For any input a^m the final state of $\mathcal{M}_{k,n}$ is

$$|v_m\rangle = \frac{1}{\sqrt{k}} \sum_{j=1}^{k} \left(\cos\left(m\frac{2\pi}{p_j}\right)|q_j^0\rangle + \sin\left(m\frac{2\pi}{p_j}\right)|q_j^1\rangle\right).$$

For any yes-instance, m is multiple of N and so each $m\frac{2\pi}{p_j}$ will be a multiple of 2π. Then, the final state is in a superposition of only the accepting states, i.e.

$$|v_m\rangle = \frac{1}{\sqrt{k}} \sum_{j=1}^{k} |q_j^0\rangle,$$

and so the input is accepted with probability 1.

For any no-instance, on the other hand, it holds that $(m \mod p_j)$ is in $\left[\frac{p_j}{8}, \frac{3p_j}{8}\right] \cup \left[\frac{5p_j}{8}, \frac{7p_j}{8}\right]$ for at least $\frac{2k}{3}$ different p_j's from the set $P_{k,n}$. If p_j is one of them, then its contribution to the overall rejecting probability is given by

$$\frac{1}{\sqrt{k}} \sin^2\left(m\frac{2\pi}{p_j}\right)$$

which takes its minimum value $\frac{1}{2k}$ when $(m \mod p_j)$ is equal to one of the border. Since there are at least $\frac{2k}{3}$ of them, the overall rejecting probability is at least $\frac{1}{3}$. □

Theorem 7. *Any bounded-error PFA solving the promise problem* $\mathsf{L}^{k,n}$ *needs* $\Omega(k(n+k)\log n)$ *states. (See [10] for the details)*

Theorem 8. *For any $n > 0$, there is a $O(k(n+k)\log(n+k))$-state PFA $\mathcal{P}_{k,n}$ solving the promise problem* $\mathsf{L}^{k,n}$ *with one-sided error bound $\frac{1}{3}$.*

Proof. Let $\mathcal{P}_{k,n}$, shortly $\mathcal{P}$, be $(Q, \{a\}, \{A\}, v_0, Q_a)$, where

- $Q = \{q_{i,j} \mid i = 1,\ldots,k, j = 0,\ldots,p_{n+i-1} - 1\}$ and $p_n,\ldots,p_{n+k-1}$ are the primes from the set $P_{k,n}$,
- v_0 is the initial probabilistic state such that the automaton is in the state $q_{i,0}$ with the probability $\frac{1}{k}$ for each $i = 1,\ldots,k$, and,
- $Q_a = \{q_{i,0} \mid i = 1,\ldots,k\}$.

The transitions of $\mathcal{P}$ are deterministic: after reading each letter, it switches from state $q_{i,j}$ to $q_{i,j+1 \pmod{p_{n+i-1}}}$. In fact, $\mathcal{P}$ executes k copies of DFAs with equal probability. The aim of the i-th DFA is to determine whether the length of the input is equivalent to zero in mod p_{n+i-1}. By construction it is clear that $\mathcal{P}$ accepts any yes-instance with the probability 1 and any no-instance with probability at most $\frac{1}{3}$.

The number of states is $|Q| = p_n + \cdots + p_{n+k-1}$. It is known [7] that the n-th prime number p_n satisfies $p_n = \Theta(n\log(n))$ and so

$$|Q| = \sum_{x=n}^{n+k-1} p_x \leq O(k(n+k)\log(n+k)).$$

□

Now, we give a lower and upper bound for DFAs.

Theorem 9. *For any $n > 0$, any DFA solving the promise problem $L^{k,n}$ needs $\Omega(n \log(n))^{\frac{k}{3}}$ states. (See [10] for the details)*

Theorem 10. *For any $n > 0$, there is a $O((n + \frac{k}{3}) \log(n + \frac{k}{3}))^{\frac{k}{3}}$-state DFA $\mathcal{D}_{k,n}$ solving the promise problem $L^{k,n} \in F_n$. (See [10] for the details)*

	DFA	PFA	QFA
lower bounds	$\Omega(n \log n)^{\frac{k}{3}}$	$\Omega(k(n + k) \log n)$	1
upper bounds	$O((n + \frac{k}{3}) \log(n + \frac{k}{3}))^{\frac{k}{3}}$	$O(k(n + k) \log(n + k))$	$2k$

Fig. 1. The summary of upper and lower state bounds for $L^{k,n}$

We give the summary of the results in Figure 1. The bounds for DFAs and PFAs are almost tight and currently we do not know any better bound for QFAs. Moreover, if we pick $n = 2^k$, then we obtain an exponential gap between QFAs and PFAs. On the other hand, if we pick $n = k$, then we obtain an exponential gap between PFAs and DFAs.

References

1. Ablayev, F., Gainutdinova, A., Khadiev, K., Yakaryılmaz, A.: Very narrow quantum OBDDs and width hierarchies for classical OBDDs. In: Jürgensen, H., Karhumäki, J., Okhotin, A. (eds.) DCFS 2014. LNCS, vol. 8614, pp. 53–64. Springer, Heidelberg (2014)
2. Ablayev, F., Gainutdinova, A.: On the lower bounds for one-way quantum automata. In: Nielsen, M., Rovan, B. (eds.) MFCS 2000. LNCS, vol. 1893, pp. 132–140. Springer, Heidelberg (2000)
3. Ablayev, F., Gainutdinova, A.: Complexity of quantum uniform and nonuniform automata. In: De Felice, C., Restivo, A. (eds.) DLT 2005. LNCS, vol. 3572, pp. 78–87. Springer, Heidelberg (2005)
4. Ablayev, F.M., Gainutdinova, A., Karpinski, M., Moore, C., Pollett, C.: On the computational power of probabilistic and quantum branching program. Information Computation **203**(2), 145–162 (2005)
5. Ambainis, A., Freivalds, R.: 1-way quantum finite automata: strengths, weaknesses and generalizations. In: FOCS 1998, pp. 332–341 (1998)
6. Ambainis, A., Yakaryılmaz, A.: Superiority of exact quantum automata for promise problems. Information Processing Letters **112**(7), 289–291 (2012)
7. Apostol, T.M.: Introduction to Analytic Number Theory. Springer (1976)
8. Bianchi, M.P., Mereghetti, C., Palano, B.: Complexity of promise problems on classical and quantum automata. In: Calude, C.S., Freivalds, R., Kazuo, I. (eds.) Gruska Festschrift. LNCS, vol. 8808, pp. 161–175. Springer, Heidelberg (2014)
9. Condon, A., Lipton, R.J.: On the complexity of space bounded interactive proofs (extended abstract). In: FOCS 1989, pp. 462–467 (1989)
10. Gainutdinova, A., Yakaryilmaz, A.: Unary probabilistic and quantum automata on promise problems. Technical Report arxiv.org/abs/1502.01462, arXiv (2015)

11. Geffert, V., Yakaryılmaz, A.: Classical automata on promise problems. In: Jürgensen, H., Karhumäki, J., Okhotin, A. (eds.) DCFS 2014. LNCS, vol. 8614, pp. 126–137. Springer, Heidelberg (2014)
12. Goldreich, O.: On promise problems: a survey. In: Goldreich, O., Rosenberg, A.L., Selman, A.L. (eds.) Theoretical Computer Science. LNCS, vol. 3895, pp. 254–290. Springer, Heidelberg (2006)
13. Gruska, J., Qiu, D., Zheng, S.: Generalizations of the distributed Deutsch-Jozsa promise problem. Technical report, arXiv (2014). arXiv:1402.7254
14. Gruska, J., Qiu, D., Zheng, S.: Potential of quantum finite automata with exact acceptance. Technical Report arXiv:1404.1689 (2014)
15. Hirvensalo, M.: Quantum automata with open time evolution. International Journal of Natural Computing 1(1), 70–85 (2010)
16. Kemeny, J.G., Snell, J.L.: Finite Markov Chains. Van Nostrand, Princeton (1960)
17. Klauck, H.: On quantum and probabilistic communication: las vegas and one-way protocols. In: STOC 2000, pp. 644–651 (2000)
18. Mereghetti, C., Palano, B., Pighizzini, G.: Note on the succinctness of deterministic, nondeterministic, probabilistic and quantum finite automata. Theoretical Informatics and Applications 35(5), 477–490 (2001)
19. Moore, C., Crutchfield, J.P.: Quantum automata and quantum grammars. Theoretical Computer Science 237(1–2), 275–306 (2000)
20. Murakami, Y., Nakanishi, M., Yamashita, S., Watanabe, K.: Quantum versus classical pushdown automata in exact computation. IPSJ Digital Courier 1, 426–435 (2005)
21. Nakanishi, M.: Quantum pushdown automata with a garbage tape. In: Italiano, G.F., Margaria-Steffen, T., Pokorný, J., Quisquater, J.-J., Wattenhofer, R. (eds.) SOFSEM 2015-Testing. LNCS, vol. 8939, pp. 352–363. Springer, Heidelberg (2015)
22. Nakanishi, M., Yakaryılmaz, A.: Classical and quantum counter automata on promise problems (2014). (arXiv:1412.6761) (Accepted to CIAA2015)
23. Nielsen, M.A., Chuang, I.L.: Quantum Computation and Quantum Information. Cambridge University Press (2000)
24. Rashid, J., Yakaryılmaz, A.: Implications of quantum automata for contextuality. In: Holzer, M., Kutrib, M. (eds.) CIAA 2014. LNCS, vol. 8587, pp. 318–331. Springer, Heidelberg (2014)
25. Salomaa, A., Soittola, M.: Automata-Theoretic Aspects of Formal Power Series. Texts and monographs in computer science. Springer-Verlag (1978)
26. Cem Say, A.C., Yakaryılmaz, A.: Quantum finite automata: a modern introduction. In: Calude, C.S., Freivalds, R., Kazuo, I. (eds.) Gruska Festschrift. LNCS, vol. 8808, pp. 208–222. Springer, Heidelberg (2014)
27. Watrous, J.: Encyclopedia of Complexity and System Science. In: Quantum computational complexity (chapter). Springer (2009). arXiv:0804.3401
28. Yakaryılmaz, A., Cem Say, A.C.: Languages recognized by nondeterministic quantum finite automata. Quantum Information and Computation 10(9&10), 747–770 (2010)
29. Yakaryılmaz, A., Cem Say, A.C.: Unbounded-error quantum computation with small space bounds. Information and Computation 279(6), 873–892 (2011)
30. Zheng, S., Gruska, J., Qiu, D.: On the state complexity of semi-quantum finite automata. In: Dediu, A.-H., Martín-Vide, C., Sierra-Rodríguez, J.-L., Truthe, B. (eds.) LATA 2014. LNCS, vol. 8370, pp. 601–612. Springer, Heidelberg (2014)
31. Zheng, S., Qiu, D., Gruska, J., Li, L., Mateus, P.: State succinctness of two-way finite automata with quantum and classical states. Theoretical Computer Science 499, 98–112 (2013)

Generalizations of Code Languages
with Marginal Errors

Yo-Sub Han[1($\boxtimes$)], Sang-Ki Ko[1], and Kai Salomaa[2]

[1] Department of Computer Science, Yonsei University, 50, Yonsei-Ro, Seoul,
Seodaemun-Gu 120-749, Republic of Korea
{emmous,narame7}@cs.yonsei.ac.kr
[2] School of Computing, Queen's University, Kingston, ON K7L 3N6, Canada
ksalomaa@cs.queensu.ca

Abstract. We study k-prefix-free, k-suffix-free and k-infix-free languages that generalize prefix-free, suffix-free and infix-free languages by allowing marginal errors. For example, a string x in a k-prefix-free language L can be a prefix of up to k different strings in L. Namely, a code (language) can allow some marginal errors. We also define finitely prefix-free languages in which a string x can be a prefix of finitely many strings. We present efficient algorithms that determine whether or not a given regular language is k-prefix-free, k-suffix-free or k-infix-free, and analyze their runtime. Lastly, we establish the undecidability results for (linear) context-free languages.

Keywords: Codes · Marginal errors · Decision algorithms · Undecidability · Regular languages · Context-free languages

1 Introduction

Codes are useful in many areas including information processing, data compression, cryptography and information transmission [2,12]. Many researchers have studied various codes such as *prefix codes*, *suffix codes* and *infix codes*. Since a code is a set of strings—a language, a code property defines a subset of languages preserving the corresponding property. For instance, for regular langauges, the prefix-freeness defines a proper subset of regular languages, *prefix-free regular languages*. Recently, Kari et al. [13] considered the problem of deciding whether or not a given regular language is maximal with respect to certain combined types of code properties.

There are different applications based on different properties subfamilies [1,3, 4,7,9,11]. For example, Huffman codes [11] are prefix-free languages and useful for lossless data compression; Han [6] proposed an efficient pattern matching algorithm for the prefix-free regular expressions based on the prefix-freeness of the input pattern. Infix-free languages have been used in text searching [3] and computing forbidden words [1,4]. Han et al. [8] observed the structural properties of FAs for infix-free regular languages and designed a decision algorithm.

© Springer International Publishing Switzerland 2015
I. Potapov (Ed.): DLT 2015, LNCS 9168, pp. 264–275, 2015.
DOI: 10.1007/978-3-319-21500-6_21

Table 1. The upper bound for the time complexity of decision algorithms for the generalizations of prefix-, suffix- and infix-free regular languages, where n is the size of FAs

	DFA	NFA
prefix-free	$O(n)$	$O(n^2)$
suffix-free	$O(n^2)$	$O(n^2)$
infix-free	$O(n^2)$	$O(n^2)$
k-prefix-free	$O(n^2 \cdot \log k)$	PSPACE-complete
k-suffix-free	PSPACE-complete	PSPACE-complete
k-infix-free	PSPACE-hard	PSPACE-hard
finitely prefix-free	$O(n)$	$O(n^2)$
finitely suffix-free	$O(n^2)$	$O(n^2)$
finitely infix-free	$O(n^2)$	$O(n^2)$

When people design a code in practice—for instance, designing a DNA code by experiments [14, 16]—it may not always be successful. There may be a few number of undesired code words in the resulting code. This motivates us to examine a relaxed version of codes; in other words, we allow some marginal errors in the code. We generalize prefix-free, suffix-free and infix-free languages and define new codes, *k-prefix-free, k-suffix-free and k-infix-free* languages. Recall that there is no prefix of any other string in the language L if L is prefix-free [2]. In a k-prefix-free language L, we allow at most k strings in L to have another string in L as a prefix. We also introduce *finitely prefix-free, finitely suffix-free* and *finitely infix-free* languages in which we allow a finite number of such strings.

Given a family $\mathcal{L}$ of languages and a language L, the *decision problem* of L with respect to $\mathcal{L}$ is to decide whether or not L belongs to $\mathcal{L}$. For the decision problems of finitely prefix-free, finitely suffix-free and finitely infix-free regular languages, we use the *state-pair graph* used for the decision problems of prefix-free, suffix-free and infix-free regular languages [5]. Table 1 summarizes the time complexity of the prior algorithms and our algorithms based on the state-pair graph. Interestingly, we have a polynomial time algorithm only for deciding the k-prefix-freeness of a language recognized by a DFA. We prove that the complexity for determining whether or not an NFA is k-prefix-free is PSPACE-complete. For both k-suffix-free cases, it is already PSPACE-hard for DFAs. For the decision problems of finitely prefix-free, suffix-free and infix-free regular languages, we present polynomial algorithms based on the state-pair graph construction. We also establish the undecidability results for context-free languages by the reduction from the Post's Correspondence Problem [17].

We define some basic notions in Section 2. We define the generalizations of prefix-free, suffix-free and infix-free regular languages in Section 3 and observe the hierarchy between the proposed subfamilies. In Section 4, we present decision algorithms and complexity results for regular languages. We also establish undecidability result for context-free languages in Section 5.

2   Preliminaries

We briefly present definitions and notations. We refer to the books [10,18] for more knowledge on automata theory. For more details on coding theory, refer to Berstel and Perrin [2] or Jürgensen and Konstantinidis [12].

Let Σ be a finite alphabet and Σ^* be the set of all strings over Σ. A language over Σ is any subset of Σ^*. The symbol $\emptyset$ denotes the empty language, the symbol λ denotes the null string and Σ^+ denotes $\Sigma^* \setminus \{\lambda\}$. Let $|w|$ be the length of w.

For strings x, y and z, we say that x is a *prefix* of y and z is a *suffix* of y if $y = xz$. For strings x, y, w and z, we say that z is an infix of y if $y = xzw$. We define a language L to be prefix-free if a string $x \in L$ is not a prefix of any other strings in L. Similarly, we define a language L to be suffix-free (or infix-free) if a string $x \in L$ is not a suffix (or infix) of any other strings in L. The reversal of a string w is denoted w^R and the reversal of a language L is $L^R = \{w^R \mid w \in L\}$.

A *nondeterministic finite automaton with λ-transitions* (λ-NFA) is a five-tuple $A = (Q, \Sigma, \delta, s, F)$, where Q is a finite set of states, Σ is a finite alphabet, δ is a multi-valued transition function from $Q \times (\Sigma \cup \{\lambda\})$ into 2^Q, $s \in Q$ is the initial state and $F \subseteq Q$ is the set of final states. By an NFA, we mean a nondeterministic automaton without λ-transitions, that is, A is an NFA if δ is a function from $Q \times \Sigma$ into 2^Q. The automaton A is *deterministic* (a DFA) if δ is a single-valued function $Q \times \Sigma \to Q$. The language $L(A)$ recognized by A is the set of strings w such that some sequence of transitions spelling out w takes the initial state of A to a final state. We define the size $|A|$ of A to be $|Q| + |\delta|$. For $q \in Q$, we denote by A_q the NFA that is obtained from A by replacing the initial state s with q. For $q, p \in Q$, we also denote by $A_{q,p}$ the NFA that is as A except has q as the initial state and p as the only final state. We assume that an FA A has only useful states; that is, each state appears on some path from the initial state to some final state.

3 k-prefix-free, k-suffix-free and k-infix-freeness

We consider generalizations of prefix-free, suffix-free and infix-free languages. First, we define a generalization of prefix-free languages called the *k-prefix-free languages*.

Definition 1. *We define a language L to be k-prefix-free if there is no string x in L that is a prefix of more than k strings in $L \setminus \{x\}$.*

Recently, Konitzer and Simon [15] defined the *maximum activity level* of a language L as follows: Given a language L, the maximum activity level l_L of L is

$$l_L = \sup\{l : (\exists w_1, \ldots, w_l \in \Sigma^+, \forall = 1, \ldots, l : w_1 \cdots w_l \in L)\}.$$

Since we consider the number of strings in L that have a common prefix from L and the activity level is defined as the number of proper prefixes of a string in L, the k-prefix-freeness is a different concept from the activity level of a language.

From the definition, we can say that prefix-free languages are in fact, 0-prefix-free. We define similar generalizations for suffix-free and infix-free languages as follows:

Definition 2. *We define a language L to be k-suffix-free (or k-infix-free) if there is no string x in L that is a suffix (or an infix) of more than k strings in $L \setminus \{x\}$.*

We have the following observations from the definition. Note that similar statements hold for suffix-free and infix-free languages.

Observation 1. *If a language L is k-prefix-free, then L is $(k+n)$-prefix-free for all $n \geq 1$.*

Observation 2. *If a language L is prefix-free, then L is k-prefix-free for all $k \geq 1$.*

We also define the following code properties from the k-prefix (suffix, infix)-freeness.

Definition 3. *We define a language L to be minimally k-prefix (suffix, infix)-free if L is k-prefix (suffix, infix)-free but not $(k-1)$-prefix (suffix, infix)-free.*

Definition 4. *We define a language L to be finitely prefix-free (suffix-free or infix-free) if L is k-prefix-free (k-suffix-free or k-infix-free) for a constant k.*

4   Decision Algorithms

We determine whether or not a regular language given by a DFA is k-prefix-free or finitely prefix-free. First, we consider decision problems where k is given as part of the input in Section 4.1 and consider the problem of deciding whether the property holds for some finite value of k in Section 4.2.

4.1   For k-prefix-free, k-suffix-free and k-infix-free Regular Languages

Given a DFA A, we can determine whether or not $L(A)$ is k-prefix-free or finitely prefix-free by examining the following properties:

Lemma 1. *Given a DFA $A = (Q, \Sigma, \delta, s, F)$, $L(A)$ is k-prefix-free if and only if there is no final state $f \in F$ satisfying the following conditions:*

$$L(A_{s,f}) \neq \emptyset \quad and \quad |L(A_f) \setminus \{\lambda\}| > k. \tag{1}$$

Proof. ($\Longrightarrow$) We prove that $L(A)$ is k-prefix-free if there is no final state f satisfying Condition (1). Assume that there is such a final state f satisfying the condition. This means that there is a string w spelled out by the path from s to f and

$$|L(A_f) \setminus \{\lambda\}| > k.$$

Since there are more than k different strings that can be computed after w, there exist more than k strings in $L(A)$ that contain w as a proper prefix. Therefore, $L(A)$ is not k-prefix-free. We have a contradiction.

($\Longleftarrow$) We prove that there is no final state f satisfying Condition (1) if $L(A)$ is k-prefix-free. Assume that $L(A)$ is not k-prefix-free. This implies that there exist a string w in L and more than k strings—let P_w denote the set of these strings—that contain w as a prefix in $L(A)$. Since A is a DFA, there is a unique final state f that we reach by reading w and all the accepting paths for strings in P_w must pass through f. This guarantees that $|L(A_f) \setminus \{\lambda\}| > k$ since $|P_w| > k$. Therefore, we have a contradiction.

Lemma 2. *Let A be a DFA (or an NFA) over an alphabet Σ with n states such that $L(A)$ is minimally k-prefix-free. Then*

$$k \leq |\Sigma| \cdot \frac{|\Sigma|^{n-1} - 1}{|\Sigma| - 1}.$$

Conversely, there exists an n-state DFA A over Σ that is minimally $|\Sigma| \cdot \frac{|\Sigma|^{n-1}-1}{|\Sigma|-1}$-prefix-free.

From Lemma 1, we can design a quadratic algorithm for deciding whether or not a language recognized by a DFA is k-prefix-free.

Theorem 1. *Given a DFA A and $k \in \mathbb{N}$, we can determine whether or not $L(A)$ is k-prefix-free in $O(n^2 \cdot \log k)$ time, where $n = |A|$.*

Note that the family of k-prefix-free languages form a proper hierarchy with respect to k. For example, if a regular language L is k-prefix-free, then L is not necessarily $(k - 1)$-prefix-free. This leads us to a new problem that determines whether or not a regular language L is minimally k-prefix-free.

Theorem 2. *Given a DFA $A = (Q, \Sigma, \delta, s, F)$, we can find $k \in \mathbb{N}$ such that $L(A)$ is minimally k-prefix-free in $O(n^3)$ time, where $n = |A|$.*

Proof. For an arbitrary final state f of A, if A_f is not acyclic, then we know that $L(A)$ is not k-prefix-free for any $k \in \mathbb{N}$. Otherwise, based on the property of Lemma 1, it is possible to compute in polynomial time the cardinality of $L(A_f)$, for each final state f, and pick the maximum value M. Then $L(A)$ is minimally M-prefix-free. Since M is at most $O(|\Sigma|^{|Q|})$ by Lemma 2, the runtime is

$$O(n^2 \cdot \log |\Sigma|^n) = O(n^3)$$

by assuming $|\Sigma|$ as a constant. $\qquad\square$

Now we consider the case when the input is specified by an NFA instead of a DFA.

Lemma 3. *Given an NFA $A = (Q, \Sigma, \delta, s, F)$, $L(A)$ is k-prefix-free if and only if there is no set $Q' \subseteq Q$ of states satisfying the following conditions:*

(i) $Q' \cap F \neq \emptyset,$

(ii) $\displaystyle\bigcap_{q \in Q'} L(A_{s,q}) \neq \emptyset$ *and* $\displaystyle\left| \bigcup_{q \in Q'} L(A_q) \setminus \{\lambda\} \right| > k.$

Proof. ($\Longrightarrow$) We prove that $L(A)$ is k-prefix-free if there is no set $Q' \subseteq Q$ of states satisfying the two conditions (i) and (ii). Assume that there is such a set Q' satisfying all these conditions. Then, there exists a string $w \in L(A)$ such that $Q' \subseteq \delta(s, w)$ by the conditions. Furthermore, we have more than k non-empty strings that can be spelled out by a path from a state in Q' to a final state in F because of condition (ii). This implies that $L(A)$ has more than k strings that contain w as a proper prefix. Therefore, $L(A)$ is not k-prefix-free.

($\Longleftarrow$) We prove that if $L(A)$ is k-prefix-free, then there is no set Q' of states satisfying these three conditions. Assume that $L(A)$ is not k-prefix-free. This implies that there exist a string w in L and more than k strings having w as a prefix in $L(A)$. Since A accepts w, there should exist a final state f such that $f \in \delta(s, w)$. We denote the set of all states that can be reached by reading w from the initial state s in A by P. Obviously, f is in P. Since there are more than k strings that have w as a proper prefix, there are also more than k non-empty strings that can be spelled out by paths from states in P to some final states. Since P satisfies conditions (i) and (ii), we have a contradiction. $\qquad\square$

Now we show that the problem of determining whether or not a regular language accepted by an NFA is k-prefix-free is PSPACE-complete.

Theorem 3. *Given an NFA A and $k \in \mathbb{N}$, it is PSPACE-complete to determine whether or not $L(A)$ is k-prefix-free.*

We also establish that given an NFA A, the problem of finding a non-negative integer k where $L(A)$ is minimally k-prefix-free is also PSPACE-complete.

Theorem 4. *Given an NFA A, the problem of finding $k \in \mathbb{N}$ where $L(A)$ is minimally k-prefix-free is PSPACE-complete.*

Now we consider the k-suffix-free case. The properties that characterize the k-suffix-freeness of a language recognized by a DFA turn out to be similar to the properties that characterize the k-prefix-freeness of an NFA. Intuitively, suffix-freeness corresponds to the prefix-freeness of the reversed language and reversing the transitions of a DFA makes it nondeterministic. We present necessary and sufficient conditions for DFAs to accept k-suffix-free regular languages as follows:

Lemma 4. *Given an NFA $A = (Q, \Sigma, \delta, s, F)$, $L(A)$ is k-suffix-free if and only if there is no set $Q' \subseteq Q$ of states satisfying the following conditions:*

$$\left| \bigcup_{q \in Q'} L(A_{s,q}) \setminus \{\lambda\} \right| > k \quad and \quad \bigcap_{q \in Q' \cup \{s\}} L(A_q) \neq \emptyset. \tag{2}$$

Proof. ($\Longrightarrow$) We prove that $L(A)$ is k-suffix-free if there is a set $Q' \subseteq Q$ of states satisfying Condition (2). Assume that there is a set $Q' \subseteq Q$ of states satisfying the condition. From the condition

$$\bigcap_{q \in Q' \cup \{s\}} L(A_q) \neq \emptyset,$$

we know that there is a common string $w \in L(A)$ that is also spelled out by paths from the states in Q' to one of the final states. Since the cardinality of the set of strings spelled out by the paths from the initial state to the states in Q' is greater than k, there are more than k strings that have w as a suffix. Since the string w is also in $L(A)$, $L(A)$ is not k-suffix-free. We have a contradiction.

($\Longleftarrow$) We prove that there is no set $Q' \subseteq Q$ of states satisfying Condition (2) if $L(A)$ is k-suffix-free. Assume that $L(A)$ is not k-suffix-free. This implies that there exist a string $w \in L(A)$ and more than k strings containing w as a suffix in $L(A)$. Assume that there exist $k_1 > k$ strings that contain w as a suffix and denote the strings by $w_1, w_2, w_3, \ldots, w_{k_1}$. For all strings $w_i, 1 \leq i \leq k_1$, we can decompose w_i into $w_i'w$ since they have the common suffix w. Then, the strings from w_1' to w_{k_1}' should be distinct and non-empty since $w_1, w_2, w_3, \ldots, w_{k_1}$ are distinct and properly containing w as a suffix. Let us denote the set of states that are reachable from the initial state s by reading $w_1', w_2', \ldots, w_{k_1}'$ by $P \subseteq Q$. Therefore, the following inequality holds:

$$\left| \bigcup_{q \in P} L(A_{s,q}) \setminus \{\lambda\} \right| > k.$$

Moreover, there should be at least one path from each state in P and the initial state to one of the final states of A spelling out the string w since A accepts the strings $w, w_1, w_2, w_3, \ldots, w_{k_1}$ that are in fact, $\lambda \cdot w, w_1' \cdot w, w_2' \cdot w, \ldots, w_{k_1}' \cdot w$. Therefore, the following inequality also holds:

$$\bigcap_{q \in P \cup \{s\}} L(A_q) \neq \emptyset.$$

Now we have a contradiction since there is a set $P \subseteq Q$ of states satisfying Condition (2). $\qquad\square$

We discuss the computational complexity of the decision problem for k-suffix-free regular languages. Interestingly, it turns out to be much more complicated than the k-prefix-free case even for DFAs.

Theorem 5. *Given a DFA (or an NFA) A and $k \in \mathbb{N}$, it is* PSPACE-*complete to determine whether or not $L(A)$ is k-suffix-free.*

Theorem 6. *Given a DFA (or an NFA) A, the problem of finding $k \in \mathbb{N}$ where $L(A)$ is minimally k-suffix-free is* PSPACE-*complete.*

Moreover, it is immediate from the k-suffix-free case that the problem of determining whether or not a regular language $L(A)$ is k-infix-free is also PSPACE-hard.

Corollary 1. *Given a DFA (or an NFA) A and $k \in \mathbb{N}$, it is PSPACE-hard to determine whether or not $L(A)$ is k-infix-free. Given a DFA (or an NFA) A, the problem of finding $k \in \mathbb{N}$ where $L(A)$ is minimally k-infix-free is PSPACE-hard.*

Proof. We show that the problem of determining whether or not $L(A)$ is k-infix-free is PSPACE-hard by the reduction from the k-suffix-free case. Let $L \subseteq \Sigma^*$ be a regular language. Then, we can check the k-suffix-freeness of L by checking the k-infix-freeness of $L \cdot \{\$\}$, where $\$ \notin \Sigma$ is a new symbol.

It remains open to show that the problem of determining whether or not $L(A)$ is k-infix-free is in PSPACE.

4.2   For Finitely Prefix-free, Finitely Suffix-free and Finitely Infix-free Regular Languages

We establish the following corollary in a similar way to the proof of Lemma 1.

Corollary 2. *Given a DFA $A = (Q, \Sigma, \delta, s, F)$, $L(A)$ is finitely prefix-free if and only if there is no final state $f \in F$ satisfying the following conditions:*

$$L(A_{s,f}) \neq \emptyset \quad and \quad |L(A_f)| = \infty.$$

By Corollary 2, we can determine whether or not a regular language is finitely prefix-free in linear time when the regular language is given by a DFA.

Theorem 7. *Given a DFA A, we can determine whether or not $L(A)$ is finitely prefix-free in $O(n)$ time, where $n = |A|$.*

For the NFA case, on the other hand, we need to use algorithms based on the *state-pair graphs* [5] to determine whether or not a regular language accepted by an NFA is finitely prefix-free. For example, for DFAs, we can easily decide the prefix-freeness (including k-prefix-freeness and finitely prefix-freeness) by observing the accepting path after the final states since all strings having the same prefix should share the same path to spell out the prefix. However, in NFAs, we cannot guarantee such properties. There can be several completely disjoint paths even for a single string. Han [5] showed that it is possible to determine whether or not a regular language given by an NFA is prefix-free, suffix-free or infix-free using the *state-pair graph* in quadratic time in the size of an input.

Definition 5 (Han [5]). *Given an NFA $A = (Q, \Sigma, \delta, s, F)$, we define the state-pair graph $G_A = (V, E)$ of A, where V is a set of nodes and E is a set of labeled edges, as follows:*

- $V = \{(i, j) \mid i, j \in Q\}$ *and*
- $E = \{((i, j), a, (x, y)) \mid x \in \delta(i, a), y \in \delta(j, a) \text{ and } a \in \Sigma \}.$

Then, we have $|G_A| \leq |Q|^2 + |\delta|^2$; namely $|G_A| = O(|A|^2)$.

Without loss of generality, we assume that A has one final state f since we can always make any NFA to have only one final state by introducing a new final state f and making f to be the target state of all final states by a λ-transition.

Lemma 5. *Given an NFA $A = (Q, \Sigma, \delta, s, f)$ and $q, p \in Q$, $L(A)$ is finitely prefix-free if and only if there is no path labeled by a string from (s, s) to (f, p) in G_A satisfying $|L(A_{p,f})| = \infty$.*

Proof. ($\Longrightarrow$) We first prove that $L(A)$ is finitely prefix-free if there is no such path. Assume that there is a such path labeled by a string w. By the assumption, the cardinality of $L(A_{p,f})$ is infinite. This implies that A accepts an infinite number of strings having w as a proper prefix. Since $w \in L(A)$ by assumption, $L(A)$ is not finitely prefix-free.
($\Longleftarrow$) We prove that if $L(A)$ is finitely prefix-free there is no such path. Assume that $L(A)$ is not finitely prefix-free. This implies that there are an infinite number of strings having a prefix w and w is accepted by A. Consider a set Q' of states that are reachable by reading w from the initial state of A. Since there are an infinite number of strings having w as a prefix, there should be an infinite number of distinct accepting paths starting from the states in Q'. If $L(A_{p,f})$ is finite for all $p \in Q'$, it is impossible to have an infinite number of accepting paths from the states in Q'. Thus at least one state $p \in Q'$ should have an infinite number of accepting paths. Now we have a contradiction. $\square$

From the properties of the state-pair graph observed in Lemma 5, we can determine whether or not $L(A)$ is finitely prefix-free by exploring the existence of such paths in the state-pair graph. Since the size of the state-pair graph is quadratic in the size of the given NFA, the time complexity of the algorithm is also quadratic.

Theorem 8. *Given an NFA $A = (Q, \Sigma, \delta, s, f)$, we can determine whether or not $L(A)$ is finitely prefix-free in $O(n^2)$ time, where $n = |A|$.*

Proof. First, we check whether or not $L(A_q)$ is finite for each $q \in Q$. This procedure takes $O(n^2)$ time since we can check the existence of a cycle of an NFA in linear time by depth-first traversal. Then, we explore the state-pair graph to check whether or not there is a path from (s, s) to (f, p) in the state-pair graph G_A when $L(A_p)$ is infinite. As the size of G_A is quadratic in the size of A, the algorithm takes $O(n^2)$ time. $\square$

Note that the decision algorithm for finitely infix-freeness is relatively more complicated than the finitely prefix-free case.

Lemma 6. *Given an NFA $A = (Q, \Sigma, \delta, s, f)$ and $q, p \in Q$, $L(A)$ is finitely infix-free if and only if there is no path labeled by a string from (s, q) to (f, p) where $f \neq p$ in G_A satisfying $|L(A_{s,q}) \cup L(A_{p,f})| = \infty$.*

Proof. ($\Longrightarrow$) We prove that $L(A)$ is finitely infix-free if there is no such path. Assume that there is a such path labeled by a string w. Then, w is in $L(A)$ and

$L(A_{q,p})$. Since the cardinality of $L(A_{s,q}) \cup L(A_{p,f})$ is infinite by assumption, the cardinality of $L(A_{s,q})$ or $L(A_{p,f})$ should be infinite. We can easily see that there are an infinite number of strings containing w as an infix in $L(A)$. Therefore, $L(A)$ is not finitely infix-free. We have a contradiction.

($\Longleftarrow$) Here we prove that if $L(A)$ is finitely infix-free, then there is no such path. Assume that $L(A)$ is not finitely infix-free. This means that for a string $w \in L(A)$, we have an infinite number of strings in $L(A)$ containing w as an infix. Consider all state pairs q, p where $w \in L(A_{q,p})$. If $L(A_{s,q}) \cup L(A_{p,f})$ is finite for all state pairs, the number of strings containing w as an infix cannot be infinite. Therefore, $L(A_{s,q}) \cup L(A_{p,f})$ should be infinite. We conclude the proof.   □

Even though the decision algorithm for finitely infix-freeness is a bit more complicated than the finitely prefix-free case, we still have a quadratic algorithm for the problem as follows:

Theorem 9. *Given an NFA* $A = (Q, \Sigma, \delta, s, f)$, *we can determine whether or not* $L(A)$ *is finitely infix-free in* $O(n^2)$ *time, where* $n = |A|$.

Proof. We first construct a state-pair graph G_A of A. Before exploring the existence of the paths, we first check whether or not $L(A_{s,q})$ and $L(A_{q,f})$ is finite. This procedure takes $O(n^2)$ time.

Once we finish the checking procedure, for all states $q \in Q$ except s, we run the depth-first search from (s, q) to (f, p) to check whether or not there exists a such path. If there exists a such path, then we check whether or not $L(A_{s,q}) \cup L(A_{p,f})$ is finite by simply checking whether or not both languages are finite. Then, for all states $q \in Q$ except f, we run the depth-first search from (s, s) to (f, q) to check whether or not there exists a such path. Since the depth-first search takes $O(|V| + |E|)$ time when we are given a graph $G = (V, E)$, we can check whether or not there exists a path in $O(n^2)$ time. Overall, our algorithm takes $O(n^2)$ time.   □

We note that the finitely suffix-freeness can be checked in a converse way to the finitely prefix-free case.

Lemma 7. *Given an NFA* $A = (Q, \Sigma, \delta, s, f)$ *and* $q \in Q$, $L(A)$ *is finitely suffix-free if and only if there is no path labeled by a string from* (s, q) *to* (f, f) *in* G_A *satisfying* $|L(A_{s,q})| = \infty$.

Proof. The proof is immediate from Lemma 5. We just exchange the initial state s and the final state f from A, and reverse the directions of all transitions. Then the new NFA accepts $L(A)^R$—the reverse of $L(A)$. Then, we can decide whether or not $L(A)$ is finitely suffix-free by testing the finitely prefix-freeness of $L(A)^R$.   □

Since the finitely suffix-freeness can be tested by reversing the NFA and testing the finitely prefix-freeness, we establish the following result.

Theorem 10. *Given an NFA* A, *we can determine whether or not* $L(A)$ *is finitely suffix-free* $O(n^2)$ *time, where* $n = |A|$.

5   Undecidability Results for (Linear) Context-free Languages

It is known that given a (linear) context-free language L, it is undecidable whether or not L is prefix-free [12]. We establish similar undecidability results for the code properties considered here.

Theorem 11. *There is no algorithm that determines whether or not a given linear language L is k-prefix (suffix, infix)-free.*

Theorem 12. *There is no algorithm that determines whether or not a given linear language L is finitely prefix (suffix, infix)-free.*

6   Conclusions

We have introduced an extension of prefix-free, suffix-free and infix-free languages by allowing marginal errors. We have defined k-prefix-free, k-suffix-free and k-infix-free languages. For example, a k-prefix-free language L can have at most k strings containing any other string w in L as a prefix. We, then, have considered more general versions of k-prefix-free, k-suffix-free and k-infix-free languages allowing a finite number of such strings.

We have examined the time complexity of the decision problems for these subfamilies. Given a DFA, we have a quadratic algorithm for the decision problem of k-prefix-freeness. However, we have shown that it is PSPACE-complete to determine whether or not a regular language given by an NFA is k-prefix-free. For the k-suffix-free and k-infix-free cases, we have shown that it is PSPACE-hard for both DFAs and NFAs. We have designed polynomial time algorithms based on the state-pair graph construction that decide whether or not a regular language (given by an NFA) is finitely prefix- (respectively, suffix- or infix-) free. We also have established the undecidability results for (linear) context-free languages. In future, we plan to investigate the state complexity of these new subfamilies of regular languages.

References

1. Béal, M.P., Crochemore, M., Mignosi, F., Restivo, A., Sciortino, M.: Computing forbidden words of regular languages. Fundamenta Informaticae **56**(1,2), 121–135 (2002)
2. Berstel, J., Perrin, D.: Theory of codes. Academic Press, Inc. (1985)
3. Clarke, C.L.A., Cormack, G.V.: On the use of regular expressions for searching text. ACM Transactions on Programming Languages and Systems **19**(3), 413–426 (1997)
4. Crochemore, M., Mignosi, F., Restivo, A.: Automata and forbidden words. Information Processing Letters **67**(3), 111–117 (1998)
5. Han, Y.S.: Decision algorithms for subfamilies of regular languages using state-pair graphs. Bulletin of the European Association for Theoretical Computer Science **93**, 118–133 (2007)

6. Han, Y.S.: An improved prefix-free regular-expression matching. International Journal of Foundations of Computer Science **24**(5), 679–687 (2013)
7. Han, Y.S., Salomaa, K., Wood, D.: Intercode regular languages. Fundamenta Informaticae **76**(16), 113–128 (2007)
8. Han, Y.S., Wang, Y., Wood, D.: Infix-free regular expressions and languages. International Journal of Foundations of Computer Science **17**(2), 379–393 (2006)
9. Han, Y.S., Wang, Y., Wood, D.: Prefix-free regular languages and pattern matching. Theoretical Computer Science **389**(1–2), 307–317 (2007)
10. Hopcroft, J.E., Motwani, R., Ullman, J.D.: Introduction to Automata Theory, Languages, and Computation, 3rd edn. Addison-Wesley Longman Publishing Company Incorporated (2006)
11. Huffman, D.: A method for the construction of minimum-redundancy codes. Proceedings of the IRE **40**(9), 1098–1101 (1952)
12. Jürgensen, H., Konstantinidis, S.: Codes. Word, Language, Grammar, Handbook of Formal Languages **1**, 511–607 (1997)
13. Kari, L., Konstantinidis, S., Kopecki, S.: On the maximality of languages with combined types of code properties. Theoretical Computer Science **550**, 79–89 (2014)
14. Kari, L., Mahalingam, K.: DNA Codes and Their Properties. In: Mao, C., Yokomori, T. (eds.) DNA12. LNCS, vol. 4287, pp. 127–142. Springer, Heidelberg (2006)
15. Konitzer, M., Simon, H.U.: DFA with a Bounded Activity Level. In: Dediu, A.-H., Martín-Vide, C., Sierra-Rodríguez, J.-L., Truthe, B. (eds.) LATA 2014. LNCS, vol. 8370, pp. 478–489. Springer, Heidelberg (2014)
16. Marathe, A., Condon, A.E., Corn, R.M.: On combinatorial dna word design. Journal of Computational Biology **8**(3), 201–219 (2001)
17. Post, E.L.: A variant of a recursively unsolvable problem. Bulletin of the American Mathematical Society **52**(4), 264–268 (1946)
18. Wood, D.: Theory of Computation. Harper & Row (1987)

Minimal Reversible Deterministic Finite Automata

Markus Holzer, Sebastian Jakobi, and Martin Kutrib[(✉)]

Institut für Informatik, Universität Giessen,
Arndtstr. 2, 35392 Giessen, Germany
{holzer,sebastian.jakobi,kutrib}@informatik.uni-giessen.de

Abstract. We study reversible deterministic finite automata (REV-DFAs), that are partial deterministic finite automata whose transition function induces an injective mapping on the state set for every letter of the input alphabet. We give a structural characterization of regular languages that can be accepted by REV-DFAs. This characterization is based on the absence of a forbidden pattern in the (minimal) deterministic state graph. Again with a forbidden pattern approach, we also show that the minimality of REV-DFAs among all equivalent REV-DFAs can be decided. Both forbidden pattern characterizations give rise to NL-complete decision algorithms. In fact, our techniques allow us to construct the minimal REV-DFA for a given minimal DFA. These considerations lead to asymptotic upper and lower bounds on the conversion from DFAs to REV-DFAs. Thus, almost all problems that concern uniqueness and the size of minimal REV-DFAs are solved.

1 Introduction

Reversibility is a fundamental principle in physics. Since abstract computational models with discrete internal states may serve as prototypes of computing devices which can be physically constructed, it is interesting to know whether these abstract models are able to obey physical laws. The observation that loss of information results in heat dissipation [16] strongly suggests to study computations without loss of information. Many different formal models have been studied from this point of view. The reversibility of a computation means in essence that every configuration has a unique successor configuration and a unique predecessor configuration. For example, reversible Turing machines have been introduced in [4], where it turned out that every Turing machine can be simulated by a reversible one—for improved simulation constructions see [3,20]. Since Rice's theorem shows that any non-trivial property on languages accepted by (reversible) Turing machines is undecidable, it is reasonable from a practical perspective to study reversibility in devices of lower computational capacity. On the opposite end of the automata hierarchy, reversibility has been studied for finite automata [1,6,8,11,17,21], pushdown automata [13], queue automata [15], and even multi-head finite automata [2,14,18,19].

© Springer International Publishing Switzerland 2015
I. Potapov (Ed.): DLT 2015, LNCS 9168, pp. 276–287, 2015.
DOI: 10.1007/978-3-319-21500-6_22

Originally, reversible deterministic finite automata have been introduced and studied in the context of algorithmic learning theory in [1]; see also [11]. Later this concept was generalized in [21] and [17]. Almost all of these definitions agree on the fact that the transition function induces a partial injective mapping for every letter. Nevertheless, there are subtle differences. For instance, in [1] a partial deterministic finite automaton (DFA) M is defined to be reversible if M and the dual of M, that is the automaton that is obtained from M by reversing all transitions and interchanging initial and final states, are both deterministic. In particular, this definition implies that for reversible DFAs in the sense of [1] only one final state is allowed; hence these devices were called *bideterministic* in [21]. Since there are regular languages that are not accepted by any DFA with a sole accepting state, by definition, there are non-reversible regular languages in this setting. Then the definition of reversibility has been extended in [21]. Now multiple accepting as well as multiple initial states are allowed. So, reversible DFAs in the sense of [21] may have limited nondeterminism plugged in from the outside world at the outset of the computation. But still, these devices turn out to be less powerful than general (possibly irreversible) finite automata. An example is the regular language a^*b^* which is shown [21] to be not acceptable by any reversible DFA. In the same paper it is proved that for a given DFA the existence of an equivalent reversible finite automaton can be decided in polynomial time. A further generalization of reversibility to quasi-reversibility, which even allows nondeterministic transitions was introduced in [17]—see also [6]. Different aspects of reversibility for classical automata are discussed in [12]. In view of these results natural questions concern the uniqueness and the size of a minimal reversible DFA in terms of the size of the equivalent minimal DFA. For the latter question, in [8], a lower bound of $\Omega(1.001^n)$ states has been obtained which, in turn, raises the question for the construction of a minimal reversible DFA from a given (minimal) DFA. The construction problem has partially been solved in [6,17], where so-called quasi-reversible automata are constructed. However, these quasi-reversible DFAs may themselves be exponentially more succinct than the minimal reversible DFAs. In fact, the witness automata in [8] are already quasi-reversible.

This is the starting point of our investigation. For our definition of reversibility we stick to standard definitions. That is, partial DFAs with a unique initial state and potentially multiple accepting states. Then such an automaton is *reversible* if the transition function induces an injective mapping on the state set for every letter. These basic definitions are given in the next section together with an introductory example. For these reversible DFAs (REV-DFAs) we are able to solve the question on uniqueness and size of minimal representations almost completely. Section 3 is devoted to develop a method to decide the reversibility of a given regular language. While the notion of reversibility proposed in [21] is also decidable in polynomial time by an argument on the syntactic monoid of the language under consideration, here we obtain a structural characterization of regular languages that can be accepted by REV-DFAs in terms of their minimal DFAs. By this characterization an NL-complete decidability algorithm is shown, which is based on checking for the absence of forbidden patterns in the state graph. Then in Section 4 we turn to the minimality of REV-DFAs. First a structural characterization of minimal REV-DFAs is given. Again, this characterization allows to establish an NL-complete algorithm that decides whether a given DFA is already a minimal REV-DFA among all equivalent REV-DFAs. A further result is the construction of a minimal REV-DFA out of a given DFA that accepts a reversible language. Finally, this method is used to reconsider the example given in [8] and to improve the lower bound derived there to its maximum. Then we give a new family of binary witness languages that yield a better lower bound in order of Φ^n, where Φ is the golden ratio. This bound can be increased by larger alphabets, it has a limit of $\Omega(2^{n-1})$ as $|\Sigma|$ tends to infinity. Finally, our results allow to determine an upper bound of 2^{n-1} states for the conversion of DFAs to minimal REV-DFAs, even for arbitrary alphabet sizes.

2   Preliminaries

An *alphabet* Σ is a non-empty finite set, its elements are called *letters* or *symbols*. We write Σ^* for the *set of all words* over the finite alphabet Σ.

We recall some definitions on finite automata as contained, for example, in [7]. A *deterministic finite automaton* (DFA) is a system $M = \langle S, \Sigma, \delta, s_0, F \rangle$, where S is the finite set of *internal states*, Σ is the alphabet of *input symbols*, $s_0 \in S$ is the *initial state*, $F \subseteq S$ is the set of *accepting states*, and $\delta \colon S \times \Sigma \to S$ is the partial *transition function*. Note, that here the transition function is not required to be *total*. The *language accepted* by M is $L(M) = \{\, w \in \Sigma^* \mid \delta(s_0, w) \in F \,\}$, where the transition function is recursively extended to $\delta \colon S \times \Sigma^* \to S$. By $\delta^R \colon S \times \Sigma \to 2^S$, with $\delta^R(q, a) = \{\, p \in S \mid \delta(p, a) = q \,\}$, we denote the *reverse* transition function of δ. Similarly, also δ^R can be extended to words instead of symbols. Two devices M and M' are said to be *equivalent* if they accept the same language, that is, $L(M) = L(M')$. In this case we simply write $M \equiv M'$.

Let $M = \langle S, \Sigma, \delta, s_0, F \rangle$ be a DFA accepting the language L. The set of words $R_{M,q} = \{\, w \in \Sigma^* \mid \delta(q, w) \in F \,\}$ refers to the *right language* of the state q in M. In case $R_{M,p} = R_{M,q}$, for some states $p, q \in S$, we say that p and q are *equivalent* and write $p \equiv_M q$. The equivalence relation $\equiv_M$ partitions the state set S of M into equivalence classes, and we denote the equivalence class of $q \in S$ by $[q] = \{\, p \in S \mid p \equiv_M q \,\}$. Equivalence can also be defined between states of different automata: two states p and q of DFAs M and, respectively, M' are *equivalent*, denoted by $p \equiv q$, if $R_{M,p} = R_{M',q}$.

A state $p \in S$ is *accessible* in M if there is a word $w \in \Sigma^*$ such that $\delta(s_0, w) = p$, and it is *productive* if there is a word $w \in \Sigma^*$ such that $\delta(p, w) \in F$. If p is both accessible and productive then we say that p is *useful*. In this paper we only consider automata with all states useful. Let M and M' be two DFAs with $M \equiv M'$. Observe that if p is a useful state in M, then there exists a useful state p' in M', with $p \equiv p'$. A DFA is *minimal* (among all DFAs) if there does not exist an equivalent DFA with fewer states. It is well known that a DFA is minimal if and only if all its states are useful and no pair of states is equivalent.

Next we define reversible DFAs. Let $M = \langle S, \Sigma, \delta, s_0, F \rangle$ be a DFA. A state $r \in S$ is said to be *irreversible* if there are two distinct states p and q in S and a letter $a \in \Sigma$ such that $\delta(p, a) = r = \delta(q, a)$. Then a DFA is *reversible* if it does not contain any irreversible state. In this case the automaton is said to be a *reversible* DFA (REV-DFA). Equivalently the DFA M is reversible, if every letter $a \in \Sigma$ induces an *injective partial mapping* from S to itself *via* the mapping $\delta_a : S \to S$ with $p \mapsto \delta(p, a)$. In this case, the reverse transition function δ^R can then be seen as a (partial) injective function $\delta^R : S \times \Sigma \to S$. Notice that if p and q are two distinct states in a REV-DFA, then $\delta(p, w) \neq \delta(q, w)$, for all words $w \in \Sigma^*$. Finally, a REV-DFA is *minimal* (among all REV-DFAs) if there is no equivalent REV-DFA with a smaller number of states.

Example 1. Consider the finite language $L = \{aa, ab, ba\}$. The minimal DFA and a REV-DFA for this language are shown in Figure 1. Obviously, the minimal DFA is *not* reversible, since it contains the irreversible state 3. Moreover, it is also easy to see that the REV-DFA shown is minimal. Here minimality is meant with respect to all equivalent REV-DFAs. Note that redirecting the b-transition connecting state 1 and 3 in the REV-DFA to become a transition from state 1 to 4 results in a minimal REV-DFA as well.   □

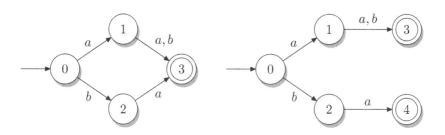

Fig. 1. The minimal DFA (left) and a minimal REV-DFA (right) for the finite language $L = \{aa, ab, ba\}$. Thus, L is a reversible language.

Finally we need some notations on computational complexity theory. We classify problems on REV-DFAs with respect to their computational complexity. Consider the complexity class NL which refers to the set of problems accepted by nondeterministic logspace bounded Turing machines.

To describe some of our algorithms we make use of nondeterministic space bounded oracle Turing machines, where the oracle tape is written deterministically. This oracle mechanism is known as RST-relativization in the literature [22]. If L is a set, we denote by $\mathsf{NL}^{\langle L \rangle}$ the class of languages accepted by nondeterministic logspace bounded RST oracle Turing machines with L oracle, and if C is a family of language, then $\mathsf{NL}^{\langle C \rangle} = \bigcup_{L \in C} \mathsf{NL}^{\langle L \rangle}$. Note that whenever C is a subset of NL, then $\mathsf{NL}^{\langle C \rangle} \subseteq \mathsf{NL}$. This is due to the well-known fact that NL is closed under complementation [10,23], that is, $\mathsf{NL} = \mathsf{coNL}$, where coNL is the set of complements of languages from NL.

Further, hardness and completeness are always meant with respect to deterministic logspace bounded reducibility, unless otherwise stated.

3   Deciding the Reversibility of a Regular Language

We consider the problem to decide whether a given regular language is reversible, that is, it is accepted by a REV-DFA. Observe, that the minimal DFA for a language need not be reversible, although the language is accepted by a REV-DFA. This is seen by Example 1. Checking reversibility for the notion of [1] is trivial, because it boils down to verify the reversibility of the minimal DFA for the language, which must have a unique final state. Hence, the language from Example 1 is *not* reversible in the sense of [1]. On the other hand, the notion of reversibility proposed in [21] is also decidable, but by a more involved argument on the syntactic monoid of the language under consideration. We prove the following structural characterization of regular languages that can be accepted by REV-DFAs in terms of their minimal DFAs. The conditions of the characterization are illustrated in Figure 2.

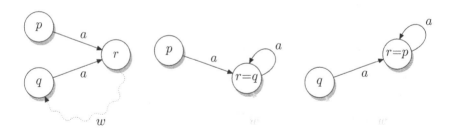

Fig. 2. The "forbidden pattern" of Theorem 2: the language accepted by a minimal DFA M can be accepted by a REV-DFA if and only if M does not contain the structure depicted on the left. Here the states p and q must be distinct, but state r could be equal to state p or state q. The situations where $r = q$ or $r = p$ are shown in the middle and on the right, respectively—here the word w and its corresponding path are grayed out because they are not relevant: in the middle, the word w that leads from r to q is not not relevant since it can be identified with the a-loop on state $r = q$. Also on the right hand side, word w is not important because we can simply interchange the roles of the states q and $r = p$.

Theorem 2. *Let* $M = \langle S, \Sigma, \delta, s_0, F \rangle$ *be a* minimal *deterministic finite automaton. The language* $L(M)$ *can be accepted by a* reversible *deterministic finite automaton if and only if there do not exist useful states* $p, q \in S$, *a letter* $a \in \Sigma$, *and a word* $w \in \Sigma^*$ *such that* $p \neq q$, $\delta(p, a) = \delta(q, a)$, *and* $\delta(q, aw) = q$.

By this characterization it is now easy to see that, e.g., both languages a^*ba^* and b^*ab^* are reversible, but their union is *not* reversible—obviously this union is a reversible language in the sense of [21]. Next we prove Theorem 2 by the upcoming two lemmata,

Lemma 3. *Let $M = \langle S, \Sigma, \delta, s_0, F \rangle$ be a deterministic finite automaton. If there exist useful states $p, q \in S$, a letter $a \in \Sigma$ and a word $w \in \Sigma^*$ such that $p \not\equiv q$, $\delta(p, a) \equiv \delta(q, a)$, and $\delta(q, aw) \equiv q$, then the language $L(M)$ cannot be accepted by a reversible deterministic finite automaton.*

Proof. Assume $M' = \langle S', \Sigma, \delta', s_0', F' \rangle$ is a REV-DFA with $L(M') = L(M)$, then of course $s_0' \equiv s_0$. Since the states p and q are useful, there must also be states $p', q' \in S'$ with $p' \equiv p$ and $q' \equiv q$. Thus, the relations $p' \not\equiv q'$, $\delta'(p', a) \equiv \delta'(q', a)$, and $\delta'(q', aw) \equiv q'$ must also hold in the REV-DFA M'. Let us now consider the sequence of states $\delta'(p', (aw)^i)$, for $i \geq 0$. From the equivalences $\delta'(p', a) \equiv \delta'(q', a)$ and $\delta'(q', aw) \equiv q'$, we conclude $\delta'(p', (aw)^i) \equiv q'$, for all $i \geq 1$. Thus, except for the first state p', all states of the above sequence are equivalent to q'. Notice that state p' cannot be equivalent to the other states of the sequence since $p' \not\equiv q'$. Since the number of states of M' must be finite, there must be a loop in the considered state sequence. This means that there must be integers $k \geq 0$ and $\ell \geq 1$ such that $\delta(p', (aw)^k) = \delta(p', (aw)^{k+\ell})$, and such that all states in the sequence $\delta(p', (aw)^0)$, $\delta(p', (aw)^1)$, $\ldots$, $\delta(p', (aw)^{k+\ell-1})$ are pairwise distinct. In fact we know that $k \geq 1$ because $\delta'(p', (aw)^{k+\ell}) \equiv q'$ cannot even be equivalent to state $\delta'(p', (aw)^0) = p'$. But now we have found two distinct states $\delta'(p', (aw)^{k-1})$ and $\delta'(p', (aw)^{k+\ell-1})$ that both map to the same state $\delta'(p', (aw)^k)$ on reading the input aw. This is a contradiction to M' being reversible, hence $L(M)$ cannot be accepted by any REV-DFA. □

When considering only minimal DFAs, the equivalences between states in Lemma 3 become equalities, so we obtain one implication of Theorem 2. Now let us prove that also the reverse implication is true. The idea how to make a given DFA reversible is very intuitive: as long as there is an irreversible state, copy this state and all states reachable from it, and distribute the incoming transitions to the new copies. The absence of the "forbidden pattern" ensures that this procedure eventually comes to an end.

For the proof of our next result we use the following notion. The state set S of a DFA $M = \langle S, \Sigma, \delta, s_0, F \rangle$ can be partitioned into *strongly connected components*: such a component is an inclusion maximal subset $C \subseteq S$ such that for all pairs of states $(p, q) \in C \times C$ there is a word $w \in \Sigma^*$ leading from p to q. Notice that also a single state q may constitute a strongly connected component, even if there is no looping transition on q.

Lemma 4. *Let $M = \langle S, \Sigma, \delta, s_0, F \rangle$ be a minimal deterministic finite automaton. If there do not exist useful states $p, q \in S$, a letter $a \in \Sigma$ and a word $w \in \Sigma^*$ such that $p \neq q$, $\delta(p, a) = \delta(q, a)$, and $\delta(q, aw) = q$, then the language $L(M)$ can be accepted by a reversible deterministic finite automaton.*

Proof. We show how to convert the DFA M into an equivalent REV-DFA. First we build a topological order $\preceq$ of the strongly connected components of M, such that if $C_1 \preceq C_2$, for two such components C_1 and C_2, then no state in C_1 can be reached from a state in C_2. Consider a minimal (with respect to $\preceq$) strongly connected component C_k that contains an irreversible state—if no such component exists then the automaton is reversible. To determine the number of necessary copies of C_k, compute

$$\alpha = \max\{\, \big|\delta^R(r,a)\big| \mid r \in C_k, a \in \Sigma \,\}. \tag{1}$$

Now we replace the component C_k by α copies of C_k and redistribute all incoming transitions among these copies, such that no state in the copies of C_k is the target of two or more transitions on the same letter. Notice that all transitions that witness the irreversibility of states in C_k come from outside of C_k, because if there were states $p, q, r \in S$ and a letter $a \in \Sigma$ with $\delta(p,a) = \delta(q,a) = r$ and $q, r \in C_k$ then M would have the "forbidden pattern" since $\delta(q, aw) = q$ for some $w \in \Sigma^*$. Therefore, the copies of C_k do not contain irreversible states.

Since also the transitions from states in the component C_k to states outside of C_k are copied, of course previously reversible states directly "behind" the copies of C_k could now become irreversible. However, in this way we only introduce irreversible states in components that are of higher rank in the topological order $\preceq$. Moreover, the obtained automaton is still equivalent to the original one. Therefore the described procedure can be applied iteratively, each time enlarging the minimal $\preceq$-rank of components that contain irreversible states, which eventually leads to a reversible DFA for $L(M)$.   □

For an example explaining the previous construction in further detail we refer to the upcoming Example 9—there all strongly connected components are singleton sets, but it is easy to see how the construction works for larger size components as well. Now we have proven Theorem 2. In fact, we will later see that the automaton constructed in the proof of Lemma 4 even is a *minimal* REV-DFA.

It can be shown that the regular language reversibility problem—given a DFA M, decide whether $L(M)$ is accepted by any REV-DFA—is NL-complete. The idea of the proof is to decide in NL whether a given DFA $M = \langle S, \Sigma, \delta, s_0, F \rangle$ accepts a *non-reversible* language with the help of Theorem 2, by witnessing the forbidden pattern depicted in Figure 2. Since NL is closed under complementation the containment of the reversibility problem within NL follows.

Theorem 5. *The regular language reversibility problem is NL-complete.*   □

4   Minimal Reversible Deterministic Finite Automata

We recall that it is well know that the minimal DFA accepting a given regular language is unique up to isomorphism. So there is the natural question asking for the relations between minimality and reversibility. It turned out that in this connection the different notions of reversibility do matter. For instance, Example 1 already shows that minimal REV-DFAs are not unique (even not up to isomorphism) in general. In [21] it is mentioned that a language L is accepted by a bideterministic finite automaton if and only if the minimal finite automaton of L is reversible and has a unique final state. This answers the question about the notion of reversibility in [1]. However, for the other notions of reversibility considered, the *minimal reversible* finite automaton for some language can be exponentially larger than the minimal automaton. In [8] finite witness languages are given that require $6n + 1$ states for a minimal DFA, but $\Omega(1.001^n)$ states for a minimal *reversible* DFA. Before we turn to determine the exact number of states for this example as well as an improved lower bound, first we derive a structural characterization of minimal REV-DFAs.

Theorem 6. *Let* $M = \langle S, \Sigma, \delta, s_0, F \rangle$ *be a reversible deterministic finite automaton with all states useful. Then M is minimal if and only if for every equivalence class* $[q_1] = \{q_1, q_2, \ldots, q_n\}$ *in S, with $n > 1$, there exists a word $w \in \Sigma^+$ such that $\delta^R(q_i, w)$ is defined for $1 \leq i \leq n$, and $\delta^R(q_k, w) \not\equiv \delta^R(q_\ell, w)$, for some k and ℓ with $1 \leq k, \ell \leq n$.* □

With the characterization of minimal REV-DFAs as stated in the previous theorem we are ready to prove that deciding minimality for these devices is NL-complete, and thus computationally not too complicated.

Theorem 7. *Deciding whether a given deterministic finite automaton M is already a minimal reversible deterministic finite automaton is NL-complete.*

Proof. Due to limited space, we only prove containment in NL, which seems the more interesting here than NL-hardness. Let the DFA $M = \langle S, \Sigma, \delta, s_0, F \rangle$ be given. We will prove minimality with respect to all REV-DFAs using Theorem 6. Our algorithm uses the following oracle subroutines:

(i) Is state p from the DFA M useful?
(ii) Is $p \equiv_M q$ in the DFA M?
(iii) Is $|[p]| = k$ in the DFA M?—besides M and p the problem instance contains also k in binary.

It is not hard to see that these problems and their complements can also be solved on a nondeterministic Turing machine in logspace.

Now our algorithm proceeds as follows: first the Turing machine verifies that the input is a REV-DFA, by inspecting all states and checking that for every letter a there is at most one a-transition leaving and entering the state. If this is not the case the Turing machine halts and rejects. Next, it is checked whether all states are useful. Here RST oracle queries are used. If this is not the case, the computation halts and rejects. Otherwise, we start verifying the conditions given in Theorem 6. To this end we cycle through all states $q \in S$—note that we already know that all these states are useful. Then we determine the size of $|[q]|$. This is done by cycling through all k with $1 \leq k \leq |S|$ and asking our oracle subroutine whether $|[q]| = k$ holds in M. If $k = 1$ nothing has to be done and the algorithm proceeds with the next q. Otherwise, let $k > 1$, and the algorithm has to verify the property stated in Theorem 6. Therefore we nondeterministically guess a word $w = av$ in reversed order on the fly letter by letter. In case $v = b_1 b_2 \cdots b_m$ with $b_i \in \Sigma$, for $1 \le i \le m$, then the machine guesses b_m, $b_{m-1}, \ldots, b_1$ and a in this order. Then for the letter b_m we deterministically compute $q' = \delta^R(q, b_m)$ and verify (i) that $|[q']| = k$ and (ii) that $\delta^R(p, b_m)$ is defined for every state p in $[q]$. Notice that in this case, every state from the equivalence class $[q']$ enters the equivalence class $[q]$ on input b_m. Again, both questions can be answered with the help of oracles on a RST oracle Turing machine. Then we continue the backward computation of M with state q' and the letter b_{m-1} proceeding as just described above. This step by step backward computation continues until we reach state q'' with the next to last guessed letter b_1. Finally, reading letter a backward must result in a situation that the condition of Theorem 6 is fulfilled. This means that $\delta^R(q'', a)$ is defined and (i) results in an equivalence class that is strictly smaller than k and (ii) moreover, $\delta^R(p, a)$ is defined for every state p in $[q'']$. As above these questions are answered with the help of the oracles described above. The Turing machine halts and rejects if any of these oracle questions is not answered appropriately. Then the equivalence class $[q]$ satisfies the condition of Theorem 6 *via* the witness $w = av$, and the Turing machine proceeds with the next q in order.

If we have found witnesses for all equivalence classes $[q]$, for all states q in S, then Turing machine halts and accepts. Otherwise, it halts and rejects. It is not hard to see that the described algorithm can be implemented on a nondeterministic logspace bounded RST oracle Turing machine. Thus, we can decide minimality of REV-DFAs in $\mathsf{NL}^{\langle \mathsf{NL} \rangle} = \mathsf{NL}$.   □

A closer look on the construction of a REV-DFA from a given minimal DFA in the proof of Lemma 4 reveals that the constructed automaton satisfies the condition given in Theorem 6, and thus, is a minimal REV-DFA.

Lemma 8. *Let $M = \langle S, \Sigma, \delta, s_0, F \rangle$ be a minimal deterministic finite automaton and $M' = \langle S', \Sigma, \delta', s_0', F' \rangle$ the reversible deterministic finite automaton constructed from M as in the proof of Lemma 4. Then M' is a minimal reversible deterministic finite automaton.*   □

Now we are prepared to derive lower bounds on the number of states for minimal reversible DFAs. The currently best known lower bound $\Omega(1.001^n)$ originates in [8]. It relies on the $2n$-fold concatenation L^{2n} of the finite language $L = \{aa, ab, ba\}$—see Figure 3. Using our technique for constructing a minimal REV-DFA, one can derive the exact number of states of a minimal REV-DFA for the language L^{2n}, which is $2^{2n+2} - 3$. Since the minimal DFA for L^{2n} has $6n + 1$ states, the blow-up in the number of states is in the order of $2^{n/3} = (\sqrt[3]{2})^n$, which is approximately 1.259^n. In our next example we present a better lower bound which is related to the Fibonacci numbers, and thus is approximately 1.618^n, the golden ratio Φ to the power of n.

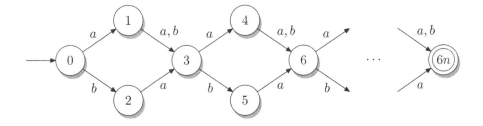

Fig. 3. The minimal DFA accepting the language L^{2n}, for $n \geq 1$

Example 9. Let $n \geq 3$ and consider the DFA $M_n = \langle S, \Sigma, \delta, s_0, F \rangle$ with state set $S_n = \{1, 2, \ldots, n\}$, initial state $s_0 = 1$, final state $F_n = \{n\}$, and transition function δ_n given through:

$$\delta_n(s, a) = \begin{cases} s+1 & \text{if } s \leq n-1 \text{ and } s \text{ is odd,} \\ s+2 & \text{if } s \leq n-2 \text{ and } s \text{ is even,} \end{cases}$$

$$\delta_n(s, b) = \begin{cases} s+2 & \text{if } s \leq n-2 \text{ and } s \text{ is odd,} \\ s+1 & \text{if } s \leq n-1 \text{ and } s \text{ is even.} \end{cases}$$

Figure 4 shows an example of the automaton M_n for $n = 6$. Notice that no transitions are defined in state n, and only one transition is defined in state $n - 1$. Clearly, the DFA M_n is minimal, but not reversible. However, since the language $L(M_n)$ is finite, one readily sees that it can be accepted by a REV-DFA.

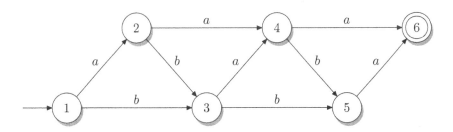

Fig. 4. The minimal DFA M_n, for $n = 6$, where the minimal REV-DFA needs $\sum_{i=1}^{n} F_i = F_{n+2} - 1$ states

Let us apply the construction from the proof of Lemma 4 to construct an equivalent REV-DFA, which, by Lemma 8 is a minimal REV-DFA. The topological order $\preceq$ of the strongly connected components of M_n clearly is the natural order $1 \preceq 2 \preceq \cdots \preceq n$. States 1 and 2 do not need to be copied, but we need two copies of state 3 because of its two predecessor states 1 and 2 by letter b. Then we need three copies of state 4 because of its three predecessors by letter a, namely state 2 and two copies of state 3. It is clear how this continues: every state s of M_n with $s \geq 3$ has two predecessors $s - 1$ and $s - 2$ either on letter a (if s is even) or letter b (if s is odd). Therefore the number of copies of state s is the sum of the number of copies of $s - 1$ and those of $s - 2$. Since we start with one copy of state 1 and one copy of state 2, the number of copies of a state $s \in S_n$ in the minimal REV-DFA for $L(M_n)$ is exactly F_s, the s-th Fibonacci number. Therefore the number of states of the minimal REV-DFA is $\sum_{i=1}^{n} F_n$. This is equal to $F_{n+2} - 1$. From the closed form

$$F_n = \frac{1}{\sqrt{5}} \cdot \left(\frac{1 + \sqrt{5}}{2}\right)^n - \frac{1}{\sqrt{5}} \cdot \left(\frac{1 - \sqrt{5}}{2}\right)^n$$

and the fact that $\left(\frac{1 - \sqrt{5}}{2}\right)^n$ tends to zero, for large n, we see that state blow-up when transforming M into an equivalent REV-DFA is in the order of $\left(\frac{1 + \sqrt{5}}{2}\right)^n$, that is, approximately 1.618^n. ☐

Thus, we have shown the following theorem.

Theorem 10. *For every n with $n \geq 3$ there is an n-state DFA M_n over a binary input alphabet accepting a reversible language, such that any equivalent REV-DFA needs at least $\Omega(\Phi^n)$ states with $\Phi = (1 + \sqrt{5})/2$, the golden ratio.* ☐

It is worth mentioning that the lower bound of Example 9 is for a binary alphabet. It can be increased at the cost of more symbols. For a k-ary alphabet one can derive the lower bound from the k-ary Fibonacci function $F_n = F_{n-1} + F_{n-2} + \cdots + F_{n-k}$. For $k = 3$ the lower bound is of order 1.839^n and for $k = 4$ it is of order 1.927^n. For growing alphabet sizes the bound asymptotically tends to 2^{n-1}, that is, $\Omega(2^{n-1})$.

Finally, our techniques allow us to determine an upper bound of 2^{n-1} states for the conversion from DFAs to equivalent REV-DFAs, even for arbitrary alphabet sizes.

Theorem 11. *Let M be a minimal deterministic finite automaton with n states, that accepts a reversible language. Then a minimal reversible deterministic finite automaton for $L(M)$ has at most 2^{n-1} states.* ☐

References

1. Angluin, D.: Inference of reversible languages. J. ACM **29**, 741–765 (1982)
2. Axelsen, H.B.: Reversible Multi-head Finite Automata Characterize Reversible Logarithmic Space. In: Dediu, A.-H., Martín-Vide, C. (eds.) LATA 2012. LNCS, vol. 7183, pp. 95–105. Springer, Heidelberg (2012)
3. Axelsen, H.B., Glück, R.: A simple and efficient universal reversible turing machine. In: Dediu, A.-H., Inenaga, S., Martín-Vide, C. (eds.) LATA 2011. LNCS, vol. 6638, pp. 117–128. Springer, Heidelberg (2011)

4. Bennett, C.H.: Logical reversibility of computation. IBM J. Res. Dev. **17**, 525–532 (1973)
5. Cho, S., Huynh, D.T.: The parallel complexity of finite-state automata problems. Inform. Comput. **97**, 1–22 (1992)
6. García, P., Vázquez de Parga, M., López, D.: On the efficient construction of quasi-reversible automata for reversible languages. Inform. Process. Lett. **107**, 13–17 (2008)
7. Harrison, M.A.: Introduction to Formal Language Theory. Addison-Wesley (1978)
8. Héam, P.C.: A lower bound for reversible automata. RAIRO Inform. Théor. **34**, 331–341 (2000)
9. Holzer, M., Jakobi, S.: Minimal and Hyper-Minimal Biautomata. In: Shur, A.M., Volkov, M.V. (eds.) DLT 2014. LNCS, vol. 8633, pp. 291–302. Springer, Heidelberg (2014)
10. Immerman, N.: Nondeterministic space is closed under complement. SIAM J. Comput. **17**, 935–938 (1988)
11. Kobayashi, S., Yokomori, T.: Learning approximately regular languages with reversible languages. Theoret. Comput. Sci. **174**, 251–257 (1997)
12. Kutrib, M.: Aspects of Reversibility for Classical Automata. In: Calude, C.S., Freivalds, R., Kazuo, I. (eds.) Gruska Festschrift. LNCS, vol. 8808, pp. 83–98. Springer, Heidelberg (2014)
13. Kutrib, M., Malcher, A.: Reversible Pushdown Automata. In: Dediu, A.-H., Fernau, H., Martín-Vide, C. (eds.) LATA 2010. LNCS, vol. 6031, pp. 368–379. Springer, Heidelberg (2010)
14. Kutrib, M., Malcher, A.: One-Way Reversible Multi-head Finite Automata. In: Glück, R., Yokoyama, T. (eds.) RC 2012. LNCS, vol. 7581, pp. 14–28. Springer, Heidelberg (2013)
15. Kutrib, M., Malcher, A., Wendlandt, M.: Reversible queue automata. In: Non-Classical Models of Automata and Applications (NCMA 2014). books@ocg.at, vol. 304, pp. 163–178. Austrian Computer Society, Vienna (2014)
16. Landauer, R.: Irreversibility and heat generation in the computing process. IBM J. Res. Dev. **5**, 183–191 (1961)
17. Lombardy, S.: On the Construction of Reversible Automata for Reversible Languages. In: Widmayer, P., Triguero, F., Morales, R., Hennessy, M., Eidenbenz, S., Conejo, R. (eds.) ICALP 2002. LNCS, vol. 2380, pp. 170–182. Springer, Heidelberg (2002)
18. Morita, K.: Two-way reversible multi-head finite automata. Fund. Inform. **110**, 241–254 (2011)
19. Morita, K.: A Deterministic Two-Way Multi-head Finite Automaton Can Be Converted into a Reversible One with the Same Number of Heads. In: Glück, R., Yokoyama, T. (eds.) RC 2012. LNCS, vol. 7581, pp. 29–43. Springer, Heidelberg (2013)
20. Morita, K., Shirasaki, A., Gono, Y.: A 1-tape 2-symbol reversible Turing machine. Trans. IEICE **E72**, 223–228 (1989)
21. Pin, J.E.: On reversible automata. In: Latin 1992: Theoretical Informatics. LNCS, vol. 583, pp. 401–416. Springer (1992)
22. Ruzzo, W.L., Simon, J., Tompa, M.: Space-bounded hierarchies and probabilistic computations. J. Comput. System Sci. **28**, 216–230 (1984)
23. Szelepcsényi, R.: The method of forced enumeration for nondeterministic automata. Acta Inform. **26**, 279–284 (1988)
24. Wagner, K., Wechsung, G.: Computational Complexity. Reidel (1986)

Multi-sequential Word Relations

Ismaël Jecker[✉] and Emmanuel Filiot

Université Libre de Bruxelles, Brussel, Belgium
ijecker@gmaiil.com

Abstract. Rational relations are binary relations of finite words that are realised by non-deterministic finite state transducers (NFT). A particular kind of rational relations is the sequential functions. Sequential functions are the functions that can be realised by input-deterministic transducers. Some rational functions are not sequential. However, based on a property on transducers called the twinning property, it is decidable in PTIME whether a rational function given by an NFT is sequential. In this paper, we investigate the generalisation of this result to multi-sequential relations, i.e. relations that are equal to a finite union of sequential functions. We show that given an NFT, it is decidable in PTIME whether the relation it defines is multi-sequential, based on a property called the fork property. If the fork property is not satisfied, we give a procedure that effectively constructs a finite set of input-deterministic transducers whose union defines the relation. This procedure generalises to arbitrary NFT the determinisation procedure of functional NFT.

Finite transducers extend finite automata with output words on transitions. Any successful computation (called run) of a transducer defines an output word obtained by concatenating, from left to right, the words occurring along the transitions of that computation. Since transitions are non-deterministic in general, there might be several successful runs on the same input word u, and hence several output words associated with u. Therefore, finite transducers can define binary relations of finite words, the so-called class of rational relations [6,9,14]. Unlike finite automata, the equivalence problem, i.e. whether two transducers define the same relation, is undecidable [10]. This has motivated the study of different subclasses of rational relations, and their associated definability problems: given a finite transducer T, does the relation $[\![T]\!]$ it defines belong to a given class $\mathcal{C}$ of relations? We survey the most important known subclasses of rational relations.

Rational Functions. An important subclass of rational relations is the class of *rational functions*. It enjoys decidable equivalence and moreover, it is decidable whether a transducer is functional, i.e. defines a function. This latter result was first shown by Schützenberger with polynomial space complexity [16] and the complexity has been refined to polynomial time in [5,11].

I. Jecker—Author supported by the ERC inVEST (279499) project.

E. Filiot—F.R.S.-FNRS research associate (chercheur qualifié).

I. Potapov (Ed.): DLT 2015, LNCS 9168, pp. 288–299, 2015.
DOI: 10.1007/978-3-319-21500-6_23

A subclass of rational functions which enjoys good algorithmic properties with respect to evaluation is the class of *sequential functions*. Sequential functions are those functions defined by finite transducers whose underlying input automaton is deterministic (called sequential transducers). Some rational functions are not sequential. E.g., over the alphabet $\Sigma = \{a, b\}$, the function f_{swap} mapping any word of the form $u\sigma$ to σu, where $u \in \Sigma^*$ and $\sigma \in \Sigma$, is rational but not sequential, because finite transducers process input words from left-to-right, and therefore any transducer implementing that function should guess non-deterministically the last symbol of $u\sigma$. Given a functional transducer, it is decidable whether it defines a sequential function [7], based on a structural property of finite transducers called the *twinning property*. This property can be decided in PTIME [5] and therefore, deciding whether a functional transducer defines a sequential function is in PTIME. If the twinning property holds, one can "determinise" the original transducer into an equivalent sequential transducer.

It turns out that many examples of rational functions from the literature which are not sequential are *almost* sequential, in the sense that they are equal to a finite union of sequential functions. Such functions are called *multi-sequential*. For instance, the function f_{swap} is multi-sequential, as $f_{swap} = f_a \cup f_b$, where f_a is the partial sequential function mapping all words of the form ua to au (and similarly for f_b). Some rational functions are not multi-sequential, such as functions that are iterations of non-sequential functions. E.g., the function mapping $u_1 \# u_2 \# \ldots \# u_n$ to $f_{swap}(u_1) \ldots f_{swap}(u_n)$ for some separator $\#$, is not multi-sequential. Multi-sequential functions have been considered by Choffrut and Schützenberger in [8], where it is shown that multi-sequentiality for functional transducers is a decidable problem.

Contribution. We generalise the result of [8] to rational *relations* and investigate its complexity. Our main result shows that, given a finite transducer, it is decidable in PTIME whether the relation it defines is multi-sequential, i.e. is a finite union of sequential functions. Our procedure is based on a simple characterisation of non multi-sequential relations by means of a structural property, called the fork property, on finite transducers. We show that a finite transducer defines a multi-sequential relation iff it is unforked, i.e., does not satisfy the fork property. We define a procedure that decomposes into a union of sequential transducers any finite unforked transducer.

Related Works. As already mentioned, for the particular case of rational functions defined by unambiguous transducers, decidability of multi-sequentiality was shown in [8] (without complexity result). We generalise this result to arbitrary rational relations and show PTIME complexity. Restricted to functions, our decomposition procedure can handle ambiguous functional transducers without extra exponential blow-up (removing ambiguity is worst-case exponential in state-complexity). Although we apply it in a more general setting than [8], the fork property is easily seen to be equivalent to that of [8]. It was also recently used to characterise unambiguous (min,+) automata over reals that define multi-sequential functions [3] (see also [2] for a generalisation to larger classes).

Table 1. Definability problems for rational relations given by finite transducers

sequentiality	multi-sequentiality (for functions)	functionality	multi-sequentiality	finite-valuedness
PSPACE [7]	DECIDABLE [8]	PSPACE [16]	PTIME (**this paper**)	PTIME [15,17]
PTIME [5]	PTIME (**this paper**)	PTIME [5,11]		

Finite-valued rational relations are rational relations such that any input word has at most k images by the relation, for a fixed constant k. Equivalence of finite-valued rational relations is decidable [11], finite-valuedness (existence of such a k) is decidable in PTIME for rational relations, and any k-valued rational relation can be expressed as a union of k rational functions [15,17, 18]. This result can be used to decide multi-sequentiality as follows. Given a rational relation R, we begin by deciding its finite-valuedness. If it is not finite-valued, it is not multi-sequential. If it is, there exist finitely many functional transducers such that their union recognises R. If all of them are unforked, then the functions they define are multi-sequential [8], and so is R. If any of them is forked, then R is not multi-sequential (this can be proved similarly to Lemma 11). However, our main result proves that those steps are unnecessary, as it is enough to check the fork property on the initial transducer.

Finally, finitely-sequential relations have been considered in [1]. They correspond to relations that can be realised by an input-deterministic transducer whose accepting states can, at the end of the run, output additional words from a finite set. Such relations are weaker than multi-sequential relations.

The known and new results of this paper are summarised in Table 23. Due to space restriction, proofs are only sketched. We refer the reader to [13] for an extended version with proofs.

1   Rational Word Relations

We denote by $\mathbb{N}$ the set of natural numbers $\{0, 1, \dots\}$, and by $\mathcal{P}(A)$ the set of subsets of a set A, and by $\mathcal{P}_f(A)$ the set of finite subsets of A.

Words and Delays. Let Σ be a (finite) alphabet of symbols. The elements of the free monoid Σ^* are called words over Σ. The length of a word w is written $|w|$. The free monoid Σ^* is partially ordered by the prefix relation $\leq$. We denote by Σ^{-1} the set of symbols σ^{-1} for all $\sigma \in \Sigma$. Any word $u \in (\Sigma \cup \Sigma^{-1})^*$ can be reduced into a unique irreducible word $\overline{u}$ by the equations $\sigma\sigma^{-1} = \sigma^{-1}\sigma = \epsilon$ for all $\sigma \in \Sigma$. Let G_Σ be the set of irreducible words over $\Sigma \cup \Sigma^{-1}$. The set G_Σ equipped with concatenation $u.v = \overline{uv}$ is a group, called the free group over Σ. We denote by u^{-1} the inverse of u. E.g. $(a^{-1}bc)^{-1} = c^{-1}b^{-1}a$. The *delay* between two words v and w is the element $\Delta(v, w) := v^{-1}w \in G_\Sigma$. E.g., $\Delta(ab, acd) = b^{-1}cd$.

Finite Automata. A (finite) automaton over a monoid M is a tuple $\mathcal{A} = (Q, E, I, T)$ where Q is the finite set of states, $I \subset Q$ is the set of initial states,

$T \subset Q$ is the set of final states, and $E \subset Q \times M \times Q$ is the finite set of edges, or transitions, labelled by elements of the monoid M.

For all transitions $e = (q_1, m, q_2) \in E$, q_1 is called the source of e, q_2 its target and m its label. A run of an automaton is a sequence of transitions $r = e_1 \ldots e_n$ such that for every $1 \le i < n$, the target of e_i is equal to the source of e_{i+1}. We write $p \xrightarrow{m}_A q$ (or just $p \xrightarrow{m} q$ when it is clear from the context) to mean that there is a run $e_1 \ldots e_n$ such that p is the source of e_1, q the target of e_n, and m is the product of the labels of the e_i. A run is called accepting if its source is an initial state and its target is a final state. An automaton is called trim if each of its states occurs in at least one accepting run. It is well-known that any automaton can be trimmed in polynomial time. The language recognised by an automaton over a monoid M is the set of elements of M labelling its accepting runs. A Σ-automaton is an automaton over the free monoid Σ^* such that each edge is labelled by a single element σ of Σ. A Σ-automaton is called deterministic if it has a single initial state, and for all $q \in Q$ and $\sigma \in \Sigma$, there exists at most one transition labelled σ of source q.

Given two automata A_1 and A_2 over a monoid M, their disjoint union $A_1 \cup A_2$ is defined as the disjoint union of their set of states, initial states, final states and transitions. It recognises the union of their respective languages.

Finite Transducers. Let Σ and Γ two alphabets. A (finite) transducer $\mathcal{T}$ from Σ^* to Γ^* is a tuple (Q, E, I, T, f_T) such that (Q, E, I, T) is a finite automaton over the monoid $\Sigma^* \times \Gamma^*$, called the underlying automaton of $\mathcal{T}$, such that $E \subseteq Q \times \Sigma \times \Gamma^* \times Q$, and $f_T : T \to \Gamma^*$ is the final output function. In this paper, the input and output alphabets are always denoted by Σ and Γ. Hence, we just use the terminology transducer instead of transducer from Σ^* to Γ^*.

A run (resp. accepting run) of $\mathcal{T}$ is a run (resp. accepting run) of its underlying automaton. We write $p \xrightarrow{u|v} q$ instead of $p \xrightarrow{(u,v)} q$. The relation recognised by $\mathcal{T}$ is the set $[\![\mathcal{T}]\!]$ of pairs $(v, w f_T(t))$ such that $i \xrightarrow{v|w} t$ for $i \in I$ and $t \in T$. A transducer $\mathcal{T}$ is functional if $[\![\mathcal{T}]\!]$ is a function. It is called trim if its underlying automaton is trim. The input automaton of a transducer is the Σ-automaton obtained by projecting the labels of its underlying automaton on their first component. A transducer is called unambiguous if its input automaton has at most one accepting run for each word it accepts. Given a transducer $\mathcal{T}$, we denote by M_T the maximal integer $|v|$, $v \in \Gamma^*$, such that v labels a transition of $\mathcal{T}$ or $v = f_T(q)$ for some $q \in Q$.

A rational transducer is defined as a transducer, except that its transitions are labeled in $\Sigma^* \times \Gamma^*$. Rational transducers are strictly more expressive than transducers and define the class of rational relations. E.g., by using loops $q \xrightarrow{\epsilon|w} q$, a word can have infinitely many images by a rational relation. If there is no such loop, it is easily shown that rational transducers are equivalent to transducers[1].

---

[1] Transducers are sometimes called real-time in the literature, and rational transducers just transducers [14]. To avoid unecessary technical difficulties, we establish our main result for (real-time) transducers, but, as shown in Remark 9, it still holds for rational transducers.

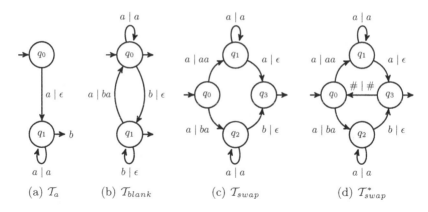

Fig. 1. Finite transducers

2   Multi-sequential Relations

In this section, we define multi-sequential relations, and give a decidable property on transducers that characterise them.

2.1   From Sequential Functions to Multi-sequential Relations

A transducer $\mathcal{T} = (Q, E, I, T, f_T)$ is *sequential* if its input automaton is deterministic. A function $f : A^* \to B^*$ is *sequential* if it can be realised by a sequential transducer[2], i.e. $f = [\![\mathcal{T}]\!]$ for some sequential transducer $\mathcal{T}$.

Let $\Sigma = \Gamma = \{a, b\}$. Fig. 1 depicts transducers that implement sequential and non-sequential functions. All states which is the target of a source-less arrow are initial, and those which are the source of an arrow without target, or whose target is a word, are accepting. The function f_a that maps any word of the form a^n, $n > 0$, to $a^{n-1}b$, is sequential. It is realised by the transducer of Fig. 1(a). The function f_{blank} replaces each block of consecutive b by a single b, and deletes the trailing b's. E.g. $f_{blank}(abbbab) = aba$. It is sequential and defined by the sequential transducer of Fig. 1(b). The function f_{swap} maps any word of the form $a^n \sigma$ to σa^n, for $\sigma \in \Sigma$. It is not sequential, because the transducer has to guess the last symbol σ. It can be defined by the transducer of Fig. 1(c).

Sequential functions have been characterised by a structural property of the transducers defining them, called the *twinning property*. Precisely, a trim transducers with initial state q_0 is twinned iff for all states q_1, q_2, all words $u, v \in \Sigma^*$ and $u_1, v_1, u_2, v_2 \in \Gamma^*$, if $q_0 \xrightarrow{u|u_1} q_1 \xrightarrow{v|v_1} q_1$ and $q_0 \xrightarrow{u|u_2} q_2 \xrightarrow{v|v_2} q_2$, then $\Delta(u_1, u_2) = \Delta(u_1 v_1, u_2 v_2)$. E.g., by taking $u = v = a$, $u_1 = aa$, $u_2 = ba$ and $v_1 = v_2 = a$, it is easy to see that the transducer of Fig.1(c) is not twinned.

---

[2] These functions are sometimes called subsequential in the literature. We follow the terminology of [14].

Theorem 1 ([5,7]). *Let $\mathcal{T}$ be a trim transducer.*

1. *$\mathcal{T}$ is twinned iff $[\![\mathcal{T}]\!]$ is sequential.*
2. *It is decidable in PTIME whether a trim transducer is twinned.*

The following result is a folklore result which states that the difference (the delay) between the outputs of two input words is linearly bounded by the difference of their input words.

Proposition 2. *Let $\mathcal{D}$ be a sequential transducer. For all pairs $(u_1, v_1), (u_2, v_2) \in [\![\mathcal{D}]\!], |\Delta(v_1, v_2)| \leq M_{\mathcal{D}}(|\Delta(u_1, u_2)| + 2)$.*

Multi-sequential relations. The function f_{swap} is not sequential, but it is multi-sequential, in the sense that it is the union of two sequential functions f_1, f_2 such that f_1 is the restriction of f_{swap} to words in a^+a and f_2 its restriction to words in a^+b. Precisely:

Definition 3 (Multi-sequential relations). *A relation $R \subseteq \Sigma^* \times \Gamma^*$ is multi-sequential if there exist k sequential functions $f_1, \ldots, f_k$ such that $R = \bigcup_{i=1}^{k} f_i$.*

The *multi-sequentiality problem* asks, given a transducer $\mathcal{T}$, whether $[\![\mathcal{T}]\!]$ is multi-sequential. It should be clear that the answer to this problem is not always positive. Indeed, even some rational functions are not multi-sequential. It is the case for instance for the function f_{swap}^* that maps any word of the form $u_1 \# u_2 \# \ldots \# u_n$ to $f_{swap}(u_1) \# f_{swap}(u_2) \# \ldots \# f_{swap}(u_n)$, where $u_i \in \Sigma^*$ and $\# \notin \Sigma$ is a fresh symbol. This function is rational, as it can be defined by the transducer of Fig. 1(d). In this paper, we investigate the intrinsic reasons making a rational relation like f_{swap} multi-sequential and a rational relation like f_{swap}^* not. In particular, we define a weaker variant of the twinning property that characterises the multi-sequential relations by structural properties of the transducers which define them.

2.2 Fork Property

Definition 4. *Let $\mathcal{T}$ be a trim transducer and q_1, q_2 be two states of $\mathcal{T}$. We say that q_1 forks to q_2 if there exist words $u, v \in \Sigma^*$ and $u_1, u_2, v_1, v_2 \in \Gamma^*$, such that $q_1 \xrightarrow{u|u_1} q_1 \xrightarrow{v|v_1} q_1 \xrightarrow{u|u_2} q_2 \xrightarrow{v|v_2} q_2$, or graphically*

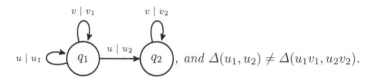

and $\Delta(u_1, u_2) \neq \Delta(u_1 v_1, u_2 v_2)$.

$\mathcal{T}$ satisfies the fork property if there are two states q_1 and q_2 of $\mathcal{T}$ such that q_1 forks to q_2. $\mathcal{T}$ is forked if it satisfies the fork property, otherwise it is unforked.

Remark 5. Any transducer satisfying the twinning property is unforked. Indeed, suppose that $\mathcal{T}$ satisfies the twinning property. We show that in any pattern depicted in Definition 4, we immediately get $\Delta(u_1, u_2) = \Delta(u_1v_1, u_2v_2)$. Indeed, since $\mathcal{T}$ is trim, there exist words $(x, x') \in \Sigma^* \times \Gamma^*$ such that $q_0 \xrightarrow{x|x'} q_1$, where q_0 is the initial state of $\mathcal{T}$. Then, we have $q_0 \xrightarrow{xu|x'u_1} q_1 \xrightarrow{v|v_1} q_1$ and $q_0 \xrightarrow{xu|x'u_2} q_2 \xrightarrow{v|v_2} q_2$. Since $\mathcal{T}$ satisfies the twinning property, we get that $\Delta(x'u_1, x'u_2) = \Delta(x'u_1v_1, x'u_2v_2)$, and therefore $\Delta(u_1, u_2) = \Delta(u_1v_1, u_2v_2)$.

Theorem 6 (Choffrut, Schützenberger [8]). *Let $\mathcal{T}$ be a (functional) unambiguous trim transducer. Then $[\![\mathcal{T}]\!]$ is multi-sequential iff $\mathcal{T}$ is unforked.*

2.3   Main Result

We generalise Theorem 6 to arbitrary transducers, and show the fork property to be decidable in polynomial time.

Theorem 7 (Main Result). *Let $\mathcal{T}$ be a trim transducer.*

1. *$[\![\mathcal{T}]\!]$ is multi-sequential iff $\mathcal{T}$ is unforked.*
2. *It is decidable in PTIME whether a trim transducer is forked.*

Deciding the fork property is done with a reversal-bounded counter machine, whose emptiness is known to be decidable in PTIME [12] (see [13]). The proof of Theorem 7.1 is done via two lemmas, Lemma 10 and 12, that are shown in the rest of this paper. An immediate consequence of this theorem and the fact that any transducer can be trimmed in polynomial time, is the following corollary:

Corollary 8. *It is decidable in PTIME whether a transducer defines a multi-sequential relation.*

Remark 9. Theorem 7 is also true when $\mathcal{T}$ is a rational transducer. Indeed, if $[\![\mathcal{T}]\!]$ is multi-sequential, then it is finite-valued, and therefore there is no loop of the form $q \xrightarrow{\epsilon|w} q$. Then $\mathcal{T}$ can be transformed into an equivalent real-time transducer, while preserving the fork property. Conversely, if $\mathcal{T}$ is unforked, then there is no loop of the form $q \xrightarrow{\epsilon|w} q$, otherwise by taking $q_1 = q_2 = q$, $u = v = u_1 = v_1 = \epsilon$ and $u_2 = v_2 = w$ in the definition of the fork property, one would raise a contradiction. As before, one can transform $\mathcal{T}$ into a real-time transducer while preserving the fork property.

Lemma 10. *Let $\mathcal{T}$ be a trim transducer. If $\mathcal{T}$ is unforked, then $[\![\mathcal{T}]\!]$ is multi-sequential.*

The proof of this lemma is the goal of Sec. 3 which provides a procedure that decomposes a transducer $\mathcal{T}$ into a union of sequential transducers. This procedure generalises to relations the determinisation procedure of functional transducers. In particular, it is based on a subset construction extended with delays, and a careful analysis of the strongly connected components of $\mathcal{T}$.

The following lemma is a key result to prove the other direction of Theorem 7.1. It states that the fork property is preserved by transducer equivalence, and therefore is independent from the transducer that realises the relation.

Lemma 11. *Let $\mathcal{T}_1, \mathcal{T}_2$ be two trim transducers.*

1. *If $[\![\mathcal{T}_1]\!] \subseteq [\![\mathcal{T}_2]\!]$ and $\mathcal{T}_1$ is forked, then $\mathcal{T}_2$ is forked.*
2. *If $[\![\mathcal{T}_1]\!] = [\![\mathcal{T}_2]\!]$, then $\mathcal{T}_1$ is forked iff $\mathcal{T}_2$ is forked.*

Proof. Clearly, 2 is a consequence of 1. Let us prove 1 by contradiction. Since $\mathcal{T}_1$ is forked, there exist two states q_1 and q_2 of $\mathcal{T}_1$, an accepting run $q_0 \xrightarrow{x|x'}_{\mathcal{T}_1}$ $q_1 \xrightarrow{u|u_2}_{\mathcal{T}_1} q_2 \xrightarrow{y|y'}_{\mathcal{T}_1} q_f$ and loops $q_1 \xrightarrow{u|u_1}_{\mathcal{T}_1} q_1 \xrightarrow{v|v_1}_{\mathcal{T}_1} q_1$ and $q_2 \xrightarrow{v|v_2}_{\mathcal{T}_1} q_2$ such that for every $n \in \mathbb{N}$, $|\Delta(u_1 v_1^n, u_2 v_2^n)| \geq n$ (see [13]).

Suppose that $\mathcal{T}_2$ is *unforked*. Then, by Lemma 10, there exist sequential transducers $\mathcal{D}_1, \dots, \mathcal{D}_k$ such that $[\![\mathcal{T}_2]\!] = \bigcup_{i=1}^k [\![\mathcal{D}_i]\!]$. Let $M = max\{|u|, |v|, |y|\}$. Let b be the maximal $M_{\mathcal{D}_i}$, $1 \leq i \leq k$. Let $r = 4b(k+1)(M+1)$. For every $1 \leq i \leq k+1$, consider the pair (w_i, w_i') in $[\![\mathcal{T}_1]\!]$ defined by

$$w_i = xv^{r^k} uv^{r^{k-1}} \dots uv^{r^{i+1}} uv^{r^i} y, \qquad w_i' = x'v_1^{r^k} u_1 v_1^{r^{k-1}} \dots u_1 v_1^{r^{i+1}} u_2 v_2^{r^i} y'.$$

Since $[\![\mathcal{T}_1]\!] \subseteq [\![\mathcal{T}_2]\!]$, and $(w_i, w_i') \in [\![\mathcal{T}_1]\!]$, we have $(w_i, w_i') \in [\![\mathcal{T}_2]\!]$, hence there exists $1 \leq j \leq k$ such that $(w_i, w_i') \in [\![\mathcal{D}_j]\!]$. As there are k different $\mathcal{D}_j$ and $k+1$ pairs (w_i, w_i'), there exist $k \geq i_1 > i_2 \geq 0$ such that $(w_{i_1}, w_{i_1}'), (w_{i_2}, w_{i_2}') \in [\![\mathcal{D}_j]\!]$. By Proposition 2, $|\Delta(w_{i_1}', w_{i_2}')| \leq b(|\Delta(w_{i_1}, w_{i_2})| + 2)$. However,

$$\begin{aligned}
&b(|\Delta(w_{i_1}, w_{i_2})| + 2) \\
&= b|y^{-1} uv^{r^{i_1-1}} \dots uv^{r^{i_2}} y| + 2b \\
&\leq b|uv^{r^{i_1-1}} \dots uvy| + 2b \\
&= i_1 b|u| + b|y| + \frac{r^{i_1}-1}{r-1} b|v| + 2b \\
&\leq b(k+1)(M+1) + 2bM(r^{i_1-1}-1) \\
&\leq \frac{r}{4} + \frac{r^{i_1}-1}{4} < \frac{r^{i_1}}{2} \\
&< \frac{r^{i_1}}{2}
\end{aligned}$$

$$\begin{aligned}
&|\Delta(w_{i_1}', w_{i_2}')| \\
&= |(u_2 v_2^{r^{i_1}} y')^{-1} u_1 v_1^{r^{i_1}} \dots u_1 v_1^{r^{i_2+1}} u_2 v_2^{r^{i_2}} y'| \\
&\geq |(u_2 v_2^{r^{i_1}})^{-1} u_1 v_1^{r^{i_1}}| \\
&\quad - |u_1 v_1^{r^{i_1-1}} \dots u_1 v_1^{r^{i_2+1}} u_2 v_2^{r^{i_2}} y'| \\
&\geq r^{i_1} - b|y| - b|uv^{r^{i_1-1}} \dots uvy| \\
&> r^{i_1} - \frac{r^{i_1}}{2} \\
&= \frac{r^{i_1}}{2}
\end{aligned}$$

which is a contradiction. $\qquad\square$

Lemma 12. *Let $\mathcal{T}$ be a trim transducer. If $[\![T]\!]$ is multi-sequential, then T is unforked.*

Proof. If $[\![T]\!]$ is multi-sequential, then $\mathcal{T}$ is equivalent to a transducer $\mathcal{T}'$ given as a union of k sequential transducers $\mathcal{D}_i$ for some $k \geq 0$ with disjoint sets of states. Clearly, if each $\mathcal{D}_i$ is unforked, then so is $\mathcal{T}'$. Since the $\mathcal{D}_i$ are sequential, they satisfy the twinning property, and therefore they are unforked by Remark 5. Hence, $\mathcal{T}'$ is unforked. By Lemma 11 and since $\mathcal{T}'$ and $\mathcal{T}$ are equivalent, $\mathcal{T}$ is also unforked. $\qquad\square$

The following result implies that, in order to show that a rational relation is not multi-sequential, it suffices to exhibit a function contained in that relation, which is not multi-sequential.

Corollary 13. *Let R be a rational relation, and f a rational function such that $f \subseteq R$ and f is not multi-sequential. Then R is not multi-sequential.*

Proof. We assume that R and f are defined by transducers $\mathcal{T}$ and $\mathcal{T}_f$. The result still holds for rational transducers, for the same reasons as the one explained in Remark 9. Since f is not multi-sequential, by Theorem 7.1, $\mathcal{T}_f$ is forked. Since $f \subseteq R$, by Lemma 11 it implies that $\mathcal{T}$ is forked, and hence not multi-sequential, again by Theorem 7.1. $\qquad\square$

3   Decomposition Procedure

In this section, we show how to decompose an unforked transducer into a union of sequential transducers, via a series of constructions. For simplicity, we sometimes consider *multi-transducers*, i.e. transducers such that the function f_T maps any final state to a finite set of output words. Let $\mathcal{T} = (Q, E, I, T, f_T)$ be a transducer. Let $\sim\, \subseteq Q^2$ defined by $q_1 \sim q_2$ if q_1 and q_2 are strongly connected, i.e. if there exist a run from q_1 to q_2 and a run from q_2 to q_1. The equivalence classes of $\sim$ are called the strongly connected components (SCC) of $\mathcal{T}$. An edge of $\mathcal{T}$ is called *transient* if its source and target are in distinct SCC, or equivalently, if there exist no run from its target to its source. The condensation of $\mathcal{T}$ is the directed acyclic graph $\Psi(\mathcal{T})$ whose vertices are the SCC of $\mathcal{T}$ and whose edges are the transient edges of $\mathcal{T}$. A transducer is called *separable* if it has a single initial state and any two edges of same source and same input symbol are transient.

Split. Let $\mathcal{T} = (Q, E, I, T, f_T)$ be a transducer. Let P be the paths of the condensation $\Psi(\mathcal{T})$ starting in an SCC containing an initial state. Note that P is finite as $\Psi(\mathcal{T})$ is a DAG. For each path $p \in P$, let $\mathcal{T}_p$ be the subtransducer of $\mathcal{T}$ obtained by removing all the transient edges of $\mathcal{T}$ but the ones occurring in p. The split of $\mathcal{T}$ is the transducer $\mathsf{split}(\mathcal{T}) = \bigcup_{p \in P} \mathcal{T}_p$. Clearly,

Lemma 14. *The transducer $\mathsf{split}(\mathcal{T})$ is equivalent to $\mathcal{T}$, i.e. $[\![\mathsf{split}(\mathcal{T})]\!] = [\![\mathcal{T}]\!]$.*

If $\mathcal{T}$ is separable, then $\mathsf{split}(\mathcal{T})$ is a decomposition of $\mathcal{T}$ into sequential transducers. Since any multi-transducer can be transformed into an equivalent union of transducers over the same underlying automaton while preserving separability, we get the following result:

Lemma 15. *Let $\mathcal{T}$ be a separable multi-transducer with a single initial state. Then $[\![\mathcal{T}]\!]$ is multi-sequential.*

Determinisation. We recall the determinisation procedure for transducers, for instance presented in [4]. It extends the subset construction with delays between output words, and outputs the longest common prefix of all the output words produced on transitions on the same input symbol, and keep the remaining suffixes

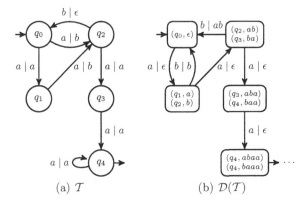

(a) $\mathcal{T}$ (b) $\mathcal{D}(\mathcal{T})$

Fig. 2. Non determinisable transducer

(delays) in the states. Precisely, let $\mathcal{T} = (Q, E, I, T, f_T)$ be a trim transducer. For every $U \in \mathcal{P}_f(Q \times \Gamma^*)$, for every $\sigma \in \Sigma$, let

$R_{U,\sigma} = \{(q, w) \in Q \times \Gamma^* | \exists (p, u) \in U, \exists (p, \sigma | v, q) \in E \text{ s.t. } w = uv\}$,
$w_{U,\sigma}$ be the largest common prefix of the words $\{w | \exists q \in Q \text{ s.t. } (q, w) \in R_{U,\sigma}\}$,
$P_{U,\sigma} = \{(q, w) | (q, w_{U,\sigma}w) \in R_{U,\sigma}\}$.

Let $\bar{\mathcal{D}}(\mathcal{T})$ be the infinite-state multi-transducer over the set of states $\mathcal{P}_f(Q \times \Gamma^*)$, with set of edges $\{(U, \sigma | w_{U,\sigma}, P_{U,\sigma}) | U \in \mathcal{P}_f(Q \times \Gamma^*), \sigma \in \Sigma\}$, initial state $U_0 = I \times \{\epsilon\}$, set of final states $\{U \in \mathcal{P}_f(Q \times \Gamma^*) | U \cap (T \times \Gamma^*) \neq \emptyset\}$, and final output relation that maps each final state U to $\{w f_T(q) | q \in T \text{ and } (q, w) \in U\}$. Note that $\bar{\mathcal{D}}(\mathcal{T})$ has a deterministic (potentially infinite) input-automaton.

The determinisation of $\mathcal{T}$, written $\mathcal{D}(\mathcal{T})$, is the trim part of $\bar{\mathcal{D}}(\mathcal{T})$. The transducer $\mathcal{D}(\mathcal{T})$ is equivalent to $\mathcal{T}$ (see [13]). It is well-known that $\mathcal{D}(\mathcal{T})$ is a (finite) sequential transducer iff $\mathcal{T}$ satisfies the twinning property.

Example 16. Fig. 2(a) depicts an unforked transducer that does not satisfy the twinning property. As a consequence, $\mathcal{D}(\mathcal{T})$ is infinite (a part of $\mathcal{D}(\mathcal{T})$ can be seen on Fig. 2(b)). The non satisfaction of the twinning property is witnessed by the two runs $q_0 \xrightarrow{aaaa|abaa} q_4 \xrightarrow{a|a} q_4$ and $q_0 \xrightarrow{aaaa|baaa} q_4 \xrightarrow{a|a} q_4$. Note that these runs do not create an instance of the fork property. The idea of the next construction, called the weak determinisation, is to keep some well-chosen non-deterministic transitions, and reset the determinisation whenever it definitively leaves an SCC (the SCC $\{q_0, q_1, q_2\}$ in this example). We explain this procedure when there is a single initial state, as any transducer can be easily transformed into a finite union of transducers with single initial states.

Weak Determinisation. Let $\mathcal{T} = (Q, E, I, T, f_T)$ be a trim transducer with a single initial state. For every $U \in \mathcal{P}_f(Q \times \Gamma^*)$, let the rank n_U of U be the set containing all the SCC of $\mathcal{T}$ accessible from the states q components of an element of U. The multi-transducer $\bar{\mathcal{W}}(\mathcal{T})$ is obtained from $\bar{\mathcal{D}}(\mathcal{T})$ by splitting the edges

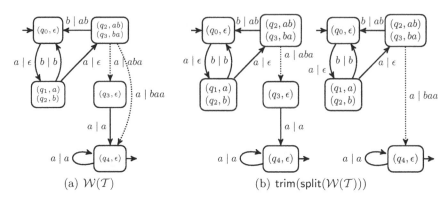

(a) $\mathcal{W}(\mathcal{T})$ (b) trim(split($\mathcal{W}(\mathcal{T})$))

Fig. 3. Weak determinisation and split

that do not preserve the rank, as follows. If $(U, u|v, U')$ is an edge of $\bar{\mathcal{D}}(\mathcal{T})$ such that $n_{U'}$ is strictly included in n_U, it is removed, and replaced by the set of edges $\{(U, u|vw, \{(q, \epsilon)\})|(q, w) \in U'\}$. Any pair of distinct edges of the form $U \xrightarrow{a|v_1} U_1$ and $U \xrightarrow{a|v_2} U_2$ in $\bar{\mathcal{W}}(\mathcal{T})$ have necessarily been created by this transformation, as everything stays input-deterministic otherwise. Therefore, since the rank strictly decreases ($n_{U_2} \subsetneq n_U$ and $n_{U_1} \subsetneq n_U$) and can never increase in $\bar{\mathcal{W}}(\mathcal{T})$, there is no run from U_2 to U, nor from U_1 to U in $\bar{\mathcal{W}}(\mathcal{T})$, and the two edges are transient. As a consequence,

Lemma 17. *The infinite transducer $\bar{\mathcal{W}}(\mathcal{T})$ is separable.*

The weak determinisation of $\mathcal{T}$, written $\mathcal{W}(\mathcal{T})$, is the trim part of $\bar{\mathcal{W}}(\mathcal{T})$.

Proposition 18. *$\mathcal{W}(\mathcal{T})$ and $\mathcal{T}$ are equivalent. Moreover, if $\mathcal{T}$ is unforked, $\mathcal{W}(\mathcal{T})$ is finite, and it is a multi-transducer.*

The main idea of the proof is that as long as the fork property is not satisfied, the length of the words present in the states of $\mathcal{W}(\mathcal{T})$ can be bounded (see [13]).

Proof of Lemma 10. We can finally prove that every unforked transducer is multi-sequential. Let $\mathcal{T} = (Q, E, I, T, f_T)$ be an unforked transducer. Then $\mathcal{T}$ is equivalent to $\bigcup_{i \in I} \mathcal{T}_i$, where $\mathcal{T}_i$ is the transducer obtained by keeping only i as initial state. Given $i \in I$, as we just saw, $\mathcal{W}(\mathcal{T}_i)$ is a transducer equivalent to $\mathcal{T}_i$. Moreover, as $\bar{\mathcal{W}}(\mathcal{T}_i)$ is separable, so is $\mathcal{W}(\mathcal{T}_i)$, hence, by Lemma 15, it is multi-sequential. The desired result follows.

Example 19. Let us illustrate the weak determinisation on the transducer $\mathcal{T}$ of Fig. 2(a). Consider the determinisation $\mathcal{D}(\mathcal{T})$ of $\mathcal{T}$ of Fig. 2(b). When it is in state $U_1 = \{(q_2, ab), (q_3, ba)\}$, on input a, it moves to state $U_2 = \{(q_3, aba), (q_4, baa)\}$, definitely leaving the SCC $\{q_0, q_1, q_2\}$ of $\mathcal{T}$ (the rank n_{U_2} of U_2 is strictly included in the rank n_{U_1} of U_1). As a result, this transition is removed from $\bar{\mathcal{D}}(\mathcal{T})$, and replaced by the transitions $U_2 \xrightarrow{a|aba} \{(q_3, \epsilon)\}$ and $U_2 \xrightarrow{a|baa} \{(q_4, \epsilon)\}$. The resulting transducer $\mathcal{W}_\mathcal{T}$ is depicted on Fig. 3(a) (where the new transitions are dotted). Fig. 3(b) shows how the latter transducer is split into a union.

References

1. Allauzen, C., Mohri, M.: Finitely subsequential transducers. Int. J. Found. Comput. Sci. **14**(6), 983–994 (2003)
2. Bala, S.: Which Finitely Ambiguous Automata Recognize Finitely Sequential Functions? In: Chatterjee, K., Sgall, J. (eds.) MFCS 2013. LNCS, vol. 8087, pp. 86–97. Springer, Heidelberg (2013)
3. Bala, S., Koniński, A.: Unambiguous Automata Denoting Finitely Sequential Functions. In: Dediu, A.-H., Martín-Vide, C., Truthe, B. (eds.) LATA 2013. LNCS, vol. 7810, pp. 104–115. Springer, Heidelberg (2013)
4. Béal, M.-P., Carton, O.: Determinization of transducers over finite and infinite words. Theoretical Computer Science **289**(1), 225–251 (2002)
5. Béal, M.-P., Carton, O., Prieur, C., Sakarovitch, J.: Squaring transducers: an efficient procedure for deciding functionality and sequentiality. Theoretical Computer Science **292**(1), 45–63 (2003)
6. Berstel, J.: Transductions and context-free languages. http://www-igm.univ-mlv.fr/berstel/ (December 2009)
7. Choffrut, C.: Une caractérisation des fonctions séquentielles et des fonctions soussuentielles en tant que relations rationnelles. TCS **5**(3), 325–337 (1977)
8. Choffrut, C., Schützenberger, M.P.: Décomposition de fonctions rationnelles. In: STACS, pp. 213–226 (1986)
9. Elgot, C.C., Mezei, J.E.: On relations defined by generalized finite automata. IBM Journal of Research and Development **9**, 47–68 (1965)
10. Griffiths, T.V.: The unsolvability of the equivalence problem for lambda-free nondeterministic generalized machines. Journal of the ACM **15**(3), 409–413 (1968)
11. Gurari, E.M., Ibarra, O.H.: A note on finite-valued and finitely ambiguous transducers. Theory of Computing Systems **16**(1), 61–66 (1983)
12. Ibarra, O.H.: Reversal-bounded multicounter machines and their decision problems. Journal of the ACM **25**(1), 116–133 (1978)
13. Jecker, I., Filiot, E.: Multi-sequential word relations. CoRR, abs/1504.03864 (2015)
14. Sakarovitch, J.: Elements of Automata Theory. Cambridge University Press (2009)
15. Sakarovitch, J., de Souza, R.: Lexicographic decomposition of k-valued transducers. Theory of Computing Systems **47**(3), 758–785 (2010)
16. Schützenberger, M.P.: Sur les relations rationnelles. In Automata Theory and Formal Languages. LNCS, vol. 33, pp. 209–213 (1975)
17. Weber, A.: On the valuedness of finite transducers. Acta Informatica **27**(8), 749–780 (1989)
18. Weber, A.: Decomposing finite-valued transducers and deciding their equivalence. SIAM Journal on Computing **22**(1), 175–202 (1993)

The Boundary of Prefix-Free Languages

Jozef Jirásek[1] and Galina Jirásková[2]([⊠])

[1] Institute of Computer Science, Faculty of Science, P.J. Šafárik University,
Jesenná 5, 04001 Košice, Slovakia
jozef.jirasek@upjs.sk

[2] Mathematical Institute, Slovak Academy of Sciences,
Grešákova 6, 04001 Košice, Slovakia
jiraskov@saske.sk

Abstract. We investigate the boundary operation on the class of prefix-free regular languages. We show that if a prefix-free language is recognized by a deterministic finite automaton of n states, then its boundary is recognized by a deterministic automaton of at most $(n-1) \cdot 2^{n-4} + n + 1$ states. We prove that this bound is tight, and to describe worst-case examples, we use a three-letter alphabet. Next we show that the tight bound for boundary on binary prefix-free languages is $2n - 2$, and that in the unary case, the tight bound is $n - 2$.

1 Introduction

Boundary is a combined operation on languages defined to be $\mathrm{bd}(L) = L^* \cap (L^c)^*$, where L^* is the Kleene star of L, and L^c is the complement of L [1,8,9]. We investigated the boundary operation on regular languages in [6]. We proved that the boundary of a regular language recognized by a deterministic finite automaton (DFA) of n states can be accepted by a deterministic automaton of at most $3/8 \cdot 4^n + 2^{n-2} - 2 \cdot 3^{n-2} - n + 2$ states. We also proved that this upper bound is tight, and to describe worst-case examples, we used a five-letter alphabet. Then we showed that this bound cannot be met by any quaternary language, and we obtained asymptotically tight bound $\Theta(4^n)$ in the binary case. This completely solved Open Problem 15 in [9].

In this paper, we study the boundary operation on the class of prefix-free regular languages. A language is prefix-free if it does not contain two distinct strings, one of which is a prefix of the other. In prefix codes, like variable-length Huffman codes or country calling codes, there is no codeword that is a proper prefix of any other codeword. Hence a receiver can identify each codeword without any special marker between words. Motivated by prefix codes, the class of prefix-free regular languages has been recently investigated [2,3,4,5].

A minimal DFA recognizes a prefix-free language if and only if it has exactly one final state from which only the empty string is accepted [5]. Next, the Kleene

J. Jirásek—Research supported by VEGA grant 1/0142/15.

G. Jirásková—Research supported by VEGA grant 2/0084/15.

I. Potapov (Ed.): DLT 2015, LNCS 9168, pp. 300–312, 2015.
DOI: 10.1007/978-3-319-21500-6_24

star of an n-state DFA prefix-free language is accepted by a DFA which has at most n states. Moreover, a DFA for the complement of a prefix-free language has a specific structure — all but one states are final, and there is a final state which has a loop on every input symbol. We use these two facts to get the exact complexity of the boundary operation.

First, we prove that the boundary of an n-state DFA prefix-free language can be accepted by a DFA of at most $(n-1) \cdot 2^{n-4} + n + 1$ states. We also prove that this upper bound is tight, and to describe worst-case examples, we use a three-letter alphabet. Then, we show that this upper bound cannot be met by any binary prefix-free language. Moreover, we get the tight bound $2n - 2$ in the binary case, and the tight bound $n - 2$ in the unary case. Thus, our results show that the complexity of the boundary operation on prefix-free languages is given by an exponential function for every alphabet with at least three symbols, but it is given by a linear function otherwise.

2   Preliminaries

We assume that the reader is familiar with basic concepts of regular languages and finite automata. For details or unexplained notions we refer to [7,10,11].

A *nondeterministic finite automaton* (NFA) is a tuple $M = (Q, \Sigma, \cdot, s, F)$, where Q is a finite non-empty set of states, Σ is an input alphabet, $\cdot : Q \times \Sigma \to 2^Q$ is the transition function which can be extended to the domain $2^Q \times \Sigma^*$ in a natural way, $s \in Q$ is the initial state, and $F \subseteq Q$ is the set of final states. The *language accepted by NFA* M is $L(M) = \{w \in \Sigma^* \mid s \cdot w \cap F \neq \emptyset\}$.

An NFA A is a *deterministic finite automaton* (DFA), if $|q \cdot a| = 1$ for each q in Q and each a in Σ; then we write $q \cdot a = q'$ instead of $q \cdot a = \{q'\}$.

The *state complexity* of a regular language L, $\mathrm{sc}(L)$, is the number of states in the minimal DFA recognizing the language L. It is well-known that a DFA is minimal if all of its states are reachable and pairwise distinguishable.

Every NFA $A = (Q, \Sigma, \cdot, s, F)$ can be converted to an equivalent deterministic automaton $A' = (2^Q, \Sigma, \cdot', \{s\}, F')$, where $R \cdot' a = R \cdot a$ and $F' = \{R \in 2^Q \mid R \cap F \neq \emptyset\}$. We call the resulting DFA A' the *subset automaton* of the NFA A.

The *product automaton* for the intersection of two languages recognized by DFAs $A = (Q_A, \Sigma, \cdot_A, s_A, F_A)$ and $B = (Q_B, \Sigma, \cdot_B, s_B, F_B)$, is the DFA $A \times B = (Q_A \times Q_B, \Sigma, \cdot, (s_A, s_B), F)$, where $(p, q) \cdot a = (p \cdot_A a, q \cdot_B a)$ and $F = F_A \times F_B$.

If L is a regular language accepted by a DFA $A = (Q, \Sigma, \cdot, s, F)$, then we can construct an NFA N for the language L^* from the DFA A as follows: First, we add a new initial and final state q_0 going on an input a to $\{s \cdot a\}$ if $s \cdot a \notin F$, and to $\{s \cdot a, s\}$ if $s \cdot a \in F$. Then, for each state q and each symbol a such that $q \cdot a \in F$, we add the transition from q to s on a. To get an NFA for L^{c*}, we first interchange the final and non-final states in the DFA A, to get a DFA A^c for the language L^c, and then we construct an NFA N' for the language L^{c*} from the DFA A^c as described above. Let D and D' be the subset automata of the NFAs N and N', respectively. Then the boundary of L, that is, the language $L^* \cap L^{c*}$, is accepted by the product automaton $D \times D'$.

3 Construction of DFA for Boundary on PF Languages

If L is a prefix-free language, then we can simplify the construction of a DFA for $L^* \cap L^{c*}$. First, if a prefix-free language is accepted by a 1-state or a 2-state minimal DFA, then it is empty or it equals $\{\varepsilon\}$, respectively, and its boundary is equal to $\{\varepsilon\}$. Therefore in what follows, we assume that $sc(L) \geq 3$. Moreover, the minimal DFA for L has exactly one final state which goes to the dead state on every input symbol [5].

Thus let L be a prefix-free language accepted by a minimal DFA $A = (Q, \Sigma, \cdot, s, \{f\})$, where $Q = \{1, 2, \ldots, n-3\} \cup \{s, f, d\}$, and d is the dead state; if $n = 3$, then $Q = \{s, f, d\}$.

We can construct a DFA D for the language L^* from the DFA A by making the state f initial, and by replacing every transition $f \xrightarrow{a} d$ by the transition $f \xrightarrow{a} s \cdot a$ [5]. Formally, we have $D = (Q, \Sigma, \circ, f, \{f\})$, where

$$q \circ a = \begin{cases} q \cdot a, & \text{if } q \neq f, \\ s \cdot a, & \text{if } q = f. \end{cases} \tag{1}$$

The language L^c is accepted by the DFA $A^c = (Q, \Sigma, \cdot, s, Q \setminus \{f\})$. Since we assume that $|Q| \geq 3$, the initial state s is final in A^c. Next, the state d is final in A^c, and it goes to itself on every input symbol. It follows that to get an NFA N' for the language L^{c*}, we do not need to add a new initial state, and moreover, we do not need to add any new transition $q \xrightarrow{a} s$ if $q \cdot a = d$. Formally, we have $N' = (Q, \Sigma, \bullet, s, Q \setminus \{f\})$, where

$$q \bullet a = \begin{cases} \{q \cdot a\}, & \text{if } q \cdot a \in \{f, d\} \\ \{q \cdot a, s\}, & \text{if } q \cdot a \in \{s, 1, 2, \ldots, n-3\}. \end{cases} \tag{2}$$

Since we added only those transitions that are added also in a general construction for L^*, we have $L(N') \subseteq L^{c*}$. Let us show that $L^{c*} \subseteq L(N')$. The empty string is in $L(N')$. Next, if a non-empty string w is in L^{c*}, then $w = w_1 w_2 \cdots w_k$, where $k \geq 1$ and $w_i \in L^c$ for $i = 1, 2, \ldots, k$. If each w_i is accepted in A^c in a state q_i which is different from d, then w is accepted by N' since we have $s \xrightarrow{w_1} s \xrightarrow{w_2} s \xrightarrow{w_3} \cdots \xrightarrow{w_{k-1}} s \xrightarrow{w_k} q_k$ in N'. Let there be an i such that w_i is accepted by A^c in the state d, and assume that i is the smallest such index. Then in the NFA N', we have the following accepting computation on w: $s \xrightarrow{w_1} s \xrightarrow{w_2} s \xrightarrow{w_3} \cdots \xrightarrow{w_{i-1}} s \xrightarrow{w_i} d \xrightarrow{w_{i+1}w_{i+2}\cdots w_k} d$. Hence N' accepts L^{c*}.

Let D' be the subset automaton of the NFA N'. Then the language $L^* \cap L^{c*}$ is accepted by the product automaton $D \times D'$ whose states are pairs (q, T), where $q \in Q$ and $T \subseteq Q$. The initial state of the product automaton is the pair $(f, \{s\})$. A state (q, T) is final if $q = f$ and $T \cap \{s, 1, 2, \ldots, n-3, d\} \neq \emptyset$. Next, on input a, a state (q, T) goes to the state $(q \circ a, T \bullet a)$, where

$$q \circ a = \begin{cases} q \cdot a & \text{if } q \neq f, \\ s \cdot a & \text{if } q = f, \end{cases} \quad T \bullet a = \begin{cases} T \cdot a, & \text{if } T \cdot a \subseteq \{f, d\}, \\ T \cdot a \cup \{s\}, & \text{if } T \cdot a \cap \{s, 1, \ldots, n-3\} \neq \emptyset. \end{cases}$$

4   State Complexity of Boundary on PF Languages

Let us start with an upper bound on the state complexity of the boundary operation on prefix-free languages. We will need the following observation.

Proposition 1. *Let* $q \in Q$, $S \subseteq \{s, 1, 2, \ldots, n-3, f\}$, $\emptyset \neq T \subseteq \{s, 1, 2, \ldots, n-3\}$. *Then in the product automaton* $D \times D'$ *described in Section 3,*
  (i) $(q, S \cup \{d\})$ *is equivalent to* $(q, \{d\})$;
  (ii) $(f, \{f\})$ *is equivalent to* $(s, \{d\})$;
  (iii) $(q, T \cup \{f\})$ *is equivalent to* $(q, \{d\})$.

Lemma 2 (Upper Bound). *Let* L *be a prefix-free language with* $\mathrm{sc}(L) = n$, *where* $n \geq 3$. *Then* $\mathrm{sc}(L^* \cap L^{c*}) \leq (n-1) \cdot 2^{n-4} + n + 1$.

Proof. Let $D \times D'$ be the product automaton for $L^* \cap L^{c*}$ from Section 3. Let
  $\mathcal{R}_1 = \{(q, \{d\}) \mid q \in Q\}$,
  $\mathcal{R}_2 = \{(q, T) \mid q \in \{s, 1, \ldots, n-3\}, T \subseteq \{s, 1, \ldots, n-3\}, \text{ and } \{s, q\} \subseteq T\}$.
Let (q, T) be a state of $D \times D'$ which is reached from the initial state $(f, \{s\})$ by a non-empty string w. We will show that either (q, T) is equivalent to a state in $\mathcal{R}_1$, or (q, T) is in $\mathcal{R}_2$. The proof is by induction on $|w|$.

The base case is $w = a$ for some a in Σ. On symbol a, the initial state $(f, \{s\})$ goes to the state $(f \circ a, \{s\} \bullet a) = (s \cdot a, \{s\} \bullet a)$. We have three cases:

  (*i*) If $s \cdot a = d$, then $(s \cdot a, \{s\} \bullet a) = (d, \{d\})$, which is a state in $\mathcal{R}_1$.
  (*ii*) If $s \cdot a = f$, then $(s \cdot a, \{s\} \bullet a) = (f, \{f\})$. By Proposition 1, the state $(f, \{f\})$ is equivalent to the state $(s, \{d\})$ which is in $\mathcal{R}_1$.
  (*iii*) If $s \cdot a \in \{s, 1, \ldots, n-3\}$, then $(s \cdot a, \{s\} \bullet a) = (s \cdot a, \{s \cdot a, s\})$ which is a state in $\mathcal{R}_2$.

Assume that our claim holds for all states that are reachable by strings of length k, and let $w = va$, where $a \in \Sigma$, be a string of length $k + 1$. Then, by the induction hypothesis, the state (q, T) that is reached after reading v is either equivalent to a state $(p, \{d\})$ in $\mathcal{R}_1$, or it is in $\mathcal{R}_2$. In the first case, on symbol a, the state $(p, \{d\})$ goes to the state $(p \circ a, \{d\})$ which is in $\mathcal{R}_1$. In the second case, we have $q \in \{s, 1, 2, \ldots, n-3\}$, $T \subseteq \{s, 1, 2, \ldots, n-3\}$, and $\{s, q\} \subseteq T$, therefore on symbol a, the state (q, T) goes to the state $(q \cdot a, T \bullet a)$. We have three cases:

  (*i*) If $q \cdot a = d$, then $(q \cdot a, T \bullet a) = (d, T \bullet a)$, and $d \in T \bullet a$ since $q \in T$. By Proposition 1, the resulting state is equivalent to $(d, \{d\})$.
  (*ii*) If $q \cdot a = f$, then $(q \cdot a, T \bullet a) = (f, T \bullet a)$, and $f \in T \bullet a$ since $q \in T$. By Proposition 1, the resulting state is equivalent to $(s, \{d\})$ if $T \bullet a = \{f\}$, and it is equivalent to $(f, \{d\})$ otherwise.
  (*iii*) If $q \cdot a \in \{s, 1, \ldots, n-3\}$, then $(q \cdot a, T \bullet a) = (q \cdot a, T \cdot a \cup \{s\})$. Now, if $d \in T \cdot a$ or $f \in T \cdot a$, then $(q \cdot a, T \cdot a \cup \{s\})$ is equivalent to $(q \cdot a, \{d\})$ by Proposition 1. Otherwise, $T \cdot a \cup \{s\} \subseteq \{s, 1, 2, \ldots, n-3\}$ and $\{s, q \cdot a\} \subseteq T \cdot a \cup \{s\}$, so the resulting state $(q \cdot a, T \cdot a \cup \{s\})$ is in $\mathcal{R}_2$.

Thus, including the initial state $(f, \{s\})$, the product automaton has at most $|R_2| + |R_1| + 1 = (n-1) \cdot 2^{n-4} + n + 1$ reachable and distinguishable states.   □

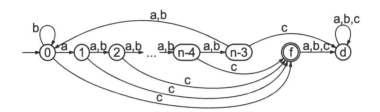

Fig. 1. The DFA of a ternary prefix-free witness for boundary meeting the upper bound $(n-1) \cdot 2^{n-4} + n + 1$

Now we prove that the upper bound given in Lemma 2 is tight. Our witness languages are defined over a three-letter alphabet.

Lemma 3 (Lower Bound). *Let $n \geq 3$. There exists a ternary prefix-free language with $\mathrm{sc}(L) = n$ such that $\mathrm{sc}(L^* \cap L^{c*}) = (n-1) \cdot 2^{n-4} + n + 1$.*

Proof. If $n = 3$, then the language b^*a over the ternary alphabet $\{a, b, c\}$ meets the upper bound $(n-1) \cdot 2^{n-4} + n + 1 = 5$.

Let $n \geq 4$. Let L be the ternary prefix-free language accepted by the DFA $A = (Q, \Sigma, \cdot, 0, \{f\})$ shown in Fig. 1 Construct the product automaton $D \times D'$ as described in Section 3; the DFA D and the NFA N' are shown in Fig. 2 and 3, respectively. Let us show that the product automaton has $(n-1) \cdot 2^{n-4} + n + 1$ reachable and pairwise distinguishable states.

We first prove reachability. The initial state of the product automaton is the state $(f, \{0\})$. Next we have

$$(f, \{0\}) \xrightarrow{c} (f, \{f\}) \xrightarrow{c} (f, \{d\}) \xrightarrow{b} (0, \{d\}),$$

$$(0, \{d\}) \xrightarrow{a^i} (i, \{d\}) \text{ for } i = 1, 2, \ldots, n - 3, \text{ and}$$

$$(n - 3, \{d\}) \xrightarrow{c} (d, \{d\}),$$

thus each state $(q, \{d\})$ with $q \in Q$ is reachable. Now we are going to prove the reachability of each state (i, T) such that $0 \leq i \leq n - 3$, $T \subseteq \{0, 1, \ldots, n - 3\}$, and $\{0, i\} \subseteq T$. The proof is by induction on the size of T. The basis, $|T| \leq 2$, holds since we have

$$(f, \{0\}) \xrightarrow{b} (0, \{0\}) \xrightarrow{a} (1, \{0, 1\}),$$

$$(1, \{0, 1\}) \xrightarrow{b^{i-1}} (i, \{0, i\}) \text{ for } i = 2, 3, \ldots, n - 3,$$

$$(n - 3, \{0, n - 3\}) \xrightarrow{a} (0, \{0, 1\}),$$

$$(0, \{0, 1\}) \xrightarrow{b^{i-1}} (0, \{0, i\}) \text{ for } i = 2, 3, \ldots, n - 3.$$

Now assume that $2 \leq k \leq n - 3$, and that for each i with $0 \leq i \leq n - 3$ and each subset T of $\{0, 1, \ldots, n - 3\}$ with $|T| = k$ and $\{0, i\} \subseteq T$, the state (i, T) is

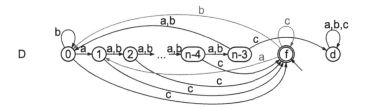

Fig. 2. The DFA D for the language L^*

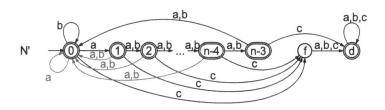

Fig. 3. The NFA N' for the language L^{c*}

reachable in the product automaton. Let $0 \le i \le n-3$. Let $T = \{0, j_1, j_2, \ldots, j_k\}$, where $1 \le j_1 < j_2 < \cdots < j_k \le n-3$, be a set of size $k+1$ such that $\{0, i\} \subseteq T$. Our aim is to show that the state (i, T) is reachable. Consider two cases:

(i) First, let $i \ne 0$. Then, since $i \in T$, we have $i = j_\ell$ for some ℓ with $1 \le \ell \le k$. Take $T' = \{0, j_2 - j_1, \ldots, j_k - j_1\}$. Then $|T'| = k$ and $\{0, j_\ell - j_1\} \subseteq T'$. It follows that $(j_\ell - j_1, T')$ is reachable by the induction hypothesis. Next,

$$(j_\ell - j_1, T') \xrightarrow{a} (j_\ell - j_1 + 1, \{0, 1, j_2 - j_1 + 1, \ldots, j_k - j_1 + 1\})$$
$$\xrightarrow{b^{j_1 - 1}} (j_\ell, \{0, j_1, j_2, \ldots, j_k\}) = (j_\ell, T) = (i, T),$$

and therefore, the state (i, T) is reachable.

(ii) Now, let $i = 0$. Take $T' = \{0, j_2 - j_1, \ldots, j_k - j_1, n - 3\}$. Then $|T'| = k+1$ and $\{0, n - 3\} \subseteq T'$, so the state $(n - 3, T')$ is reachable as shown in case (i); notice that $n - 3 \ne 0$ since $n \ge 4$. Next we have

$$(n - 3, T') \xrightarrow{a} (0, \{0, 1, j_2 - j_1 + 1, \ldots, j_k - j_1 + 1\})$$
$$\xrightarrow{b^{j_1 - 1}} (0, \{0, j_1, j_2, \ldots, j_k\}) = (0, T),$$

and therefore, the state $(0, T)$ is reachable.

Hence we have shown the reachability of $(n-1) \cdot 2^{n-4} + n + 1$ states in the product automaton $D \times D'$. Now we prove distinguishability. Denote by $\mathcal{R}$ the family of all the reachable states, that is,

$$\mathcal{R} = \{(f, \{0\})\} \cup \{(q, \{d\}) \mid q \in Q\}$$
$$\cup \{(i, T) \mid 0 \le i \le n-3, T \subseteq \{0, 1, \ldots, n-3\}, \{0, i\} \subseteq T\}.$$

Let (p, S) and (q, T) be two distinct states in $\mathcal{R}$. Consider several cases:

(i)   Let $(p, S) = (f, \{0\})$. The state $(f, \{0\})$ is accepting, so it is distinguishable from every rejecting state (q, T) with $q \ne f$. Moreover, it can be distinguished from $(f, \{d\})$ by c since c is rejected from $(f, \{0\})$ and accepted from $(f, \{d\})$.

(ii) Let $(p, S) = (d, \{d\})$. The state $(d, \{d\})$ is the only dead state since every other state (q, T) with $q \ne d$ accepts the string $b^n cc$; notice that $q \cdot b^n = 0$ if $q \in \{0, 1, \ldots, n-3\}$. Thus $(d, \{d\})$ is distinguishable from any other state.

(iii) Let $(p, S) = (f, \{d\})$. The state $(f, \{d\})$ is accepting, so it is distinguishable from every rejecting state (q, T) with $0 \le q \le n-3$.

(iv) Let $(p, S) = (i, \{d\})$ and $(q, T) = (j, \{d\})$ where $0 \le i < j \le n-3$. Then $a^{n-3-j}c$ is rejected from $(j, \{d\})$ and accepted from $(i, \{d\})$ since we have

$$(j, \{d\}) \xrightarrow{a^{n-3-j}} (n-3, \{d\}) \xrightarrow{c} (d, \{d\}), \text{ and}$$
$$(i, \{d\}) \xrightarrow{a^{n-3-j}} (n-3-(j-i), \{d\}) \xrightarrow{c} (f, \{d\}).$$

(v)   Let $(p, S) = (i, \{d\})$ with $0 \le i \le n-3$. Let $0 \le q \le n-3$, $T \subseteq \{0, 1, \ldots, n-3\}$, and $\{0, q\} \subseteq T$. Then $b^n c$ is accepted from $(i, \{d\})$ and rejected from (q, T) since we have

$$(i, \{d\}) \xrightarrow{b^n} (0, \{d\}) \xrightarrow{c} (f, \{d\}), \text{ and}$$
$$(q, T) \xrightarrow{b^n} (0, \{0\}) \xrightarrow{c} (f, \{f\}).$$

(vi) Let $0 \le p, q \le n-3$, $S, T \subseteq \{0, 1, \ldots, n-3\}$, $\{0, p\} \subseteq S$, $\{0, q\} \subseteq T$, and let $p < q$. Then by $a^{n-3-q}c$, the state (q, T) goes to the dead state while (p, S) goes to a non-dead state since we have

$$(q, T) \xrightarrow{a^{n-3-q}} (n-3, T \bullet a^{n-3-q}) \xrightarrow{c} (d, \{d\}) \text{ since } (n-3) \in T \bullet a^{n-3-q},$$
$$(p, S) \xrightarrow{a^{n-3-q}} (n-3-(q-p), S \bullet a^{n-3-q}) \xrightarrow{c} (f, S \bullet a^{n-3-q}c).$$

(vii) Finally, let $0 \le p, q \le n-3$, $S, T \subseteq \{0, 1, \ldots, n-3\}$, $\{0, p\} \subseteq S$, $\{0, q\} \subseteq T$, and $p = q$. Then $S \ne T$. Without loss of generality, there is a state i with $i \ge 1$ and $i \ne p$ such that $i \in S$ and $i \notin T$. Let us show that $a^{n-3-i}c$ is accepted from (p, S) and rejected from (q, T). Since $i \in S$ and $i \notin T$, we have $(n-3) \in S \bullet a^{n-3-i}$ and $T \bullet a^{n-3-i} \subseteq \{0, 1, \ldots, n-4\}$. Let $p' = (p + n - 3 - i) \bmod (n-2)$. Since $i \ne p$, we have $p' \ne n-3$. Therefore,

$$(p, S) \xrightarrow{a^{n-3-i}} (p', S \bullet a^{n-3-i}) \xrightarrow{c} (f, \{d\}),$$
$$(p, T) \xrightarrow{a^{n-3-i}} (p', T \bullet a^{n-3-i}) \xrightarrow{c} (f, \{f\}).$$ □

As a corollary of the two lemmata above, we get the following result.

Theorem 4 (Boundary on Prefix-Free Languages: State Complexity).
Let L be a prefix-free language over an alphabet Σ with $\mathrm{sc}(L) = n$. If $n \in \{1, 2\}$, then $\mathrm{sc}(L^* \cap L^{c*}) = 2$. If $n \geq 3$, then $\mathrm{sc}(L^* \cap L^{c*}) \leq (n-1) \cdot 2^{n-4} + n + 1$, and the bound is tight if $|\Sigma| \geq 3$. □

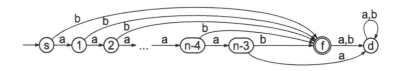

Fig. 4. The DFA of a binary prefix-free language L with $\mathrm{sc}(L^* \cap L^{c*}) = 2n - 2$

4.1   Binary Case

Now we consider the binary case. Our aim is to show that the upper bound $(n-1) \cdot 2^{n-4} + n + 1$ given in Lemma 2 cannot be met by any binary prefix-free language. Moreover, we will show that the tight bound for boundary on binary prefix-free languages is $2n - 2$. Let us start with a lower bound.

Lemma 5 (Binary Case: Lower Bound). *Let $n \geq 3$. There exists a binary prefix-free language L with $\mathrm{sc}(L) = n$ such that $\mathrm{sc}(L^* \cap L^{c*}) = 2n - 2$.*

Proof. Let L be the binary prefix-free language accepted by the DFA A shown in Fig. 4; if $n = 3$, then we have $s \cdot a = d$ and $s \cdot b = f$. Construct the product automaton $D \times D'$ as described in Section 3. The product automaton has $2n - 2$ reachable states as shown in Fig. 5. All these states are pairwise distinguishable. Hence $\mathrm{sc}(L^* \cap L^{c*}) = 2n - 2$. □

Now we are going to show that $2n - 2$ is also an upper bound on the state complexity of the boundary operation on binary prefix-free languages. We will also show that the language L accepted by the DFA shown in Fig. 4 is the only binary prefix-free language, up to renaming the input symbols, with $\mathrm{sc}(L) = n$ and $\mathrm{sc}(L^* \cap L^{c*}) = 2n - 2$. First, we need the following technical lemma.

Lemma 6. *Let $n \geq 4$ and $1 \leq k \leq n - 3$. Let L be a binary prefix-free language accepted by a DFA $A = (\{s, 1, 2, \ldots, n-3, f, d\}, \{a, b\}, \cdot, s, \{f\})$, in which*
$$s \cdot a = 1 \text{ and } i \cdot a = i + 1 \text{ for } i = 1, 2, \ldots, k - 1;$$
$$s \cdot b = f \text{ and } i \cdot b = f \text{ for } i = 1, 2, \ldots, k - 1.$$
If $a^k b \in L^c$, then $L^* \cap L^{c*} = L^* \setminus \{a^i b \mid 0 \leq i \leq k - 1\}$, and $\mathrm{sc}(L^* \cap L^{c*}) \leq n + k$.

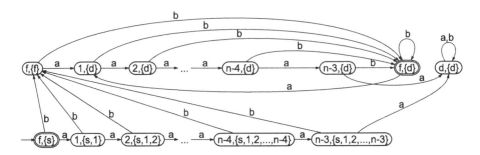

Fig. 5. The reachable states of the product automaton for $L^* \cap L^{c*}$, where L is accepted by the DFA in Fig. 4

Proof. Notice that we have

   (1) $a \in L^c$ and $a^k b \in L^c$;

   (2) $a^i b \notin L^c$ for $i = 0, 1, 2, \ldots, k - 1$;

   (3) if $w \neq \varepsilon$ and $0 \leq i \leq k - 1$, then $a^i b w \in L^c$.

It follows that every string bw with $w \neq \varepsilon$ is in L^c, so it is also in L^{c*}. Next, every string in a^* is in L^{c*}. Let $u = a^\ell bw$ where $\ell \geq 1$.

If $w \neq \varepsilon$, then u can be partitioned as $u = a \cdot a \cdot \cdots \cdot a \cdot bw$, so $u \in L^{c*}$. Let $w = \varepsilon$. If $\ell \geq k$, then u can be partitioned as $u = a \cdot \cdots \cdot a \cdot a^k b$, so $u \in L^{c*}$. If $\ell \leq k - 1$, then $a^\ell b \notin L^c$, and since $a^i b \notin L^c$ for $i = 0, 1, \ldots, k - 1$, we have $a^\ell b \notin L^{c*}$. Hence we have $L^{c*} = \Sigma^* \setminus \{a^i b \mid 0 \leq i \leq k - 1\}$, and therefore $L^* \cap L^{c*} = L^* \setminus \{a^i b \mid 0 \leq i \leq k - 1\}$.

Now let us construct an $(n + k)$-state DFA for $L^* \cap L^{c*}$. We start with the DFA D for the language L^*. Recall that the DFA D can be obtained from the DFA A by making the state f initial, and by redirecting the transitions on a and b from the state f to the states $s \cdot a$ and $s \cdot b$, respectively. The resulting DFA D, shown in Fig. 6, has at most n states and accepts L^*.

To get a DFA B for $L^* \cap L^{c*} = L^* \setminus \{a^i b \mid 0 \leq i \leq k - 1\}$ from the DFA D, we first add a new initial and final state s', and new non-final states $1', 2', \ldots, (k-1)'$. Next, we add the transitions on a and b for the new states as follows:

$$s' \xrightarrow{a} 1' \xrightarrow{a} 2' \xrightarrow{a} \cdots \xrightarrow{a} (k-1)' \xrightarrow{a} k,$$
$$s' \xrightarrow{b} s \text{ and } i' \xrightarrow{b} s \text{ for } i = 1, 2, \ldots, k - 1;$$

the transitions are shown in Fig 7. Notice that the state s is not reachable in the DFA D for the language L^*. Next, we have $s \cdot a = f \circ a$ in the DFA D. Finally, the strings a^i with $1 \leq i \leq k - 1$ are rejected in the DFA D. It follows that the DFA B accepts the language $L^* \cap L^{c*} = L^* \setminus \{a^i b \mid 0 \leq i \leq k - 1\}$. Hence $\mathrm{sc}(L^* \cap L^{c*}) \leq n + k$. $\qquad \square$

Now we are ready to get a tight bound on the state complexity of boundary on binary prefix-free languages.

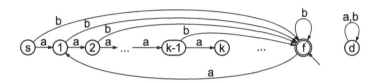

Fig. 6. The DFA D for the language L^*

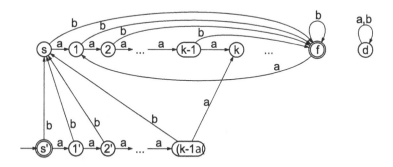

Fig. 7. The DFA B for the language $L^* \cap L^{c*} = L^* \setminus \{a^i b \mid 0 \le i \le k - 1\}$

Theorem 7 (Binary Case: Tight Bound). *Let $n \ge 4$. Let L be a binary prefix-free regular language with $\mathrm{sc}(L) = n$. Then $\mathrm{sc}(L^* \cap L^{c*}) \le 2n - 2$. Next, there exists exactly one binary prefix-free language, up to renaming the input symbols, such that $\mathrm{sc}(L) = n$ and $\mathrm{sc}(L^* \cap L^{c*}) = 2n - 2$, and this language is accepted by the DFA shown in Fig. 4.*

Proof. Let $n \ge 4$. Let L be a binary prefix-free regular language accepted by a minimal DFA $A = (\{s, 1, 2, \ldots, n - 3, f, d\}, \{a, b\}, \cdot, s, \{f\})$.

If both a and b are in L^c, then $L^{c*} = \{a, b\}^*$ and $L^* \cap L^{c*} = L^*$. Therefore in this case we have $\mathrm{sc}(L^* \cap L^{c*}) = \mathrm{sc}(L^*) \le n$.

Without loss of generality, we assume that $a \in L^c$ and $b \notin L^c$. Then we must have $s \cdot b = f$ and since we need to reach the states $1, 2, \ldots, n - 3$, without loss of generality, let $s \cdot a = 1$.

Now assume by induction on k ($2 \le k \le n - 3$), that to get sufficiently large complexity of boundary of L, we have defined the transitions on a and b in states $s, 1, 2, \ldots, k - 1$ as in Lemma 6, that is, as follows:

   $s \cdot a = 1$ and $i \cdot a = i + 1$ for $i = 1, 2, \ldots, k - 1$,
   $s \cdot b = f$ and $i \cdot b = f$ for $i = 1, 2, \ldots, k - 1$.
If $a^k b \in L^c$, then by Lemma 6, we have $\mathrm{sc}(L^* \cap L^{c*}) \le n + k \le 2n - 3$. To get a larger complexity, we must have $a^k b \notin L^c$, so $k \cdot b = f$. Now consider two cases:

(1) Let $k \leq n - 4$. Then we still have to reach states $k + 1, k + 2, \ldots, n - 3$ so, without loss of generality, we must have $k \cdot a = k + 1$, and we continue by the next step of induction.

(2) Let $k = n - 3$. Then we have already reached all states in $\{1, 2, \ldots, n-3\}$. For defining the transition on a in the state $n - 3$, we have three possibilities:

(2a) Let $(n-3) \cdot a \in \{s, 1, 2, \ldots, n-3\}$. Then, all states in $\{s, 1, 2, \ldots, n-3\}$ of the DFA A would be equivalent, so we would have $sc(L) < n$ since $n \geq 4$.

(2b) Let $(n - 3) \cdot a = f$. Then, by a similar argument as in Lemma 6, $sc(L^* \cap L^{c*}) \leq n + (n - 2)$. However, in the DFAs D and B, described in the proof of Lemma 6, the state d would not be reachable, which would result in a $(2n - 3)$-state DFA B for $L^* \cap L^{c*}$.

(2c) Let $(n - 3) \cdot a = d$. Then the DFA A is the same as the DFA shown in Fig. 4, and $sc(L^* \cap L^{c*}) = 2n - 2$.   □

4.2   Unary Case

Recall that a unary language is prefix-free if it is empty, or if it contains exactly one string. The following observation follows immediately from this fact.

Proposition 8. *Let L be a unary prefix-free language with $sc(L) = n$. Then*

$$sc(L^* \cap L^{c*}) = \begin{cases} 2, & \text{if } n = 1 \text{ or } n = 2, \\ 3, & \text{if } n = 3, \\ n - 2, & \text{if } n \geq 4. \end{cases}$$

Proof. If $n = 1$, then $L = \emptyset$. If $n = 2$, then $L = \varepsilon$. In both cases, we get $L^* \cap L^{c*} = \varepsilon$, so $sc(L^* \cap L^{c*}) = 2$.

If $n = 3$, then $L = a$, and $L^* \cap L^{c*} = a^* \setminus \{a\}$, so $sc(L^* \cap L^{c*}) = 3$.

If $n \geq 4$, then $L = a^{n-2}$. Since $a \in L^c$, we have $L^{c*} = a^*$. Hence $L^* \cap L^{c*} = (a^{n-2})^*$, so $sc(L^* \cap L^{c*}) = n - 2$.   □

We summarize the results of our paper in the following theorem.

Theorem 9 (Boundary on Prefix-Free Languages). *Let $f_k(n)$ be the state complexity of the boundary operation on prefix-free languages over a k-letter alphabet defined to be*

$$f_k(n) = \max\{sc(L^* \cap L^{c*}) \mid L \subseteq \Sigma^*, |\Sigma| = k, sc(L) = n, \text{ and } L \text{ is prefix-free}\}.$$

Then

(i) $f_k(1) = f_k(2) = 2$;

(ii) $f_k(n) = (n - 1) \cdot 2^{n-4} + n + 1$ if $k \geq 3$ and $n \geq 3$;

(iii) $f_2(n) = 2n - 2$, if $n \geq 3$;

(iv) $f_1(n) = \begin{cases} 3, & \text{if } n = 3, \\ n - 2, & \text{if } n \geq 4. \end{cases}$

Proof. (*i*) If L is a prefix-free language with $\mathrm{sc}(L) = 1$ or $\mathrm{sc}(L) = 2$, then $L = \emptyset$ or $L = \varepsilon$, respectively. In both cases, we have $L^* \cap L^{c*} = \varepsilon$, so $\mathrm{sc}(L^* \cap L^{c*}) = 2$.

(*ii*) The state complexity of boundary on prefix-free languages over an alphabet of at least three symbols is given by Theorem 4.

The tight bounds in (*iii*) and (*iv*) are given by Theorem 7 and Proposition 8, respectively; direct computations in the case of $n = 3$ give two binary witness languages b and a^*b meeting the bound $2n - 2 = 4$. $\qquad\square$

5   Conclusions

We investigated the boundary operation, defined as $\mathrm{bd}(L) = L^* \cap (L^c)^*$, on the class of prefix-free regular languages. We proved that if a prefix-free language is recognized by an n-state deterministic finite automaton, then its boundary is recognized by a deterministic automaton of at most $(n-1) \cdot 2^{n-4} + n + 1$ states. We also proved that this upper bound is tight. To describe worst-case examples, we used a three-letter alphabet.

Then we showed that the tight bound on the state complexity of boundary on binary prefix-free languages is $2n - 2$. This is quite an interesting result which not only shows that a ternary alphabet is optimal for defining witness languages meeting the upper bound $(n - 1) \cdot 2^{n-4} + n + 1$ for boundary, but it also gives the exact complexity of boundary on binary prefix-free languages. Finally, we obtained the tight bound $n - 2$ for boundary on unary prefix-free languages.

Hence we proved that the state complexity of the boundary operation on the class of prefix-free regular languages is given by an exponential function for an alphabet which contains at least three symbols, and it is given by a linear function otherwise.

References

1. Brzozowski, J.A., Grant, E., Shallit, J.: Closures in formal languages and Kuratowski's theorem. Internat. J. Found. Comput. Sci. **22**(2), 301–321 (2011)
2. Eom, H.-S., Han, Y.-S., Salomaa, K.: State Complexity of k-Union and k-Intersection for Prefix-Free Regular Languages. In: Jurgensen, H., Reis, R. (eds.) DCFS 2013. LNCS, vol. 8031, pp. 78–89. Springer, Heidelberg (2013)
3. Eom, H., Han, Y., Salomaa, K., Yu, S.: State complexity of combined operations for prefix-free regular languages. In: Paun, G., Rozenberg, G., Salomaa, A. (eds.) Discrete Mathematics and Computer Science, pp. 137–151 (2014)
4. Han, Y., Salomaa, K., Wood, D.: Nondeterministic state complexity of basic operations for prefix-free regular languages. Fundam. Inform. **90**(1–2), 93–106 (2009)
5. Han, Y., Salomaa, K., Wood, D.: Operational state complexity of prefix-free regular languages. In: Ésik, Z., Fülöp, Z. (eds.) Automata, Formal Languages, and Related Topics, pp. 99–115. University of Szeged, Hungary, Institute of Informatics (2009)
6. Jirásek, J., Jirásková, G.: On the boundary of regular languages. Theoret. Comput. Sci. **578**, 42–57 (2015)

7. Maslov, A.: Estimates of the number of states of finite automata. Soviet Mathematics Doklady **11**, 1373–1375 (1970)
8. Salomaa, A., Salomaa, K., Yu, S.: State complexity of combined operations. Theoret. Comput. Sci. **383**(2–3), 140–152 (2007)
9. Shallit, J.: Open problems in automata theory and formal languages. https://cs. uwaterloo.ca/shallit/Talks/open10r.pdf
10. Sipser, M.: Introduction to the theory of computation. PWS Publishing Company, Boston (1997)
11. Yu, S., Zhuang, Q., Salomaa, K.: The state complexities of some basic operations on regular languages. Theoret. Comput. Sci. **125**(2), 315–328 (1994)

A Connected 3-State Reversible Mealy Automaton Cannot Generate an Infinite Burnside Group

Ines Klimann[1], Matthieu Picantin[1]([⊠]), and Dmytro Savchuk[2]

[1] University Paris Diderot, Sorbonne Paris Cité, LIAFA, UMR 7089 CNRS,
F-75013 Paris, France
{klimann,picantin}@liafa.univ-paris-diderot.fr
[2] Department of Mathematics and Statistics, University of South Florida,
4202 E Fowler Ave, Tampa, FL 33620-5700, USA
savchuk@usf.edu

Abstract. The class of automaton groups is a rich source of the simplest examples of infinite Burnside groups. However, no such examples have been constructed in some classes, as groups generated by non reversible automata. It was recently shown that 2-state reversible Mealy automata cannot generate infinite Burnside groups. Here we extend this result to connected 3-state reversible Mealy automata, using new original techniques. The results rely on a fine analysis of associated orbit trees and a new characterization of the existence of elements of infinite order.

Keywords: Burnside groups · Reversible mealy automata · Automaton groups

1 Mealy Automata and the General Burnside Problem

In 1902, Burnside has introduced a question which would become highly influential in group theory [6]:

Is a finitely generated group whose elements have finite order necessarily finite?

This problem is now known as the *General Burnside Problem*. A group is commonly called a *Burnside* group if it is finitely generated and all its elements have finite order.

In 1964, Golod and Shafarevich [14] were the first ones to give a negative answer to the general Burnside problem and around the same time Glushkov suggested that groups generated by automata could serve as a different source of counterexamples [12]. In 1972, Aleshin gave an answer as a subgroup of an automaton group [2], and then in 1980, Grigorchuk exhibited the first and the

---

This work was partially supported by the French *Agence Nationale de la Recherche*, through the Project **MealyM** ANR-JS02-012-01. The third author was partially supported by the New Researcher Grant from the USF Internal Awards Program.

I. Potapov (Ed.): DLT 2015, LNCS 9168, pp. 313–325, 2015.
DOI: 10.1007/978-3-319-21500-6_25

simplest by now example of an infinite Burnside automaton group [15]. Since then many infinite Burnside automaton groups have been constructed [4,16,18, 24]. Even by now, the simplest examples of infinite Burnside groups are still automaton groups.

All the examples of infinite Burnside automaton groups in the literature happen to be generated by non-reversible Mealy automata (where all the letters do not act as permutations on the stateset). It was proved in [19] that a 2-state reversible Mealy automaton cannot generate an infinite Burnside group, but the techniques were strongly based on the size of the stateset. Here we address this problem for a larger class, the connected 3-state reversible automata, and prove:

Theorem 1. *A connected 3-state reversible Mealy automaton cannot generate an infinite Burnside group.*

To this end we develop new techniques, centered on the orbit tree of the dual of a Mealy automaton. In particular, we give a new characterization of the existence of elements of infinite order in a reversible automaton semigroup (Theorem 17). Although not stated in full generality, such a characterization has already been applied successfully in various situations (see [13,20]). We hope that these techniques could be further extended to attack similar problems for bigger automata.

The class of automaton groups is very interesting from an algorithmic point of view. Even though the word problem is decidable, most of other basic algorithmic questions, including finiteness, order, and conjugacy problems, are either known to be undecidable, or their decidability is unknown. For example, it was proved recently that the order problem is undecidable for automaton semigroups [11] and for groups generated by, so called, asynchronous Mealy automata [5]. But this problem still remains open for the class of all automaton groups. There are many partial methods to find elements of infinite order in such groups, but the class of reversible automata is known as the class for which most of these algorithms do not work or perform poorly. The proof of Theorem 1 is completely constructive and gives a uniform algorithm to produce many elements of infinite order in infinite groups generated by 3-state invertible reversible automata. Unfortunately, it does not provide an algorithm that can determine if the group itself is infinite, however, it is known that any invertible reversible not bireversible automaton generates an infinite group [1].

The paper is organized as follows. In Section 2, we recall basics about automaton groups and rooted trees. In Sections 3 and 4, we introduce crucial constructions: the labeled orbit tree of a Mealy automaton and its self-liftable subtrees. Section 5 is devoted to the new characterization of the existence of elements of infinite order. Section 6 contains the proof of our main result (Theorem 1).

2   Basic Notions

2.1   Groups Generated by Mealy Automata

We first recall the formal definition of an automaton. A *(finite, deterministic, and complete) automaton* is a triple $(Q, \Sigma, \delta = (\delta_i \colon Q \to Q)_{i \in \Sigma})$, where the *stateset* Q and the *alphabet* Σ are non-empty finite sets, and the δ_i are functions.

A *Mealy automaton* is a quadruple $(Q, \Sigma, \delta, \rho)$, such that (Q, Σ, δ) and (Σ, Q, ρ) are both automata. In other terms, a Mealy automaton is a complete, deterministic, letter-to-letter transducer with the same input and output alphabet.

The graphical representation of a Mealy automaton is standard, see Figure 1 left.

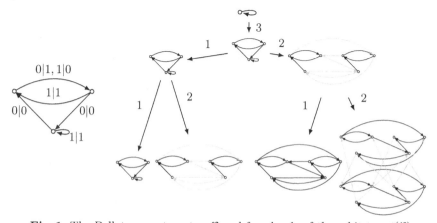

Fig. 1. The Bellaterra automaton $\mathcal{B}$ and four levels of the orbit tree $\mathsf{t}(\mathcal{B})$

A Mealy automaton $(Q, \Sigma, \delta, \rho)$ is *invertible* if the functions ρ_x are permutations of Σ and *reversible* if the functions δ_i are permutations of Q.

In a Mealy automaton $\mathcal{A} = (Q, \Sigma, \delta, \rho)$, the sets Q and Σ play dual roles. So we may consider the *dual (Mealy) automaton* defined by $\mathfrak{d}(\mathcal{A}) = (\Sigma, Q, \rho, \delta)$. Obviously, a Mealy automaton is reversible if and only if its dual is invertible.

Let $\mathcal{A} = (Q, \Sigma, \delta, \rho)$ be a Mealy automaton. Each state $x \in Q$ defines a mapping from Σ^* into itself recursively defined by:

$$\forall i \in \Sigma, \ \forall \mathbf{s} \in \Sigma^*, \qquad \rho_x(i\mathbf{s}) = \rho_x(i)\rho_{\delta_i(x)}(\mathbf{s}) \, .$$

The image of the empty word is itself. The mapping ρ_x for each $x \in Q$ is length-preserving and prefix-preserving. We say that ρ_x is the function *induced* by x. For $\mathbf{x} = x_1 \cdots x_n \in Q^n$ with $n > 0$, set $\rho_{\mathbf{x}} \colon \Sigma^* \to \Sigma^*, \rho_{\mathbf{x}} = \rho_{x_n} \circ \cdots \circ \rho_{x_1}$.

Denote dually by $\delta_i \colon Q^* \to Q^*, i \in \Sigma$, the functions induced by the states of $\mathfrak{d}(\mathcal{A})$. For $\mathbf{s} = s_1 \cdots s_n \in \Sigma^n$ with $n > 0$, set $\delta_{\mathbf{s}} \colon Q^* \to Q^*, \delta_{\mathbf{s}} = \delta_{s_n} \circ \cdots \circ \delta_{s_1}$.

The semigroup of mappings from Σ^* to Σ^* generated by $\{\rho_x, x \in Q\}$ is called the *semigroup generated by* $\mathcal{A}$ and is denoted by $\langle \mathcal{A} \rangle_+$. When $\mathcal{A}$ is invertible, the functions induced by its states are permutations on words of the same length and thus we may consider the group of mappings from Σ^* to Σ^* generated by $\{\rho_x, x \in Q\}$. This group is called the *group generated by* $\mathcal{A}$ and is denoted by $\langle \mathcal{A} \rangle$.

Let us recall some known results that will be used in our proofs.

Proposition 2 (see, for example, [1]). *An invertible Mealy automaton generates a finite group if and only if it generates a finite semigroup.*   □

Proposition 3 ([1, 22, 23]). *A Mealy automaton generates a finite (semi)group if and only if so does its dual.*   □

2.2 Terminology on Trees

Throughout this paper, we will use different sorts of labeled trees. Here we set up some terminology that are common for all of them.

All our trees are rooted, *i.e.* with a selected vertex called the *root*. We will visualize the trees traditionally as growing down from the root. A *path* is a (possibly infinite) sequence of adjacent edges without backtracking from top to bottom. A path is said to be *initial* if it starts at the root of the tree. A *branch* is an infinite initial path. The initial vertex of a non-empty path $\mathbf{e}$ is denoted by $\top(\mathbf{e})$ and its terminal vertex by $\bot(\mathbf{e})$ whenever the path is finite.

The *level of a vertex* is its distance to the root and the *level of an edge* or a *path* is the level of its initial vertex. For V a vertex in a tree $\mathcal{T}$, by *section of* $\mathcal{T}$ *at* V (denoted by $\mathcal{T}_{|V}$) we mean the subtree of $\mathcal{T}$ with root V consisting of all those vertices of $\mathcal{T}$ that are descendant of V. Additionally, for $\mathbf{e}$ an initial finite path of $\mathcal{T}$, we also call $\mathcal{T}_{|\bot(\mathbf{e})}$ the *section of* $\mathcal{T}$ *at* $\mathbf{e}$ and denote it by $\mathcal{T}_{|\mathbf{e}}$.

If the edges of a rooted tree are labeled by elements of some finite set, the *label* of a (possibly infinite) path is the ordered sequence of labels of its edges.

3 Connected Components of the Powers of an Automaton

In this section we detail the basic properties of the connected components of the powers of a reversible Mealy automaton. The link between these components is central in our construction.

Let $\mathcal{A} = (Q, \Sigma, \delta, \rho)$ be a reversible Mealy automaton. By reversibility, all the connected components of its underlying graph are strongly connected. Consider the powers of $\mathcal{A}$: for $n > 0$, its *n-th power* $\mathcal{A}^n$ is the Mealy automaton

$$\mathcal{A}^n = \left(Q^n, \Sigma, (\delta_i \colon Q^n \to Q^n)_{i \in \Sigma}, (\rho_{\mathbf{x}} \colon \Sigma \to \Sigma)_{\mathbf{x} \in Q^n} \right) \ .$$

By convention, $\mathcal{A}^0$ is the trivial automaton on the alphabet Σ.

As $\mathcal{A}$ is reversible, so are its powers and the connected components of $\mathcal{A}^n$ coincide with the orbits of the action of $\langle \mathfrak{d}(\mathcal{A}) \rangle$ on Q^n.

Definition 4. *The* connection degree $\lambda(\mathcal{A})$ *of* $\mathcal{A}$ *is the largest n such that $\mathcal{A}^n$ is connected: it is 0 if $\mathcal{A}$ is disconnected, and infinite if all its powers are connected.*

Since $\mathcal{A}$ is reversible, there is a very particular connection between the connected components of $\mathcal{A}^n$ and those of $\mathcal{A}^{n+1}$ as highlighted in [19]. More precisely, take a connected component $\mathcal{C}$ of some $\mathcal{A}^n$, and let $\mathbf{u} \in Q^n$ be a state of $\mathcal{C}$. Take also $x \in Q$ a state of $\mathcal{A}$, and let $\mathcal{D}$ be the connected component of $\mathcal{A}^{n+1}$ containing the state $\mathbf{u}x$. Then, for any state $\mathbf{v}$ of $\mathcal{C}$, there exists a state of $\mathcal{D}$ prefixed with $\mathbf{v}$:

$$\exists \mathbf{s} \in \Sigma^* \mid \delta_{\mathbf{s}}(\mathbf{u}) = \mathbf{v} \quad \text{and so} \quad \delta_{\mathbf{s}}(\mathbf{u}x) = \mathbf{v}\delta_{\rho_{\mathbf{u}}(\mathbf{s})}(x) \ .$$

Furthermore, if $\mathbf{u}y$ is a state of $\mathcal{D}$, for some state $y \in Q$ different from x, then $\delta_{\mathbf{s}}(\mathbf{u}x)$ and $\delta_{\mathbf{s}}(\mathbf{u}y)$ are two different states of $\mathcal{D}$ prefixed with $\mathbf{v}$, because of the reversibility of $\mathcal{A}^{n+1}$: the transition function $\delta_{\rho_{\mathbf{u}}(\mathbf{s})}$ is a permutation. Hence $\mathcal{D}$ can be seen as consisting of several copies of $\mathcal{C}$ and $\#\mathcal{C}$ divides $\#\mathcal{D}$. They have the same size if and only if, for each state $\mathbf{u}$ of $\mathcal{C}$ and any different states $x, y \in Q$, $\mathbf{u}x$ and $\mathbf{u}y$ cannot simultaneously lie in $\mathcal{D}$. If from a connected component $\mathcal{C}$ of $\mathcal{A}^n$, we obtain several connected components of $\mathcal{A}^{n+1}$, we say that $\mathcal{C}$ *splits up*.

The connected components of the powers of a Mealy automaton and the finiteness of the generated group or of a monogenic subgroup are closely related, as shown in the following propositions (obtained also independently in [8]).

Proposition 5. *A reversible Mealy automaton generates a finite group if and only if the connected components of its powers have bounded size.*

Proposition 6. *Let $\mathcal{A} = (Q, \Sigma, \delta, \rho)$ be an invertible reversible Mealy automaton and let $\mathbf{u} \in Q^+$ be a non-empty word. The following conditions are equivalent:*

(i) $\rho_{\mathbf{u}}$ has finite order,
(ii) the sizes of the connected components of $(\mathbf{u}^n)_{n \in \mathbb{N}}$ are bounded,
(iii) there exists a word $\mathbf{v}$ such that the sizes of the connected components of $(\mathbf{v}\mathbf{u}^n)_{n \in \mathbb{N}}$ are bounded,
(iv) for any word $\mathbf{v}$, the sizes of the connected components of $(\mathbf{v}\mathbf{u}^n)_{n \in \mathbb{N}}$ are bounded.

Proof. (ii)$\Rightarrow$(iii), (iv)$\Rightarrow$(ii), and (iv)$\Rightarrow$(iii) are immediate.

(i)$\Rightarrow$(ii) is a direct consequence of Proposition 5: let k be the order of $\rho_{\mathbf{u}}$; it means that $\mathbf{u}^k$ acts as the identity, and so do all the states of its connected component (as it is in fact strongly connected by reversibility of the automaton). By Proposition 5, the connected components of the $(\mathbf{u}^{kn})_n$ have bounded size, which leads to (ii).

(iii)$\Rightarrow$(i): for each n, denote by $\mathcal{C}_n$ the connected component of $\mathbf{v}\mathbf{u}^n$. As the sizes of these components are bounded, the sequence $(\mathcal{C}_n)_n$ admits a subsequence whose all elements are the same, up to state numbering. Within this subsequence, there are two elements such that two different words in $\mathbf{v}\mathbf{u}^*$ name the same state, say $\mathbf{v}\mathbf{u}^p$ and $\mathbf{v}\mathbf{u}^q$, which means that $\rho_{\mathbf{v}\mathbf{u}^p} = \rho_{\mathbf{v}\mathbf{u}^q}$, and $\rho_{\mathbf{u}}$ has finite order.

(ii)$\Rightarrow$(iv): the size of the connected component of $\mathbf{v}\mathbf{u}^n$ is at most $\#\Sigma^{|\mathbf{v}|}$ times the size of the connected component of $\mathbf{u}^n$. $\qquad\square$

4   The Labeled Orbit Tree

In this section, we build a tree capturing the links between the connected components of consecutive powers of a Mealy automaton. See an example in Figure 1.

Let $\mathcal{A} = (Q, \Sigma, \delta, \rho)$ be an invertible reversible Mealy automaton. Consider the tree with vertices the connected components of the powers of $\mathcal{A}$, and the incidence relation built by adding an element of Q: for any $n \geq 0$, the connected component of $\mathbf{u} \in Q^n$ is linked to the connected component(s) of $\mathbf{u}x$, for any $x \in Q$. This tree is called the *orbit tree* of $\mathfrak{d}(\mathcal{A})$ [10,17]. It can be seen as the quotient of the tree Q^* under the action of the group $\langle \mathfrak{d}(\mathcal{A}) \rangle$.

We label any edge $\mathcal{C} \to \mathcal{D}$ of the orbit tree by the ratio $\frac{\#\mathcal{D}}{\#\mathcal{C}}$, which is always an integer by the reversibility of $\mathcal{A}$. We call this labeled tree the *labeled orbit tree* of $\mathfrak{d}(\mathcal{A})$. In [10], in the definition of the labeled orbit tree, each vertex is labeled by the size of the associated connected component, which encodes exactly the same information as our relative labeling. We denote by $\mathfrak{t}(\mathcal{A})$ the labeled orbit tree of $\mathfrak{d}(\mathcal{A})$. Note that for each vertex of $\mathfrak{t}(\mathcal{A})$ the sum of the labels of all edges going down from this vertex always equals to the number of states in $\mathcal{A}$.

Let $\mathbf{u}$ be a (possibly infinite) word over Q. The *path of* $\mathbf{u}$ in the orbit tree $\mathfrak{t}(\mathcal{A})$ is the unique initial path going from the root through the connected components of the prefixes of $\mathbf{u}$; $\mathbf{u}$ can be called *a representative* of this initial path.

Definition 7. *Let e and f be two edges in the orbit tree $\mathfrak{t}(\mathcal{A})$. We say that e is liftable to f if each word of $\perp(e)$ admits some word of $\perp(f)$ as a suffix.*

Consider $\mathbf{u}$ in $\top(e)$ and its suffix $\mathbf{v}$ in $\top(f)$: any state $x \in Q$ such that $\mathbf{u}x \in \perp(e)$ satisfies $\mathbf{v}x \in \perp(f)$. Informally, "e liftable to f" means that what can happen after $\top(e)$ by following e can also happen after $\top(f)$ by following f. This condition is equivalent to a weaker one:

Lemma 8. *Let $\mathcal{A}$ be a reversible Mealy automaton, and let e and f be two edges in the orbit tree $\mathfrak{t}(\mathcal{A})$. If there exists a word of $\perp(e)$ which admits a word of $\perp(f)$ as suffix, then e is liftable to f.*

Proof. Let $\mathcal{A} = (Q, \Sigma, \delta, \rho)$. Assume $\mathbf{uv} \in \perp(e)$ with $\mathbf{v} \in \perp(f)$. By reversibility, for any word $\mathbf{w}$ in the connected component $\perp(e)$, there exists $\mathbf{s} \in \Sigma^*$ satisfying $\mathbf{w} = \delta_{\mathbf{s}}(\mathbf{uv})$, which can also be written $\mathbf{w} = \delta_{\mathbf{s}}(\mathbf{u})\delta_{\mathbf{t}}(\mathbf{v})$ with $\mathbf{t} = \rho_{\mathbf{u}}(\mathbf{s})$. Hence the suffix $\delta_{\mathbf{t}}(\mathbf{v})$ of $\mathbf{w}$ belongs to the connected component $\perp(f)$ of $\mathbf{v}$.   □

Obviously if e is liftable to f, then f is closer to the root of the orbit tree. The fact that an edge is liftable to another one reflects a deeper relation stated below. The following lemma is one of the key observations.

Lemma 9. *Let e and f be two edges in the orbit tree $\mathfrak{t}(\mathcal{A})$. If e is liftable to f, then the label of e is less than or equal to the label of f.*

Proof. Since e is liftable to f, each word in $\perp(e)$ has a form $\mathbf{vu}x$ for some $\mathbf{u}x \in \perp(f)$. Suppose that $\mathbf{vu}x$ and $\mathbf{vu}y$ are in the same connected component: there exists $\mathbf{s} \in \Sigma^*$ such that $\delta_{\mathbf{s}}$ moves $\mathbf{vu}x$ to $\mathbf{vu}y$. In this case, $\rho_{\mathbf{v}}(\mathbf{s})$ moves $\mathbf{u}x$ to $\mathbf{u}y$.

Thus, the number of children of **vu** in the connected component of **vu**x (which is equal to the label of e) is less than or equal to the number of children of **u** in the connected component of **u**x (which is equal to the label of f). $\qquad\qquad\square$

The notion of liftability can be generalized to paths:

Definition 10. *Let* $\mathbf{e} = (e_i)_{i\in I}$ *and* $\mathbf{f} = (f_i)_{i\in I}$ *be two paths of the same (possibly infinite) length in the orbit tree* $\mathsf{t}(\mathcal{A})$. *The path* $\mathbf{e}$ *is liftable to the path* $\mathbf{f}$ *if, for any* $i \in I$, *the edge* e_i *is liftable to the edge* f_i.

As each word $\mathbf{u} \in Q^*$ is a state in a connected component of $\mathcal{A}^{|\mathbf{u}|}$, we can notice the following fact which is crucial for all our forthcoming proofs.

Lemma 11. *Let* $\mathbf{e}$ *be a path at level* k *in the orbit tree* $\mathsf{t}(\mathcal{A})$. *Then, for any* $\ell < k$, $\mathbf{e}$ *is liftable to some path at level* ℓ. *In particular,* $\mathbf{e}$ *is liftable to some initial path.*

Proposition 12. *If for some* n, *a connected component of* $\mathcal{A}^n$ *does not split up, then the connection degree of* $\mathcal{A}$ *is at least* $n + 1$.

Proof. Suppose that $\mathcal{A}$ has m states. If an edge at level n in the orbit tree $\mathsf{t}(\mathcal{A})$ has label m, by Lemma 11, it is liftable to some edge at any level above n, and by Lemma 9 this edge is labeled by m. Now, by going from top to bottom, we can conclude that there is only one edge at each level above $n + 1$ in $\mathsf{t}(\mathcal{A})$. $\qquad\square$

Proposition 13. *If all edges coming down from the only connected component at vertex* $\mathcal{A}^{\lambda(\mathcal{A})}$ *are labeled by 1, the group generated by* $\mathcal{A}$ *is finite.*

Definition 14. *Let* $\mathcal{A}$ *be a reversible Mealy automaton and* $\mathfrak{r}$ *be a (possibly infinite) path or subtree of* $\mathsf{t}(\mathcal{A})$. *For* $k > 0$, $\mathfrak{r}$ *is* k-self-liftable *whenever any path in* $\mathfrak{r}$ *starting at level* $i+k$ *is liftable to a path in* $\mathfrak{r}$ *starting at level* i, *for any* $i \geq 0$. *A path or a subtree is* self-liftable *if it is* k-self-liftable *for some* $k > 0$.

5   Existence of Elements of Infinite Order

Here we provide a new characterization of the existence of elements of infinite order in the semigroup generated by a reversible Mealy automaton $\mathcal{A}$ in terms of path properties of the associated orbit tree $\mathsf{t}(\mathcal{A})$.

Let us start with a straightforward observation: any periodic word $\mathbf{u}$ over the stateset of a reversible Mealy automaton $\mathcal{A}$ is the representative of a $|\mathbf{u}|$-self-liftable branch in the orbit tree $\mathsf{t}(\mathcal{A})$. The following gives the converse.

Proposition 15. *Let* $\mathcal{A}$ *be a reversible Mealy automaton with stateset* Q. *Every self-liftable branch in the orbit tree* $\mathsf{t}(\mathcal{A})$ *admits a periodic representative, i.e. of the form* $\mathbf{u}^\omega$ *for some* $\mathbf{u} \in Q^+$.

Proof. Let $\mathbf{e} = e_0 e_1 e_2 \cdots$ be a k-self-liftable branch in $\mathfrak{t}(\mathcal{A})$. By self-liftability, it is enough to prove that $\mathbf{e}$ admits an ultimately periodic representative: removing first k letters in any representative of $\mathbf{e}$ produces another representative.

Put the natural partial order on $\mathbb{N}^k$: $(f_1, \ldots, f_k) \preceq (g_1, \ldots, g_k) \Leftrightarrow \forall i, \ f_i \leq g_i$.

By Lemma 9 an infinite sequence $\{L_n = (l_{nk}, l_{nk+1}, \ldots, l_{nk+k-1})\}_{n \geq 0}$ over $\mathbb{N}^k$, where l_j is the label of the edge e_j, is $\preceq$-decreasing and, thus, ultimately constant: let N be a number satisfying $L_n = L_{n+1}$ for all $n \geq N$. For each word $\mathbf{u}$ representing a vertex in $\mathbf{e}$, consider the set

$$F(\mathbf{u}) = \{\mathbf{v} \colon \mathbf{uv} \text{ is a representative of a vertex in } \mathbf{e}\}$$

of words that can follow $\mathbf{u}$ without leaving the path $\mathbf{e}$. Then for each $n \geq N$ and each word $\mathbf{wu}$ in $\perp(e_{n+k})$ with $|\mathbf{w}| = k$ we have

$$F(\mathbf{wu}) = F(\mathbf{u}). \tag{1}$$

Indeed, by k-self-liftability of $\mathbf{e}$ we get $F(\mathbf{wu}) \subset F(\mathbf{u})$. On the other hand, the cardinalities of these sets are equal to

$$l_{n+k+1} l_{n+k+2} \cdots l_{n+k+|v|} \quad \text{and} \quad l_{n+1} l_{n+2} \cdots l_{n+|v|}$$

respectively. But these two numbers are equal due to the choice of $n \geq N$.

Now we can construct an ultimately periodic representative of $\mathbf{e}$ as follows. Choose words $\mathbf{w}_1, \mathbf{w}_2, \ldots$ of length kM for some M such that $kM > N$ arbitrarily in such a way that $\mathbf{w}_1 \mathbf{w}_2 \cdots \mathbf{w}_i$ is a representative of a vertex in $\mathbf{e}$. Since there is only finite number of choices for $\mathbf{w}_j$'s, there will be $1 < i < j$ such that $\mathbf{w}_i = \mathbf{w}_j$. For $\mathbf{u} = \mathbf{w}_1 \mathbf{w}_2 \cdots \mathbf{w}_{i-1}$ and $\mathbf{v} = \mathbf{w}_{i+1} \mathbf{w}_{i+2} \cdots \mathbf{w}_{j-1}$, applying (1) twice, we get

$$F(\mathbf{uw}_i \mathbf{vw}_i) = F(\mathbf{w}_i) = F(\mathbf{uw}_i).$$

Therefore, the word $\mathbf{uw}_i (\mathbf{vw}_i)^\omega$ is an ultimately periodic representative of $\mathbf{e}$. $\quad\square$

Definition 16. *Any branch labeled by a word not suffixed by 1^ω is called* active.

Theorem 17. *The semigroup generated by an invertible reversible automaton $\mathcal{A}$ admits elements of infinite order if and only if the orbit tree $\mathfrak{t}(\mathcal{A})$ admits an active self-liftable branch.*

6   The Connected 3-State Case

We study here the case where $\mathcal{A}$ is a connected invertible reversible 3-state Mealy automaton, which means that the orbit tree $\mathfrak{t}(\mathcal{A})$ has a unique edge adjacent to the root, labeled by 3. We prove that if $\mathcal{A}$ generates an infinite group, then the orbit tree $\mathfrak{t}(\mathcal{A})$ admits a (necessarily unique) active 1-self-liftable branch, more precisely a branch labeled by either 3^ω or $3^n 2^\omega$. The conclusion comes then from Theorem 17: $\mathcal{A}$ cannot generate an infinite Burnside group.

Note that there exist disconnected 3-state invertible reversible Mealy automata with no active 1-self-liftable branch in the associated orbit tree.

6.1   Reducing the Scope

If the connection degree of such an automaton $\mathcal{A}$ is infinite, the semigroup $\langle \mathcal{A} \rangle_+$ is free of rank 3 [19, Prop. 14], and $\mathcal{A}$ cannot generate an infinite Burnside group. **So from now on, we assume** $0 < \lambda(\mathcal{A}) < \infty$. The orbit tree $\mathfrak{t}(\mathcal{A})$ has a prefix linear part until the level $\lambda(\mathcal{A})$ and, below this level, all the vertices split up.

Definition 18. *Let* $\mathbf{i}$ *be a (possibly infinite) word over an alphabet F. For $j \in F$, a j-block $\mathbf{j}$ of $\mathbf{i}$ is a maximal factor of $\mathbf{i}$ in $j^* \cup \{j^\omega\}$, that is, $\mathbf{i} = \mathbf{kjl}$ holds, where the last letter of $\mathbf{k}$ and the first letter of $\mathbf{l}$, if not empty, are not j.*

Lemma 19. *If the lengths of the 2-blocks in the orbit tree $\mathfrak{t}(\mathcal{A})$ are not bounded, then $\mathfrak{t}(\mathcal{A})$ admits a branch labeled by $3^{\lambda(\mathcal{A})} 2^\omega$. If the lengths of the 2-blocks are bounded with supremum N, then $\mathfrak{t}(\mathcal{A})$ admits an initial path labeled by $3^{\lambda(\mathcal{A})} 2^N$ (and none labeled by $3^{\lambda(\mathcal{A})} 2^{N+1}$). In both cases, the branch (or path) is unique.*

Proof. As there is at most one path starting at $\mathcal{A}^{\lambda(\mathcal{A})}$ with a maximal prefix in 2^ω because the stateset has size 3, Lemma 11 leads to the conclusion. □

By Proposition 12, no edge can be labeled by 3 below the connection degree of $\mathcal{A}$. On the other hand, the case when all edges going down from the unique vertex on the level $\lambda(\mathcal{A})$ are labeled by 1 does not produce infinite Burnside groups by to Proposition 13.

From now on, we assume that the only connected component at vertex $\mathcal{A}^{\lambda(\mathcal{A})}$ splits up in two connected components. Below we will put the emphasis on the larger one.

6.2   Self-liftable Subtrees and $\mathfrak{s}$-Words

Let $\mathcal{A}$ be a 3-state invertible reversible Mealy automaton with the stateset Q satisfying the above conditions.

Definition 20. *At level $\lambda(\mathcal{A})$ there are two edges, say e_1 labeled by 1 and e_2 labeled by 2. Denote by $\mathfrak{s}$ the restriction of the tree $\mathfrak{t}(\mathcal{A})$ containing the linear part until the level $\lambda(\mathcal{A})$ and then all the edges which are liftable to e_2.*

It is straightforward to see that $\mathfrak{s}$ is the maximal 1-self-liftable subtree of $\mathfrak{t}(\mathcal{A})$ containing e_2 but not containing e_1. Let $W(\mathfrak{s})$ be the set of all words over Q representing the initial paths of $\mathfrak{s}$ (we will call them $\mathfrak{s}$-*words*). Any edge of such a path below level $\lambda(\mathcal{A}) + 1$ is liftable to e_2. In particular, $W(\mathfrak{s})$ is prefix-stable and hence can be seen as a rooted tree whose edges are labeled by elements of Q (called also *orbital tree* in [20]). Note that $W(\mathfrak{s})$ is also suffix-stable.

The edge e_2 and the tree $\mathfrak{s}$ allow us to structure our vision of the orbit tree $\mathfrak{t}(\mathcal{A})$.

Lemma 21. *Each vertex of $\mathfrak{t}(\mathcal{A})$ below the level $\lambda(\mathcal{A}) + 1$ is the initial vertex of one edge which is not liftable to e_2, and either one or two edges which are.* □

Lemma 22. *For any two $\mathfrak{s}$-words $\mathbf{u}$ and $\mathbf{v}$, there exists an infinite word $\mathbf{r} \in Q^\omega$ such that $\mathbf{ur}$ and $\mathbf{vr}$ are $\mathfrak{s}$-words.*

Lemma 23. *Let $x \in Q$ and $\mathbf{u} \in Q^{\curlywedge(\mathcal{A})}$. The set of length $\curlywedge(\mathcal{A})$ suffixes of all words in the connected component of $x\mathbf{u}$ is the whole $Q^{\curlywedge(\mathcal{A})}$.*

Proof. Since $\mathcal{A}$ is reversible and $\mathcal{A}^{\curlywedge(\mathcal{A})}$ is connected, for each $\mathbf{v} \in Q^{\curlywedge(\mathcal{A})}$, there exists $\mathbf{s} \in \Sigma^*$ such that $\delta_{\mathbf{s}}(\mathbf{u}) = \mathbf{v}$. By the invertibility of $\mathcal{A}$, $\mathbf{t} = \rho_x^{-1}(\mathbf{s})$ is well defined and we have: $\delta_{\mathbf{t}}(x\mathbf{u}) = \delta_{\mathbf{t}}(x)\delta_{\rho_x(\mathbf{t})}(\mathbf{u}) = \delta_{\mathbf{t}}(x)\delta_{\mathbf{s}}(\mathbf{u}) = \delta_{\mathbf{t}}(x)\mathbf{v}.$ □

Proposition 24. *For any $\mathfrak{s}$-word $\mathbf{u}$, there are infinitely many edges in $W(\mathfrak{s})_{|\mathbf{u}}$ labeled by each state of the automaton.*

Proof. Denote the stateset $Q = \{x, y, z\}$ and let $\mathbf{u}$ be an $\mathfrak{s}$-word such that no edge of $W(\mathfrak{s})_{|\mathbf{u}}$ is labeled by z (we have $|\mathbf{u}| \geq \curlywedge(\mathcal{A})$, otherwise it is impossible).

As each word of $W(\mathfrak{s})$ can be extended in $W(\mathfrak{s})$ by two different states, x and y belong to $W(\mathfrak{s})_{|\mathbf{u}}$. By induction: $\{x, y\}^* \subseteq W(\mathfrak{s})_{|\mathbf{u}}$; and, as $W(\mathfrak{s})$ is suffix-closed: $\{x, y\}^* \subseteq W(\mathfrak{s})$. Let $\mathbf{v} \in \{x, y\}^{\curlywedge(\mathcal{A})-1}$: $x\mathbf{v}$ and $y\mathbf{v}$ are $\mathfrak{s}$-words and $x\mathbf{v}x$, $x\mathbf{v}y$, $y\mathbf{v}x$, and $y\mathbf{v}y$ are in $\perp(e_2)$. Hence $x\mathbf{v}z$ and $y\mathbf{v}z$ have length $\curlywedge(\mathcal{A}) + 1$ and belong to $\perp(e_1)$, which is of size $3^{\curlywedge(\mathcal{A})}$.

By Lemma 23, the connected component of $x\mathbf{v}z$ has at least $3^{\curlywedge(\mathcal{A})}$ words with different suffixes starting from position 2. Therefore, as $y\mathbf{v}z$ is also in this component, this latter has size at least $3^{\curlywedge(\mathcal{A})} + 1$. Contradiction. □

6.3 Cyclic $\mathfrak{s}$-words and Elements of Infinite Order

In this subsection, we exhibit a family of words whose induced actions have finite bounded orders. Then we prove that each word admits a bounded power which induces the same action as some word in this family.

Definition 25. *A word over Q is a cyclic $\mathfrak{s}$-word if all its powers are $\mathfrak{s}$-words (equivalently, if it is an $\mathfrak{s}$-word viewed as a cyclic word).*

Note that the existence of such cyclic $\mathfrak{s}$-words is ensured by the simple fact that any $\mathfrak{s}$-word of length $\curlywedge(\mathcal{A}) \times (1 + \#Q^{\curlywedge(\mathcal{A})})$ admits a cyclic $\mathfrak{s}$-word as a factor.

Proposition 26. *If the lengths of the 2-blocks in $\mathfrak{t}(\mathcal{A})$ are bounded by N, then any edge at level $\curlywedge(\mathcal{A}) + N$ or below in $\mathfrak{s}$ is followed by three edges.*

Proof. From Lemma 19, there is a unique initial path labeled by $3^{\curlywedge(\mathcal{A})}2^N$ (and none labeled by $3^{\curlywedge(\mathcal{A})}2^{N+1}$), call it $\mathbf{e}$: $\perp(\mathbf{e})$ is the initial vertex of 3 edges, two of which belong to $\mathfrak{s}$ by Lemma 21. By Lemmas 11 and 21, each path of $\mathfrak{s}$ is either a prefix of $\mathbf{e}$ or prefixed with $\mathbf{e}$. Suppose that an edge of $\mathfrak{s}$ below level $|\mathbf{e}| + 1$ is labeled by 2 and consider an initial branch $\mathbf{f}$ in $\mathfrak{s}$ which minimizes the length of the 1-block from $\perp(\mathbf{e})$: its label has prefix $3^{\curlywedge(\mathcal{A})}2^N1^k2$ for some $k > 0$. Consider the (non-initial) path in $\mathfrak{t}(\mathcal{A})$ obtained from $\mathbf{f}$ by erasing its first edge: by Lemma 11, it is liftable to an initial path, say $\mathbf{g}$. As $\mathbf{f}$ is in $\mathfrak{s}$, so is $\mathbf{g}$, hence $\mathbf{g}$

and $\mathbf{e}$ coincide until level $\curlywedge(\mathcal{A}) + N$; so the label of $\mathbf{g}$ has prefix $3^{\curlywedge(\mathcal{A})}2^N$. By Lemmas 11 and 9, it has also a prefix whose label is greater than or equal to (coordinatewise) $3^{\curlywedge(\mathcal{A})-1}2^N1^k2$. Hence the label of $\mathbf{g}$ has a prefix greater than or equal to $3^{\curlywedge(\mathcal{A})}2^N1^{k-1}2$, contradicting to the choice of $\mathbf{f}$.   □

Proposition 27. *If the lengths of the 2-blocks in* $\mathsf{t}(\mathcal{A})$ *are bounded, every cyclic* s*-word induces an action of finite order, bounded by a uniform constant.*

Proof. Let $\mathbf{u}$ be a cyclic s-word and n be an integer such that $|\mathbf{u}^n| > \curlywedge(\mathcal{A})$: $\mathbf{u}^n$ is an s-word. So, by Proposition 26, the label of the path of $\mathbf{u}^\omega$ is ultimately 1 and, by Proposition 5, the action induced by $\mathbf{u}$ has finite order, bounded by a constant which depends on $\curlywedge(\mathcal{A})$ (more precisely on the number of output labelings of $\bot(\mathbf{e})$, where $\mathbf{e}$ is the path of $\mathsf{t}(\mathcal{A})$ defined in the proof of Proposition 26).   □

Proposition 28. *If the 2-blocks in* $\mathsf{t}(\mathcal{A})$ *are bounded, every non-empty word over* Q *admits a non-empty bounded power equivalent to some cyclic* s*-word.*

Proof. Let $\curlywedge = \curlywedge(\mathcal{A})$ be the connection degree of $\mathcal{A}$. For a (possibly infinite) s-word $\mathbf{w}$, let $\mathrm{fact}_\curlywedge(\mathbf{w})$ denote the (finite) set of its length $\curlywedge$ factors.

Consider an infinite s-word $\mathbf{u}$, that we assume to be *maximal* in the sense that there is no other infinite s-word $\mathbf{u}'$ satisfying $\mathrm{fact}_\curlywedge(\mathbf{u}) \subsetneq \mathrm{fact}_\curlywedge(\mathbf{u}')$. Let fix a finite prefix $\mathbf{v}$ of $\mathbf{u}$ satisfying $\mathrm{fact}_\curlywedge(\mathbf{u}) = \mathrm{fact}_\curlywedge(\mathbf{v})$. From the maximality assumption on $\mathbf{u}$, we deduce that each word $\mathbf{w}$ such that $\mathbf{vw}$ is an s-word satisfies $\mathrm{fact}_\curlywedge(\mathbf{w}) \subseteq \mathrm{fact}_\curlywedge(\mathbf{v})$. We will refer to this property as Property ($\natural$).

Let $a_1 a_2 \ldots a_n \in Q^+$. By Proposition 24, a_1 appears infinitely many often in the tree $W(\mathsf{s})_{|\mathbf{v}}$. Therefore there is some word $\mathbf{u_0}$ satisfying $\mathbf{u_0}a_1 \in W(\mathsf{s})_{|\mathbf{v}}$. The goal is to build a word $\mathbf{u_1}$ satisfying $\mathbf{u_0}a_1\mathbf{u_1}a_2 \in W(\mathsf{s})_{|\mathbf{v}}$. If $\mathbf{u_0}a_1a_2 \in W(\mathsf{s})_{|\mathbf{v}}$, then take the empty word for $\mathbf{u_1}$. Otherwise, as in Figure 2, choose some word $\mathbf{v_0}$ satisfying $\mathbf{v_0}a_2 \in W(\mathsf{s})_{|\mathbf{v}}$. By Lemma 22 and Property ($\natural$), there exists a word $\mathbf{r} \in \mathrm{fact}_\curlywedge(\mathbf{v})$ such that both $\mathbf{u_0}a_1\mathbf{r} \in W(\mathsf{s})_{|\mathbf{v}}$ and $\mathbf{v_0}\mathbf{r} \in W(\mathsf{s})_{|\mathbf{v}}$ hold. Let $\mathbf{v_r}$ be a word such that $\mathbf{v_r}\mathbf{r}$ is a prefix of $\mathbf{v}$ and let $\mathbf{u_1'}$ be the word satisfying $\mathbf{v_r}\mathbf{u_1'} = \mathbf{vv_0}$. Then $\mathbf{u_1'}$ is a cyclic s-word, since its length $\curlywedge$ prefix $\mathbf{r}$ satisfies $\mathbf{v_r}\mathbf{u_1'}\mathbf{r} \in W(\mathsf{s})$.

By Proposition 27, $\mathbf{u_1'}$ has finite order q. Set $\mathbf{u_1} = \mathbf{u_1'}^q$: $a_1\mathbf{u_1}a_2 \in W(\mathsf{s})_{|\mathbf{vu_0}}$.

The same method produces words $(\mathbf{u_i})_{1 \leq i \leq n}$ of length at least $\curlywedge$ that induce the trivial action and such that the word $\mathbf{w}^{(0)} = a_1\mathbf{u_1}a_2 \cdots a_{n-1}\mathbf{u_{n-1}}a_n\mathbf{u_n}$ satisfies $\mathbf{w}^{(0)}a_1 \in S_{1|\mathbf{vu_0}}$ and induces the same action as $a_1 \cdots a_n$. For $i \geq 0$, we analogously define a word $\mathbf{w}^{(i+1)}$ inducing the same action as $a_1 \cdots a_n$ such that $\mathbf{w}^{(i+1)}a_1 \in S_{1|\mathbf{vu_0}\mathbf{w}^{(0)}\ldots\mathbf{w}^{(i)}}$. Eventually, there exist $\mathbf{i} < \mathbf{j}$ such that $\mathbf{u_1}^{(i)}$ and $\mathbf{u_1}^{(j+1)}$ have the same prefix of length $\curlywedge$, hence $\mathbf{w}^{(i)} \cdots \mathbf{w}^{(j)}$ is a cyclic s-word and induces the same action as $(a_1 \cdots a_n)^{\mathbf{j}-\mathbf{i}}$. Note that $\mathbf{j} - \mathbf{i}$ is bounded by a constant depending on $\curlywedge(\mathcal{A})$ and $\#Q$.   □

Corollary 29. *If* $\mathcal{A}$ *generates an infinite group, the orbit tree* $\mathsf{t}(\mathcal{A})$ *admits an active self-liftable branch, which is labeled either by* 3^ω *or by* 3^n2^ω *for some* n.

Proof. If the lengths of the 2-blocks are bounded, any non-empty word has a non-empty bounded power which is equivalent to a cyclic s-word by Proposition 28,

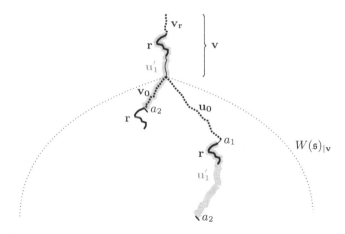

Fig. 2. Proof of Proposition 28: building a word $\mathbf{u}'_1$ satisfying $\mathbf{u}_0 a_1 \mathbf{u}'_1 a_2 \in W(\mathfrak{s})_{|\mathbf{v}}$

so the order of the action it induces is finite and bounded by a constant from Propositions 27 and 28. From Zelmanov's solution to the restricted Burnside problem (see [7] for a simpler proof in our framework) the group $\langle \mathcal{A} \rangle$ is finite, which contradicts the hypothesis. Proposition 26 leads to the conclusion.   □

Our main result (Theorem 1) immediately follows now from Theorem 17. Note that in the case when the group is infinite the proof of the existence of elements of infinite order is completely constructive. Indeed, such elements are constructed from Proposition 15 where all bounds are known. In particular, Theorem 1 allows us to detect some infinite order elements, undetectable by the existing packages FR [3] and automgrp [21] for GAP system [9], dedicated to automaton (semi)groups.

References

1. Akhavi, A., Klimann, I., Lombardy, S., Mairesse, J., Picantin, M.: On the finiteness problem for automaton (semi)groups. Int. J. Algebr. Comput. 22(6) (2012)
2. Alešin, S.V.: Finite automata and the Burnside problem for periodic groups. Mat. Zametki **11**, 319–328 (1972)
3. Bartholdi, L.: FR functionally recursive groups, self-similar groups, GAP package for computation in self-similar groups and semigroups, V. 2.2.1 (2015)
4. Bartholdi, L., Šuniḱ, Z.: On the word and period growth of some groups of tree automorphisms. Comm. Algebra **29–11**, 4923–4964 (2001)
5. Belk, J., Bleak, C.: Some undecidability results for asynchronous transducers and the Brin-Thompson group $2V$. ArXiv:1405.0982
6. Burnside, W.: On an unsettled question in the theory of discontinuous groups. Quart. J. Math. **33**, 230–238 (1902)
7. D'Angeli, D., Rodaro, E.: Freeness of automata groups vs boundary dynamics. ArXiv:1410.6097v2
8. D'Angeli, D., Rodaro, E.: A geometric approach to (semi)-groups defined by automata via dual transducers. Geometriae Dedicata **174**(1), 375–400 (2015)

9. The GAP Group: GAP Groups, Algorithms, and Programming (2015)
10. Gawron, P.W., Nekrashevych, V.V., Sushchansky, V.I.: Conjugation in tree automorphism groups. Internat. J. Algebr. Comput. **11**–**5**, 529–547 (2001)
11. Gillibert, P.: The finiteness problem for automaton semigroups is undecidable. Internat. J. Algebr. Comput. **24**–**1**, 1–9 (2014)
12. Gluškov, V.M.: Abstract theory of automata. Uspehi Mat. Nauk **16**(5), 3–62 (1961)
13. Godin, T., Klimann, I., Picantin, M.: On Torsion-Free Semigroups Generated by Invertible Reversible Mealy Automata. In: Dediu, A.-H., Formenti, E., Martín-Vide, C., Truthe, B. (eds.) LATA 2015. LNCS, vol. 8977, pp. 328–339. Springer, Heidelberg (2015)
14. Golod, E.S., Shafarevich, I.: On the class field tower. Izv. Akad. Nauk SSSR Ser. Mat. **28**, 261–272 (1964)
15. Grigorchuk, R.: On Burnside's problem on periodic groups. Funktsional. Anal. i Prilozhen **14**(1), 53–54 (1980)
16. Grigorchuk, R.: Degrees of growth of finitely generated groups and the theory of invariant means. Izv. Akad. Nauk SSSR Ser. Mat. **48**(5), 939–985 (1984)
17. Grigorchuk, R., Savchuk, D.: Ergodic decomposition of group actions on rooted trees. In: Proc. of Steklov Inst. of Math. (to appear, 2015)
18. Gupta, N., Sidki, S.: On the Burnside problem for periodic groups. Math. Z. **182**–**3**, 385–388 (1983)
19. Klimann, I.: Automaton semigroups: The two-state case. Theor. Comput. Syst., 1–17 (special issue STACS 2013) (2014)
20. Klimann, I., Picantin, M., Savchuk, D.: Orbit automata as a new tool to attack the order problem in automaton groups. ArXiv:1411.0158v2
21. Muntyan, Y., Savchuk, D.: automgrp automata groups, GAP package for computation in self-similar groups and semigroups, V. 1.2.4 (2014)
22. Nekrashevych, V.: Self-similar groups, Mathematical Surveys and Monographs, vol. 117. American Mathematical Society, Providence (2005)
23. Savchuk, D., Vorobets, Y.: Automata generating free products of groups of order 2. J. Algebra **336**–**1**, 53–66 (2011)
24. Sushchansky, V.I.: Periodic permutation p-groups and the unrestricted Burnside problem. DAN SSSR. **247**(3), 557–562 (1979). (in Russian)

Path Checking for MTL and TPTL
over Data Words

Shiguang Feng[1]([⊠]), Markus Lohrey[2], and Karin Quaas[1]

[1] Institut für Informatik, Universität Leipzig, Leipzig, Germany
[2] Department für Elektrotechnik und Informatik,
Universität Siegen, Siegen, Germany
shig.feng@gmail.com

Abstract. Precise complexity results are derived for the model checking problems for MTL and TPTL on (in)finite data words and deterministic one-counter machines. Depending on the number of register variables and the encoding of constraint numbers (unary or binary), the complexity is P-complete or PSPACE-complete. Proofs can be found in the long version [10].

1 Introduction

Linear time temporal logic (LTL) is nowadays of the main logical formalisms for describing system behaviour Triggered by real time applications, various timed extensions of LTL have been invented. Two of the most prominent examples are MTL (metric temporal logic) [13] and TPTL (timed propositional temporal logic) [2]. In MTL, the operators next (X) and until (U) are indexed by time intervals. For instance, the formula $p \, U_{[2,3)} \, q$ holds at time t, if there is a time $t' \in [t+2, t+3)$, where q holds, and p holds during the interval $[t, t')$. TPTL is a more powerful logic that is equipped with a freeze formalism. It uses register variables, which can be set to the current time value and later these register variables can be compared with the current time value. For instance, the above MTL-formula $p \, U_{[2,3)} \, q$ is equivalent to the TPTL-formula $x.(p \, U \, (q \wedge 2 \leq x < 3))$. Here, the constraint $2 \leq x < 3$ should be read as: The difference of the current time value and the value stored in x is in the interval $[2, 3)$. In this paper, we always use the discrete semantics (opposed to the continuous semantics), where formulae are interpreted over (in)finite timed sequences $(P_0, d_0)(P_1, d_1) \ldots$, where the d_i are time stamps and the P_i are sets of atomic propositions.

The freeze mechanism from TPTL has also received attention in connection with data words. A data word is a finite or infinite sequence $(P_0, d_0)(P_1, d_1) \ldots$ of the above form, where we do not require the data values d_i to be monotonic, and we speak of *non-monotonic data words*. As for TPTL, freezeLTL can store the current data value in a register x. But in contrast to TPTL, the value of x can only be compared for equality with the current data value.

S. Feng—The author is supported by the German Research Foundation (DFG), GRK 1763.

© Springer International Publishing Switzerland 2015
I. Potapov (Ed.): DLT 2015, LNCS 9168, pp. 326–339, 2015.
DOI: 10.1007/978-3-319-21500-6_26

Satisfiability and model checking for MTL, TPTL and freezeLTL have been studied intensively in the past [2–4,7,8,15–17]. For model checking freezeLTL the authors of [8] consider one-counter machines (OCM) as a mechanism for generating infinite non-monotonic data words, where the data values are the counter values along the unique computation path. Whereas freezeLTL model checking for non-deterministic OCM is Σ_1^1-complete, the problem becomes PSPACE-complete for deterministic OCM [8].

In this paper, we study MTL and TPTL over non-monotonic data words. The latter logic extends both freezeLTL over non-monotonic data words and TPTL over monotonic data words: As for freezeLTL, data values are natural numbers that can vary arbitrarily over time. In contrast to the latter, one can express that the difference of the current data value and the value stored in a register belongs to a certain interval, whereas freezeLTL only allows to say that this difference is zero. Applications for TPTL over non-monotonic data values can be seen in areas, where data streams of discrete values have to be analyzed and the focus is on the dynamic variation of the values (e.g. streams of discrete sensor data or stock charts). Recently, it has been shown that (in contrast to the monotonic setting [1]) in the non-monotonic setting, TPTL is strictly more powerful than MTL [5].

We investigate the complexity of model checking problems for TPTL over non-monotonic data words. These data words can be either finite or infinite periodic; in the latter case the data word is specified by an initial part, a period, and an offset number, which is added to the data values in the period after each repetition of the period. For periodic words without data values (i.e., ω-words of the form uv^ω), the complexity of LTL model checking (also known as LTL path checking) belongs to $\mathsf{AC}^1(\mathsf{LogDCFL})$ (a subclass of NC) [14]. This result solved a long standing open problem. For finite monotonic data words, the same complexity bound has been shown for MTL in [4].

We show that the latter result of [4] is quite sharp in the following sense: Path checking for MTL over non-monotonic (finite or infinite) data words as well as path checking for TPTL with one register variable over monotonic (finite or infinite) data words is P-complete. Moreover, path checking for TPTL (with an arbitrary number of register variables) over finite as well as infinite periodic data words becomes PSPACE-complete. We also show that PSPACE-hardness already holds (i) for the fragment of TPTL with only two register variables and (ii) for full TPTL, where all interval borders are encoded in unary (the latter result can be shown by a straightforward adaptation of the PSPACE-hardness proof in [8]). These results yield a rather complete picture on the complexity of path checking for MTL and TPTL, see Fig. 5.

2   Temporal Logics over Data Words

Let $\mathcal{P}$ be a finite set of *atomic propositions*. A *data word* over $\mathcal{P}$ is a finite or infinite sequence $(P_0, d_0)(P_1, d_1) \cdots$ of pairs from $2^{\mathcal{P}} \times \mathbb{N}$. It is *monotonic* (*strictly monotonic*), if $d_i \leq d_{i+1}$ ($d_i < d_{i+1}$) for all appropriate i. It is *pure*, if $P_i = \emptyset$ for all $i \geq 0$. A pure data word is just written as a sequence of natural numbers. We denote with $(2^{\mathcal{P}} \times \mathbb{N})^*$ and $(2^{\mathcal{P}} \times \mathbb{N})^\omega$, respectively, the set of finite and infinite, respectively, data words over $\mathcal{P}$. The *length* of a data word u is denoted by $|u|$, where we set $|u| = \infty$ for the case that u is infinite. For the data word $u = (P_0, d_0)(P_1, d_1) \cdots$, we use the notations $u[i] = (P_i, d_i)$, $u[: i] = (P_0, d_0)(P_1, d_1) \cdots (P_i, d_i)$, $u[i :] = (P_i, d_i)(P_{i+1}, d_{i+1}) \cdots$, and $u_{+k} = (P_0, d_0 + k)(P_1, d_1 + k) \cdots$, where $k \in \mathbb{N}$. We use $u_1 u_2$ to denote the concatenation of two data words u_1 and u_2, where u_1 has to be finite. For finite data words u_1, u_2 and $k \in \mathbb{N}$, let

$$u_1(u_2)^\omega_{+k} = u_1 u_2 (u_2)_{+k}(u_2)_{+2k}(u_2)_{+3k} \cdots .$$

For complexity considerations, the encoding of the data values and the offset number k (in an infinite data word) makes a difference. We speak of *unary* (resp., *binary*) encoded data words if all these numbers are given in unary (resp., binary) encoding.

The set of formulae of the logic MTL is built up from $\mathcal{P}$ by Boolean connectives, the *next* and the *until* modality using the following grammar, where $p \in \mathcal{P}$ and $I \subseteq \mathbb{Z}$ is an interval with endpoints in $\mathbb{Z} \cup \{-\infty, +\infty\}$:

$$\varphi ::= p \mid \neg\varphi \mid \varphi \wedge \varphi \mid \mathsf{X}_I \varphi \mid \varphi \mathsf{U}_I \varphi$$

Formulae of MTL are interpreted over data words. Let $w = (P_0, d_0)(P_1, d_1) \cdots$ be a data word, and let $i \leq |w|$. We define the *satisfaction relation for* MTL inductively as follows (we omit the obvious cases for $\neg$ and $\wedge$):

- $(w, i) \models p$ if and only if $p \in P_i$
- $(w, i) \models \mathsf{X}_I \varphi$ if and only if $i + 1 \leq |w|$, $d_{i+1} - d_i \in I$ and $(w, i + 1) \models \varphi$
- $(w, i) \models \varphi_1 \mathsf{U}_I \varphi_2$ if and only if there exists a position j with $i \leq j \leq |w|$, $(w, j) \models \varphi_2$, $d_j - d_i \in I$, and $(w, t) \models \varphi_1$ for all $t \in [i, j)$.

We say that a data word *satisfies* an MTL-formula φ, written $w \models \varphi$, if $(w, 0) \models \varphi$. We use the following standard abbreviations: $\varphi_1 \vee \varphi_2 := \neg(\neg\varphi_1 \wedge \neg\varphi_2)$, $\varphi_1 \rightarrow \varphi_2 := \neg\varphi_1 \vee \varphi_2$, $\mathsf{true} := p \vee \neg p$, $\mathsf{false} := \neg\mathsf{true}$, $\mathsf{F}_I \varphi := \mathsf{true} \mathsf{U}_I \varphi$, $\mathsf{G}_I \varphi := \neg\mathsf{F}_I \neg\varphi$.

Next we define formulae of the logic TPTL. For this, let V be a countable set of *register variables*. The set of TPTL-formulae is given by the following grammar, where $p \in \mathcal{P}$, $x \in V$, $c \in \mathbb{Z}$, and $\sim \in \{<, \leq, =, \geq, >\}$:

$$\varphi ::= p \mid x \sim c \mid \neg\varphi \mid \varphi \wedge \varphi \mid \mathsf{X}\varphi \mid \varphi \mathsf{U}\varphi \mid x.\varphi \tag{1}$$

We use the same syntactical abbreviations as for MTL. The fragment freezeLTL is obtained by restricting $\sim$ in (1) to $=$. Ordinary LTL is obtained by disallowing the use of register variables. Given $r \geq 1$, we use TPTL^r (resp., freezeLTL^r) to denote the fragment of TPTL (resp., freezeLTL) that uses at most r different register variables.

A *register valuation* ν is a function from V to $\mathbb{Z}$. Given a register valuation ν, a data value $d \in \mathbb{Z}$, and a variable $x \in V$, we define the register valuations $\nu + d$

and $\nu[x \mapsto d]$ as follows: $(\nu + d)(y) = \nu(y) + d$ for every $y \in V$, $(\nu[x \mapsto d])(y) = \nu(y)$ for every $y \in V \setminus \{x\}$, and $(\nu[x \mapsto d])(x) = d$.

Let $w = (P_0, d_0)(P_1, d_1) \cdots$ be a data word, let ν be a register valuation, and let $i \in \mathbb{N}$. The satisfaction relation for TPTL is inductively defined in a similar way as for MTL; we only give the definitions for the new formulae:

- $(w, i, \nu) \models X\varphi$ if and only if $i + 1 \leq |w|$ and $(w, i + 1, \nu) \models \varphi$
- $(w, i, \nu) \models \varphi_1 U \varphi_2$ if and only if there exists a position j with $i \leq j \leq |w|$, $(w, j, \nu) \models \varphi_2$, and $(w, t, \nu) \models \varphi_1$ for all $t \in [i, j)$
- $(w, i, \nu) \models x.\varphi$ if and only if $(w, i, \nu[x \mapsto d_i]) \models \varphi$
- $(w, i, \nu) \models x \sim c$ if and only if $d_i - \nu(x) \sim c$.

Note that $x \sim c$ does not mean that the current value $v = \nu(x)$ of x satisfies $v \sim c$, but expresses that $d_i - v \sim c$, where d_i is the current data value. We say that a data word w satisfies a TPTL-formula φ, written $w \models \varphi$, if $(w, 0, \bar{0}) \models \varphi$, where $\bar{0}$ denotes the valuation that maps all variables to the initial data value d_0.

For complexity considerations, it makes a difference, whether the numbers c in constraints $x \sim c$ are binary or unary encoded, and similarly for the interval borders in MTL. We write TPTL_u^r, TPTL_u, MTL_u (resp., TPTL_b^r, TPTL_b, MTL_b) if we want to emphasize that numbers are encoded in unary (resp., binary) notation. The *length* of a (TPTL or MTL) formula ψ, denoted by $|\psi|$, is the number of symbols occurring in ψ.

3   Path Checking Problems for TPTL and MTL

In this section, we study the path checking problems for our logics over data words. Data words can be (i) finite or infinite, (ii) monotonic or non-monotonic, (iii) pure or non-pure, and (iv) unary encoded or binary encoded. For one of our logics L and a class of data words C, we consider the *path checking problem for* L *over* C. It asks whether for a given data word $w \in C$ and a given formula $\varphi \in L$, $w \models \varphi$ holds.

3.1   Upper Bounds

In this section we prove our upper complexity bounds. All bounds hold for non-monotonic and non-pure data words (and we will not mention this explicitly in the theorems). But we have to distinguish whether (i) data words are unary or binary encoded, and (ii) whether data words are finite or infinite. For the most general path checking problem (TPTL_b over infinite binary encoded data words) we can devise an alternating polynomial time algorithm (and hence a polynomial space algorithm). The only technical difficulty is to bound the position in the infinite data word and the values of the register valuation, so that they can be stored in polynomial space, see [10] for details.

Theorem 1. *Path checking for* TPTL_b *over infinite binary encoded data words is in* PSPACE.

If the number of register variables is fixed and all data values are unary encoded, then the alternating Turing-machine in the proof of Theorem 1 works in logarithmic space. Since $\mathsf{ALOGSPACE} = \mathsf{P}$, we obtain the following statement for (i). For (ii) we show that an infinite *binary* encoded monotonic data word can be replaced by an infinite *unary* encoded data word, which allows to apply (i).

Theorem 2. *For every fixed $r \in \mathbb{N}$, path checking for TPTL_u^r over (i) infinite unary encoded data words or (ii) infinite binary encoded monotonic data words is in P.*

Actually, for finite data words, we obtain a polynomial time algorithm also for binary encoded data words (assuming again a fixed number of register variables):

Theorem 3. *For every fixed $r \in \mathbb{N}$, path checking for TPTL_b^r over finite binary encoded data words is in P.*

For infinite data words we have to reduce the number of register variables to one in order to get a polynomial time complexity for binary encoded numbers:

Theorem 4. *Path checking for TPTL_b^1 over infinite binary encoded data words is in P.*

For the proof of Theorem 4 we need the following two lemmas.

Lemma 5. *For a given LTL-formula ψ, words $u_1, \ldots, u_k, u \in (2^{\mathcal{P}})^*$ and binary encoded numbers $N_1, \ldots, N_k \in \mathbb{N}$, the question whether $u_1^{N_1} u_2^{N_2} \cdots u_k^{N_k} u^\omega \models \psi$ holds, belongs to P (actually, $\mathsf{AC}^1(\mathsf{LogDCFL})$).*

The crucial point is that for all finite words $u, v \in (2^{\mathcal{P}})^*$, every infinite word $w \in (2^{\mathcal{P}})^\omega$ and every number $N \geq |\psi|$, we have $uv^N w \models \psi$ if and only if $uv^{|\psi|} w \models \psi$. This can be shown by using the Ehrenfeucht-Fraïssé game for LTL from [9]. Hence, one can replace all exponents N_i by small numbers of size at most $|\psi|$. Then, one can use a polynomial time algorithm (or $\mathsf{AC}^1(\mathsf{LogDCFL})$ algorithm) for LTL path checking [14].

Lemma 6. *Path checking for TPTL_b-formulae, which do not contain subformulae of the form $x.\theta$ for a register variable x, over infinite binary encoded data words is in P (in fact, $\mathsf{AC}^1(\mathsf{LogDCFL})$).*

Proof. We reduce the question, whether $w \models \psi$ in logspace to an instance of the succinct LTL path checking problem from Lemma 5. Let $w = u_1(u_2)_{+k}^\omega$ and let $w[i] = (P_i, d_i) \in 2^{\mathcal{P}} \times \mathbb{N}$. Let $n_1 = |u_1|$ and $n_2 = |u_2|$. We can assume that only one register variable x appears in ψ (since we do not use the freeze construct $x.()$ in ψ all register variables remain at the initial value d_0).

In order to construct an LTL-formula from ψ, it remains to eliminate occurrences of constraints $x \sim c$ in ψ. W.l.o.g. all constraints are of the form $x < c$ or $x > c$. Let $x \sim_1 c_1, \ldots, x \sim_m c_m$ be a list of all constraints that appear in ψ. We introduce for every $1 \leq j \leq m$ a new atomic proposition p_j and let $\mathcal{P}' = \mathcal{P} \cup \{p_1, \ldots, p_m\}$. Let ψ' be obtained from ψ by replacing every occurrence of $x \sim_j c_j$ by p_j, and let $w' \in (2^{\mathcal{P}'})^\omega$ be the ω-word with $w'[i] = P_i \cup \{p_j \mid 1 \le j \le m, d_i - d_0 \sim_j c_j\}$. Clearly $w \models \psi$ if and only if $w' \models \psi'$. We will show that the word w' can be written in the form considered in Lemma 5.

First of all, we can write w' as $w' = u'_1 u'_{2,0} u'_{2,1} u'_{2,2} \cdots$, where $|u'_1| = n_1$ and $|u'_{2,i}| = n_2$. The word u'_1 can be computed in logspace by evaluating all constraints at all positions of u_1. Moreover, every word $u'_{2,i}$ is obtained from u_2 (without the data values) by adding the new propositions p_j at the appropriate positions. Consider the equivalence relation $\equiv$ on $\mathbb{N}$ with $a \equiv b$ if and only if $u'_{2,a} = u'_{2,b}$. The crucial observations are that (i) every equivalence class of $\equiv$ is an interval, and (ii) the index of $\equiv$ is bounded by $1 + n_2 \cdot m$ (one plus length of u_2 times number of constraints). To see this, consider a position $0 \le i \le n_2 - 1$ in the word u_2 and a constraint $x \sim_j c_j$ ($1 \le j \le m$). Then, the truth value of "proposition p_j is present at the ith position of $u'_{2,x}$" switches (from true to false or from false to true) at most once when x grows. The reason for this is that the data value at position $n_1 + i + n_2 \cdot x$ is $d_{n_1 + i + n_2 \cdot x} = d_{n_1 + i} + k \cdot x$ for $x \ge 0$, i.e., it grows monotonically with x. Hence, the truth value of $d_{n_1 + i} + k \cdot x - d_0 \sim_j c_j$ switches at most once, when x grows. So, we get at most $n_2 \cdot m$ many "switching points" in $\mathbb{N}$ which produce at most $1 + n_2 \cdot m$ many intervals.

Let $I_1, \ldots, I_l$ be a list of all $\equiv$-classes (intervals), where $a < b$ whenever $a \in I_i$, $b \in I_j$ and $i < j$. The borders of these intervals can be computed in logspace using arithmetic on binary encoded numbers (addition, multiplication and division with remainder can be carried out in logspace on binary encoded numbers [12]). Hence, we can compute in logspace the lengths $N_i = |I_i|$ of the intervals, where $N_l = \omega$. Also, for all $1 \le i \le l$ we can compute in logspace the unique word v_i such that $v_i = u'_{2,a}$ for all $a \in I_i$. Hence, $w' = u'_1 v_1^{N_1} \cdots v_l^{N_l}$. We can now apply Lemma 5. □

Proof of Theorem 4. Consider an infinite binary encoded data word $w = u_1 (u_2)^\omega_{+k}$ and a TPTL^1_b-formula ψ. Let $n = |u_1| + |u_2|$. We check in polynomial time whether $w \models \psi$. A TPTL-formula φ is closed if every occurrence of a register variable x in φ appears within a subformula of the form $x.\theta$. The following two claims are straightforward:

Claim 1: If φ is closed, then for all valuations ν, ν', $(w, i, \nu) \models \varphi$ iff $(w, i, \nu') \models \varphi$.

Claim 2: If φ is closed and $i \ge |u_1|$, then for every valuation ν, $(w, i, \nu) \models \varphi$ iff $(w, i + |u_2|, \nu) \models \varphi$.

By Claim 1 we can write $(w, i) \models \varphi$ for $(w, i, \nu) \models \varphi$. It suffices to compute for every (necessarily closed) subformula $x.\varphi$ of ψ the set of all positions $i \in [0, n-1]$ such that $(w, i) \models x.\varphi$, or equivalently $w[i :] \models \varphi$. We do this in a bottom-up process. Consider a subformula $x.\varphi$ of ψ and a position $i \in [0, n-1]$. We have to check whether $w[i :] \models \varphi$. Let $x.\varphi_1, \ldots, x.\varphi_l$ be all maximal (with respect to the subformula relation) subformulae of φ of the form $x.\theta$. We can assume that for every $1 \le s \le l$ we have already determined the set of positions $j \in [0, n-1]$ such that $(w, j) \models x.\varphi_s$. We can therefore replace every subformula $x.\varphi_s$ of φ by a new atomic proposition p_s and add in the data words u_1 (resp., u_2) the proposition

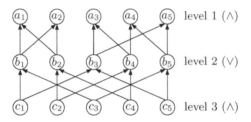

Fig. 1. An SAM2-circuit

p_s to all positions j (resp., $j - |u_1|$) such that $(w, j) \models x.\varphi_s$, where $j \in [0, n-1]$. Here, we make use of Claim 2. We denote the resulting formula and the resulting data word with φ' and $w' = u'_1(u'_2)^{\omega}_{+k}$, respectively. Next, it is easy to compute from u'_1 and u'_2 new finite data words v_1 and v_2 such that $v_1(v_2)^{\omega}_{+k} = w'[i :]$: If $i < |u'_1|$ then we take $v_1 = u'_1[i :]$ and $v_2 = u'_2$. If $|u'_1| \leq i \leq n - 1$, then we take $v_1 = \varepsilon$ and $v_2 = u'_2[i :](u'_2[: i - 1] + k)$. Finally, using Lemma 6 we can check in polynomial time whether $w'[i :] \models \varphi'$. $\square$

3.2 Lower Bounds

We prove several P-hardness and PSPACE-hardness results for path checking.

P-Hardness. We prove our P-hardness results by a reduction from a restricted version of the Boolean circuit value problem. A *synchronous alternating monotone circuit with fanin 2 and fanout 2* (briefly, SAM2-circuit) is a Boolean circuit divided into levels $1, \ldots, l$ ($l \geq 2$) such that the following properties hold:

- All wires go from a gate in level $i + 1$ to a gate from level i ($1 \leq i \leq l - 1$).
- All output gates are in level 1 and all input gates are in level l, and the latter are labelled with input bits. Moreover, there is a distinguished output gate on level 1.
- All gates in the same level $1 \leq i \leq l - 1$ are of the same type ($\wedge$ or $\vee$) and the levels alternate between $\wedge$-levels and $\vee$-levels.
- All gates except the output gates have fanout 2 and all gates except the input gates have fanin 2. The two input gates for a gate at level $i \leq l - 1$ are different.

By the restriction to fanin 2 and fanout 2, we know that each level contains the same number of gates. Fig. 1 shows an example of an SAM2-circuit (the node names a_i, b_i, c_i will be needed later). The circuit value problem for SAM2-circuits (i.e., the question whether the distinguished output gate of a given SAM2-circuit evaluates to 1), which is called SAM2CVP, is P-complete [11].

Recall that finite path checking for MTL (a fragment of TPTL^1) over monotonic data words is in the parallel complexity class $\mathsf{AC}^1(\mathsf{LogDCFL})$ [4]. We will show that for both (i) MTL_u over non-monotonic data words and (ii) TPTL^1_u over monotonic data words the path checking problem becomes P-hard (and hence P-complete).

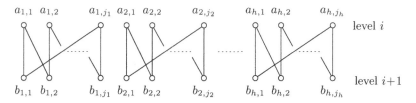

Fig. 2. The induced subgraph between level i and $i + 1$

Theorem 7. *Path checking for* MTL_u *over finite unary encoded pure data words is* P*-hard.*

Proof. We reduce from SAM2CVP. Let α be the input circuit. We first encode each two consecutive levels of α into a data word, and combine these data words into a data word w, which is the encoding of the whole circuit. Then we construct a formula ψ such that $w \models \psi$ if and only if α evaluates to 1. The data word w that we are constructing contains gate names of α (and some copies of the gates) as atomic propositions. These propositions will be only needed for the construction. At the end, we can remove all propositions from the data word w and hence obtain a pure data word. The whole construction can be done in logspace. The reader might look at the example in [10], where the construction is carried out for the circuit from Fig. 1.

Let α be an SAM2-circuit with $l \geq 2$ levels and n gates in each level. By the restriction to fanin 2 and fanout 2, the induced undirected subgraph which contains the nodes in level i and $i + 1$ ($1 \leq i < l$) consists of several cycles; see Fig. 2. For instance, for the circuit in Fig. 1 the number of cycles between level 1 and 2 (resp., 2 and 3) is 2.

We can enumerate in logspace the gates of level i and $i + 1$ such that they occur in the order shown in Fig. 2. For this, let $a_1, \ldots, a_n$ (resp., $b_1, \ldots, b_n$) be the nodes in level i (resp., $i + 1$) in the order in which they occur in the input description. We start with a_1 and enumerate the nodes in the cycle containing a_1 (from a_1 we go to the smaller neighbor among $b_1, \ldots, b_n$, then the next node on the cycle is uniquely determined since the graph has degree 2). Thereby we store the current node in the cycle and the starting node a_1. As soon as we return to a_1, we know that the first cycle is completed. To find the next cycle, we search for the first node from $a_2, \ldots, a_n$ that is not reachable from a_1 (reachability in undirected graphs is in logspace), and continue this way.

So, assume that the nodes in layer i and $i + 1$ are ordered as in Fig. 2. In particular, we have h cycles. For each $1 \leq t \leq h$, we add a new node $a'_{t,1}$ (resp., $b'_{t,1}$) after a_{t,j_t} (resp., b_{t,j_t}). Then we replace the edge $(a_{t,j_t}, b_{t,1})$ by the edge $(a_{t,j_t}, b'_{t,1})$ ($1 \leq t \leq h$). In this way we obtain the graph from Fig. 3. Again, the construction can be done in logspace by adding the new nodes and new edges once a cycle was completed in the enumeration procedure from the previous paragraph.

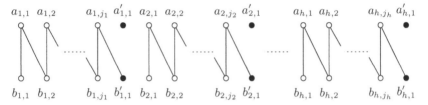

Fig. 3. The graph obtained from the induced subgraph

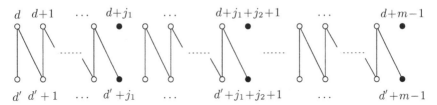

Fig. 4. Labeling the new graph

By adding dummy nodes, we can assume that for every $1 \leq i \leq l - 1$, the subgraph between level i and $i + 1$ has the same number (say h) of cycles. We still denote by n the number of nodes in each level. Thus, after the above step we have $m = n + h$ nodes in each level. Let $d = (i - 1) \cdot 2m$ and $d' = d + m$. In Fig. 3, we label the nodes in level i (resp., $i + 1$) with the numbers $d, d + 1, \ldots, d + m - 1$ (resp. $d', d' + 1 \ldots, d' + m - 1$) in this order, see Fig. 4. By this labeling, the difference between two connected nodes in level i and level $i + 1$ is always m or $m + 1$. So we can use the modality $\mathsf{F}_{[m,m+1]}$ (resp., $\mathsf{G}_{[m,m+1]}$) to jump from an $\vee$-gate (resp., $\wedge$-gate) in level i to a successor gate in level $i + 1$. We now obtain in logspace the data word $w_i = w_{i,1} w_{i,2}$, where

$$
w_{i,1} = \begin{cases}
(a_{1,1}, d)(a_{1,2}, d + 1) \cdots (a_{1,j_1}, d + j_1 - 1) \\
(a_{2,1}, d + j_1 + 1)(a_{2,2}, d + j_1 + 2) \cdots (a_{2,j_2}, d + j_1 + j_2) \cdots \\
(a_{h,1}, d + \sum_{t=1}^{h-1} j_t + h - 1)(a_{h,2}, d + \sum_{t=1}^{h-1} j_t + h) \cdots (a_{h,j_h}, d + m - 2)
\end{cases}
$$

$$
w_{i,2} = \begin{cases}
(b_{1,1}, d') \cdots (b_{1,j_1}, d' + j_1 - 1)(b'_{1,1}, d' + j_1) \\
(b_{2,1}, d' + j_1 + 1) \cdots (b_{2,j_2}, d' + j_1 + j_2)(b'_{2,1}, d' + j_1 + j_2 + 1) \cdots \\
(b_{h,1}, d' + \sum_{t=1}^{h-1} j_t + h - 1) \cdots (b_{h,j_h}, d' + m - 2)(b'_{h,1}, d' + m - 1)
\end{cases}
$$

which is the encoding of the wires between level i and level $i + 1$ from Fig. 4. Note that the new nodes $a'_{1,1}, a'_{2,1}, \ldots, a'_{h,1}$ in level i of the graph in Fig. 3 do not occur in $w_{i,1}$.

Suppose now that all data words w_i ($1 \leq i \leq l - 1$) are constructed. We then combine them to obtain the data word w for the whole circuit as follows.

Suppose that

$$w_{i,2} = (\tilde{b}_1, y_1) \cdots (\tilde{b}_m, y_m) \text{ and } w_{i+1,1} = (b_1, z_1) \cdots (b_n, z_n).$$

Note that every $\tilde{b}_i$ is either one of the b_j or b'_j (the copy of b_j). Let

$$v_{i+1,1} = (\tilde{b}_1, z'_1) \cdots (\tilde{b}_m, z'_m),$$

where the data values z'_i are determined as follows: If $\tilde{b}_i = b_j$ or $\tilde{b}_i = b'_j$, then $z'_i = z_j$. Then, the data word w is $w = w_{1,1} w_{1,2} v_{2,1} w_{2,2} \cdots v_{l-1,1} w_{l-1,2}$.

Let us explain the idea. Consider a gate a_j of level $2 \leq i \leq l-1$, and assume that level i consists of $\vee$-gates. Let b_{j_1} and b_{j_2} (from level $i+1$) be the two input gates for a_j. In the above data word $v_{i,1}$ there is a unique position where the proposition a_j occurs, and possibly a position where the copy a'_j occurs. If both positions exist, then they carry the same data value. Let us point to one of these positions. Using an MTL formula, we want to branch (existentially) to the positions in the factor $v_{i+1,1}$, where the propositions $b_{j_1}, b'_{j_1}, b_{j_2}, b'_{j_2}$ occur (where b'_{j_1} and b'_{j_2} possibly do not exist). For this, we use the modality $F_{[m,m+1]}$. By the construction, this modality branches existentially to positions in the factor $w_{i,2}$, where the propositions $b_{j_1}, b'_{j_1}, b_{j_2}, b'_{j_2}$ occur. Then, using the iterated modality X^m (which is an abbreviation for m copies of the MTL-modality $X_{\mathbb{Z}}$), we jump to the corresponding positions in $v_{i+1,1}$.

In the above argument, we assumed that $2 \leq i \leq l-1$. If $i = 1$, then we can argue similarly, if we assume that we are pointing to the unique a_j-labeled position of the prefix $w_{1,1}$ of w. Now consider level $l-1$. Suppose that

$$w_{l-1,2} = (\tilde{d}_1, v_1) \ldots (\tilde{d}_m, v_m).$$

Let $d_1, \ldots, d_n$ be the original gates of level l, which all belong to $\{\tilde{d}_1, \ldots, \tilde{d}_m\}$, and let $x_i \in \{0,1\}$ be the input value for gate d_i. Define

$$I = \{j \mid j \in [1,m], \exists i \in [1,n] : \tilde{d}_j \in \{d_i, d'_i\}, x_i = 1\}. \tag{2}$$

Let the designated output gate be the kth node in level 1. We construct the MTL-formula $\psi = X^{k-1}\varphi_1$, where φ_i $(1 \leq i \leq l-1)$ is defined inductively as follows:

$$\varphi_i = \begin{cases} F_{[m,m+1]}X^m\varphi_{i+1} & \text{if } i < l-1 \text{ and level } i \text{ is a } \vee\text{-level,} \\ G_{[m,m+1]}X^m\varphi_{i+1} & \text{if } i < l-1 \text{ and level } i \text{ is a } \wedge\text{-level,} \\ F_{[m,m+1]}(\bigvee_{j\in I} X^{m-j}\neg X\,\text{true}) & \text{if } i = l-1 \text{ and level } i \text{ is a } \vee\text{-level,} \\ G_{[m,m+1]}(\bigvee_{j\in I} X^{m-j}\neg X\,\text{true}) & \text{if } i = l-1 \text{ and level } i \text{ is a } \wedge\text{-level.} \end{cases}$$

Note that the formula $\neg X\,\text{true}$ is only true in the last position of a data word. Suppose data word w is the encoding of the circuit. From the above consideration, it follows that $w \models \psi$ if and only if the circuit α evaluates to 1. Note that we do not use any propositional variables in the formula ψ. So we can ignore the propositional part in the data word w to get a pure data word.   □

Note that the above construction uses non-monotonic data words. This is unavoidable since finite path checking for MTL over monotonic data words is in NC [4]. On the other hand, for the extension TPTL_u^1 of MTL_u we can show, using again a reduction from SAM2CVP (see [10]), P-hardness also for monotonic data words:

Theorem 8. *Path checking for* TPTL_u^1 *over finite unary encoded strictly monotonic pure data words is* P*-hard.*

PSPACE-Hardness. In [10], we prove three PSPACE lower bounds, which complete our complexity picture. The first one is shown by a reduction from QBF, whereas the latter two results are shown by a reduction from a quantified variant of the subset sum problem [19].

Theorem 9. *Path checking for* TPTL_u *over finite unary encoded strictly monotonic pure data words is* PSPACE*-hard.*

Theorem 10. *Path checking for* TPTL_b^2 *over the infinite strictly monotonic pure data word* $w = 0(1)_{+1}^\omega = 0, 1, 2, 3, 4, \ldots$ *is* PSPACE*-hard.*

Theorem 11. *Path checking for* $\mathsf{freezeLTL}^2$ *(and hence* TPTL_u^2*) over infinite binary encoded pure data words is* PSPACE*-hard.*

Recall from Theorem 2 that for every fixed r, path checking for TPTL_u^r over infinite binary encoded monotonic data words can be solved in polynomial time. Hence, Theorem 11 shows that monotonicity is important for Theorem 2.

3.3   Summary of the Results

Figure 5 collects our complexity results for path checking problems (here the superscript $< \infty$ is a place holder for any number $r \geq 2$). Whether data words are pure or not does not influence the complexity in all cases. Moreover, for finite data words, the complexity does not depend upon the encoding of data words (unary or binary) and the fact whether data words are monotonic or non-monotonic. On the other hand, for infinite data words, these distinctions influence the complexity: For binary and non-monotonic data words we get another picture than or unary encoded or (quasi-)monotonic data words. Note that for MTL_b and MTL_u the complexity is P-complete for all classes of data words (since MTL translates in logspace into TPTL^1).

One may also study the complexity of path checking problems for various fragments of MTL and TPTL. In this context, it is interesting to note that all lower bounds already hold for the corresponding unary fragments (where the until-operator is replaced by F and G) with only one exception: Our proof for Theorem 11 in [10] for $\mathsf{freezeLTL}^2$ needs the until operator. It is not clear, whether path checking for the unary fragment of $\mathsf{freezeLTL}^2$ over infinite binary encoded data words is still PSPACE-complete.

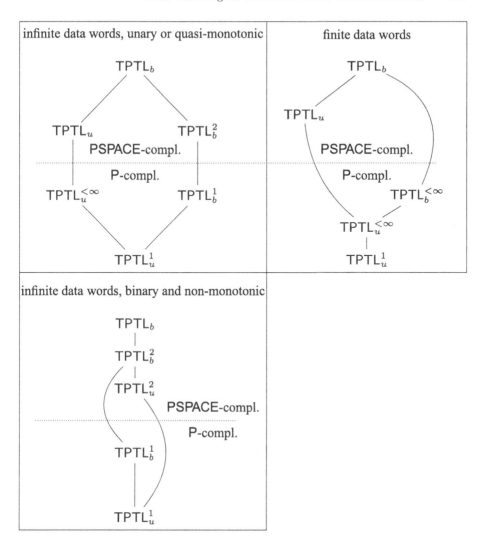

Fig. 5. Complexity results for path checking

Our complexity results for infinite unary encoded data words also hold for *deterministic one-counter machines (DOCMs)*, see [10] for a precise definition. A DOCM produces in general an infinite data word, where the sequence of atomic propositions is the sequence of states of the machine, and the sequence of data values is the sequence of counter values produced by the DOCM (the DOCM can block in which case it produces a finite data word). It is an easy observation that the data word produced by a DOCM $\mathcal{A}$ is periodic in case it is infinite, and one can in fact compute in logspace from $\mathcal{A}$ two unary encoded finite data words u_1 and u_2 and a unary encoded number k such that $u_1(u_2)_{+k}^{\omega}$ is the data word produced by $\mathcal{A}$, see also [8, Lemma 9]. For this it is crucial that the counter can be incremented or decremented in each step by at most one (or, more general, a unary encoded number). This, in turn implies that for each of the logics L considered in this paper, the model checking problem for L over DOCM (i.e., the question, whether a given formula $\varphi \in$ L holds in the data word produced by a given DOCM) is equivalent with respect to logspace reductions to the path checking problem for L over infinite unary encoded data words. Hence, the upper left diagram from Figure 5 also shows the complexity results for TPTL model checking over DOCM. In particular we strengthen the third author's recent decidability result for model checking non-monotonic TPTL over DOCMs [18]. Our results also generalizes the PSPACE-completeness result for freezeLTL over DOCMs from [8].

References

1. Alur, R., Henzinger, T.A.: Real-Time Logics: Complexity and Expressiveness. Inf. Comput. **104**(1), 35–77 (1993)
2. Alur, R., Henzinger, T.A.: A really temporal logic. J. ACM **41**(1), 181–204 (1994)
3. Bouyer, P., Larsen, K.G., Markey. Model checking one-clock priced timed automata. Log. Meth. Comput. Sci. **4**(2) (2008)
4. Bundala, D., Ouaknine, J.: On the complexity of temporal-logic path checking. In: Esparza, J., Fraigniaud, P., Husfeldt, T., Koutsoupias, E. (eds.) ICALP 2014, Part II. LNCS, vol. 8573, pp. 86–97. Springer, Heidelberg (2014)
5. Carapelle, C., Feng, S., Gil, O.F., Quaas, K.: On the expressiveness of TPTL and MTL over ω-data words. In: Proc. AFL 2014. EPTCS, vol. 151, pp. 174–187 (2014)
6. Carapelle, C., Feng, S., Fernández Gil, O., Quaas, K.: Satisfiability for MTL and TPTL over non-monotonic data words. In: Dediu, A.-H., Martín-Vide, C., Sierra-Rodríguez, J.-L., Truthe, B. (eds.) LATA 2014. LNCS, vol. 8370, pp. 248–259. Springer, Heidelberg (2014)
7. Demri, S., Lazić, R.: LTL with the freeze quantifier and register automata. ACM Trans. Comput. Log. **10**(3) (2009)
8. Demri, S., Lazić, R., Sangnier, A.: Model checking memoryful linear-time logics over one-counter automata. Theor. Comput. Sci. **411**(22–24), 2298–2316 (2010)
9. Etessami, K., Wilke, T.: An until hierarchy and other applications of an Ehrenfeucht-Fraïssé game for temporal logic. Inf. Comput. **160**(1–2), 88–108 (2000)
10. Feng, S., Lohrey, M., Quaas, K.: Path-Checking for MTL and TPTL, arXiv.org 1412.3644 (2014)
11. Greenlaw, R., Hoover, H.J., Ruzzo, W.L.: Limits to Parallel Computation: P-completeness Theory. Oxford University Press (1995)
12. Hesse, W., Allender, E., Barrington, D.A.M.: Uniform constant-depth threshold circuits for division and iterated multiplication. J. Comput. System Sci. **65**, 695–716 (2002)
13. Koymans, R.: Specifying real-time properties with metric temporal logic. Real-Time Systems **2**(4), 255–299 (1990)
14. Kuhtz, L., Finkbeiner, B.: Efficient parallel path checking for linear-time temporal logic with past and bounds. Log. Meth. Comput. Sci. **8**(4) (2012)
15. Laroussinie, F., Markey, N., Schnoebelen, P.: On model checking durational kripke structures. In: Nielsen, M., Engberg, U. (eds.) FOSSACS 2002. LNCS, vol. 2303, pp. 264–279. Springer, Heidelberg (2002)

16. Ouaknine, J., Worrell, J.B.: On metric temporal logic and faulty turing machines. In: Aceto, L., Ingólfsdóttir, A. (eds.) FOSSACS 2006. LNCS, vol. 3921, pp. 217–230. Springer, Heidelberg (2006)
17. Ouaknine, J., Worrell, J.: On the decidability and complexity of metric temporal logic over finite words. Log. Meth. Comput. Sci. 3(1) (2007)
18. Quaas, K.: Model checking metric temporal logic over automata with one counter. In: Dediu, A.-H., Martín-Vide, C., Truthe, B. (eds.) LATA 2013. LNCS, vol. 7810, pp. 468–479. Springer, Heidelberg (2013)
19. Travers, S.: The complexity of membership problems for circuits over sets of integers. Theor. Comput. Sci. 369(1), 211–229 (2006)

On Distinguishing $\mathbf{NC}^1$ and $\mathbf{NL}$

Andreas Krebs, Klaus-Jörn Lange$^{(\boxtimes)}$, and Michael Ludwig

WSI - University of Tübingen, Sand 13, 72076 Tübingen, Germany
{krebs,lange,ludwigm}@informatik.uni-tuebingen.de

Abstract. We obtain results within the area of dense completeness, which describes a close relation between families of formal languages and complexity classes. Previously we were able show that this relation exists between counter languages and **NL** but not between the regular languages and $\mathbf{NC}^1$.

We narrow the gap between the regular languages and the counter languages by considering visibly counter languages. It turns out that they are not densely complete for $\mathbf{NC}^1$. At the same time we found a restricted counter automaton model which is densely complete for **NL**.

Besides counter automata we show more positive examples in terms of L-systems.

1 Introduction

Turing machines are the key model for computation and the most general as well. A consequence however is for example the undecidability of the word problem. Two of the major areas of theory can be understood as different branches originating from the concept of the Turing machine. One branch limits Turing machines in terms of resources like space and time which led to what we know as complexity theory, the study of complexity classes. In the other branch we are in a way limiting the functionality of Turing machines resulting e.g. in pushdown or finite automata. We want to name the objects of the second branch *families of formal languages*. It turned out that complexity classes and families of formal languages have very different properties but they are also connected in many ways. The present work is a contribution to understanding the relationship between complexity classes and families of formal languages. We hope that this leads to new insights to complexity as it is much harder to analyze compared to families of formal languages.

The term *complexity class* is clear, the term *family of formal languages* however needs clarification. Certainly one can interpret *formal language* in a way, that every subset of Σ^* is a formal language but we understand it, as outlined above, as languages which are accepted or generated by certain objects like automata or grammars. Finding a final definition of what we want to consider a family of formal languages is part of our ongoing work.

The regular languages represent a very basic example of a large abundance of families of formal languages, coined by pumping theorems and built on that decision properties, which distinguish them from complexity classes. Nevertheless, most of them exhibit very close relationships to complexity classes.

© Springer International Publishing Switzerland 2015
I. Potapov (Ed.): DLT 2015, LNCS 9168, pp. 340–351, 2015.
DOI: 10.1007/978-3-319-21500-6_27

General algorithms in terms of Turing machines or circuit families are immune to a combinatorial or algebraic analysis. This makes families of formal languages interesting as they contain problems complete for complexity classes and thus in their word problems exhibit close connections to complexity theory. At the same time they are restricted in a way which makes them open for a combinatorial and even algebraic analysis.

It indeed is often the case that a family of formal languages $\mathcal{F}$ is complete for a complexity class $\mathcal{C}$ in the sense that $\mathcal{F}$ is contained in $\mathcal{C}$, and $\mathcal{F}$ contains a $\mathcal{C}$-complete problem. Examples for this situation abound in circuit complexity, e.g., with the regular languages and NC^1, or the context-free languages and SAC^1. Strengthening this link, the notion of dense completeness [KL12] further requires that each $C \in \mathcal{C}$ corresponds to a formal language $F \in \mathcal{F}$ of the same complexity, i.e. such that C and F are reducible to each other.

While it is usual to have a complete family of formal languages corresponding to some complexity class, the picture is different for dense completeness. Up to now we only found dense completeness in non-deterministic classes. Also the proofs heavily rely on non-determinism. Our working hypothesis is that only non-deterministic classes have a densely complete family of formal languages. The first examples established in [KL12] are the following:

- The index languages are densely complete for NP.
- The context-free languages are densely complete for SAC^1.
- The nondeterministic one-counter languages are densely complete for NL.
- The regular languages are **not** densely complete for NC^1.

We are especially interested in non-denseness results. In the instance of the regular languages the proof relies on the gap result of [BCST92]. In [KLL15] we were able to derive a corresponding gap result for the family of visibly counter languages, which we will use in this paper to show unconditionally that even the visibly counter languages are not densely complete in NC^1.

Our results lead to an interesting situation: We have a counter-based family which is densely complete for NL and one which is not densely complete for NC^1. The next logical step of course would be to find out whether the deterministic one-counter languages are densely complete for L.

A different direction is to look at the visibly pushdown languages which are also NC^1 complete [Dym88]. We cannot rule out the possibility that NC^1 contains indeed a dense family of formal languages. However it seems easier to show non-denseness for NC^1. On the other hand it seems easy to show denseness results for non-deterministic classes. We present examples of families of formal languages which are densely complete for certain complexity classes in terms of L-systems. This underlines our assumption that dense completeness captures an inherent and important property.

Due to space restriction, we omit some of the proofs.

We thank the anonymous referees.

2   Families of Formal Languages

In this section we recollect some notions and results of classes which we will call *families of formal languages*. They have in commen that the complexities of their word problems typically range between $\mathbf{AC^0}$ and $\mathbf{NP}$. In contrast to complexity classes they exhibit pumping or iteration properties which lead to the decidability of emptiness and finiteness of their members, but typically not to that of equivalence or universality.

At present, we have no finished definition of a family $\mathcal{F}$ to be a family of formal languages. What we assume as minimum requirements are:

Recursive Presentability: There is recursively enumerable set $R \subseteq \Sigma^*$ and a mapping ϕ from R into the powerset of Σ^*. Each $x \in R$ represents a language generating device, e.g.: a grammar or an automaton, which generates the language $\phi(x)$ and we have $\mathcal{F} = \{\phi(x) | x \in R\}$.

Decidabilities: The emptiness and the finiteness of elements of $\mathcal{F}$, i.e. the sets $\{x \in R | \phi(x) \neq \emptyset\}$ and $\{x \in R | \phi(x)$ is finite$\}$, are decidable.

Closure Properties: $\mathcal{F}$ is constructively closed under intersection with regular sets and under inverse morphisms. These closures are constructively in the sense, that from $x \in R$, morphism h, or given finite automaton A a $y \in R$ is computable, such that $\phi(y) = \phi(x) \cap L(A)$ resp. $h^{-1}(\phi(x))$.

Unfortunately, we can construct language families, which we do not regard as family of formal languages, but which in fact fulfill these properties. Thus we go by well-known examples like the family of regular set, that of context-free languages and some of their various subfamilies, and some families of context-free Lindenmayer languages, which all fulfill the requirements mentioned above.

Notation. We fix a finite alphabet Σ. By Σ^* we denote the set of words over Σ. A language over Σ is a subset of Σ^*. For $w \in L \subseteq \Sigma^*$, by $|w|$ we denote the length of w and ϵ is the word of length 0. For $A \subseteq \Sigma$ we denote by $|w|_A$ the number positions in w having a letter in A. For a word, w_i is the letter in position i.

Regular Languages. The regular languages are a prime example for a family of formal languages and indeed fulfill all the requirements we proposed for something being called a family of formal languages. They can be defined in terms of finite automata (DFA and NFA), logic (MSO), and algebra (finite syntactic monoid).

Context-Free Languages. Context-free languages (CFL) correspond to those accepted by pushdown automata (PDA). The deterministic variant is strictly weaker. The same goes for counter languages. Here only one stack symbol may be pushed by the accepting automaton. We write NOCA for non-deterministic one-counter automata and DOCA for the deterministic variant.

Visibly Pushdown Languages. Context-free languages are accepted by pushdown automata. A way to restrict pushdown automata has received much attention in the last ten years. The visibility restriction for pushdown automata leads to the class of visibly pushdown languages (a.k.a. input-driven pushdown languages), short: VPL. Here, the input symbol determines the stack operation, i.e. if a symbol is pushed or popped. This leads to a partition of Σ into call, return and internal letters: $\Sigma = \Sigma_{call} \cup \Sigma_{ret} \cup \Sigma_{int}$. Then $\hat{\Sigma} = (\Sigma_{call}, \Sigma_{ret}, \Sigma_{int})$ is a visibly alphabet. In the rest of the paper we always assume that there is a visibly alphabet for Σ if we speak about VPL.

We define a function $\Delta : \Sigma^* \to \mathbb{Z}$ which gives us the *height* of a word by $\Delta(w) = |w|_{\Sigma_{call}} - |w|_{\Sigma_{ret}}$. Each word w over a visibly alphabet can be assigned its *height profile* w^Δ, which is a map $\{0, \ldots, |w|\} \to \mathbb{Z}$ with $w^\Delta(i) = \Delta(w_1 \cdots w_i)$. Mehlhorn [Meh80] and independently also Alur and Madhusudan [AM04] introduced *input-driven* or *visibly pushdown automata* (VPDA). In these automata the input letter determines the kind of stack operation. We omit a formal definition for VPDA here, as we are actually interested in a more restricted model.

The family of languages which are accepted by some VPDA is called VPL. This family enjoys many constructive closure and decision properties.

In [BLS06], a reasonable restriction of VPDA was introduced by visibly counter automata (VCA). That is a counter automaton which obeys the visibly restriction. In [BLS06] this model was used as a tool for showing a certain problem concerning VPL to be decidable. In particular they showed that given a VPDA, it is decidable if the language is accepted by some VCA. The following definition exhibits a natural m which we will call threshold. It allows the automaton to have a limited access to the current stack height.

Definition 1 (m-VCA). *An m-VCA $\mathcal{A}$ over $\hat{\Sigma} = (\Sigma_{call}, \Sigma_{ret}, \Sigma_{int})$ is a tuple $\mathcal{A} = (Q, q_0, F, \hat{\Sigma}, \delta_0, \ldots, \delta_m)$, where $m \geq 0$ is the threshold, Q is the set of states, q_0 the initial state, F the set of final states, and $\delta_i \colon Q \times \Sigma \to Q$ are the transition functions.*

A configuration is an element of $Q \times \mathbb{N}$. Note that m-VCAs, similar to VPDAs, can only recognize words where the height profile is non-negative. All other words are rejected. An m-VCA $\mathcal{A}$ performs the following transition when a letter $\sigma \in \Sigma$ is read: $(q, k) \xrightarrow{\sigma} (\delta_{\min(m,k)}(q, \sigma), k + \Delta(\sigma))$. Then $w \in L(\mathcal{A})$ iff $(q_0, 0) \xrightarrow{w} (f, \delta(w))$ for $f \in F$.

The class of the visibly counter languages (VCL) contains the languages recognized by an m-VCA for some m.

As previously argued, VPL has many nice properties and is still expressible enough for many applications. The class VCL is even simpler and can function as an intermediate step if we want to extend results form the regular domain to, say, VPL.

Lindenmayer Systems. The models we looked at so far are automata-based. Lindenmayer introduced a formal rewriting system similar to grammars whose purpose was to model growth of plants. The main difference is that each leaf in the derivation tree of L-Systems has to have the same depth in contrast to ordinary grammars. One can see the resulting objects as fractals. Besides describing biological processes, L-systems have gotten applied in other fields like computer graphics. L-systems have also found their way in the theory of formal languages. Refer e.g. Rozenberg and Salomaa [RS80].

We call a map $h\colon \Sigma \to 2^{\Sigma^*}$ a *substitution* if $h(\epsilon) = \epsilon$ and $h(uv) = h(u)h(v)$.

Definition 2. *The following are L-systems:*

- *An 0L system is a tuple $G = (\Sigma, h, w)$ where Σ is the alphabet, $h\colon \Sigma \to 2^{\Sigma^*}$ a substitution and $w \in \Sigma^*$ is a word we call axiom. The language of G is $L(G) = \bigcup_k h^k(w)$.*
- *An E0L system is a tuple $G = (\Sigma, h, w, \Delta)$ where (Σ, h, w) is a 0L system and $\Delta \subseteq \Sigma$ is a set of terminals. The language of G is $L(G) = \bigcup_k h^k(w) \cap \Delta^*$.*
- *An ET0L system is a tuple $G = (\Sigma, H, w, \Delta)$, where H is a finite set of substitutions and for every $h \in H$, (Σ, h, w) is a 0L system. The language of G is $L(G) = \{x\Delta^* \mid \exists k \in \mathbb{N} \exists h_1, \ldots, h_k \in H \colon x \in h_1(h_2(\ldots h_k(w)\ldots))\}$.*
- *An ET0L system $G = (\Sigma, H, w, \Delta)$ is an EDT0L system if for all $h \in H$ and for all $a \in \Sigma$ holds that $|h(a)| = 1$, i.e. h is an endomorphism.*
- *An EDT0L system $G = (\Sigma, H, w, \Delta)$ is an ED0L system, if $|H| = 1$.*

3   Complexity Classes

In complexity theory we are interested in the amount of resources needed for solving the word problem. Turing machines are the standard model resulting in the resource measures time, space and (non-)determinism.

Using the logarithmic space bound, we get the nondeterministic class **NL** and using polynomial time bound, we get **NP**.

Circuits. When considering very low complexity classes, other models of computations are needed. A circuit is a directed acyclic graph where the nodes are labeled with Boolean functions unless it is an input node. A word over $\{0,1\}$ is accepted by the circuit if the output gate results to 1, whereas the result is computed in the obvious way. We only consider $\{0,1\}$ as an input alphabet; other alphabets can be simulated. If we want to accept languages we naturally want to accept words ob arbitrary length. To achieve this we speak of families of circuits $(C_n)_{n \in \mathbb{N}}$. Here there is one circuit of each input length. If there is some resource-bounded machine computing $(C_n)_{n \in \mathbb{N}}$, we speak of uniformity. If we do not require such a machine, we say, that the circuit family is non-uniform. If not stated otherwise, circuit families are assumed to be non-uniform

Typical complexity measures in circuits are size, depth, fan-in of the gates, type of the gates and uniformity. We get for example the following classes:

- **AC^i**: circuits of polynomial size, depth in $\mathcal{O}(\log^i(n))$, unbounded fan-in and Boolean gates.
- **ACC^i**: circuits of polynomial size, depth in $\mathcal{O}(\log^i(n))$, unbounded fan-in and Boolean and modulo gates. If we want to emphasize the modulus k, we denote this by using the notation **ACC^i_k**.
- **TC^i**: circuits of polynomial size, depth in $\mathcal{O}(\log^i(n))$, unbounded fan-in and threshold gates.

- **NC^i**: circuits of polynomial size, depth in $\mathcal{O}(\log^i(n))$, bounded fan-in and Boolean gates.
- **SAC^i**: circuits of polynomial size, depth in $\mathcal{O}(\log^i(n))$, semi-unbounded fan-in and Boolean gates.

In particular we are interested in the following classes: $\mathbf{AC^0} \subset \mathbf{ACC^0} \subseteq \mathbf{TC^0} \subseteq \mathbf{NC^1} \subseteq \mathbf{L} \subseteq \mathbf{NL} \subseteq \mathbf{SAC^1} \subseteq \mathbf{P}$. Note that the classes $\mathbf{AC^0}$ and $\mathbf{ACC^0}$ are seperated. All other inclusions are unknown whether they are strict. Refer e.g. to [Vol99].

Complexities of Families of Formal Languages. The various families mentioned so far, have the following connections to complexity classes in terms of completeness results. When we say that a family $\mathcal{F}$ is complete for $\mathcal{C}$, we mean that both $\mathcal{F} \subset \mathcal{C}$ and that $\mathcal{F}$ contains a $\mathcal{C}$-complete problem.

- The *ETOL*-languages are **NP**-complete.
- Both the context-free languages and the *EOL*-languages are **SAC^1**-complete.
- Both the nondeterministic counter languages and the *EDTOL*-languages are **NL**-complete.
- the regular languages, the visibly counter languages, and the visibly pushdown languages are **NC^1**-complete.
- The *EDOL*-languages are **AC^0**-complete (which is rather the case because of the **AC^0** reductions we chose).

Dense Completeness. Finally we state the definition of dense completeness as it is introduced in [KL12]. In our setting we use many-one-reductions. If a language A is reducible to B we write $A \leq B$. If $A \leq B$ and $B \leq A$ we write $A \approx_m B$ and say A and B are many-one-equivalent. The reductions we use are all DLOGTIME-uniform $\mathbf{AC^0}$ reductions, so we write e.g. $A \approx_m^{\mathbf{AC^0}} B$.

Definition 3. *Let $\mathcal{F}$ and $\mathcal{C}$ be sets of languages. We say $\mathcal{F}$ is densely complete in $\mathcal{C}$ if*

- *$\mathcal{F} \subseteq \mathcal{C}$ and*
- *for all $C \in \mathcal{C}$ there exists a language $F \in \mathcal{F}$ such that $C \approx_m^{\mathbf{AC^0}} F$.*

The gist of the definition is that we can say that a family of formal languages is densely complete in some complexity class. However in the definition we do not require $\mathcal{F}$ and $\mathcal{C}$ to be restricted in any way. Like that we get transitivity of the dense completeness property which is desirable.

4 Negative Instances for NC^1

As all the examples we know, where a densely complete family of formal languages exists correspond to a (more or less) non-deterministic complexity class, it is rather interesting to consider deterministic classes. For its closeness to **L** we consider $\mathbf{NC^1}$ as a deterministic class, and hence would like to show that there is no densely complete family of formal languages in it.

Our approach to prove that a family of formal languages is not densely complete will show that the class is too sparse inside of $\mathbf{NC^1}$. We have shown this before for the regular languages using finite monoids, and it was not clear this approach would ever work for any family of formal languages outside the regular languages. Using a result from a recent paper [KLL15] we are now able to break this barrier and show a family of formal languages outside the regular language that is not densely complete in $\mathbf{NC^1}$. In this section we will show that VCL is not densely complete in $\mathbf{NC^1}$. This also narrows the gap between the counter languages which are densely complete in a complexity class and the regular languages which are not densely complete in a complexity class to counter languages vs. visibly counter languages.

Theorem 4. *The visibly counter languages are not densely complete in* $\mathbf{NC^1}$.

Proof. We show the statement by contradiction. The contradiction will be achieved by using two facts:

- Ladner's theorem in the generalized version by Vollmer [Vol90] shows us how to get arbitrarily long lists of languages L_i for which $L_i \leq L_j$ iff $i \leq j$. That means that complexity classes in a way have infinitely many ascending levels of complexity inside.
- Visibly counter languages (and regular languages as well) exhibit a kind of dichotomy [KLL15] when it comes to the membership of a language to $\mathbf{AC^0}$. Either a visibly counter language is in $\mathbf{AC^0}$ or it is hard for a proper superclass of $\mathbf{AC^0}$.

So it seems likely that those to facts contradict each other and this is what we will prove.

Assume VCL to be densely complete in $\mathbf{NC^1}$. Then for all languages L in $\mathbf{NC^1}$ there exists a language V in VCL such that L and V are many-one-equivalent under $\mathbf{AC^0}$-reductions.

The language PARITY= $\{w \in \{0,1\} \mid |w| \equiv 0 \pmod 2\}$ is not in $\mathbf{AC^0}$ but $\mathbf{ACC_2^0}$-complete [FSS84]. Using Ladner's theorem we choose L to be a language whose complexity lies strictly between $\mathbf{AC^0}$ and $\mathbf{ACC_2^0}$, i.e. $L \leq$ PARITY, PARITY $\not\leq L$ and $L \notin \mathbf{AC^0}$ [Hås86]. The construction for L basically takes a subset of PARITY by only allowing certain word lengths.

Having L we look at V, which must be many-one-equivalent to L:

- If V is in $\mathbf{AC^0}$ then we have a contradiction, since L is not in $\mathbf{AC^0}$.
- If V is not in $\mathbf{AC^0}$ then according to [KLL15], we have to consider again two cases for the two different reasons a visibly counter languages can be outside $\mathbf{AC^0}$. One reason is that the height behavior is too complex, resulting in V being $\mathbf{TC^0}$-hard. The other reason concerns the regular part of the language and is related to the case for regular languages, resulting in V being $\mathbf{ACC_k^0}$-hard for some k. So in both cases, V is hard for a proper superclass.
  - V is hard for $\mathbf{TC^0}$. This is a contradiction, since L is not $\mathbf{ACC_2^0}$-hard.

- V is hard for $\mathbf{ACC}_k^0$-hard. If k is even then again we have the contradiction because L is not $\mathbf{ACC}_2^0$-hard. If k is odd then this implied $\mathbf{ACC}_k^0 \subseteq \mathbf{ACC}_2^0$. Due to [Smo87] we know that this is contradictory also. □

Since the regular languages are visibly counter languages and $\mathbf{NC^1}$-complete, we get the following.

Corollary 5. *The regular languages are not densely complete in* $\mathbf{NC^1}$.

This completes a proof from the previous paper on dense completeness [KL12]. The statement for the non-denseness of the regular languages was true, but the proof was incomplete by not considering the case that the syntactic monoid of a regular language in AC^0 might contain in fact a nontrivial group.

As we saw in the proof, we used Ladner's theorem on the one hand and some kind of dichotomy on the other. Up to now we do not know any other proof strategy for showing that a formal language class is not densely complete in some complexity class.

5   Positive Instances for NL

We introduce a restricted counter-based automaton model being densely complete in **NL**. It can then be used to demonstrate that a certain type of L-system is also densely complete in **NL**.

The automaton model we introduce is a non-deterministic counter automaton with the restriction that once the automaton performs a pop action on the stack, it has to pop until it is empty. See figure 1. We call it a sweeping counter automaton (SCA) and the corresponding family of formal languages we call SCL.

Definition 6. *A nondeterministic sweeping counter automaton (SCA) is a tuple* $\mathcal{A} = (Q^\uparrow, Q^\downarrow, \Sigma, q_0, F, \delta, \delta_0)$, *where* $Q = Q^\uparrow \,\dot{\cup}\, Q^\downarrow$ *and* $Q^\uparrow$ *is a set of push-states,* $Q^\downarrow$ *a set of pop-states,* $q_0 \in Q^\uparrow$ *is the initial state,* $F \subseteq Q^\downarrow$ *a set of final states and* $\delta \subseteq (Q^\uparrow \times \Sigma \times Q) \cup (Q^\downarrow \times \Sigma \times Q^\downarrow)$ *and* $\delta_0 \subseteq Q^\downarrow \times \Sigma \times Q$ *are transition functions. The transition* δ_0 *is applied if the counter is 0 and* δ *is applied otherwise.*

A configuration of a SCA $\mathcal{A}$ is an element of $Q \times \mathbb{N}$. The transition relations δ and δ_0 take an input word and define a run through configurations. The initial configuration is $(q_0, 0)$. Further if $q \in Q^\uparrow$ and $k > 0$ then $(q, k) \overset{a \in \Sigma}{\rightarrow} (\delta(q, a), k+1)$ and $(q, k) \overset{a \in \Sigma}{\rightarrow} (\delta(q, a), k - 1)$ in the case of $q \in Q^\downarrow$. If $k = 0$ then $(q, k) \overset{a \in \Sigma}{\rightarrow} (\delta_0(q, a), k')$ where $k'=0$ iff $\delta_0(q, a) \in Q^\downarrow$; otherwise $k' = 1$. Then: $L(\mathcal{A}) = \{w \in \Sigma^* \mid (q_0, 0) \overset{w}{\rightarrow} (f, 0), f \in F\}$. Note that there are some similar ways to define a SCA but the present definition serves our purpose.

SCL is closed under union, intersection, Kleene star and inverse homomorphisms but not under complement. The regular languages are contained in SCL. Further the decidabilities of NOCA translate to SCL. SCA cannot be determinized.

stack height

$w \in \Sigma^*$

Fig. 1. Characterizing stack height behavior for an SCA

Theorem 7. SCL *is densely complete for* **NL**.

We can use this result to prove denseness of the L-system EDT0L. It is sufficient to show that a densely complete family is a subset of EDT0L and that EDT0L lies within **NL**. The latter is known to be true [RS80].

Lemma 8. SCL $\subseteq$ *EDT0L*

Proof. We are given an SCA $\mathcal{A} = (Q^\uparrow, Q^\downarrow, \Sigma, q_0, F, \delta, \delta_0)$ and construct an EDT0L system $G = (\Sigma_G, H, w, \Delta)$ with $\Delta = \Sigma$. For convenience we assume $\epsilon \notin L(\mathcal{A})$. We set $\Sigma_G = \{(q_1, q_2) \mid q_1, q_2 \in Q^\downarrow\} \cup \{(q, *) \mid q \in Q^\downarrow\} \cup \Delta$.

The idea is to let G generate a letter for each push-pop-cycle of the SCA and then extend each of those letters to the actual word the automaton reads. Hence we set $w = (q_0, *)$. We need a first set of substitutions for extending one more push-pop-cycle. Hence

$$h_1^q((q', *)) = (q', q)(q, *)$$

for all $q' \in Q^\downarrow$. In general for all substitutions: On letters not specified the map is the identity. For the final state we need the following variant:

$$h'^q_1((q', *)) = (q', q)$$

for all $q \in F$. Next we have to build the word as the counter in- and decreases.

$$h_2^{a,b}((q, q')) = a(q'', q''')b$$

if $q \in Q$, $q', q''' \in Q^\downarrow$, $q'' \in Q^\uparrow$ and $q' \in \delta(q''', b)$. Further if $q \in Q^\downarrow$ it must hold that $q'' \in \delta_0(q, a)$ and $q'' \in \delta(q, a)$ else. Finally if the final stack height is reached we have to terminate: $h_3((q, q)) = \epsilon$. Now H is the set of all substitutions we just described.

We verify that the construction is correct. Each word $w = w_{(1)} \ldots w_{(k)}$ is accepted by $\mathcal{A}$ where every $w_{(i)}$ corresponds to one push-pop-cycle of the automaton. If $\mathcal{A}$ is in q after $w_{(1)}$ then we use the derivation $(q_0, *) \rightarrow (q_0, q)(q, *)$. If $w_{(1)} = aw'b$, then we can apply $h_2^{a,b}$ and so on. It is easy to see that we can derivate as such: $(q_0, *) \rightarrow (q_0, q)(q, *) \rightarrow \cdots \rightarrow w_{(1)}(q, *)$. After that we can proceed with $w_{(2)}$ etc. until the whole word is derived. Conversely every word we get by the grammar is also a word accepted by the automaton. The only thing we have to note is that if we derivate like this $(q_0, *) \rightarrow \cdots \rightarrow (q_0, q)(q, q')(q', q'')(q'', *)$ instead of building the word for each cycle first, we just get additional ways to derivate words.   □

In conclusion we get the result:

Theorem 9. *EDT0L is densely complete for* **NL**.

6   More Densely Complete L-Systems

An easy case is ED0L, which is densely complete for **AC^0** because of the reductions. Further since CFL $\subseteq$ E0L $\subseteq$ **SAC^1**, we get that E0L is densely complete for **SAC^1**.

In the previous section we showed that EDT0L is densely complete for **NL**. It is natural to ask whether e.g. the **NP**-complete L-system ET0L is in fact densely complete for **NP**. Using the so-called checking-stack pushdown automata characterization of ET0L [vL76], we can translate proofs from [KL12] to this case. There we showed that PDA are densely complete for **SAC^1**.

Checking-stack pushdown automata (CS-PDA) are a well researched model [vL76, RS80] in the context of L-systems. A CS-PDA is basically a PDA equipped with a checking stack. This additional stack is nondeterministically filled with some word and after that the checking stack is read only. Further the head for the checking stack is synchronized with the normal stack. This enables the automaton to perform tasks, a usual pushdown automaton cannot do. The checking stack can be used for synchronization of different parts of the computation. E.g. the language $\{ww \mid w \in \Sigma^*\}$ is an easy example which is accepted by an CS-PDA, but by no PDA.

For PDA we have the two-way-variant 2-PDA and also we can use k input heads, which is denoted by $2 - \text{PDA}(k)$. We always restrict such automata to polynomial time and write $2 - \text{PDA}(k)_{\text{poly-time}}$. Using the same notation we get the corresponding CS-PDA models.

Theorem 10 ([vL76]). *ET0L equals CS-PDA and is* **NP**-*complete*.

Also by [vL75] it is easy to see that:

Lemma 11. *2-CS-PDA(k) equals* **NP**.

The proof strategy is very similar to the case of PDA and **SAC^1**. We first show that going from two-way to one-way preserves denseness and then that the number of heads can be reduced without losing denseness. In conclusion we get:

Theorem 12. *ET0L is densely complete for* **NP**.

7   Discussion

In this paper we primarily inspected aspects of the relationship between NC^1 and NL, using the notion of dense completeness. We developed further the theory of dense completeness and gathered evidence that strengthens our confidence that dense completeness captures an essential property of nondeterministic complexity.

Dense completeness might offer new angles to separate complexity classes. For instance if NC^1 has no densely complete family of formal languages, we know the class to be strictly contained in NL. We could also try to show that every complete family of formal languages in NL is already densely complete. We conjecture that dense completeness is a feature of nondeterministic classes. Of course this should be hard to show since such a result would separate determinism from nondeterminism. To obtain such arguments we need a formal definition of what a family of formal languages should be. In the present work we exhibited one possibility, which is our first contribution.

Our second contribution is showing concrete denseness and non-denseness results. We want to narrow the gap between NC^1 and NL. Hence we want to find large families in NC^1 and show that they are not densely complete. In NL we want to achieve the opposite and find small families being densely complete. For the NC^1 case we performed this with VCL. In the NL case we proposed (to our best knowledge) new variant of counter languages, namely sweeping counter languages. As a byproduct we could show the L-system of EDT0L to be densely complete in NL and ET0L to be densely complete in NP. Also E0L is densely complete for SAC^1 and E0L for AC^0.

There are some interesting questions arising from this work which we will pursue. One of them: Can we show that the deterministic counter languages are not densely complete in L? We are also interested in showing non-denseness of the visibly pushdown languages for NC^1. This would relate rather to the context-free languages which are densely complete for SAC^1.

The only way we know to show non-denseness is using some kind of *dichotomy* of the family of formal languages regarding AC^0. The regular languages for instance are either in AC^0 or hard for ACC_0^k. Not having this dichotomy is actually a weaker property than dense completeness. It is worth investigating how denseness is related to this dichotomy.

The dichotomy for VCL relies on two properties: The regular part must be quasi-aperiodic and the height behavior of the language must be simple. The second one is interesting if we compare this to our new found SCA. In the visibly case, the property of being sweeping would be sufficient for the language to have simple stack behavior. So in the case of visibly sweeping automata, only the regular part determines whether the language is in AC^0. Hence we will continue investigating height behavior properties of counter languages.

References

AM04. Alur, R., Madhusudan, P.: Visibly pushdown languages. In: Babai, L. (ed.) Proceedings of the 36th Annual ACM Symposium on Theory of Computing, June 13–16, pp. 202–211. ACM, Chicago (2004)

BCST92. Barrington, D.A.M., Compton, K.J., Straubing, H., Thérien, D.: Regular Languages in NC^1. J. Comput. Syst. Sci. **44**(3), 478–499 (1992)

BLS06. Bárány, V., Löding, C., Serre, O.: Regularity problems for visibly pushdown languages. In: Durand, B., Thomas, W. (eds.) STACS 2006. LNCS, vol. 3884, pp. 420–431. Springer, Heidelberg (2006)

Dym88. Dymond, P.W.: Input-Driven Languages are in log n Depth. Inf. Process. Lett. **26**(5), 247–250 (1988)

FSS84. Furst, M.L., Saxe, J.B., Sipser, M.: Parity, Circuits, and the Polynomial-Time Hierarchy. Mathematical Systems Theory **17**(1), 13–27 (1984)

Hås86. Håstad, J.: Almost optimal lower bounds for small depth circuits. In: Hartmanis, J. (ed.) Proceedings of the 18th Annual ACM Symposium on Theory of Computing, May 28–30, pp. 6–20. ACM, Berkeley (1986)

KL12. Krebs, A., Lange, K.-J.: Dense completeness. In: Yen, H.-C., Ibarra, O.H. (eds.) DLT 2012. LNCS, vol. 7410, pp. 178–189. Springer, Heidelberg (2012)

KLL15. Krebs, A., Lange, K.-L., Ludwig, M.: Visibly counter languages and constant depth circuits. In: Mayr, E.W., Ollinger, N. (eds) 32nd International Symposium on Theoretical Aspects of Computer Science, STACS 2015, March 4–7. LIPIcs, vol. 30, pp. 594–607. Schloss Dagstuhl - Leibniz-Zentrum fuer Informatik, Garching (2015)

Meh80. Mehlhorn, K.: Pebbling moutain ranges and its application of DCFL-Recognition. In: de Bakker, J.W., van Leeuwen, J. (eds) Proceedings of the Automata, Languages and Programming, 7th Colloquium, Noordweijkerhout, July 14–18, The Netherland. LNCS, vol. 85, pp. 422–435. Springer, Heidelberg (1980)

RS80. Rozenberg, G., Salomaa, A.: Mathematical Theory of L Systems. Academic Press Inc., Orlando (1980)

Smo87. Smolensky, R.: Algebraic methods in the theory of lower bounds for boolean circuit complexity. In: Aho, A.V. (ed.) Proceedings of the 19th Annual ACM Symposium on Theory of Computing, pp. 77–82. ACM, New York (1987)

vL75. van Leeuwen, J.: The Membership Question for ET0L-Languages is Polynomially Complete. Inf. Process. Lett. **3**(5), 138–143 (1975)

vL76. van Leeuwen, J.: Variations of a new machine model. In: 17th Annual Symposium on Foundations of Computer Science, October 25–27, pp. 228–235. IEEE Computer Society, Texas (1976)

Vol90. Vollmer, H.: The gap-language-technique revisited. In: Börger, E., Böuning, H.K., Richter, M.M., Schönfeld, W. (eds) CSL 1990. LNCS, vol. 533, pp. 389–399. Springer, Heidelberg (1990)

Vol99. Vollmer, H.: Introduction to circuit complexity - a uniform approach. Texts in theoretical computer science. Springer (1999)

Surminimisation of Automata

Victor Marsault[(✉)]

Telecom-ParisTech, 46 rue Barrault, 75013 Paris, France
marsault@telecom-paristech.fr

Abstract. We introduce the notion of *surminimisation* of a finite deterministic automaton; it consists in performing a transition relabelling before executing the minimisation and it produces an automaton smaller than a sole minimisation would. While the classical minimisation process preserves the accepted language, the surminimisation process preserves its underlying ordered tree only. Surminimisation induces on languages and on Abstract Rational Numeration Systems (ARNS) a transformation that we call *label reduction*. We prove that all positional numeration systems are label-irreducible and that an ARNS and its label reduction are very close, in the sense that converting the integer representations from one system into the other is done by a simple Mealy machine.

1 Introduction

The classical notion of minimisation (*cf.* for instance [7]) is a transformation of deterministic finite automata and is associated with the automaton equivalence. Two automata are equivalent if they accept the same language L and minimising any automaton accepting L produces the automaton accepting L with the fewest amount of states. Hence, the invariant of minimisation is the accepted language.

In this article, we assume that the alphabet (of every automaton) is equipped with a total order. We then define another automaton transformation called *surminimisation* that produces an automaton with fewer states than the one resulting from a sole minimisation. The invariant of this new transformation is no longer the accepted language, but its underlying ordered tree.

For each state p of a given automaton, the order on the alphabet induces a (total) order on the outgoing transitions of p: a transition is smaller if it is labelled by a smaller letter. The *surminimisation* process consists in two steps. First, it relabels the outgoing transitions of each state p, such that their order is preserved: the smallest transition is relabelled by the letter 0, the second smallest is relabelled by 1, and so on. The second step simply consists in a minimisation.

Surminimisation induces on automata an equivalence relation that we call *T-equivalence*: two automata are T-equivalent if their surminimisations are isomorphic. Moreover, the surminimisation process is idempotent[1], hence each T-equivalence class features a canonical representative computed by surminimising any member of the class.

---

[1] Two successive surminimisations produce the same result as only one.

© Springer International Publishing Switzerland 2015
I. Potapov (Ed.): DLT 2015, LNCS 9168, pp. 352–363, 2015.
DOI: 10.1007/978-3-319-21500-6_28

We then lift T-equivalence to regular languages (over ordered alphabets). We prove that if two trim automata are equivalent, then their surminimisations are equivalent as well. Hence, we say that two languages are T-equivalent if they are accepted by two T-equivalent (trim) automata and we call label reduction of a regular language L, the language accepted by the surminimisation of any trim automaton accepting L.

A regular language over an ordered alphabet is nothing else than an Abstract Regular Numeration System (ARNS, *cf.* [9]). It consists in ordering a language L by the *radix*, or *genealogical* order: a longer word is always genealogically greater than a shorter word, and the genealogical ordering of two words of equal length coincides with their lexicographical ordering. The representation of an integer n in the ARNS L is then defined as the $(n+1)$-th word of L according to the radix order. The two notions are so close that we use for ARNS's every notion defined for regular languages (such as T-equivalence, label reduction, etc.).

Two T-equivalent ARNS's are very close, in the sense that the function converting one system into the other is realised by a Mealy machine, as stated below.

Theorem 1. *The function that maps the representation of an integer n in an ARNS into the representation of n in an T-equivalent ARNS is realised by a Mealy machine.*

The converse to Theorem 1 is false in the general case. Indeed there exist ARNS's that are not T-equivalent but such that the conversion from one to another is realised by a Mealy machine. We call *locally increasing* a Mealy machine that is locally preserving the order of letters and prove the following statement, a weak converse to Theorem 1 .

Theorem 2. *If the function that maps the representation of an integer n in an ARNS into the representation of n in another ARNS is realised by a locally-increasing Mealy machine, then the ARNS's are T-equivalent.*

ARNS's form the most general class of numeration systems. In particular, all (reasonable) *positional numeration systems* (or U-systems, *cf.* [5]) and all *Substitution Numeration Systems* (SNS, *cf.* [3]) are ARNS's.

We first prove that 0^*L is label-irreducible if L denotes the representation language of any positional numeration system. It has quite a significance when comparing the class of ARNS's to the class of label-irreducible ARNS's: the former contains the latter, but 1) brings no supplementary expressive power and 2) contains no additional concrete examples.

We also prove that every prefix-closed ARNS is T-equivalent to some SNS, by using classical transformations from substitutions into automata (*cf.* [11] or even [2]). It is known that every SNS is a prefix-closed ARNS (*cf.* [1]), and the previous results induce a weak converse to this statement: every prefix-closed ARNS is very close to some SNS (in the sense of Theorem 1).

The paper is organised as follows. In Section 2 , we define in details the notions of surminimisation, label reduction, etc. The following Section 3 is dedicated to the proof of Theorems 1 and 2 . Finally, Section 4 consists in a discussion of label reduction within numeration system theory.

2   Label Reduction and Surminimisation

For every integer k of $\mathbb{N}$, we write $[\![k]\!]$ for the set of the k smallest non-negative integers: $[\![k]\!] = \{0, 1, \ldots, k-1\}$. An *alphabet* is a set of *letters* and in the following **we consider ordered alphabets only**, that is, alphabets (implicitly) equipped with a total order, denoted by $<$. The set $[\![k]\!]$ will be considered both as an integer interval and as a digit alphabet naturally ordered by $0 < 1 < \cdots < (k-1)$.

Automata are directed labelled graphs and in the following **we consider deterministic automata only**, written as a 5-tuple $\mathcal{A} = \langle Q, A, \delta, i, F \rangle$ where Q is a finite set of *states*; A is a finite (ordered) alphabet; δ is the *transition function*, a **partial** function $Q \times A \rightarrow Q$; $i \in Q$ is called the *initial state*; and $F \subseteq Q$ is the set of *final states*. As usual, δ is extended to $Q \times A^*$ by $\delta(p, u\,a) = \delta(\delta(p, u), a)$ and we write $p \xrightarrow[\mathcal{A}]{u} p'$ if $\delta(p, u) = p'$.

The automaton $\mathcal{A}$ is said to be *trim* if each state of $\mathcal{A}$ may reach a final state and is reachable from the initial state. The language accepted by $\mathcal{A}$, denoted by $L(\mathcal{A})$, is the set of the words u such that $\delta(i, u)$ is a final state. Two automata are said *equivalent* if they accept the same language.

We also denote by $\mathsf{Out}_{\mathcal{A}}(p)$ the set of the transitions going out from p. We write $\mathsf{od}(p)$, or more often k_p, for $|\mathsf{Out}_{\mathcal{A}}(p)|$ the out-degree of the state p, and $\mathsf{od}(\mathcal{A}) = \max\{\mathsf{od}(p) \mid p \in Q\}$. For every state p of Q, the order on A induces an order on $\mathsf{Out}_{\mathcal{A}}(p)$; we enumerate $\mathsf{Out}_{\mathcal{A}}(p)$ w.r.t. this order as follows:

$$\forall i \in [\![k_p]\!] \qquad p \xrightarrow{a_i} p_i \quad \text{with } a_0 < a_1 < \cdots < a_{(k_p-1)}.$$

We call $(i+1)$-*th transition of* $\mathsf{Out}_{\mathcal{A}}(p)$ the transition $p \xrightarrow{a_i} p_i$, as defined above.

We first define the *label reduction* of an automaton. It consists in relabelling, for each state p, the transitions of $\mathsf{Out}_{\mathcal{A}}(p)$ using the alphabet $[\![k_p]\!]$ and such that the order of $\mathsf{Out}_{\mathcal{A}}(p)$ is preserved. More precisely.

Definition 1. *Let $\mathcal{A} = \langle Q, A, \delta, i, F \rangle$ be a (deterministic) automaton. We call label reduction of $\mathcal{A}$, denoted by $\mathsf{lred}(\mathcal{A})$ the automaton:*

$$\mathsf{lred}(\mathcal{A}) \;=\; \langle\, Q,\, [\![\mathsf{od}(\mathcal{A})]\!],\, \delta',\, i,\, F \,\rangle \;,$$

where δ' is such that, for every state p of Q, if $p \xrightarrow[\mathcal{A}]{a_i} p_i$ is the $(i+1)$-th transition of $\mathsf{Out}_{\mathcal{A}}(p)$ then $p \xrightarrow[\mathsf{lred}(\mathcal{A})]{i} p_i$ is a transition of $\mathsf{lred}(\mathcal{A})$.

Figure 1 shows an automaton $\mathcal{A}_1$ and its label reduction. The label-reduction process commutes with quotient (*cf.* Definition 2, below), as stated at Lemma 1.

Definition 2. *Let $\mathcal{A} = \langle Q_{\mathcal{A}}, A, \delta_{\mathcal{A}}, i_{\mathcal{A}}, F_{\mathcal{A}} \rangle$ and $\mathcal{M} = \langle Q_{\mathcal{M}}, A, \delta_{\mathcal{M}}, i_{\mathcal{M}}, F_{\mathcal{M}} \rangle$ be two automata. An automaton morphism $\phi : \mathcal{A} \rightarrow \mathcal{M}$ is a surjective function $Q_{\mathcal{A}} \rightarrow Q_{\mathcal{M}}$ meeting the three following conditions.*
1. *$\phi(i_{\mathcal{A}}) = i_{\mathcal{M}}$;*
2. *$p \xrightarrow[\mathcal{A}]{a} p'$ is a transition of $\mathcal{A}$ $\iff$ $\phi(p) \xrightarrow[\mathcal{M}]{a} \phi(p')$ is a transition of $\mathcal{M}$;*
3. *$F_{\mathcal{A}} = \phi^{-1}(F_{\mathcal{M}})$.*

(a) An automaton L_1 (b) $\mathsf{lred}(L_1)$

Fig. 1. Label reduction of an automaton L_1

In this case, $\mathcal{M}$ is called a quotient *of $\mathcal{A}$. If in addition, $\mathcal{M}$ is a quotient of another automaton $\mathcal{B}$, then $\mathcal{A}$ and $\mathcal{B}$ are said* bisimilar. *Every regular language L is canonically associated with a minimal trim automaton $\mathcal{M}_L$; it is a quotient of every trim automaton accepting L.*

Lemma 1. *Let $\mathcal{A}$ and $\mathcal{M}$ be two automata. If $\mathcal{M}$ is a quotient of $\mathcal{A}$, then $\mathsf{lred}(\mathcal{M})$ is a quotient of $\mathsf{lred}(\mathcal{A})$.*

Proof. We denote by $\phi : \mathcal{A} \to \mathcal{M}$ the automaton morphism associated with the quotient. Note that the state set of $\mathcal{A}$ and of $\mathsf{lred}(\mathcal{A})$ are identical (and similarly for $\mathcal{M}$ and $\mathsf{lred}(\mathcal{M})$), hence ϕ also maps states of $\mathsf{lred}(\mathcal{A})$ to states of $\mathsf{lred}(\mathcal{M})$; let us prove that ϕ is an automaton morphism from $\mathsf{lred}(\mathcal{A})$ to $\mathsf{lred}(\mathcal{M})$.

Let p be a state of $\mathsf{lred}(\mathcal{A})$. We enumerate the outgoing transitions of p in $\mathcal{A}$ as follows: $\forall i \in [\![k_p]\!]$, $p \xrightarrow{a_i} p_i$ with $a_0 < a_1 < \cdots < a_{(k_p-1)}$, where $k_p = \mathrm{od}(p)$. It follows that the enumeration of the outgoing transitions of $\phi(p)$ in $\mathcal{M}$ are $\forall i \in [\![k_p]\!]$, $\phi(p) \xrightarrow{a_i} \phi(p_i)$ with $a_0 < a_1 < \cdots < a_{(k_p-1)}$.

Hence, from Definition 1 , $\mathsf{Out}_{\mathsf{lred}(\mathcal{A})}(p)$ consists of $p \xrightarrow{i} p_i$, $\forall i \in [\![k_p]\!]$. Similarly, $\mathsf{Out}_{\mathsf{lred}(\mathcal{M})}(\phi(p))$ consists of the transitions $\phi(p) \xrightarrow{i} \phi(p_i)$, $\forall i \in [\![k_p]\!]$.

The next proposition, follows almost immediately.

Proposition 1. *Let $\mathcal{A}$ and $\mathcal{B}$ be two trim automata. If $\mathcal{A}$ and $\mathcal{B}$ are equivalent then so are $\mathsf{lred}(\mathcal{A})$ and $\mathsf{lred}(\mathcal{B})$.*

The hypothesis *trim* in Proposition 1 is crucial. Indeed the complete automaton accepting 1^* is equivalent to the trim automaton accepting 1^* whereas their label reductions are not.

In the following, we consider trim automata only. Hence, Proposition 1 allows to lift *label reduction* to regular languages: the label reduction[2] $\mathsf{lred}(L)$ of a regular language L is the language $L(\mathsf{lred}(\mathcal{A}))$ where $\mathcal{A}$ is any *trim* automaton accepting L. For instance, the label reduction of $((a + b^*)c)^*$ is $(00 + 10^*1 + 2)^*$.

Definition 3 (T-equivalent automata). *Two automata $\mathcal{A}$ and $\mathcal{B}$ are said* tree-equivalent *(or for short T-equivalent), denoted by $\mathcal{A} \stackrel{T}{\sim} \mathcal{B}$, if their label reductions are equivalent: $L(\mathsf{lred}(\mathcal{A})) = L(\mathsf{lred}(\mathcal{B}))$. Similarly, two regular languages L and K are said* T-equivalent *if their label reductions are equal.*

---

[2] Label reduction may be defined directly on language; *cf.* Remark 2 , page 362.

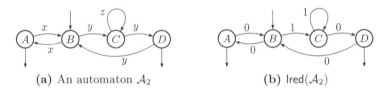

Fig. 2. Label reduction of another automaton $\mathcal{A}_2$

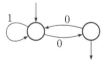

Fig. 3. The surminimisation either of $\mathcal{A}_1$ or of $\mathcal{A}_2$

Figure 2 shows the automaton $\mathcal{A}_2$ and its label reduction. This automaton is T-equivalent to the automaton $\mathcal{A}_1$ (previously shown at Figure 1a). Indeed, their respective label-reductions $\mathsf{lred}(\mathcal{A}_1)$ and $\mathsf{lred}(\mathcal{A}_2)$ are equivalent: they have the same minimisation, shown at Figure 3 . This method is a good way to decide whether two automata are T-equivalent, as formalised below.

Definition 4. *We call* surminimisation *of an automaton* $\mathcal{A}$, *the minimisation of the label reduction of* $\mathcal{A}$: $\mathsf{surmin}(\mathcal{A}) = \mathsf{minim}(\mathsf{lred}(\mathcal{A}))$.

Figure 3 shows the surminimisation either of $\mathcal{A}_1$ or of $\mathcal{A}_2$. The next proposition follows directly from the definitions; it gives both a characterisation and an efficient decision algorithm for T-equivalence.

Proposition 2. *Two automata are T-equivalent if and only if their respective surminimisations are isomorphic.*

Remark 1. Surminimisation (or label reduction) removes the *meaning* of the letters (if there is any) and retains their order only. For instance the language 0^*1^* may be described as 0*'s followed by* 1*'s* while its label reduction is $0^*+0^*10^*$ that may be described as *words with at most one* 1. In particular, surminimisation also removes the complexity due to an arbitrary choice of letters: for instance the label reduction of the language $L_3 = arbi(trary)^*$ is $0^4 \left(0^5\right)^*$ and the label reduction of[3] $\mathsf{Pre}\,(L_3)$ is 0^*; this example highlights that the question of the succinctness of surminimisation is meaningless.

The classical notions of equivalence and minimisation feature a natural invariant: they preserve the accepted language. We have already seen that T-equivalence and surminimisation do not preserve the language; however they feature another invariant: the underlying (ordered) tree.

---

[3] $\mathsf{Pre}\,(L)$ is the set of prefixes of words of L: $\mathsf{Pre}\,(L) = \{\,u \mid u\,v \in L \text{ for some word } v\,\}$.

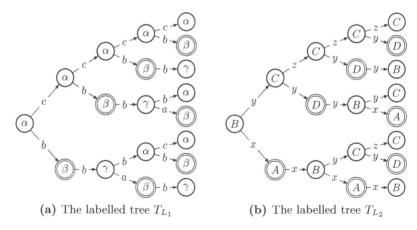

(a) The labelled tree T_{L_1} **(b)** The labelled tree T_{L_2}

Fig. 4. The unfolding of two T-equivalent automata

Definition 5. *A language L over an alphabet A may be represented as an infinite labelled tree (or infinite acyclic automaton) as follows: $T_L = (V, A, E, F)$. The vertex set is $V = \mathsf{Pre}(L)$; the edge labels are taken in the alphabet A; the edge set is $E = \{(u, a, ua) \mid ua \in V\} \subseteq V \times A \times V$; the set of final vertices $F = L$.*

If L is a regular language, an isomorphic tree may be obtained by unfolding a (trim) automaton accepting L.

Figure 4 shows the tree representations of $L_1 = L(\mathcal{A}_1)$ and $L_2 = L(\mathcal{A}_2)$; in the figure, a vertex is labelled by the corresponding state of the automaton, and is drawn with a double line if it is final. These two trees, T_{L_1} and T_{L_2}, coincide up to labelling; it is a consequence of the fact that L_1 and L_2 are T-equivalent, as stated in the next Proposition 3 . It is a direct consequence of Lemma 2 .

Proposition 3. *Let $\mathcal{A}$ and $\mathcal{B}$ be two automata. If $\mathcal{A} \overset{T}{\sim} \mathcal{B}$, then their respective unfoldings differ only by the labellings.*

Lemma 2. *Let $\mathcal{A}$ be an automaton and $\mathcal{M} = \mathsf{surmin}(\mathcal{A})$. Then the respective unfoldings of $\mathcal{A}$ and $\mathcal{M}$ differ only by the labellings.*

Proof (Sketch). Let ϕ be the automaton morphism $\mathsf{lred}(\mathcal{A}) \to \mathcal{M}$. It is also a function from $Q_{\mathcal{A}}$ to $Q_{\mathcal{M}}$ which is **not** an automaton morphism $\mathcal{A} \to \mathcal{M}$: it does not meet the condition 2 of Definition 2 . However, ϕ satisfies the next condition.

$$p \xrightarrow{\;a\;}_{\mathcal{A}} q \text{ is the } (i+1)\text{-th transition of } \mathsf{Out}_{\mathcal{A}}(p) \quad \Longleftrightarrow \quad \phi(p) \xrightarrow{\;i\;}_{\mathcal{M}} \phi(q)$$

The function ϕ may be extended to a bijection that maps vertices of $T_{L(\mathcal{A})}$ to vertices of $T_{L(\mathcal{M})}$, and satisfies an analogous condition.

3   T-equivalent Languages Define the Same ARNS

An ordered alphabet A induces two orders on words of A^*, the classical *lexicographic order* $<_{\text{lex}}$ and the *radix order* $<_{\text{rad}}$ defined as follows: $u <_{\text{rad}} v$ either if $|u| < |v|$ or if $|u| = |v|$ and $u <_{\text{lex}} v$. Ordering a language L with the radix order defines the *Abstract Numeration System* (ANS, *cf.* [9]) associated with L: every integer n is represented by the $(n+1)$-th word of L in the radix order which is denoted by $\langle n \rangle_L$. If L is a regular language, it defines an Abstract **Regular** Numeration System (ARNS). Let K be another ANS; we call *conversion function from L into K* the function that maps $\langle n \rangle_L$ to $\langle n \rangle_K$, for every integer n.

A Mealy machine is a graph labelled with pair of letters (*cf.* [6]); it is written as a 6-tuple $\mathcal{T} = \langle Q, A, B, \tau, i, F \rangle$, where Q, A, i and F are defined as in an automaton, B is the *output alphabet* and τ is a function $Q \times A \to B \times Q$, extended as usual to $Q \times A^* \to B^* \times Q$. We write $p \xrightarrow[\mathcal{T}]{u \,|\, v} q$ if $\tau(p, u) = (v, q)$ and the pair $u \,|\, v$ is said to be *accepted by $\mathcal{T}$* if in addition $p = i$ and $q \in F$. The *function realised by $\mathcal{T}$* maps u to v for all pairs $u \,|\, v$ accepted by $\mathcal{T}$. [4]

Let $\mathcal{A}$ and $\mathcal{B}$ be two (trim) automata. We now define a Mealy machine $\mathcal{A} \boxtimes \mathcal{B}$; it is a variant of the well-known automaton product (used for regular-language intersection). The underlying graphs of $\mathcal{A} \boxtimes \mathcal{B}$ and $\text{lred}(\mathcal{A}) \times \text{lred}(\mathcal{B})$ coincide, but the transitions of $\mathcal{A} \boxtimes \mathcal{B}$ are labelled using the labels of $\mathcal{A}$ and $\mathcal{B}$. [5]

Definition 6. *Let $\mathcal{A} = \langle Q_\mathcal{A}, A, \delta_\mathcal{A}, i_\mathcal{A}, F_\mathcal{A} \rangle$ and $\mathcal{B} = \langle Q_\mathcal{B}, B, \delta_\mathcal{B}, i_\mathcal{B}, F_\mathcal{B} \rangle$ be two automata. We denote by $\mathcal{A} \boxtimes \mathcal{B}$ the Mealy machine*

$$\mathcal{A} \boxtimes \mathcal{B} \;=\; \langle Q_\mathcal{A} \times Q_\mathcal{B},\, A,\, B,\, \tau,\, (i_\mathcal{A}, i_\mathcal{B}),\, F_\mathcal{A} \times F_\mathcal{B} \rangle \;,$$

where the transition function τ is defined as follows. If $p \xrightarrow[\mathcal{A}]{a_i} p_i$ and $q \xrightarrow[\mathcal{B}]{b_i} q_i$ are respectively the $(i+1)$-th transitions of $\text{Out}_\mathcal{A}(p)$ and of $\text{Out}_\mathcal{B}(q)$, then $\mathcal{A} \boxtimes \mathcal{B}$ features the transition $(p, q) \xrightarrow[\mathcal{A} \boxtimes \mathcal{B}]{a_i \,|\, b_i} (p_i, q_i)$.

We say that a state (p, q) is inconsistent *if either 1) p or q is final but the other is not; or 2) the out-degrees of p and q are not equal: $\text{od}_\mathcal{A}(p) \neq \text{od}_\mathcal{B}(q)$.*

Figure 5 shows the automaton $\mathcal{A}_1 \boxtimes \mathcal{A}_2$; in the figure, inconsistent states are drawn in dotted lines and their outgoing transitions are omitted and inaccessible but consistent states are drawn in dashed lines. We can now state Theorem 1 under the more precise following form.

Theorem 1. *Let $\mathcal{A}$ and $\mathcal{B}$ be two trim automata. If $\mathcal{A}$ and $\mathcal{B}$ are associated with two T-equivalent ARNS's L and K, then $\mathcal{A} \boxtimes \mathcal{B}$ realises the conversion function from L into K.*

---

[4]   According to transducer terminology, a Mealy machine is a pure sequential and letter-to-letter transducer *cf.* [5,12]. Mealy machines have the same expressive power as Moore Machines, also called deterministic finite automata with output (DFAO).

[5]   In the classical automaton product, transitions are matched using transition labels, hence $\text{lred}(\mathcal{A}) \times \text{lred}(\mathcal{B})$ and $\mathcal{A} \times \mathcal{B}$ have different underlying graphs.

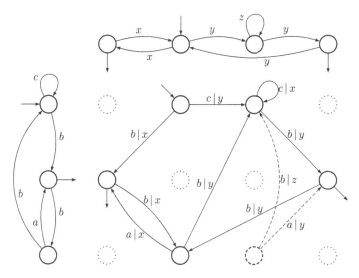

Fig. 5. The Mealy machine $\mathcal{A}_1 \boxtimes \mathcal{A}_2$

The proof of Theorem 1 breaks down into Lemmas 3 and 4. First, we give a few properties following directly from Definition 6.

Property 1. Let $\mathcal{A}$, $\mathcal{B}$ be two automata and $u \mid v$, $u' \mid v'$ be accepted by $\mathcal{A} \boxtimes \mathcal{B}$.

a. $u = u' \Leftrightarrow v = v'$.

b. $u <_{\mathrm{rad}} u' \Leftrightarrow v <_{\mathrm{rad}} v'$.

c. Let $(p,q) \xrightarrow{a\mid b} (p',q')$ be a transition of $\mathcal{A}\boxtimes\mathcal{B}$. Then $p \xrightarrow{a} p'$ is a transition of $\mathcal{A}$ and $q \xrightarrow{b} q'$ is a transition of $\mathcal{B}$.

d. Let $p \xrightarrow{a} p'$ be a transition of $\mathcal{A}$ and q be a state of $\mathcal{B}$. If (p,q) is consistent, then $(p,q) \xrightarrow{a\mid b} (p',q')$ is a transition of $\mathcal{A} \boxtimes \mathcal{B}$, for some q' and b. [6]

Lemma 3. *Let $\mathcal{A}$ and $\mathcal{B}$ be two automata. If $\mathcal{A}\boxtimes\mathcal{B}$ has no inconsistent accessible states, then it realises the conversion function from $L(\mathcal{A})$ into $L(\mathcal{B})$*

Proof. Let us denote by $\mathcal{A}'$ the trim of the input automaton of $\mathcal{A} \boxtimes \mathcal{B}$. If $\mathcal{A} \boxtimes \mathcal{B}$ has no inconsistent accessible states then it follows from Properties 1.c and 1.d that $\mathcal{A}$ is a quotient of $\mathcal{A}'$ (through the projection $(p,q) \mapsto p$). It follows that the input language of $\mathcal{A} \boxtimes \mathcal{B}$ is $L(\mathcal{A})$; symmetrically, the output language of $\mathcal{A} \boxtimes \mathcal{B}$ is $L(\mathcal{B})$. Since $\mathcal{A} \boxtimes \mathcal{B}$ realises a bijection (from Property 1.a) and preserves the order (from Property 1.b), it follows that $\mathcal{A} \boxtimes \mathcal{B}$ maps $\langle n \rangle_{\mathcal{A}}$ to $\langle n \rangle_{\mathcal{B}}$.

Lemma 4. *Let $\mathcal{A}$ and $\mathcal{B}$ be two automata. If $\mathcal{A} \overset{T}{\leadsto} \mathcal{B}$, then every inconsistent state of $\mathcal{A} \boxtimes \mathcal{B}$ is not accessible.*

---

[6] For concision, we omitted the symmetrical statement.

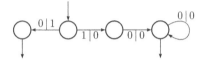

Fig. 6. A Mealy machine which is not locally increasing

Proof. Since they are T-equivalent, $\mathcal{A}$ and $\mathcal{B}$ have the same surminimisation, denoted by $\mathcal{M}$; it is a quotient both of $\mathsf{lred}(\mathcal{A})$ and of $\mathsf{lred}(\mathcal{B})$ realised by the automaton morphisms $\phi : \mathsf{lred}(\mathcal{A}) \to \mathcal{M}$ and $\psi : \mathsf{lred}(\mathcal{B}) \to \mathcal{M}$. The proof of the next claim consists in an induction over a traversal of $\mathcal{A} \boxtimes \mathcal{B}$ and is omitted here.

Claim. *Every accessible state* (p, q) *of* $\mathcal{A} \boxtimes \mathcal{B}$ *meets the condition* $\phi(p) = \psi(q)$.

Let (p, q) be an accessible state of $\mathcal{A} \boxtimes \mathcal{B}$. It follows that $\phi(p) = \psi(q)$ (from the claim), hence that p and q are both final or both non-final and that p and q have the same amount of outgoing transitions. Hence (p, q) is not an inconsistent state.

Theorem 1 follows directly from Lemmas 3 and 4. However, its converse is false in the general case: the Mealy machine shown at Figure 6 realises the conversion from $0+10^+$ into $1+00^+$, two distinct and label-irreducible languages hence not T-equivalent.

We say that a Mealy machine $\mathcal{T}$ is *locally increasing* if it locally preserves the order of labels or, more formally, if it satisfies the following condition.

$$\text{For every pair of transitions of } \mathcal{T} \qquad \begin{array}{c} p \xrightarrow{a \mid b} q \\ \\ p \xrightarrow{c \mid d} q' \end{array} \qquad a < c \Leftrightarrow b < d \qquad (1)$$

For instance, the Mealy machine $\mathcal{A} \boxtimes \mathcal{B}$ is always locally increasing whereas the one shown at Figure 6 is not: the two outgoing transitions of the initial state reverse the order of the letters.

Theorem 2. *Let* $\mathcal{T}$ *be a locally increasing Mealy machine,* $\mathcal{A}$ *and* $\mathcal{B}$ *its respective input and output automata. Then* $\mathcal{A}$ *and* $\mathcal{B}$ *are T-equivalent.*

Proof. Note that $\mathcal{A}$, $\mathcal{B}$, $\mathcal{T}$ have the same state set and that since $\mathcal{T}$ is locally increasing, then $\mathcal{B}$ is deterministic. Let p be a state of $\mathcal{A}$, $\mathcal{B}$ and $\mathcal{T}$. We enumerate the outgoing transitions of p in $\mathcal{T}$: $\forall i \in [\![k_p]\!]$ $p \xrightarrow{a_i \mid b_i} p_i$ where $k_p = \mathsf{od}_{\mathcal{T}}(p)$ and $a_0 < a_1 < \cdots < a_{k_p-1}$; since $\mathcal{T}$ is locally increasing, then $b_0 < b_1 < \cdots < b_{k_p-1}$.

We fix i in $[\![k_p]\!]$. The transitions $p \xrightarrow{a_i} p_i$ and $p \xrightarrow{b_i} p_i$ are respectively the $(i+1)$-th transitions of $\mathsf{Out}_{\mathcal{A}}(p)$ and $\mathsf{Out}_{\mathcal{B}}(p)$. It follows that both transitions are relabelled by the same digit i in $\mathsf{lred}(\mathcal{A})$ and $\mathsf{lred}(\mathcal{B})$, hence coincide. Since $\mathcal{A}$ and $\mathcal{B}$ differ only by their transition labels, $\mathsf{lred}(\mathcal{A})$ and $\mathsf{lred}(\mathcal{B})$ are isomorphic.

4   Label Reduction Within Numeration System Theory

We briefly recall here basic definitions and notations for positional numeration system; see for instance Section 2.3.3 of [5] for more details. A *basis* is a strictly increasing sequence of integers $(U_i)_{i \in \mathbb{N}}$ with $U_0 = 1$ defining the positional numeration system U. The U-evaluation function π_U maps a finite word $d_k \, d_{k-1} \cdots d_0$ over a digit alphabet to the integer $\pi_U(d_k \, d_{k-1} \cdots d_0) = \sum_{i=0}^{k} d_i \, U_i$. The Rényi greedy algorithm (*cf.* for instance [10, Chapter 7]) computes a word whose evaluation is n; it is called *the U-representation of n* and is denoted by $\langle n \rangle_U$. We also denote by $L(U)$ the language $L(U) = \{ \langle n \rangle_U \mid n \in \mathbb{N} \}$.

In the following, we always assume that the ratio U_{n+1}/U_n is bounded by an integer constant $M = \sup\{ \lceil U_{n+1}/U_n \rceil \mid n \in \mathbb{N} \}$, in which case the digits of each U-representation belong to the alphabet $A_U = [\![M]\!]$. The next two classical propositions follow.

Proposition 4 ([5, Proposition 2.3.44]). *Let u be a word of A_U^*. If u does not start with the letter 0 then $u \leq_{rad} \langle n \rangle_U$, where $n = \pi_U(u)$.*

Proposition 5 ([5, Proposition 2.3.45]). *Let n and m be two positive integers. Then $n < m$ if and only if $\langle n \rangle_U <_{rad} \langle m \rangle_U$.*

The next proposition (together with the remark following it) is the main result from this section.

Proposition 6. *Let U be a positional numeration system. If $L(U)$ is a regular language, then $0^* L(U)$ is label-irreducible.* [7]

Proof. For concision, we write $\langle \; \rangle$ and $\pi(\;)$ instead of $\langle \; \rangle_U$ and $\pi_U(\;)$ in this proof. The whole statement follows from the next claim.

Claim. Let m be an integer u, v be two words of A_U^* and $(d+1)$ be a positive digit. If $\langle m \rangle = u(d+1)v$ and $(u,d) \neq (\epsilon, 0)$, then there exists an integer n such that $\langle n \rangle = u d v'$, for some v'.

Proof of the Claim. Without loss of generality, we may assume that v is the smallest word in the radix order such that $u(d+1)v$ is the representation of an integer (Assumption $(*)$). Let $k = |v|$, we denote by n the value of $u \, d \, v$: that is $n = \pi(u \, d \, v) = (m - U_k)$. Note that (since it is a prefix of $\langle m \rangle$,) u may not start with the letter zero; since $(u, d) \neq (\epsilon, 0)$, it follows that $u d v$ does not start with the letter 0 either and applying Propositions 4 and 5 yields the two following inequations: $u \, d \, v \; \leq_{rad} \; \langle n \rangle \; <_{rad} \; u\,(d+1)\,v$.
It follows that $\langle n \rangle$ is of one of the three following forms.

- $\langle n \rangle = u \, d \, w$ with $v <_{rad} w$ and $\pi(v) = \pi(w)$. It follows that $u\,(d+1)\,v <_{rad} u\,(d+1)\,w$ and that $\pi(u\,(d+1)\,w) = m$, hence the representation of m cannot be equal to $u\,(d+1)\,v$, a contradiction to Proposition 4 .
- $\langle n \rangle = u(d+1)w$ for some $w <_{rad} v$, a contradiction to Assumption $(*)$.
- $\langle n \rangle = u \, d \, v$ yielding the proof of the claim.

---

[7] A language is said *label-irreducible* if it is equal to its label reduction.

Remark 2. Proposition 6 could be made substantially stronger if label reduction were defined directly on languages (ie. independently of automata). The label reduction of a language would be defined as $\mathsf{lred}(L) = \{f_L(u) \mid u \in L\}$ with

$$f_L(\epsilon) = \epsilon$$
$$f_L(u\,b) = f_L(u)\,g_L(u,b) \quad \text{where } g_L(u,b) = |\{ua \mid a < b \text{ and } ua \in \mathsf{Pre}\,(L)\}| \;.$$

A language L is *label-irreducible* if $\forall ub, \; ub \in \mathsf{Pre}\,(L) \;\Rightarrow\; \forall a < b, \; ua \in \mathsf{Pre}\,(L)$. The statement *every positional numeration system is label-irreducible* is then an immediate consequence of the Claim of the previous proof.

Remark 3. Let us compare the classes of ARNS's and label-irreducible ARNS's.
- Every ARNS is T-equivalent to some label-irreducible ARNS, hence from Theorem 1 , one may be converted into the other by means of a Mealy machine. It follows that both systems will share the same properties
- Within an T-equivalence class, the unique label-irreducible ARNS is associated with an automaton with the smallest amount of states.
- All concrete numeration systems seem to be label-irreducible. [8]

In the remainder of this section, we prove that every T-equivalence class contains a *Substitution Numeration System* (SNS, *cf.* [3]). It is known that an SNS is a particular ARNS (*cf.* [1]) and we use this result to give here a very brief description of SNS's as ARNS's.

Let X be an alphabet and let $\sigma : X^* \to X^*$ be a monoid morphism. We assume that σ is *prolongable on $x \in X$*, that is, 1) $\sigma(x)$ starts with an x and 2) $\lim_{n\to\infty}(|\sigma^n(x)|) = \infty$. In the following, we manipulate the alphabet B_σ (defined below) whose *letters* are *words* over the alphabet X. If u or $x_0 x_1 \cdots x_k$ denotes a word of X^*, the corresponding letter of B_σ is denoted by $[u]$ or $[x_0 x_1 \cdots x_k]$. $B_\sigma = \{ [u] \mid u \text{ is a strict prefix of } \sigma(y) \text{ for some } y \in A \}$.

The *prefix automaton* $\mathcal{A}_\sigma = \langle X, B_\sigma, \delta, x, X \rangle$ is defined as follows. Its state set is X (*ie.* the alphabet of σ), its initial state is x and all states are accepting. The transition function δ is defined such that $\mathcal{A}_\sigma$ features the transition $y \xrightarrow{\;[u]\;}_{\mathcal{A}_\sigma} z$ if and only if uz is a prefix of $\sigma(y)$. The SNS σ is then the ARNS L, where L consists of the words of $L(\mathcal{A}_\sigma)$ that do not start with the letter $[\epsilon]$.

If every ARNS is not necessarily an SNS, it is pretty close to be the case, as stated by the Corollary 1 of the next proposition. The proof of this statement is omitted here, and consists in a thorough examination of classical transformations from automaton to substitution (*cf.* [8,11]).

Proposition 7. *Every prefix-closed ARNS is T-equivalent to an SNS.*

Corollary 1. *Every prefix-closed ARNS can be converted into an SNS through a Mealy machine.*

---

[8] We call here *concrete* the numeration systems that may be defined by an evaluation function, by opposition to those defined by their representation language.

To the best of our knowledge, the only label-reducible concrete numeration systems are the *rational base numeration systems* (*cf.* Section 2.5 of [5]) and, even in this case, it happens to exist a variant which is label-irreducible (*cf.* [4]).

5   Conclusion

We introduced the notion of surminimisation of automata, a transformation producing an automata smaller than the one resulting from the classical minimisation. While minimisation preserves the language (that is, a labelled tree), surminimisation preserves the underlying unlabelled tree only.

Surminimisation induces on Abstract Regular Numeration Systems (ARNS) a transformation, called label reduction, and an equivalence relation; each equivalence class features a canonical representative: the label reduction of any element of the class. We proved that members of the same equivalence class are essentially the same (*ie.* may be converted from one into another by a Mealy machine) and, conversely, that if the conversion from one ARNS into another is realised by a locally increasing Mealy machine, then the ARNS's are T-equivalent.

Moreover, a simple verification yields that all positional numeration systems are label-irreducible. In summary, label reduction allows to simplify ARNS's without excluding any concrete cases.

Acknowledgments. The author thanks Michel Rigo and Émilie Charlier for their invitation to Liège and the discussions he had with them about this work.

References

1. Berthé, V., Rigo, M.: Odometers on regular languages. Theory Comput. Syst. **40**(1), 1–31 (2007)
2. Cobham, A.: Uniform tag sequences. Math. Systems Theory **6**, 164–192 (1972)
3. Dumont, J.-M., Thomas, A.: Digital sum problems and substitutions on a finite alphabet. Journal of Number Theory **39**(3), 351–366 (1991)
4. Frougny, C., Klouda, K.: Rational base number systems for p-adic numbers. RAIRO - Theor. Inf. and Applic. **46**(1), 87–106 (2012)
5. Frougny, C., Sakarovitch, J.: Number representation and finite automata. In: Berthé, V., Rigo, M. (eds) Combinatorics, Automata and Number Theory. Encyclopedia of Mathematics and its Applications 135, pp. 34–107. Cambridge Univ. Press (2010)
6. Mealy, G.H.: A method for synthesizing sequential circuits. Bell Syst. Tech. J. **34**, 1045–1079 (1955)
7. Hopcroft, J.E., Ullman, J.D.: Introduction to Automata Theory, Languages and Computation. Addison-Wesley (1979)
8. Lecomte, P., Rigo, M.: Abstract numeration systems. In: Berthé, V., Rigo, M. (eds.) Combinatorics, Automata and Number Theory. Encyclopedia of Mathematics and its Applications 135, pp. 108–162. Cambridge Univ. Press (2010)
9. Lecomte, P., Rigo, M.: Numeration systems on a regular language. Theory Comput. Syst. **34**, 27–44 (2001)
10. Lothaire, M.: Algebraic Combinatorics on Words. Cambridge University Press (2002)
11. Rigo, M., Maes, A.: More on generalized automatic sequences. Journal of Automata, Languages and Combinatorics **7**(3), 351–376 (2002)
12. Sakarovitch, J.: Eléments de théorie des automates. Vuibert, 2003. Corrected English translation: Elements of Automata Theory. Cambridge University Press (2009)

On the Complexity of k-Piecewise Testability and the Depth of Automata

Tomáš Masopust$^{(\boxtimes)}$ and Michaël Thomazo

TU Dresden, Dresden, Germany
{tomas.masopust,michael.thomazo}@tu-dresden.de

Abstract. For a non-negative integer k, a language is k-piecewise testable (k-PT) if it is a finite boolean combination of languages of the form $\Sigma^* a_1 \Sigma^* \cdots \Sigma^* a_n \Sigma^*$ for $a_i \in \Sigma$ and $0 \le n \le k$. We study the following problem: Given a DFA recognizing a piecewise testable language, decide whether the language is k-PT. We provide a complexity bound and a detailed analysis for small k's. The result can be used to find the minimal k for which the language is k-PT. We show that the upper bound on k given by the depth of the minimal DFA can be exponentially bigger than the minimal possible k, and provide a tight upper bound on the depth of the minimal DFA recognizing a k-PT language.

1 Introduction

A regular language is *piecewise testable* (PT) if it is a finite boolean combination of languages of the form $\Sigma^* a_1 \Sigma^* a_2 \Sigma^* \cdots \Sigma^* a_n \Sigma^*$, where $a_i \in \Sigma$ and $n \ge 0$. It is *k-piecewise testable* (k-PT) if $n \le k$. These languages were introduced by Simon in his PhD thesis [13]. Simon proved that PT languages are exactly those regular languages whose syntactic monoid is $\mathcal{J}$-trivial. He provided various characterizations of PT languages in terms of monoids, automata, etc.

In this paper, we study the *k-piecewise testability* problem, that is, to decide whether a PT language is k-PT.

NAME: k-PIECEWISETESTABILITY
INPUT: an automaton (minimal DFA or NFA) $\mathcal{A}$
OUTPUT: YES if and only if $\mathcal{L}(\mathcal{A})$ is k-piecewise testable

Note that the problem is trivially decidable, since there is only a finite number of k-PT languages over the input alphabet of $\mathcal{A}$.

It is not hard to see that if a language is k-PT, it is also $(k+1)$-PT. It was shown in [10] that if the depth of a minimal DFA recognizing a PT language is k, then the language is k-PT. However, the opposite implication does not hold. To the best of our knowledge, no efficient algorithm to find the minimal k for which a PT language is k-PT nor an algorithm to decide whether a language is k-PT has been published so far.[1]

T. Masopust—Research supported by the DFG in grant KR 4381/1-1.

M. Thomazo—Research supported by the Alexander von Humboldt Foundation.

[1] Very recently, a co-NP upper bound appeared in [5] in terms of separability.

© Springer International Publishing Switzerland 2015
I. Potapov (Ed.): DLT 2015, LNCS 9168, pp. 364–376, 2015.
DOI: 10.1007/978-3-319-21500-6_29

We first show a co-NP upper bound to decide whether a minimal DFA recognizes a k-PT language for a fixed k (Theorem 1), which results in an algorithm to find the minimal k that runs in the time single exponential with respect to the size of the DFA and double exponential with respect to the resulting k. We then provide a detailed complexity analysis for small k's. In particular, the problem is trivial for $k = 0$, decidable in deterministic logarithmic space for $k = 1$ (Theorem 2), and NL-complete for $k = 2, 3$ (Theorems 3 and 4). As a result, we obtain a PSPACE upper bound to decide whether an NFA recognizes a k-PT language for a fixed k. Recall that it is PSPACE-complete to decide whether an NFA recognizes a PT language, and it is actually PSPACE-complete to decide whether an NFA recognizes a 0-PT language (Theorem 5).

Since the depth of the minimal DFAs plays a role of an upper bound on k, we investigate the relationship between the depth of an NFA and k-piecewise testability of its language. We show that, for every $k \geq 0$, there exists a k-PT language with an NFA of depth $k - 1$ and with the minimal DFA of depth $2^k - 1$ (Theorem 7). Although it is well known that DFAs can be exponentially larger than NFAs, a by-product of our result is that all the exponential number of states of the DFA form a simple path. Finally, we investigate the opposite implication and show that the tight upper bound on the depth of the minimal DFA recognizing a k-PT language over an n-letter alphabet is $\binom{k+n}{k} - 1$ (Theorem 8).

For the missing proofs, the reader is referred to [11].

2   Preliminaries and Definitions

We assume that the reader is familiar with automata theory. To fix the notation, the cardinality of a set A is denoted by $|A|$ and the power set of A by 2^A. An alphabet Σ is a finite nonempty set. The free monoid generated by Σ is denoted by Σ^*. A word over Σ is any element of Σ^*; the empty word is denoted by ε. For a word $w \in \Sigma^*$, $\mathrm{alph}(w) \subseteq \Sigma$ denotes the set of all letters occurring in w, and $|w|_a$ denotes the number of occurrences of letter a in w. A language over Σ is a subset of Σ^*.

A *nondeterministic finite automaton* (NFA) is a quintuple $\mathcal{A} = (Q, \Sigma, \cdot, I, F)$, where Q is a finite nonempty set of states, Σ is an input alphabet, $I \subseteq Q$ is a set of initial states, $F \subseteq Q$ is a set of accepting states, and $\cdot : Q \times \Sigma \to 2^Q$ is the transition function that can be extended to the domain $2^Q \times \Sigma^*$. The language *accepted* by $\mathcal{A}$ is the set $L(\mathcal{A}) = \{w \in \Sigma^* \mid I \cdot w \cap F \neq \emptyset\}$. We usually omit $\cdot$ and write simply Iw instead of $I \cdot w$. A *path* π from a state q_0 to a state q_n under a word $a_1 a_2 \cdots a_n$, for some $n \geq 0$, is a sequence of states and input symbols $q_0 a_1 q_1 a_2 \ldots q_{n-1} a_n q_n$ such that $q_{i+1} \in q_i \cdot a_{i+1}$, for all $i = 0, 1, \ldots, n - 1$. The path π is *accepting* if $q_0 \in I$ and $q_n \in F$. We use the notation $q_0 \xrightarrow{a_1 a_2 \cdots a_n} q_n$ to denote that there exists a path from q_0 to q_n under the word $a_1 a_2 \cdots a_n$. A path is *simple* if all states of the path are pairwise different. The number of states on the longest simple path of $\mathcal{A}$ decreased by one is called the *depth* of the automaton $\mathcal{A}$, denoted by $depth(\mathcal{A})$.

The *reachability* relation $\leq$ on the set of states is defined by $p \leq q$ if there exists a word w in Σ^* such that $q \in p \cdot w$. The NFA $\mathcal{A}$ is *partially ordered* if the reachability relation $\leq$ is a partial order. For two states p and q of $\mathcal{A}$, we write $p < q$ if $p \leq q$ and $p \neq q$. A state p is *maximal* if there is no state q such that $p < q$. Partially ordered automata are also called *acyclic automata*.

The NFA $\mathcal{A}$ is *deterministic* (DFA) if $|I| = 1$ and $|q \cdot a| = 1$ for every q in Q and a in Σ. Then the transition function $\cdot$ is a map from $Q \times \Sigma$ to Q that can be extended to the domain $Q \times \Sigma^*$. Two states p_1, p_2 of a DFA are *distinguishable* if there exists a word w and states $f \in F$ and $r \in Q \setminus F$ such that $p_1 \cdot w = f$ and $p_2 \cdot w = r$, that is, w is accepted from one of them and rejected from the other. A DFA is *minimal* if all its states are reachable from the initial state and pairwise distinguishable.

The notion of confluent DFAs was introduced in [10]. Let $\mathcal{A} = (Q, \Sigma, \cdot, q_0, F)$ be a DFA and $\Gamma \subseteq \Sigma$ be a subalphabet. The DFA $\mathcal{A}$ is Γ-*confluent* if, for every state q in Q and every pair of words u, v in Γ^*, there exists a word w in Γ^* such that $(qu)w = (qv)w$. The DFA $\mathcal{A}$ is *confluent* if it is Γ-confluent for every subalphabet Γ. The DFA $\mathcal{A}$ is *locally confluent* if, for every state q in Q and every pair of letters a, b in Σ, there exists a word w in $\{a, b\}^*$ such that $(qa)w = (qb)w$.

An NFA $\mathcal{A} = (Q, \Sigma, \cdot, I, F)$ can be turned into a directed graph $G(\mathcal{A})$ with the set of vertices Q, where a pair (p, q) in $Q \times Q$ is an edge in $G(\mathcal{A})$ if there is a transition from p to q in $\mathcal{A}$. For $\Gamma \subseteq \Sigma$, we define the directed graph $G(\mathcal{A}, \Gamma)$ with the set of vertices Q by considering all those transitions that correspond to letters in Γ. For a state p, let $\Sigma(p) = \{a \in \Sigma \mid p \in p \cdot a\}$ denote the set of all letters under which the NFA $\mathcal{A}$ has a self-loop in the state p. Let $\mathcal{A}$ be a partially ordered NFA. If for every state p of $\mathcal{A}$, state p is the unique maximal state of the connected component of $G(\mathcal{A}, \Sigma(p))$ containing p, then we say that the NFA satisfies the *unique maximal state (UMS) property*.

A regular language is k-*piecewise testable* if it is a finite boolean combination of languages of the form $\Sigma^* a_1 \Sigma^* a_2 \Sigma^* \cdots \Sigma^* a_n \Sigma^*$, where $0 \leq n \leq k$ and $a_i \in \Sigma$. A regular language is *piecewise testable* if it is k-piecewise testable for some $k \geq 0$. We adopt the notation $L_{a_1 a_2 \cdots a_n} = \Sigma^* a_1 \Sigma^* a_2 \Sigma^* \cdots \Sigma^* a_n \Sigma^*$ from [10]. For two words $v = a_1 a_2 \cdots a_n$ and $w \in L_v$, we say that v is a *subword* of w, denoted by $v \preccurlyeq w$. For $k \geq 0$, let $sub_k(v) = \{u \in \Sigma^* \mid u \preccurlyeq v, |u| \leq k\}$. For words w_1, w_2, we define $w_1 \sim_k w_2$ if and only if $sub_k(w_1) = sub_k(w_2)$. If $w_1 \sim_k w_2$, we say that w_1 and w_2 are k-*equivalent*. Note that $\sim_k$ is a congruence with finite index.

Fact 1 ([13]). *Let L be a regular language, and let $\sim_L$ denote the Myhill congruence. A language L is k-PT if and only if $\sim_k \subseteq \sim_L$. Moreover, L is a finite union of $\sim_k$ classes.*

Fact 2. *Let L be a language recognized by the minimal DFA $\mathcal{A}$. The following is equivalent.*

1. *The language L is PT.*
2. *The minimal DFA $\mathcal{A}$ is partially ordered and (locally) confluent [10].*
3. *The minimal DFA $\mathcal{A}$ is partially ordered and satisfies the UMS property [15].*

3   Complexity of k-Piecewise Testability for DFAs

The k-*piecewise testability problem for DFAs* asks whether, given a minimal DFA $\mathcal{A}$, the language $L(\mathcal{A})$ is k-PT. We show that it belongs to co-NP, which can be used to compute the minimal k for which the language is k-PT in the time single exponential with respect to the size of $\mathcal{A}$ and double exponential with respect to the resulting k. For small k's we then provide precise complexity analyses.

Theorem 1. *The following problem belongs to co-NP:*

NAME: k-PIECEWISETESTABILITY
INPUT: *a minimal DFA $\mathcal{A}$*
OUTPUT: YES *if and only if $\mathcal{L}(\mathcal{A})$ is k-PT*

Proof (sketch). One first checks that the automaton $\mathcal{A}$ over Σ recognizes a PT language. If $\mathcal{L}(\mathcal{A})$ is not k-PT, then there exist two k-equivalent words w_1 and w_2. It can be shown that the length of w_1 is at most $k|\Sigma|^k$, w_1 is a subword of w_2, and w_1 and w_2 lead the automaton to two different states. In addition, it can be shown that one can choose w_2 of length at most $depth(\mathcal{A})$ bigger than the length of w_1. A polynomial certificate for non k-piecewise testability can thus be given by providing such w_1 and w_2, which are indeed of polynomial length in the size of $\mathcal{A}$ and Σ. □

If we search for the minimal k for which the language is k-PT, we can first check whether it is 0-PT. If not, we check whether it is 1-PT and so on until we find the required k. In this case, the bounds $k|\Sigma|^k$ and $k|\Sigma|^k + depth(\mathcal{A})$ on the length of words w_1 and w_2 that need to be investigated are exponential with respect to k. To investigate all the words up to these lengths then gives an algorithm that is exponential with respect to the size of the minimal DFA and double exponential with respect to the desired k.

Proposition 1. *Let $\mathcal{A}$ be a minimal DFA that is partially ordered and confluent. To find the minimal k for which the language $L(\mathcal{A})$ is k-PT can be done it time exponential with respect to the size of $\mathcal{A}$ and double exponential with respect to the resulting k.*

Theorem 1 gives an upper bound on the complexity to decide whether a language is k-PT for a fixed k. We now show that for $k \leq 3$, the complexity is much simpler.

0-*Piecewise Testability* The language $L(\mathcal{A})$ of a minimal DFA $\mathcal{A}$ over Σ is 0-PT if and only if it has a single state, that is, it recognizes either Σ^* or $\emptyset$. Thus, given a minimal DFA, it is decidable in $O(1)$ whether its language is 0-PT.

1-*Piecewise Testability* Let $\mathcal{A} = (Q, \Sigma, \cdot, q_0, F)$ be a minimal DFA. It can be shown that the language $L(\mathcal{A})$ is 1-PT if and only if (1) for every $p \in Q$ and $a \in \Sigma$, $pa = q$ implies $qa = q$, and (2) for every $p \in Q$ and $a, b \in \Sigma$, $pab = pba$. Since this property can be verified locally in the DFA, we have the following.

Theorem 2. *The problem to decide whether a minimal DFA recognizes a 1-PT language is in LOGSPACE.*

2-Piecewise Testability We show that the problem to decide whether a minimal DFA recognizes a 2-PT language is NL-complete. This coincides with the complexity to decide whether the language is PT.

Theorem 3. *The problem to decide whether a minimal DFA recognizes a 2-PT language is NL-complete.*

We prove this theorem by a sequence of lemmas.

Lemma 1. *Let* $\mathcal{A} = (Q, \Sigma, \cdot, q_0, F)$ *be a minimal DFA. For every* $k \geq 0$, *if* $w_1 \sim_k w_2$ *and* $q_0 w_1 \neq q_0 w_2$, *then there exist two words* w *and* w' *such that* $w \sim_k w'$, w' *is obtained from* w *by adding a single letter at some place, and* $q_0 w \neq q_0 w'$.

Proof. Let w_1, w_2 be such that $w_1 \sim_k w_2$ and $q_0 w_1 \neq q_0 w_2$. By [12, Theorem 6.2.6], there is w_3 such that w_1, w_2 are subwords of w_3 and $w_1 \sim_k w_2 \sim_k w_3$. Either w_1 and w_3, or w_2 and w_3, do not lead to the same state. Denote that pair by v, v' with $v \preccurlyeq v'$. Let $v = u_0, u_1, \ldots, u_n = v'$ be a sequence such that u_{j+1} is obtained from u_j by adding a letter at some place. There must be j such that u_j and u_{j+1} lead to two different states. Set $w = u_j$ and $w' = u_{j+1}$.   □

Lemma 2. *Let* $\mathcal{A} = (Q, \Sigma, \cdot, q_0, F)$ *be a minimal partially ordered and confluent DFA. The language* $L(\mathcal{A})$ *is 2-PT if and only if for every* $a \in \Sigma$ *and every states* p *such that there exists* w *with* $|w|_a \geq 1$, $pua = paua$, *for every* $u \in \Sigma^*$.

Proof. ($\Rightarrow$) By contraposition. Assume that there exists $u \in \Sigma^*$ and a state p such that $q_0 w = p$ for some $w \in \Sigma^*$ containing a and such that $pua \neq paua$. By the assumption, $w = w_1 a w_2$, for some $w_1, w_2 \in \Sigma^*$ such that $a \notin \text{alph}(w_1)$, and we want to show that $w_1 a w_2 u a \sim_2 w_1 a w_2 a u a$. However, for any $c \in \text{alph}(w_1 a w_2)$, if $ca \preccurlyeq w_1 a w_2 a u a$, then $ca \preccurlyeq w_1 a w_2 u a$. Similarly for $d \in \text{alph}(ua)$ and $ad \preccurlyeq w_1 a w_2 a u a$. Since $q_0 \cdot wua \neq q_0 \cdot waua$, the minimality of $\mathcal{A}$ gives that there exists a word v such that $wuav \in L(\mathcal{A})$ if and only if $wauav \notin L(\mathcal{A})$. Since $\sim_2$ is a congruence, $wuav \sim_2 wauav$, which violates Fact 1, hence $L(\mathcal{A})$ is not 2-PT.

($\Leftarrow$) Let w_1 and w_2 be two words such that $w_1 \sim_2 w_2$. We want to show that $q_0 w_1 = q_0 w_2$. By Lemma 1, it is sufficient to show this direction of the theorem for two words w and w' such that w' is obtained from w by adding a single letter at some place. Thus, let a be the letter, and let

$$w = a_1 \ldots a_k a_{k+1} \ldots a_n \text{ and } w' = a_1 \ldots a_k a a_{k+1} \ldots a_n$$

for $0 \leq k \leq n$. Let $w_{i,j} = a_i a_{i+1} \ldots a_j$. We distinguish two cases.

(A) Assume that a does not appear in $w_{1,k}$. Then a must appear in $w_{k+1,n}$. Consider the first occurrence of a in $w_{k+1,n}$. Then $w_{k+1,n} = u_1 a u_2$, where a does not appear in u_1. Let $B = \text{alph}(u_1 a)$. Then $B \subseteq \text{alph}(u_2)$, because if there is no a in $w_{1,k} u_1$, any subword ax, for $x \in B$, that appears in $w' = w_{1,k} a u_1 a u_2$ must also appear in the subword $a u_2$ of $w = w_{1,k} u_1 a u_2$.

Let $u_2 = x_1b_1x_2b_2x_3 \ldots x_\ell b_\ell x_{\ell+1}$, where $B = \{b_1, b_2, \ldots, b_\ell\}$ and b_j does not appear in $x_1b_1x_2 \ldots x_j$, $j = 1, 2, \ldots, \ell$. Let $v = b_1b_2 \ldots b_\ell$. Let $z \in \{q_0 \cdot w_{1,k}u_1a,$ $q_0 \cdot w_{1,k}au_1a\}$. We prove (by induction on j) that for every $j = 1, 2, \ldots, \ell$, there exists a word y_j such that $z \cdot (b_1b_2 \ldots b_j)^R y_j = z \cdot x_1b_1x_2b_2x_3 \ldots x_jb_jx_{j+1}$. Since b_1 appears in u_1, we use the assumption from the statement of the theorem to obtain $(z \cdot x_1b_1) \cdot x_2 = (z \cdot b_1x_1b_1) \cdot x_2$. Assume that it holds for $j < k$. We prove it for $j + 1$. Again, b_{j+1} appears in u_1 implies that

$$
\begin{aligned}
z \cdot x_1b_1x_2b_2x_3 \ldots x_jb_jx_{j+1}b_{j+1}x_{j+2} &= ((z \cdot x_1b_1x_2b_2x_3 \ldots x_jb_jx_{j+1})b_{j+1})x_{j+2} \\
&= ((z \cdot b_j \ldots b_2b_1y_j)b_{j+1})x_{j+2} \\
&= z \cdot b_{j+1}\underline{b_j \ldots b_2b_1y_j}b_{j+1}x_{j+2}
\end{aligned}
$$

where the second equality is by the induction hypothesis and the third is by the assumption from the statement of the theorem applied to the underlined part. Thus, in particular, there exists a word y such that $q_0 \cdot w_{1,k}u_1av^Ry = q_0 \cdot w$ and $q_0 \cdot w_{1,k}au_1av^Ry = q_0 \cdot w'$.

Finally, let $z_1 = q_0 \cdot w_{1,k}u_1a$ and $z_2 = q_0 \cdot w_{1,k}au_1a$. We prove that $z_1 \cdot v^R = z_2 \cdot v^R$, which then concludes the proof since it implies that $q_0 \cdot w = q_0 \cdot w'$. To prove this, we make use of the following claim, presented without proof.

Claim (Commutativity). For every $a, b \in \Sigma$ and every state p such that $q_0 \cdot w = p$ and a and b appear in w, $p \cdot ab = p \cdot ba$.

We can now finish the proof by induction on the length of $v^R = b_\ell \ldots b_2b_1$ by showing that the state $z_i' = z_i \cdot b_\ell \ldots b_2b_1$ has self-loops under B, $i = 1, 2$. Let $z_i \xrightarrow{b_\ell \ldots b_2b_1} z_i' = q_{i,\ell+1}b_\ell q_{i,\ell}b_{\ell-1}q_{i,\ell-1} \ldots q_{i,2}b_1q_{i,1}$ denote the path defined by the word v^R from the state z_i, $i = 1, 2$. We state the following claim without proof.

Claim. Both states z_1' and z_2' have self-loops under all letters of the alphabet B.

Thus, since no other states are reachable from z_1' and z_2' under B, and z_1' and z_2' are reachable from $q_0 \cdot w_{1,k}$ by words over B, confluency of the automaton implies that $z_1' = z_2'$, which completes the proof of part (A).

(B) If $a = a_i$ for some $i \leq k$, we consider two cases. First, assume that for every $c \in \Sigma \cup \{\varepsilon\}$, ca is a subword of $w_{1,k}a$ implies that ca is a subword of $w_{1,k}$. Then aa is a subword of $w_{1,k}$. Let $w_{1,k} = w_3aw_4$, where a does not appear in w_4. Let $q = q_0 \cdot w_3a$, and let $B = \text{alph}(w_4)$. Note that $B \subseteq \text{alph}(w_3)$, since if xa is a subword of $w_{1,k}a$, then it is also in w_3a. By the assumption of the theorem, $q = q_0 \cdot w_3a = q_0 \cdot w_3aa$, hence we get that there is a self-loop in q under a. Now, by the self-loop under a in q and commutativity (the claim above), $q \cdot w_4 = q \cdot aw_4 = q \cdot w_4a$. Thus, $q_0 \cdot w_{1,k} = q_0 \cdot w_{1,k}a$.

Second, assume that there exists c in $w_{1,k}$ such that $ca \preccurlyeq w_{1,k}a$ is not a subword of $w_{1,k}$. Then a must appear in $w_{k+1,n}$. Together, there exist $i \leq k < j$ such that $a_i = a_j = a$. By the assumption of the theorem, we obtain that $q_0 \cdot w_{1,k}aw_{k+1,j} = q_0 \cdot w_{1,k}w_{k+1,j}$, since $w_{k+1,j} = xa$, for some $x \in \Sigma^*$. This implies that $q_0 \cdot w = q_0 \cdot w'$, which completes the proof of part (B). $\square$

Lemma 3. *Let $\mathcal{A} = (Q, \Sigma, \cdot, q_0, F)$ be a DFA. Then the following is equivalent:*

1. *For every $a \in \Sigma$ and every state s such that $q_0 w = s$ for some $w \in \Sigma^*$ with $|w|_a \geq 1$, $sua = saua$, for every $u \in \Sigma^*$.*
2. *For every $a \in \Sigma$ and every state s such that $q_0 w = s$ for some $w \in \Sigma^*$ with $|w|_a \geq 1$, $sba = saba$ for every $b \in \Sigma \cup \{\varepsilon\}$.*

Lemma 4. *For every $k \geq 2$, the k-PT problem is NL-hard.*

Proof. We reduce *monotone graph accessibility (2MGAP)* [3] to the k-PT problem. An instance of 2MGAP is a graph (G, s, g), where $G = (V, E)$ is a graph with the set of vertices $V = \{1, 2, \ldots, n\}$, the source vertex $s = 1$ and the target vertex $g = n$, the out-degree of each vertex is bounded by 2 and for all edges (u, v), v is greater than u (the vertices are linearly ordered).

We construct the automaton $\mathcal{A} = (V \cup \{q_0, f_1, f_2, \ldots, f_{k-1}, d\}, \Sigma, \cdot, q_0, \{f_{k-1}\})$ as follows. For every edge (u, v), we construct a transition $u \cdot a_{uv} = v$ over a fresh letter a_{uv}. Moreover, we add the transitions $q_0 \cdot a = s$, $g \cdot a = f_1$ and $f_i \cdot a = f_{i+1}$, $i = 1, 2, \ldots, k-2$, over a fresh letter a. The automaton is deterministic, but not necessarily minimal, since some of the states may not be reachable from the initial state, or some states may be equivalent. To ensure minimality, we add, for each state $v \in V \setminus \{s\}$, new transitions from q_0 to v under fresh letters, and for each state $v \in V \setminus \{g\}$, new transitions from v to f_{k-1} under fresh letters. All undefined transitions go to the sink state d. Then the automaton $\mathcal{A}$ is deterministic and minimal, and the language $L(\mathcal{A})$ is finite.

Claim. Let w be a word over Σ. If every a from Σ appears at most once in w, that is, $|w|_a \leq 1$, then the language $\{w\}$ is 2-PT.

Proof. Since $\{w\}$ is PT, the minimal DFA is partially ordered and confluent. Then the condition of Lemma 2 is trivially satisfied, since, after the second occurrence of the same letter, the minimal DFA accepting $\{w\}$ is in the unique maximal non-accepting state.   ◇

We now show that the language $L(\mathcal{A})$ is k-PT if and only if g is not reachable from s.

Assume that g is reachable from s. Let w be a sequence of labels of such a path from s to g in $\mathcal{A}$. Then the word awa^{k-1} belongs to $L(\mathcal{A})$ and awa^k does not. However, $awa^{k-1} \sim_k awa^k$, which proves that $L(\mathcal{A})$ is not k-PT.

If g is not reachable from s, then $L(\mathcal{A}) = \{au_1, au_2, \ldots, au_\ell, u_{\ell+1}, \ldots, u_{\ell+s}\} \cup \{w_1 a^{k-1}, w_2 a^{k-1}, \ldots, w_m a^{k-1}\}$, where u_j and w_j are words over $\Sigma \setminus \{a\}$ that do not contain any letter twice. Then the first part is 2-PT by the previous claim, as well as the second part for $k = 2$. It remains to show that, for any $k \geq 3$, the second part of $L(\mathcal{A})$ is k-PT. Assume that $w_j a^{k-1} \sim_k w$, for some $1 \leq j \leq m$ and $w \in \Sigma^*$. Then $w = v_1 a v_2 a \ldots a v_k$ for some $v_1, v_2, \ldots, v_k$ such that $|v_1 \ldots v_k|_a = 0$. Since $|w_j|_a = 0$ and, for any letter c of $v_2 \cdots v_{k-1}$ (resp. v_k), the word aca (resp. $a^{k-1}c$) is a subword of $w_j a^{k-1}$, that is, of a^{k-1}, we have that $v_2 \cdots v_k = \varepsilon$, i.e., $w = v_1 a^{k-1}$. Since $w_j a^{k-1} \sim_k v_1 a^{k-1}$, we have that

$w_j a = v_1 a$; hence, $w_j a^{k-1}$ and w lead to the same state, which concludes the proof. □

Proof (of Theorem 3). To check whether a minimal DFA is *not* confluent or does *not* satisfy condition 2 of Lemma 3 can be done in NL; the reader is referred to [3] for a proof how to check confluency in NL. Since NL=co-NL [7,14], we have an NL algorithm to check 2-PT of a minimal DFA. NL-hardness then follows from the previous lemma. □

It was shown in [1] that the syntactic monoids of 1-PT languages are defined by equations $x = x^2$ and $xy = yx$, and those of 2-PT languages by equations $xyzx = xyxzx$ and $(xy)^2 = (yx)^2$. These equations can be used to achieve NL algorithms. However, our characterizations improve these results and show that, for 1-PT languages, it is sufficient to verify the equations $x = x^2$ and $xy = yx$ on letters (generators), and that, for 2-PT languages, equation $xyzx = xyxzx$ can be verified on letters (generators) up to the element y, which is a general element of the monoid. It decreases the complexity of the problems. Moreover, the partial order and (local) confluency properties can be checked instead of the equation $(xy)^2 = (yx)^2$.

3-Piecewise Testability The equations $(xy)^3 = (yx)^3$, $xzyxvxwy = xzxyxvxwy$ and $ywxvxyzx = ywxvxyxzx$ characterize the variety of 3-PT languages [1]. Non-satisfiability of any of these equations can be checked in the DFA in NL by guessing a finite number of states and the right sequences of transitions between them (in parallel, when labeled with the same labels). Thus, we have the following.

Theorem 4. *The problem to decide whether a minimal DFA recognizes a 3-PT language is NL-complete.*

k-Piecewise Testability Even though [2] provides a finite sequence of equations to define the k-PT languages over a fixed alphabet for any $k \geq 4$, the equations are more involved and it is not clear whether they can be used to obtain the precise complexity. So far, the k-PT problem can be shown to be NL-hard (for $k \geq 2$) and in co-NP, and it is open whether it tends rather to NL or to co-NP.[2]

4   Complexity of k-Piecewise Testability for NFAs

The *k-piecewise testability problem for NFAs* asks whether, given an NFA $\mathcal{A}$, the language $L(\mathcal{A})$ is k-PT. A language is 0-PT if and only if it is either empty or universal. Since the universality problem for NFAs is PSPACE-complete [4], the 0-PT problem for NFAs is PSPACE-complete. Using the same argument as in [6] then gives that, for every integer $k \geq 0$, the problem to decide whether an NFA recognizes a k-PT language is PSPACE-hard.

---

[2] See the acknowledgement for the recent development.

Since k is fixed, we can make use of the idea of Theorem 1 to decide whether an NFA recognizes a k-PT language. The length of the word w_2 is now bounded by 2^n, where n is the number of states of the NFA. Guessing the word w_2 on-the-fly then gives that the k-piecewise testability problem for NFAs is in PSPACE.

Theorem 5. *The following problem is PSPACE-complete:*

NAME: k-PIECEWISETESTABILITYNFA
INPUT: *an NFA $\mathcal{A}$*
OUTPUT: YES *if and only if $\mathcal{L}(\mathcal{A})$ is k-PT*

The problem to find the minimal k for which the language recognized by an NFA is k-PT is PSPACE-hard, since a language is PT if and only if there exists a minimal $k \geq 0$ for which it is PT.

5 Piecewise Testability and the Depth of NFAs

We generalize a result valid for DFAs to NFAs and investigate the relationship between the depth of an NFA and the minimal k for which its language is k-PT.

Recall that a regular language is PT if and only if its minimal DFA satisfies some properties that can be tested in a quadratic time, cf. Fact 2. We now show that this characterization generalizes to NFAs. An NFA $\mathcal{A}$ over an alphabet Σ is *complete* if for every state q of $\mathcal{A}$ and every letter a in Σ, the set $q \cdot a$ is nonempty, i.e., in every state, a transition under every letter is defined.

Theorem 6. *A regular language is PT if and only if there exists a complete NFA that is partially ordered and satisfies the UMS property.*

As it is PSPACE-complete to decide whether an NFA defines a PT language, it is PSPACE-complete to decide whether, given an NFA, there is an equivalent complete NFA that is partially ordered and satisfies the UMS property.

It was shown in [10] that the depth of minimal DFAs does not correspond to the minimal k for which the language is k-PT. Namely, an example of $(4\ell-1)$-PT languages with the minimal DFA of depth $4\ell^2$, for $\ell > 1$, has been presented. We now show that there is an exponential gap between the minimal k for which the language is k-PT and the depth of a minimal DFA.

Theorem 7. *For every $n \geq 2$, there exists an n-PT language that is not $(n-1)$-PT, it is recognized by an NFA of depth $n-1$, and the minimal DFA recognizing it has depth $2^n - 1$.*

Proof (sketch). For $k \geq 0$, let $\mathcal{A}_k = (I_k, \{a_0, a_1, \ldots, a_k\}, \cdot, I_k, \{0\})$ be an NFA with $I_k = \{0, 1, \ldots, k\}$ and the transition function consisting of the self-loops under a_i in all states $j > i$ and transitions under a_i from the state i to all states $j < i$ as depicted in Fig. 1.

Every NFA $\mathcal{A}_k$ has depth k. Using Theorem 6 or noticing that the reversed automata are deterministic, we can show that it accepts a $(k+1)$-PT language. It can be shown that the language is not k-PT and that its minimal DFA has depth $2^{k+1} - 1$. $\qquad\square$

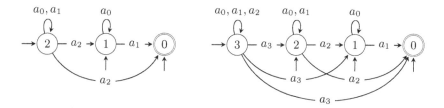

Fig. 1. Automata $\mathcal{A}_2$ and $\mathcal{A}_3$

Although it is well known that DFAs can be exponentially larger than NFAs, an interesting by-product of this result is that there are NFAs such that all the exponential number of states of their minimal DFAs form a simple path.

It could seem that NFAs are more convenient to provide upper bounds on k. However, the following simple example shows that even for 1-PT languages, the depth of an NFA depends on the size of the input alphabet. Specifically, for any alphabet Σ, the language $L = \bigcap_{a \in \Sigma} L_a$ of all words containing all letters of Σ is a 1-PT language such that any NFA recognizing it requires at least $2^{|\Sigma|}$ states and has depth $|\Sigma|$.

We now show that, given a k-PT language over an n-letter alphabet, the depth of the minimal DFA recognizing it is at most $\binom{k+n}{k} - 1$. To this end, we first investigate the following problem.

Problem 1. Let Σ be an alphabet of cardinality $n \geq 1$ and let $k \geq 1$. What is the length of a longest word, w, such that (i) $sub_k(w) = \Sigma^{\leq k} = \{v \in \Sigma^* \mid |v| \leq k\}$ and, (ii) for any two distinct prefixes w_1 and w_2 of w, $sub_k(w_1) \neq sub_k(w_2)$?

The answer is formulated in the following proposition.

Proposition 2. *Let Σ be an alphabet of cardinality n. The length of a longest word, w, satisfying the requirements of Problem 1 is given by the recursive formula $|w| = P_{k,n} = P_{k-1,n} + P_{k,n-1} + 1$, where $P_{1,m} = m = P_{m,1}$, for $m \geq 1$.*

The following two lemmas prove the proposition.

Lemma 5. *Let k and n be given, and let w' be any word over an n-letter alphabet satisfying the requirements of Problem 1. Then $|w'| \leq P_{k,n}$.*

Proof. Let $w' = w_1 z w_2$, where z is a letter, $|\operatorname{alph}(w_1)| = n-1$ and $|\operatorname{alph}(w_1 z)| = n$. Since w_1 satisfies the second requirement of Problem 1, $|w_1| \leq P_{k,n-1}$. On the other hand, since $|\operatorname{alph}(w_1 z)| = n$, any nonempty prefix of w_2 extends the set of subwords with a subword of length at least 2. Thus, w_2 cannot be longer than the longest word over Σ containing all subwords up to length $k - 1$, that is, $|w_2| \leq P_{k-1,n}$. $\qquad \square$

Lemma 6. *For any positive integers k and n, there exists a word w of length $P_{k,n}$ satisfying the requirements of Problem 1.*

Proof. Let $\Sigma_n = \{a_1, a_2, \ldots, a_n\}$ with the order $a_i < a_j$ if $i < j$. For $n = 1$ and $k \geq 1$, the word $W_{k,1} = a^k$ is of length $P_{k,1}$ and satisfies the requirements, as well as the word $W_{1,n} = a_1 a_2 \ldots a_n$ of length $P_{1,n}$ for $k = 1$ and $n \geq 1$. Assume that we have constructed the words $W_{i,j}$ of length $P_{i,j}$ for all $i < k$ and $j < n$, $W_{i,n}$ of length $P_{i,n}$ for all $i < k$, and $W_{k,j}$ of length $P_{k,j}$ for all $j < n$. We construct the word $W_{k,n}$ of length $P_{k,n}$ over Σ_n as follows: $W_{k,n} = W_{k,n-1} a_n W_{k-1,n}$.

It remains to show that $W_{k,n}$ satisfies the requirements of Problem 1. However, the set of subwords of $W_{k-1,n}$ is $\Sigma_n^{\leq k-1}$. Since $\mathrm{alph}(W_{k,n-1} a_n) = \Sigma$, we obtain that the set of subwords of $W_{k,n}$ is $\Sigma_n^{\leq k}$. Let w_1 and w_2 be two different prefixes of $W_{k,n}$. Without loss of generality, we may assume that w_1 is a prefix of w_2. If they are both prefixes of $W_{k,n-1}$, the second requirement of Problem 1 follows by induction. If w_1 is a prefix of $W_{k,n-1}$ and w_2 contains a_n, then the second requirement of Problem 1 is satisfied, because w_1 does not contain a_n. Thus, assume that both w_1 and w_2 contain a_n, that is, they both contain $W_{k,n-1} a_n$ as a prefix. Let $w_1 = W_{k,n-1} a_n w_1'$ and $w_2 = W_{k,n-1} a_n w_1' w_2'$. Since, by induction, $\mathrm{sub}_{k-1}(w_1') \subsetneq \mathrm{sub}_{k-1}(w_1' w_2')$, there exists $v \in \mathrm{sub}_{k-1}(w_1' w_2') \setminus \mathrm{sub}_{k-1}(w_1')$. Then $a_n v$ belongs to $\mathrm{sub}_k(w_2)$, but not to $\mathrm{sub}_k(w_1)$, which completes the proof.   □

It follows by induction that for any positive integers k and n

$$P_{k,n} = \binom{k+n}{k} - 1.$$

We now use this result to show that the depth of the minimal DFA recognizing a k-PT language over an n-letter alphabet is $P_{k,n}$ in the worst case.

Theorem 8. *For any natural numbers k and n, the depth of the minimal DFA recognizing a k-PT language over an n-letter alphabet is at most $P_{k,n}$. Moreover, the bound is tight for any k and n.*

Proof. Let $L_{k,n}$ be a k-PT language over an n-letter alphabet. Since $L_{k,n}$ is a finite union of $\sim_k$ classes [13], there exists F such that the $\sim_k$-canonical DFA[3] $\mathcal{A} = (Q, \Sigma, \cdot, [\varepsilon], F)$ recognizes $L_{k,n}$. The depth of $\mathcal{A}$ is $P_{k,n}$. Let $min(\mathcal{A})$ be the minimal DFA obtained from $\mathcal{A}$. Since the minimization does not increase the depth, the depth of $min(\mathcal{A})$ is at most $P_{k,n}$.

To show that the bound is tight, let w denote a fixed word of length $P_{k,n}$, which exists by Lemma 6. Consider the $\sim_k$-canonical DFA $\mathcal{A}' = (Q, \Sigma, \cdot, [\varepsilon], F)$, where $F = \{[w'] \mid w'$ is a prefix of w of even length$\}$. Then w defines a path $\pi_w = [\varepsilon] \rightarrow [w_1] \rightarrow [w_2] \ldots \rightarrow [w]$ in $\mathcal{A}'$ of length $P_{k,n}$, where w_i denotes the prefix of w of length i and accepting and non-accepting states alternate. Let $min(\mathcal{A}')$ be the minimal DFA obtained from $\mathcal{A}'$. If there were two equivalent states in π_w, then they must be of the same acceptance status. However, between any two states with the same acceptance status, there is a state with the opposite acceptance status. Thus, joining the two states creates a cycle in $min(\mathcal{A}')$, which is a contradiction with Fact 2, since the DFA $\mathcal{A}'$ recognizes a PT language.   □

---

[3] The $\sim_k$-*canonical DFA* is a DFA whose states are $\sim_k$ classes.

Finally, we present several consequences. (1) Note that it follows from the formula that $P_{k,n} = P_{n,k}$. This gives and interesting observation that increasing the length of the considered subwords has exactly the same effect as increasing the size of the alphabet. (2) Equivalently stated, Problem 1 asks what is the depth of the $\sim_k$-canonical DFA. The number of equivalence classes of $\sim_k$, i.e., the number of states, has recently been investigated in [8]. (3) It provides a precise bound on the length of w_1 of Theorem 1. However, it does not improve the statement of the theorem. (4) To provide a relationship of $P_{k,n}$ with Stirling cyclic numbers, it can be shown that, for any positive integers k and n, $P_{k,n} = \frac{1}{k!} \sum_{i=1}^{k} \begin{bmatrix} k+1 \\ i+1 \end{bmatrix} n^i$ where $\begin{bmatrix} k \\ n \end{bmatrix}$ denotes the Stirling cyclic numbers.

Acknowledgments. We would like to thank an anonymous reviewer for informing us about the unpublished manuscript [9] and its authors for providing it. It turns out that we have independently obtained two results—the bound of Theorem 8 and the co-NP bound on the k-PT problem for DFAs. Furthermore, it is shown in [9] that the k-PT problem is co-NP-complete for $k \geq 4$. It also provides a smaller bound on the length of the witnesses, which results in a single exponential algorithm to find the minimal k. On the other hand, for $k \leq 3$, that paper only says that the k-PT problem belongs to P. The authors are grateful to Sebastian Rudolph for a fruitful discussion.

References

1. Blanchet-Sadri, F.: Games, equations and the dot-depth hierarchy. Comput. Math. Appl. **18**(9), 809–822 (1989)
2. Blanchet-Sadri, F.: Equations and monoid varieties of dot-depth one and two. Theoret. Comput. Sci. **123**(2), 239–258 (1994)
3. Cho, S., Huynh, D.T.: Finite-automaton aperiodicity is PSPACE-complete. Theor. Comput. Sci. **88**(1), 99–116 (1991)
4. Garey, M.R., Johnson, D.S.: Computers and Intractability: A Guide to the Theory of NP-Completeness. W.H. Freeman (1979)
5. Hofman, P., Martens, W.: Separability by short subsequences and subwords. In: ICDT. LIPIcs, vol. 31, pp. 230–246 (2015)
6. Holub, Š., Masopust, T., Thomazo, M.: Alternating towers and piecewise testable separators. CoRR abs/1409.3943 (2014). 1409.3943
7. Immerman, N.: Nondeterministic space is closed under complementation. SIAM J. Comput. **17**(5), 935–938 (1988)
8. Karandikar, P., Kufleitner, M., Schnoebelen, P.: On the index of Simon's congruence for piecewise testability. Inform. Process. Lett. **115**(4), 515–519 (2015)
9. Klíma, O., Kunc, M., Polák, L.: Deciding k-piecewise testability (manuscript)
10. Klíma, O., Polák, L.: Alternative automata characterization of piecewise testable languages. In: Béal, M.-P., Carton, O. (eds.) DLT 2013. LNCS, vol. 7907, pp. 289–300. Springer, Heidelberg (2013)
11. Masopust, T., Thomazo, M.: On k-piecewise testability (preliminary report). CoRR abs/1412.1641 (2014), 1412.1641

12. Sakarovitch, J., Simon, I.: Subwords. In: Lothaire, M. (ed.) Combinatorics on Words, pp. 105–142. Cambridge University Press (1997)
13. Simon, I.: Hierarchies of Events with Dot-Depth One. Ph.D. thesis, Department of Applied Analysis and Computer Science. University of Waterloo, Canada (1972)
14. Szelepcsényi, R.: The method of forced enumeration for nondeterministic automata. Acta Inf. **26**(3), 279–284 (1988)
15. Trahtman, A.N.: Piecewise and local threshold testability of DFA. In: Freivalds, R. (ed.) FCT 2001. LNCS, vol. 2138, pp. 347–358. Springer, Heidelberg (2001)

Interval Exchange Words and the Question of Hof, Knill, and Simon

Zuzana Masáková[1], Edita Pelantová[1], and Štěpán Starosta[2]([⊠])

[1] Faculty of Nuclear Sciences and Physical Engineering,
Czech Technical University in Prague, Trojanova 13, 120 00 Praha 2, Czech Republic
[2] Faculty of Information Technology, Czech Technical University in Prague,
Thákurova 9, 160 00 Praha 6, Czech Republic
stepan.starosta@fit.cvut.cz

Abstract. We consider words coding non-degenerate 3 interval exchange transformation. It is known that such words contain infinitely many palindromic factors. We show that for any morphism ξ fixing such a word, either ξ or ξ^2 is conjugate to a class P morphism. By this, we provide a new family of palindromic infinite words satisfying the conjecture of Hof, Knill and Simon, as formulated by Tan.

Keywords: Palindrome · Morphism · Interval exchange

1 Introduction

Palindromes, i.e., words which read the same from the left as from the right, are source of amusement in natural languages. The reader can surely identify languages which contain the following palindromes: kuulilennuteetunneliluuk, kakak, onorarono, elevele, icipici or ailihphilia. In the theory of formal languages and stringology, palindromes and their various generalizations represent interesting objects of research. To illustrate diverse perspectives on palindromes, let us point out some recently published papers. In [2], the authors study equations on words where constants or variables are palindromes. Algorithms finding generalized repetitions, in particular finding palindromes and generalized palindromes, are studied in [5]. The definition of generalized palindromes over the alphabet $\{A, C, G, T\}$ in [7] was inspired by applications in genetics and DNA computing.

Our work is devoted to a question coming from another field, namely mathematical physics, where palindromes can also be found. The question is from paper [6], where Hof, Knill and Simon study spectra of Schrödinger operators associated to infinite sequences. These sequences are generated by a substitution over a finite alphabet. The authors show in their paper that if a sequence contains infinitely many palindromic factors (such sequences are called *palindromic*), then the associated operator has a purely singular continuous spectrum. In the same paper, the authors define a class of substitutions, called class P (see Section 5 for the exact definition) and they ask the following question: "Are there (minimal) sequences containing arbitrarily long palindromes that arise from substitutions none of which belongs to class P?" A discussion on how to transform this question into a mathematical formalism can be found in [9].

© Springer International Publishing Switzerland 2015
I. Potapov (Ed.): DLT 2015, LNCS 9168, pp. 377–388, 2015.
DOI: 10.1007/978-3-319-21500-6_30

The first result concerning class P was given by Tan in [13]. The author extended class P by morphisms conjugated to the elements of class P, since it is well-known that fixed points of conjugated morphisms have the same set of factors. This extended class is denoted by P'. In [13], it is shown that if a fixed point of a primitive substitution φ over a binary alphabet is palindromic, then the substitution φ or φ^2 belongs to class P'. The conjecture, stemming from the question of Hof, Knill and Simon, states that every (minimal) palindromic sequence is a fixed point of a morphism of class P'. It is referred to as Class P conjecture.

In [10], Labbé shows that the assumption of a binary alphabet in the theorem of Tan is essential. The author exhibited a palindromic fixed point of a morphism φ over a ternary alphabet which is not a fixed point of any morphism belonging to P'. It can be shown that the counterexample is a sequence coding a 3 interval exchange transformation with permutation (321). However, it is degenerated, i.e., it is a morphic image of a Sturmian word. For details, see Section 6.

In this paper, we study sequences that code non-degenerated 3 interval exchange transformation with the same permutation (321). It is known that such sequences are palindromic. We show that a ternary analogue of the theorem of Tan holds in this context. Let us note that another analogue is already known: in [9] it is shown for marked morphisms. These two analogues do not overlap: a morphism fixing a coding of a non-degenerate 3 interval exchange transformation is never marked.

The article is organized as follows. Section 2 contains the necessary notions. Interval exchange transformations and their properties are treated in Section 3. In Section 4 we focus on substitution invariance of words coding interval exchange transformations. The proof of the main result stated as Theorem 2 is provided in Section 5. Let us mention that the key Lemma 1 is given without proof in this contribution. The demonstration of this lemma requires some other notions. It is given in Section 5 in the full version of this paper on arXiv [12].

2   Preliminaries

Let us recall necessary notions and notation from combinatorics on words. For a basic overview we refer to [11]. An *alphabet* is a finite set of symbols, called *letters*. A *finite word* w over an alphabet $\mathcal{A}$ of length $|w| = n$ is a concatenation $w = w_0 \cdots w_{n-1}$ of letters $w_i \in \mathcal{A}$. The set of all finite words over $\mathcal{A}$ equipped with the operation of concatenation and the empty word ϵ is a monoid denoted by $\mathcal{A}^*$. For a fixed letter $a \in \mathcal{A}$, we denote by $|w|_a$ the number of occurrences of a in w, i.e., the number of indices i such that $w_i = a$. The *reversal* or *mirror image* of the word w is the word $\overline{w} = w_{n-1} \cdots w_0$. A word w for which $w = \overline{w}$ is called a *palindrome*. An *infinite word* $\mathbf{u}$ is an infinite concatenation $\mathbf{u} = u_0 u_1 u_2 \ldots \in \mathcal{A}^{\mathbb{N}}$. An infinite word $\mathbf{u} = wvvv \ldots$ with $w, v \in \mathcal{A}^*$ is said to be *eventually periodic*; it is said to be *aperiodic* if it is not of such form. We say that $w \in \mathcal{A}^*$ is a *factor* of $v \in \mathcal{A}^* \cup \mathcal{A}^{\mathbb{N}}$ if $v = w'ww''$ for some $w' \in \mathcal{A}^*$ and $w'' \in \mathcal{A}^* \cup \mathcal{A}^{\mathbb{N}}$. If $w' = \epsilon$ or $w'' = \epsilon$, then w is a *prefix* or *suffix* of v, respectively. If $v = wu$, then we write $u = w^{-1}v$ and $w = vu^{-1}$.

The set $\mathcal{L}(\mathbf{u})$ of all finite factors of an infinite word $\mathbf{u}$ is called the *language of* $\mathbf{u}$. The *factor complexity* $\mathcal{C}_\mathbf{u}$ is the function $\mathbb{N} \to \mathbb{N}$ counting the number of factors of $\mathbf{u}$ of length n. It is known that the factor complexity of an aperiodic infinite word $\mathbf{u}$ satisfies $\mathcal{C}_\mathbf{u}(n) \geq n + 1$ for all n. Aperiodic infinite words having the minimal complexity $\mathcal{C}_\mathbf{u}(n) = n + 1$ for all n are called *Sturmian words*. Since $\mathcal{C}_\mathbf{u}(1) = 2$, they are binary words. Sturmian words can be equivalently defined in many different frameworks, one of them is coding of an exchange of two intervals.

Let $\mathcal{A}$ and $\mathcal{B}$ be alphabets. Let $\varphi : \mathcal{A}^* \to \mathcal{B}^*$ be a morphism, i.e., $\varphi(wv) = \varphi(w)\varphi(v)$ for all $w, v \in \mathcal{A}^*$. We say that φ is *non-erasing* if $\varphi(b) \neq \epsilon$ for every $b \in \mathcal{A}$. The action of φ can be naturally extended to infinite words $\mathbf{u} \in \mathcal{A}^\mathbb{N}$ by setting $\varphi(\mathbf{u}) = \varphi(u_0)\varphi(u_1)\varphi(u_2)\ldots$. If $\mathcal{A} = \mathcal{B}$ and $\varphi(\mathbf{u}) = \mathbf{u}$, then $\mathbf{u}$ is said to be a *fixed point* of φ. A non-erasing morphism $\varphi : \mathcal{A}^* \to \mathcal{A}^*$ such that there is a letter $a \in \mathcal{A}$ satisfying $\varphi(a) = aw$ for some non-empty word w is called a *substitution*. Obviously, a substitution has always a fixed point, namely $\lim_{n\to\infty} \varphi^n(a)$ where the limit is taken over the product topology. Let $\mathcal{A} = \{a_1, \ldots, a_k\}$ and $\mathcal{B} = \{b_1, \ldots, b_\ell\}$. One associates to every morphism $\varphi : \mathcal{A} \to \mathcal{B}$ its *incidence matrix* $M_\varphi \in \mathbb{N}^{k \times \ell}$ defined by

$$(M_\varphi)_{ij} = |\varphi(a_i)|_{b_j}, \quad \text{for } 1 \leq i \leq k,\ 1 \leq j \leq \ell.$$

A morphism $\varphi : \mathcal{A}^* \to \mathcal{A}^*$ is said to be *primitive* if all elements of some power of its incidence matrix $M_\varphi \in \mathbb{N}^{k \times k}$ are positive.

3   Itineraries in Symmetric Exchange of Intervals

For disjoint intervals K and K' we write $K < K'$ if for $x \in K$ and $x' \in K'$ we have $x < x'$. Let J be a semi-closed interval. Consider a partition $J = J_0 \cup \cdots \cup J_{k-1}$ of J into a disjoint union of semi-closed subintervals $J_0 < J_1 < \cdots < J_{k-1}$. A bijection $T : J \to J$ is called an *exchange of k intervals with permutation π* if there exist numbers $c_0, \ldots, c_{k-1}$ such that for $0 \leq i < k$ one has

$$T(x) = x + c_i \quad \text{for } x \in J_i,\tag{1}$$

where π is a permutation of $\{0, 1, \ldots, k-1\}$ such that $T(J_{\pi^{-1}(i)}) < T(J_{\pi^{-1}(j)})$ for $i < j$. In other words, the permutation π determines the order of intervals $T(J_i)$. If π is the permutation $i \mapsto k - i + 1$, then T is called a *symmetric* interval exchange.

An exchange of intervals satisfies the *minimality condition* if the orbit of any given $\rho \in [0, 1)$, i.e, the sequence $\rho, T(\rho), T^2(\rho), T^3(\rho), \ldots$, is dense in J. An orbit can be coded by an aperiodic infinite word $\mathbf{u}_\rho = u_0 u_1 u_2 \ldots$ over the alphabet $\{0, 1, \ldots, k-1\}$ given by

$$u_n = X \quad \text{if } T^n(\rho) \in J_X \quad \text{for } X \in \{0, 1, \ldots, k-1\}.$$

The point ρ is called the *intercept* of $\mathbf{u}_\rho$. The complexity of such infinite words is known to satisfy $\mathcal{C}_{\mathbf{u}_\rho}(n) \leq (k-1)n + 1$ (see [4]). If for every $n \in \mathbb{N}$ we have

$\mathcal{C}_{\mathbf{u}_\rho}(n) = (k-1)n+1$, then the word $\mathbf{u}_\rho$ is said to be *non-degenerate*. If T satisfies the minimality condition and the coding of one of its orbits is non-degenerate, then coding of any orbit is non-degenerate and we call such T a *non-degenerate interval exchange*.

Definition 1. *Let T be an exchange of k intervals satisfying the minimality condition. Given a subinterval $I \subset J$, we define the mapping $r_I : I \to \mathbb{Z}^+ = \{1,2,3,\dots\}$ by*

$$r_I(x) = \min\{n \in \mathbb{Z}^+ : T^n(x) \in I\},$$

the so-called return time to I. The prefix of length $r_I(x)$ of the word $\mathbf{u}$ coding the orbit of the point $x \in I$ is called the I-itinerary of x and denoted $R_I(x)$. The set of all I-itineraries is denoted by $\mathrm{It}_I = \{R_I(x) : x \in I\}$. The mapping $T_I : I \to I$ defined by

$$T_I(x) = T^{r_I(x)}(x)$$

is said to be the first return map of T to I, or induced map of T on I.

Throughout the paper, when it is clear from the context, we sometimes omit the index I in r_I or R_I. It is known from Keane [8] that if T is an exchange of k intervals and $I \subset J$, then It_I has at most $k+2$ elements, and, consequently, T_I is an exchange of at most $k+2$ intervals.

Remark 1. If $I \subset J_X$, then $T(I)$ is an interval. We have

$$R \text{ is an } I\text{-itinerary} \quad \Leftrightarrow \quad X^{-1}RX \text{ is a } T(I)\text{-itinerary.}$$

Similarly, if $I \subset T(J_X)$, then $T^{-1}(I)$ is an interval and we have

$$R \text{ is an } I\text{-itinerary} \quad \Leftrightarrow \quad XRX^{-1} \text{ is a } T^{-1}(I)\text{-itinerary.}$$

We will use another fact about itineraries of an interval exchange. Without loss of generality, we consider $J = [0,1)$. The intervals J_X are left-closed right-open for all $X \in \{0,1\dots,k-1\}$. Such interval exchange T is right-continuous. Therefore, if $I = [\gamma, \delta)$, then every word $w \in \mathrm{It}_I = \{R(x) : x \in I\}$ is an I-itinerary $R(x)$ for infinitely many $x \in I$, which form an interval, again left-closed right-open.

For the rest of the section, we consider only symmetric interval exchange. In order to state a property of such interval exchanges, for an interval $K = [c,d) \subset [0,1)$ we denote $\overline{K} = [1-d, 1-c)$.

Proposition 1. *Let $T : [0,1) \to [0,1)$ be a symmetric exchange of k intervals satisfying the minimality condition. Let $I \subset [0,1)$ and let $R_1,\dots,R_m$ be the I-itineraries. The $\overline{I}$-itineraries are the mirror images of the I-itneraries, namely $\overline{R_1},\dots,\overline{R_m}$. Moreover, if*

$$[\gamma_j, \delta_j) := \{x \in I : R_I(x) = R_j\} \quad and \quad [\gamma'_j, \delta'_j) := T_I[\gamma_j, \delta_j),$$

for $j = 1,\dots,m$, then

$$\{x \in \overline{I} : R_{\overline{I}}(x) = \overline{R_j}\} = [1-\delta'_j, 1-\gamma'_j).$$

Proof. Consider the restriction of the transformation T to the set

$$S = [0,1) \setminus \{T^j(\alpha) \colon j \in \mathbb{Z}, \ \alpha \text{ is a discontinuity of } T\}.$$

Such a restriction is a bijection $S \to S$. We will show by induction that for any $i \geq 1$ and $y \in S$

$$T^{-i}(y) = 1 - T^i(1-y). \tag{2}$$

Let $y \in S$ and $j \in \{0, \ldots, k-1\}$ such that $y \in I_j$. Since T is symmetric, we have

$$1 - y \in I_j \iff y \in T(I_j).$$

The last equivalence and the definition of T imply

$$T(1-y) = 1 - y + c_j \qquad \text{and}$$
$$T^{-1}(y) = y - c_j.$$

Summing the last two equalities we obtain

$$T^{-1}(y) = 1 - T(1-y).$$

Then, using the induction hypothesis, we have for $y \in S$ that

$$T^{-(i+1)}(y) = T^{-1}\big(T^{-i}(y)\big) = 1 - T\big(1 - T^{-i}(y)\big) = 1 - T\big(T^i(1-y)\big) = 1 - T^{i+1}(1-y),$$

which proves (2).

Since $T(J_X) = \overline{J_X}$ for any letter $X \in \{0, 1, \ldots, k-1\}$, we can write for $y \in S$

$$T^{-1}(y) \in J_X \iff y \in T(J_X) \iff 1 - y \in J_X. \tag{3}$$

More generally,

$$T^{-i}(y) = T^{-1}\big(T^{-(i-1)}(y)\big) \in J_X \iff 1 - T^{-(i-1)}(y) = T^{i-1}(1-y) \in J_X, \tag{4}$$

where we have first used (3) and then (2).

Now we show that if R_j is an I-itinerary, then its mirror image $\overline{R_j}$ is an $\overline{I}$-itinerary. Consider $\rho \in (\gamma_j, \delta_j) \cap S$ and let $R_I(\rho) = a_0 a_1 \cdots a_{n-1}$ be its I-itinerary, i.e., $a_i = X$ if and only if $T^i(\rho) \in J_X$. Moreover, $T^i(\rho) \notin I$ for $1 \leq i < n$, and $T^n(\rho) \in I$. Let

$$\rho' := 1 - T^n(\rho) = 1 - T_I(\rho) \in (1 - \delta'_j, 1 - \gamma'_j) \cap S \subset \overline{I}. \tag{5}$$

By (2), we have $\rho' = T^{-n}(1-\rho)$, and therefore again by (2), $T^i(\rho') = T^{-(n-i)}(1-\rho) = 1 - T^{n-i}(\rho) \notin \overline{I}$ for $0 < i < n$. On the other hand, $T^n(\rho') = 1 - \rho \in \overline{I}$. By (4), we have for $i = 0, 1, \ldots, n-1$ that

$$J_X \ni T^i(\rho') = T^{-(n-i)}(1-\rho) \iff T^{n-i-1}(\rho) \in J_X,$$

which implies that the $\overline{I}$-itinerary of ρ' is $R_{\overline{I}}(\rho') = a_{n-1} a_{n-2} \cdots a_0$, as we wanted to show.

By right continuity of T, all points from $[1 - \delta'_j, 1 - \gamma'_j)$ have the same $\overline{I}$-itinerary as $\rho' \in (1 - \delta'_j, 1 - \gamma'_j) \cap S$. $\qquad \square$

We will be particularly interested in exchanges of three intervals. For convenience, we prefer to use for its coding the ternary alphabet $\{A, B, C\}$ instead of $\{0, 1, 2\}$. Without loss of generality let $0 < \alpha < \beta < 1$. Let $T : [0, 1) \to [0, 1)$ be given by

$$
T(x) = \begin{cases} x + 1 - \alpha & \text{if } x \in [0, \alpha) =: J_A\,, \\ x + 1 - \alpha - \beta & \text{if } x \in [\alpha, \beta) =: J_B\,, \\ x - \beta & \text{if } x \in [\beta, 1) =: J_C\,. \end{cases} \tag{6}
$$

The transformation T is an exchange of three intervals with the permutation (321). It is often called a *3iet* for short. The infinite word $\mathbf{u}_\rho$ coding the orbit of a point $\rho \in [0, 1)$ under a 3iet is called a *3iet word*.

We require that $1 - \alpha$ and β be linearly independent over $\mathbb{Q}$, which is known to be a necessary and sufficient condition for minimality of the 3iet T. Nondegeneracy of T is equivalent to the condition

$$
1 \notin (1 - \alpha)\mathbb{Z} + \beta\mathbb{Z}\,, \tag{7}
$$

see [4]. This means that the 3iet word $\mathbf{u}$ has complexity $\mathcal{C}_\mathbf{u}(n) = 2n + 1$ if and only if the parameters α and β of the corresponding 3iet T satisfy (7).

From the general result of Keane, one can derive that for a given subinterval $I \subset [0, 1)$ there exist at most five I-itineraries under a 3iet T.

Convention: For the rest of the paper, let T be a non-degenerate exchange of three intervals with the permutation (321) given by (6).

4   Substitution Invariance and Conjugation of Substitutions

Let us recall the relation of induction to a subinterval I to substitution invariance of 3iet words. Let I be an interval $I \subset [0, 1)$ such that the set It_I of I-itineraries has three elements, say R_A, R_B and R_C. For every $\rho \in I$, the infinite word $\mathbf{u}_\rho$ can be written as a concatenation of words R_A, R_B and R_C. For a letter $Y \in \{A, B, C\}$ denote $I_Y = \{x \in I : R(x) = R_Y\}$. Obviously, $I = I_A \cup I_B \cup I_C$, and the induced mapping T_I is an exchange of these three intervals. The order of the words R_A, R_B and R_C in the concatenation is determined by the iterations of $T_I(\rho)$.

Suppose that T_I is homothetic to T. Recall that mappings $f : I_f \to I_f$ and $g : I_g \to I_g$ are homothetic if there exists an affine bijection $\Phi : I_f \to I_g$ with $\Phi(x) = \lambda x + \mu$ such that

$$
\Phi f(x) = g \Phi(x) \quad \text{for all } x \in I_f\,. \tag{8}
$$

This means that f and g behave in the same way, up to a scaling factor λ and a shift by μ of the domains I_f and I_g. In other words, the graphs of the mappings f and g are the same, up to their scale and placing. The homothety of T and

T_I implies that $\Phi(J_Y) = I_Y$ for all $Y \in \{A, B, C\}$. From (8), we derive for every $k \in \mathbb{N}$ that $\Phi T^k(x) = T_I^k \Phi(x)$ for $x \in [0, 1)$, and thus $\Phi T^k(\rho) = T_I^k(\rho)$ whenever

$$\Phi(\rho) = \rho, \tag{9}$$

i.e., ρ is the homothety center. From the relation $\Phi(J_Y) = I_Y$ it follows that the k-th element in the concatenation of itineraries R_A, R_B and R_C is equal to R_Y if and only if the k-th letter in the infinite word $\mathbf{u}_\rho$ is equal to Y. This is equivalent to saying that the infinite word $\mathbf{u}_\rho$ is invariant under the substitution η given by

$$\eta(A) = R_A, \ \eta(B) = R_B, \ \eta(C) = R_C. \tag{10}$$

We conclude that the existence of an interval I with three itineraries and T_I homothetic to T leads to a substitution fixing a 3iet word whose intercept is the homothety center ρ. In fact, the converse holds, too, as shown in [1]. We summarize both statements as follows.

Theorem 1 ([1]). *Let ξ be a primitive substitution over $\{A, B, C\}$ with incidence matrix M and let T be a non-degenerate 3iet. The substitution ξ fixes the word $\mathbf{u}_\rho$ coding the orbit of a point $\rho \in [0, 1)$ under T if and only if there exists an interval $I \subset [0, 1)$ with I-itineraries $It_I = \{R_A, R_B, R_C\}$ such that T_I is homothetic to T, ρ is the homothety center, and the substitution η given by*

$$\eta = \begin{cases} \xi & \text{if no eigenvalue of } M \text{ belongs to } (-1, 0), \\ \xi^2 & \text{otherwise,} \end{cases}$$

satisfies $\eta(A) = R_A$, $\eta(B) = R_B$ and $\eta(C) = R_C$.

Let us mention that the scaling factor $\lambda \in (0, 1)$ in the homothety mapping $\Phi(x) = \lambda x + \mu$ is equal to the length of the interval $I = [\gamma, \delta)$, i.e., $\lambda = \delta - \gamma$, and the shift μ is equal to the left end-point of the interval I, namely γ. Moreover, it is related to the intercept ρ of an infinite word $\mathbf{u}_\rho$ in the following way: one has $\mu = \gamma = \rho(1 - \lambda)$, as follows from (9). In fact, λ is an eigenvalue of the incidence matrix of η. It follows from [1] that if ξ has such an eigenvalue, then the choice $\eta = \xi$ is sufficient. Otherwise, the incidence matrix of ξ^2 has such an eigenvalue.

By Theorem 1, if $\mathbf{u}_\rho$ is invariant under a substitution, we find an interval I such that T_I is homothetic to T. If $I' = T(I)$ is again an interval, then $T_{I'}$ is also homothetic to T, and the I'-itineraries change with respect to the I-itineraries, as described in Remark 1. To show the relation of the corresponding substitutions, we need the following definition.

Definition 2. *Let φ and ψ be morphisms over $\mathcal{A}^*$ and let $w \in \mathcal{A}^*$ be a word such that $w\varphi(a) = \psi(a)w$ for every letter $a \in \mathcal{A}$. The morphism φ is said to be a left conjugate of ψ and ψ is said to be a right conjugate of φ. If φ is a left or right conjugate of ψ, then we say φ is conjugate to ψ. If the only left conjugate of φ is φ itself, then φ is called the leftmost conjugate of ψ and we write $\varphi = \psi_L$. If the only right conjugate of ψ is ψ itself, then ψ is called the rightmost conjugate of φ and we write $\psi = \varphi_R$.*

Note that given a substitution ξ, its leftmost and rightmost conjugates ξ_L and ξ_R may not exist. If this happens, it can be shown that its fixed point is a periodic word. All the substitutions considered here thus possess their leftmost and rightmost conjugates.

Proposition 2. *Let $\mathbf{u}_\rho$ be a 3iet word coding the orbit of the point $\rho \in [0,1)$ under a non-degenerate 3iet T. Moreover, assume that $\mathbf{u}_\rho$ is a fixed point of a primitive substitution η such that the corresponding interval I of Theorem 1 is of length λ. Let η' be a left conjugate of η, i.e., $\eta(a)w = w\eta'(a)$ for some word $w \in \mathcal{A}^*$. The morphism η' fixes the infinite word $\mathbf{u}_{\rho'}$ with ρ' satisfying*

$$(1 - \lambda)\rho' = T^n\big((1 - \lambda)\rho\big), \qquad \text{where } n = |w|. \tag{11}$$

Moreover, the interval I' corresponding to η' by Theorem 1 satisfies $I' = T^n(I)$.

Proof. Suppose that w is a letter, i.e., $w = X \in \mathcal{A}$. Necessarily, the words $\eta(a)$ start with the letter X for all $a \in \mathcal{A}$. This means for the interval I that $I \subset J_X$. According to Remark 1, the interval $I' = T(I)$ has three I'-itineraries. Moreover, the induced mapping $T_{I'}$ is also homothetic to T. Denote $I = [\gamma, \delta)$. The homothety between the transformations T and T_I is achieved by the map $\Phi(x) = \lambda x + \gamma$. The homothety between T and $T_{I'}$ is the map $\Phi'(x) = \lambda x + T(\gamma)$. Since the intercepts ρ and ρ' are by (9) fixed by the homotheties Φ and Φ', respectively, we have

$$\Phi(\rho) = \lambda\rho + \gamma = \rho \quad \text{and} \quad \Phi'(\rho') = \lambda\rho' + T(\gamma) = \rho' \,.$$

Eliminating γ, we obtain

$$(1 - \lambda)\rho' = T(\gamma) = T\big((1 - \lambda)\rho\big) \,.$$

Since conjugation by any word w can be performed letter by letter, the proof is finished. □

We will also need to see the relation of the substitution η corresponding to the interval $I = [\gamma, \delta)$ with the substitution corresponding to the interval $\overline{I} = [1 - \delta, 1 - \gamma)$. It turns out that it is the mirror substitution of η, defined in general as follows. For a morphism $\xi : \mathcal{A} \to \mathcal{A}$, we define the morphism $\overline{\xi} : \mathcal{A} \to \mathcal{A}$ by $\overline{\xi}(a) = \overline{\xi(a)}$ for $a \in \mathcal{A}$.

Proposition 3. *Let $I \subset [0,1)$ be a left-closed right-open interval such that $\#It_I = 3$ and T_I is an exchange of three intervals with the permutation (321). The interval $\overline{I}$ satisfies $\#It_{\overline{I}} = 3$ and the induced map $T_{\overline{I}}$ is homothetic to T_I. If, moreover, T_I is homothetic to T and the substitution η corresponding to I fixes the infinite word $\mathbf{u}_\rho$, then the substitution corresponding to $\overline{I}$ is $\overline{\eta}$ and fixes the infinite word $\mathbf{u}_{\overline{\rho}}$, where $\overline{\rho} = 1 - \rho$.*

Proof. Denote $It_I = \{R_1, R_2, R_3\}$ and $I_j = \{x \in I : R_I(x) = R_j\} = [\gamma_j, \delta_j)$ for $j = 1, 2, 3$ so that $I_1 < I_2 < I_3$. By Proposition 1, the $\overline{I}$-itineraries are $\overline{R_1}, \overline{R_2}$ and $\overline{R_3}$, where

$$I'_j = \{x \in \overline{I} : R_{\overline{I}}(x) = \overline{R_j}\} = [1 - \delta'_j, 1 - \gamma'_j) \,,$$

where $[\gamma'_j, \delta'_j) = T_I[\gamma_j, \delta_j)$. Since T_I is an exchange of three intervals with the permutation (321) we have

$$T_I[\gamma_1, \delta_1) > T_I[\gamma_2, \delta_2) > T_I[\gamma_3, \delta_3)\,,$$

and therefore $I'_1 < I'_2 < I'_3$. The induced map $T_{\overline{I}}$ is therefore an exchange of three intervals I'_1, I'_2 and I'_3 with permutation (321) and since $|I'_j| = |I_j|$ for $j = 1, 2, 3$, the transformation $T_{\overline{I}}$ is homothetic to T_I.

Suppose that T_I is homothetic to the original 3iet T. By Theorem 1, there is a substitution η corresponding to the interval I and satisfying $\eta(A) = R_1$, $\eta(B) = R_2$ and $\eta(C) = R_3$. The mapping $T_{\overline{I}}$ is homothetic to T_I and thus also to T, the corresponding substitution η' satisfies $\eta'(A) = \overline{R_1}$, $\eta'(B) = \overline{R_2}$ and $\eta'(C) = \overline{R_3}$. We can see that $\eta' = \overline{\eta}$.

Let ρ be the intercept of the infinite word which is fixed by the substitution η. It is the center of homothety between T_I and T, i.e., it is the fixed point of the mapping $\Phi(x) = (\delta - \gamma)x + \gamma$. We have $\rho = (\delta - \gamma)\rho + \gamma$, which implies

$$\rho = \frac{\gamma}{1 - \delta + \gamma}\,.$$

Similarly, the intercept $\overline{\rho}$ of the substitution $\overline{\eta}$ satisfies $\overline{\rho} = (\delta - \gamma)\overline{\rho} + 1 - \delta$, whence

$$\overline{\rho} = \frac{1 - \delta}{1 - \delta + \gamma} = 1 - \rho\,. \qquad \square$$

For a finite word w, we denote by $\mathrm{Fst}(w)$ and $\mathrm{Lst}(w)$ the first and last letters of w, respectively.

Remark 2. Let η be a primitive substitution given by Theorem 1 fixing a 3iet word. Necessarily, the first and the last letters of $\eta(A)$, $\eta(B)$ and $\eta(C)$ satisfy

$$\mathrm{Fst}\big(\eta(A)\big) \leq \mathrm{Fst}\big(\eta(B)\big) \leq \mathrm{Fst}\big(\eta(C)\big) \text{ and } \mathrm{Lst}\big(\eta(A)\big) \leq \mathrm{Lst}\big(\eta(B)\big) \leq \mathrm{Lst}\big(\eta(C)\big),$$

where we consider the order $A < B < C$. The inequalities for the first letters follow from the definition of an exchange of intervals, namely from the fact that the words $\eta(A)$, $\eta(B)$ and $\eta(C)$ are given as I-itineraries. By Proposition 3, the last letters of the words $\eta(A)$, $\eta(B)$ and $\eta(C)$ are the first letters of the words $\overline{\eta(A)}$, $\overline{\eta(B)}$ and $\overline{\eta(C)}$ which proves the second set of inequalities.

For the proof of the main Theorem 2, we need the following lemma. Its proof is a technical one and necessitates some facts about the relation of morphisms fixing 3iet words and Sturmian morphisms. All the details can be found in the full version of this contribution in [12].

Lemma 1. *Let η be a primitive substitution given by Theorem 1 fixing a 3iet word. We have*

$$\Big(\mathrm{Fst}\big(\eta_L(A)\big), \mathrm{Fst}\big(\eta_L(B)\big), \mathrm{Fst}\big(\eta_L(C)\big)\Big) =$$

$$\Big(\mathrm{Lst}\big(\eta_R(A)\big), \mathrm{Lst}\big(\eta_R(B)\big), \mathrm{Lst}\big(\eta_R(C)\big)\Big) \in \{(A, B, B), (B, B, C)\}\,.$$

Corollary 1. *Let η be a primitive substitution given by Theorem 1 fixing a 3iet word $\mathbf{u}_\rho$. If $\big(\mathrm{Fst}(\eta(A)), \mathrm{Fst}(\eta(B)), \mathrm{Fst}(\eta(C))\big) = (A, B, B)$, then $\rho = \alpha$, and if $\big(\mathrm{Fst}(\eta(A)), \mathrm{Fst}(\eta(B)), \mathrm{Fst}(\eta(C))\big) = (B, B, C)$, then $\rho = \beta$.*

Proof. Let I be the interval corresponding to η such that T_I is homothetic to T. Denote $I_X = \{x \in I \colon R_I(x) = X\}$. If $\big(\mathrm{Fst}(\eta(A)), \mathrm{Fst}(\eta(B)), \mathrm{Fst}(\eta(C))\big) = (A, B, B)$, then the boundary between intervals I_A and I_B, i.e., the discontinuity point of T_I, is equal to the point α. Since T_I is homothetic to T, the homothety map Φ maps the discontinuity points of T to the discontinuity points of T_I, i.e., $\Phi(\alpha) = \alpha$. Since the fixed point of the homothety is equal to the intercept of the infinite word coded by η, we have $\rho = \alpha$. The other case is analogous. $\qquad\square$

5   Class P Conjecture for Non-degenerate 3iet

The main result of this section is Theorem 2, which states that a substitution fixing a non-degenerate 3iet word is of class P'.

Definition 3. *Let φ be a substitution over an alphabet $\mathcal{A}$. We say that φ belongs to the class P if there exists a palindrome p such that for every $a \in \mathcal{A}$ one has $\varphi(a) = pp_a$ where p_a is a palindrome. We say that φ is of class P' if it is conjugate to some morphism in class P.*

The following lemma is a generalization of a result obtained for binary alphabets by Tan [13], also shown in [9]. We provide a different proof.

Proposition 4. *Let $\varphi \colon \mathcal{A} \to \mathcal{A}$ be a non-erasing morphism. The morphism φ is conjugate to $\overline{\varphi}$ if and only if φ is of class P'.*

Proof. ($\Leftarrow$): Since φ is of class P', there exists a morphism φ' of class P which is conjugate to φ, i.e., there exists a word w such that $w\varphi(a) = \varphi'(a)w$ or $\varphi(a)w = w\varphi'(a)$ for every letter a.

We can suppose that $w\varphi(a) = \varphi'(a)w$ for every letter a as the other case is analogous. It implies $\varphi(a) = w^{-1}pp_a w$ for some palindromes p_a and p. Thus, $\overline{\varphi(a)} = \overline{w}p_a p(\overline{w})^{-1}$ for every letter a. In other words, the morphism $\overline{\varphi}$ is conjugate to $\overline{\varphi'}$. Since $\overline{\varphi'}$ is clearly conjugate to φ, we conclude that φ is conjugate to $\overline{\varphi}$.

($\Rightarrow$): Since φ is conjugate to $\overline{\varphi}$, there exists a word $w \in \mathcal{B}^*$ such that for every $a \in \mathcal{A}$, we have

$$\varphi(a)w = w\overline{\varphi(a)} \quad \text{or} \quad w\varphi(a) = \overline{\varphi(a)}w.$$

Suppose first that $\varphi(a)w = w\overline{\varphi(a)}$ holds. By Lemma 1 in [3], this implies that w is a palindrome. Let $u \in \mathcal{A}^*$ and $c \in \{\varepsilon\} \cup \mathcal{A}$ be such that $w = uc\overline{u}$. We can thus write

$$\varphi(a)uc\overline{u} = uc\overline{u}\overline{\varphi(a)}.$$

By applying $(uc)^{-1}$ from the left and $(c\overline{u})^{-1}$ from the right, we obtain for any $a \in \mathcal{A}$

$$c^{-1}u^{-1}\varphi(a)u = \overline{u}\overline{\varphi(a)}\overline{u}^{-1}c^{-1} = \overline{c^{-1}u^{-1}\varphi(a)u}.$$

This means that the word $p_a := c^{-1}u^{-1}\varphi(a)u$ is a palindrome. Set $p := c$. Denote by φ' the morphism defined for all $a \in \mathcal{A}$ by $\varphi'(a) = pp_a = u^{-1}\varphi(a)u$. Obviously, φ is conjugate to φ' which is of class P. Therefore $\varphi \in P'$.

The case $w\varphi(a) = \overline{\varphi(a)}w$ is analogous. $\qquad\square$

We are now in position to complete the proof the main theorem.

Theorem 2. *If ξ is a primitive substitution fixing a non-degenerate 3iet word, then ξ or ξ^2 belongs to class P'.*

Proof. Denote by $\eta \in \{\xi, \xi^2\}$ the substitution from Theorem 1. There exist intervals I_L and $I_R \subset [0,1)$ such that $\eta_L(A), \eta_L(B)$ and $\eta_L(C)$ are the I_L-itineraries, $\eta_R(A), \eta_R(B)$ and $\eta_R(C)$ are the I_R-itineraries, and such that T_{I_L} and T_{I_R} are 3iets homothetic to T.

Lemma 1 implies that

$$\Big(\mathrm{Fst}\big(\eta_L(A)\big), \mathrm{Fst}\big(\eta_L(B)\big), \mathrm{Fst}\big(\eta_L(C)\big)\Big) = \Big(\mathrm{Lst}\big(\eta_R(A)\big), \mathrm{Lst}\big(\eta_R(B)\big), \mathrm{Lst}\big(\eta_R(C)\big)\Big)$$

and this triple of letters equals (A, B, B) or (B, B, C). Suppose it is equal to (A, B, B). Note that by Corollary 1, η_L fixes the infinite word $\mathbf{u}_\alpha$.

According to Proposition 3, the induced transformation $T_{\overline{I_R}}$ is again homothetic to T and the corresponding substitution is $\overline{\eta_R}$. Since it is the mirror substitution to η_R, we have $\Big(\mathrm{Fst}\big(\overline{\eta_R}(A)\big), \mathrm{Fst}\big(\overline{\eta_R}(B)\big), \mathrm{Fst}\big(\overline{\eta_R}(C)\big)\Big) = (A, B, B)$. By Corollary 1, the substitution $\overline{\eta_R}$ also fixes the infinite word $\mathbf{u}_\alpha$. Since the intervals I_L and $\overline{I_R}$ are of the same length and are homothetic to the interval $[0,1)$ with the same homothety center α, necessarily $I_L = \overline{I_R}$ and thus $\overline{\eta_R} = \eta_L$. Consequently, η_R is conjugate to its mirror image. We apply Proposition 4 to finish the proof.

In case that $\big(\mathrm{Fst}\big(\eta_L(A)\big), \mathrm{Fst}\big(\eta_L(B)\big), \mathrm{Fst}\big(\eta_L(C)\big)\big) = (B, B, C)$, we proceed in a similar way. In this case, the center of the homothety of the intervals $I_L = \overline{I_R}$ and $[0,1)$ is β. $\qquad\square$

6   Comments

Recall that a substitution ξ over an alphabet $\mathcal{A}$ is called *marked* if its leftmost conjugate ξ_L and its rightmost conjugate ξ_R satisfy

$$\mathrm{Fst}\big(\xi_L(a)\big) \neq \mathrm{Fst}\big(\xi_L(b)\big) \quad \text{and} \quad \mathrm{Lst}\big(\xi_R(a)\big) \neq \mathrm{Lst}\big(\xi_R(b)\big)$$

for distinct $a, b \in \mathcal{A}$. It can be shown that if ξ is marked, then all its powers are marked. In [9], it is shown that for a marked morphism ξ with fixed point $\mathbf{u}$ having infinitely many palindromes, some power ξ^k belongs to class P'.

Our Lemma 1 shows that a substitution fixing a non-degenerated 3iet word cannot be marked. Theorem 2 thus provides a new class of substitutions satisfying the conjecture.

Let us mention that substitutions fixing degenerate 3iet words are not necessarily in class P'. In fact, a counterexample to the conjecture given by Labbé in [10] is the substitution

$$A \mapsto ABA, \ B \mapsto C, \ C \mapsto BAC,$$

which has, as a fixed point, a degenerate 3iet word coding the orbit of $\rho = \frac{2-\sqrt{2}}{4}$ under the 3iet with parameters $\alpha = \frac{1}{2}$ and $\beta = \frac{3-\sqrt{2}}{2}$.

Acknowledgements. Z.M. and E.P. acknowledge financial support by the Czech Science Foundation grant GAČR 13-03538S, Š.S. acknowledges financial support by the Czech Science Foundation grant GAČR 13-35273P.

References

1. Arnoux, P., Berthé, V., Masáková, Z., Pelantová, E.: Sturm numbers and substitution invariance of 3iet words. Integers **8** (Article A14) (2008)
2. Blondin Massé, A., Brlek, S., Garon, A., Labbé, S.: Equations on palindromes and circular words. Theoret. Comput. Sci. **412**, 2922–2930 (2011)
3. Blondin Massé, A., Brlek, S., Labbé, S.: Palindromic lacunas of the Thue-Morse word. In: Proc. GASCom 2008, pp. 53–67 (2008)
4. Ferenczi, S., Holton, C., Zamboni, L.: Structure of three-interval exchange transformations II: a combinatorial description of the tranjectories. J. Anal. Math. **89**, 239–276 (2003)
5. Gawrychowski, P., Manea, F., Nowotka, D.: Testing generalised freeness of words. In: Mayr, E.W., Portier, N. (eds.) 31st International Symposium on Theoretical Aspects of Computer Science (STACS 2014). (LIPIcs), vol. 25, pp. 337–349. Dagstuhl, Schloss Dagstuhl-Leibniz-Zentrum fuer Informatik, Germany (2014)
6. Hof, A., Knill, O., Simon, B.: Singular continuous spectrum for palindromic Schrödinger operators. Comm. Math. Phys. **174**, 149–159 (1995)
7. Kari, L., Mahalingam, K.: Watson-crick conjugate and commutative words. In: Garzon, M.H., Yan, H. (eds.) DNA 2007. LNCS, vol. 4848, pp. 273–283. Springer, Heidelberg (2008)
8. Keane, M.: Interval exchange transformations. Math. Z. **141**, 25–31 (1975)
9. Labbé, S., Pelantová, E.: Palindromic sequences generated from marked morphisms. Eur. J. Comb. **51**, 200–214 (2016)
10. Labbé, S.: A counterexample to a question of Hof, Knill and Simon. Electron. J. Comb. **21** (2014)
11. Lothaire, M.: Algebraic combinatorics on words. Encyclopedia of Mathematics and its Applications, vol. 90. Cambridge University Press (2002)
12. Masáková, Z., Pelantová, E., Starosta, Š: Interval exchange words and the question of Hof, Knill, and Simon (2015). (preprint available at) http://arxiv.org/abs/1503.03376
13. Tan, B.: Mirror substitutions and palindromic sequences. Theoret. Comput. Sci. **389**, 118–124 (2007)

State Complexity of Neighbourhoods
and Approximate Pattern Matching

Timothy Ng, David Rappaport, and Kai Salomaa[⊠]

School of Computing, Queen's University, Kingston, ON K7L 3N6, Canada
{ng,daver,ksalomaa}@cs.queensu.ca

Abstract. The neighbourhood of a language L with respect to an additive distance consists of all strings that have distance at most the given radius from some string of L. We show that the worst case (deterministic) state complexity of a radius r neighbourhood of a language recognized by an n state nondeterministic finite automaton A is $(r+2)^n$. The lower bound construction uses an alphabet of size linear in n. We show that the worst case state complexity of the set of strings that contain a substring within distance r from a string recognized by A is $(r+2)^{n-2} + 1$.

Keywords: Regular languages · State complexity · Lower bounds · Additive distance

1 Introduction

The similarity of strings is often defined using the edit distance [11,16], also known as the Levenshtein distance [14]. The edit distance is particularly useful for error-correction and error-detection applications [7–10,12]. A useful property is that the edit distance is additive with respect to concatenation of strings in the sense defined by Calude et al. [4].

If the distance of any two distinct strings of a language L is greater than r, the language L can detect up to r errors [9,11,13] (assuming the errors have unit weight). Alternatively we can consider what the shortest distance is between strings in languages L_1 and L_2, that is, what is the smallest number errors that transform a string of L_1 into a string of L_2. Calude at al. [4] showed that the neighbourhood of a regular language with respect to an additive distance is always regular. Additive quasi-distances preserve regularity as well [4]. This gives rise to the question how large is the deterministic finite automaton (DFA) needed to recognize the neigbourhood of a regular language. Informally, determining the optimal size of the DFA for the neighbourhood gives the *state complexity of error detection*. Note that since complementation does not change the size of a DFA, the size of the minimal DFA for the neighbourhood of L of radius r equals to the state complexity of the set of strings that have distance at least $r + 1$ from any string in L.

Povarov [17] showed that the Hamming neighbourhood of radius one of an n-state DFA language can be recognized by a DFA of size $n \cdot 2^{n-1} + 1$ and also

© Springer International Publishing Switzerland 2015
I. Potapov (Ed.): DLT 2015, LNCS 9168, pp. 389–400, 2015.
DOI: 10.1007/978-3-319-21500-6_31

gave a lower bound $\frac{3}{8}n \cdot 2^n - 2^{n-4} + n$ for its state complexity. Using a weighted finite automaton construction the third author and Schofield [18] gave an upper bound of $(r+2)^n$ for the neighbourhood of radius r of an n-state DFA-language. No good lower bounds are known for neighbourhoods of radius at least two.

The string matching problem consists of finding occurrences of a particular string in a text [2]. El-Mabrouk [6] considers the problem of pattern matching with r mismatches from a descriptional complexity point of view. Given a pattern P of length m and a text T, the problem is to determine whether T contains substrings of length m having characters differing from P in at most r positions, that is, substrings having Hamming distance at most r from P. For a pattern $P = a^m$ consisting of occurrences of only one character, the state complexity was shown to be $\binom{m+1}{r+1}$ [6].

The state complexity of $\Sigma^* L \Sigma^*$ was considered by Brzozowski, Jirásková, and Li [3] and was shown to have a tight bound of $2^{n-2} + 1$. A DFA recognizing $\Sigma^* L \Sigma^*$ can be viewed to solve the exact string matching problem. In the terminology of Brzozowski et al. [3], $\Sigma^* L \Sigma^*$ is a two-sided ideal and the descriptional complexity of related subregular language families was studied recently by Bordihn et al. [1].

This paper studies the descriptional complexity of neighbourhoods and of approximate string matching. As our main result we give a lower bound $(r+2)^n$ for the size of a DFA recognizing the radius r neighbourhood of an n-state regular language. The lower bound matches the previously known upper bound [18]. The bound can be reached either using a neighbourhood of an n-state DFA language with respect to an additive quasi-distance or using a neighbourhood of an n state NFA (nondeterministic finite automaton) language using an additive distance.

The lower bound constructions use an alphabet of size linear in n. A further limitation is that the (quasi-)distance associates different values to different edit operations. The precise state complexity of the edit distance with unit error costs remains open.

We also show that if L is recognized by an n-state NFA the set of strings that contain a substring within distance r from a string in L with respect to an additive (quasi-)distance is recognized by a DFA of size $(r+2)^{n-2} + 1$ and that this bound cannot be improved in the worst case. When r is zero the result coincides with the state complexity of two-sided ideals [3].

2   Preliminaries

Here we briefly recall some definitions and notation used in the paper. For all unexplained notions on finite automata and regular languages the reader may consult the textbook by Shallit [19] or the survey by Yu [20]. A survey of distances is given by Deza and Deza [5] and the notion of quasi-distance is from Calude et al. [4].

We denote by Σ a finite alphabet, Σ^* the set of words over Σ, and ε the empty word. A *nondeterministic finite automaton* (NFA) is a tuple $A = (Q, \Sigma, \delta, q_0, F)$ where Q is a finite set of states, Σ is an alphabet, δ is a multi-valued transition function $\delta : Q \times \Sigma \to 2^Q$, $q_0 \in Q$ is the initial state, and $F \subseteq Q$ is a set of final states. We extend the transition function δ to $Q \times \Sigma^* \to 2^Q$ in the usual way. A word $w \in \Sigma^*$ is *accepted* by A if $\delta(q_0, w) \cap F \neq \emptyset$ and the language recognized by A consists of all strings accepted by A. The automaton A is a *deterministic finite automaton* (DFA) if, for all $q \in Q$ and $a \in \Sigma$, $\delta(q, a)$ either consists of one state or is undefined. The DFA A is complete if $\delta(q, a)$ is defined for all $q \in Q$ and $a \in \Sigma$. Two states p and q of a DFA A are equivalent if $\delta(p, w) \in F$ if and only if $\delta(q, w) \in F$ for every string $w \in \Sigma^*$. A complete DFA A is *minimal* if each state $q \in Q$ is reachable from the initial state and no two states are equivalent. The (right) Kleene congruence of a language $L \subseteq \Sigma^*$ is the relation $\equiv_L \subseteq \Sigma^* \times \Sigma^*$ defined by setting

$$x \equiv_L y \text{ iff } [(\forall z \in \Sigma^*) \ xz \in L \Leftrightarrow yz \in L].$$

The language L is regular if and only if the index of $\equiv_L$ is finite and, in this case, the index of $\equiv_L$ is equal to the size of the minimal DFA for L [19]. The minimal DFA for a regular language L is unique. The *state complexity* of L, $sc(L)$, is the size of the minimal complete DFA recognizing L.

A function $d : \Sigma^* \times \Sigma^* \to [0, \infty)$ is a *distance* if it satisfies for all $x, y, z \in \Sigma^*$ the conditions $d(x, y) = 0$ if and only if $x = y$, $d(x, y) = d(y, x)$, and $d(x, z) \leq d(x, y) + d(y, z)$. The function d is a *quasi-distance* [4] if it satisfies conditions 2 and 3 and $d(x, y) = 0$ if $x = y$; that is, a quasi-distance between two distinct elements can be zero. In the following, unless otherwise mentioned, we consider only *integral* (quasi-)distances; that is, d is always a function $\Sigma^* \times \Sigma^* \to \mathbb{N}_0$.

The neighbourhood of a language L of radius r is the set

$$E(L, d, r) = \{x \in \Sigma^* \mid (\exists w \in L)d(x, w) \leq r\}.$$

A distance d is *finite* if for all nonnegative integers r the neighbourhood of radius r of any string with respect to d is finite. A distance d is *additive* [4] if for every factorization of a string $w = w_1 w_2$ and radius $r \geq 0$,

$$E(w, d, r) = \bigcup_{r_1 + r_2 = r} E(w_1, d, r_1) \cdot E(w_2, d, r_2).$$

A neighbourhood of a regular language with respect to an additive quasi-distance is regular [4].

The following upper bound for the state complexity of the neighbourhood of a regular language with respect to additive distances is known by [18] and by [15] for additive quasi-distances. The results are stated in terms of weighted finite automata.

Proposition 2.1 ([15,18]). *If A is an n-state NFA and d an additive quasi-distance, then for any $r \in \mathbb{N}$, $sc(E(L(A), d, r)) \leq (r + 2)^n$.*

We will use also the NFA construction for a neighbourhood due to Povarov [17]. Informally, the construction makes $r + 1$ copies of an NFA A, with each copy corresponding to a cumulative error ranging from 0 to r. A transition from a level i to a level $i' > i$ occurs when there is a transition that does not exist in A. There are $r + 1$ such copies of A to allow for at most r errors. Strictly speaking, [17] deals with additive distances but exactly the same construction works for quasi-distances.

Proposition 2.2 ([17]). *If A is an NFA with n states and d is an additive quasi-distance, then $E(L(A), d, r)$ has an NFA of size $n \cdot (r + 1)$.*

3   State Complexity of Additive Neighbourhoods

As the main result of this section we give a tight lower bound for the state complexity of a neighbourhood of a regular language given by a DFA (respectively, by an NFA) with respect to an additive quasi-distance (respectively, an additive distance).

For $n \in \mathbb{N}$ we consider an alphabet

$$\Sigma_n = \{a_1, \ldots, a_{n-1}, b_1, \ldots, b_n, c_1, \ldots, c_{n-1}\}. \tag{1}$$

For $r \in \mathbb{N}$, we define a quasi-distance $d_r : \Sigma_n^* \times \Sigma_n^* \to \mathbb{N}_0$ by the conditions:

– $d_r(a_i, a_j) = r + 1$ for $i \neq j$
– $d_r(b_i, b_j) = 1$ for $i \neq j$
– $d_r(a_i, b_j) = d_r(c_i, b_j) = r + 1$ for all $1 \leq i, j \leq n$
– $d_r(a_i, c_i) = 0$ for $1 \leq i \leq n - 1$
– $d_r(c_i, c_j) = r + 1$ for all $1 \leq i, j \leq n$
– $d_r(a_i, c_j) = r + 1$ for all $i \neq j$
– $d_r(\sigma, \varepsilon) = r + 1$ for all $\sigma \in \Sigma$.

Note that the value $d_r(\sigma, \varepsilon)$ denotes the cost of the deletion and insertion operations and that the listed substitution, insertion, and deletion operations on elements of Σ_n define a unique additive quasi-distance of Σ_n^* [4].

Lemma 3.1. *The function d_r is an additive quasi-distance.*

We define the following family of incomplete DFAs. Let $A_n = (Q_n, \Sigma_n, \delta, 1, \{n\})$ be a DFA with n states where $Q_n = \{1, \ldots, n\}$ and Σ_n is as in (1). The transition function δ is defined by setting

– $\delta(i, a_i) = i + 1$ for $1 \leq i \leq n - 1$
– $\delta(i, a_j) = i$ for $1 \leq i \leq n - 2$ and $i + 1 \leq j \leq n - 1$
– $\delta(i, b_j) = i$ for $1 \leq i \leq n - 1$ and $j = i - 1$ or $i + 1 \leq j \leq n$
– $\delta(i, c_i) = i$ for $1 \leq i \leq n - 1$

All transitions not listed are undefined. The DFA A_n is depicted in Figure 1.

The quasi-distance d_r identifies the symbols a_i and c_i, $1 \leq i \leq n$. By using two different symbols that have distance zero in our quasi-distance allows us to define A_n to be deterministic. By identifying a_i and c_i we can later modify

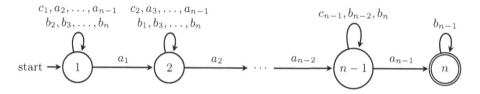

Fig. 1. The DFA A_n

the construction to give a lower bound for the neighbourhood of a language recognized by an NFA with respect to a distance (see Lemma 3.4).

To establish a lower bound for the state complexity of the neighbourhood $E(L(A_n), d_r, r)$ we define a set S of strings that are all pairwise inequivalent with respect to the Kleene congruence of the neighbourhood. First we construct an NFA $B_{n,r}$ for $E(L(A_n), d_r, r)$ and the inequivalence of the strings in S is verified using properties of $B_{n,r}$.

Suppose we have a DFA $A = (Q, \Sigma, \delta, q_0, F)$. Using Proposition 2.2 (due to [17]), an NFA $B = (Q', \Sigma, \delta', q_0', F')$ which recognizes the neighbourhood of radius r of $L(A)$ with respect to a quasi-distance d is defined by setting $Q' = Q \times \{0, \ldots, r\}$, $q_0' = (q_0, 0)$, $F' = F \times \{0, \ldots, r\}$ and the transitions of δ' for $q \in Q$, $0 \le k \le r$ and $a \in \Sigma$ are defined as

$$\delta'((q,k), a) = (\delta(q,a), k) \cup \bigcup_{b \in (\Sigma \cup \{\varepsilon\}) \setminus \{a\}} \{(\delta(q,b), k + d(a,b)) \mid k + d(a,b) \le r\}.$$

Now as described above we construct the NFA

$$B_{n,r} = (Q_n', \Sigma_n, \delta', q_0', F'), \tag{2}$$

shown in Figure 2, which recognizes the neighbourhood of $L(A_n)$ of radius r with respect to the quasi-distance d_r, where $Q_n' = Q_n \times \{0, 1, \ldots, r\}$, $q_0' = (q_0, 0)$, $F' = F \times \{0, 1, \ldots, r\}$ and the transition function δ' is defined by

- $\delta'((q,j), a_q) = \{(q,j), (q+1,j)\}$ for $1 \le q \le n-1$,
- $\delta'((q,j), a_{q'}) = \{(q,j)\}$ for all $1 \le q \le n-1$ and $q \le q' \le n-1$,
- $\delta'((q,j), b_i) = \{(q,j+1)\}$ for $1 \le q \le n$ and $i = 1, \ldots, q-2, q$,
- $\delta'((q,j), b_i) = \{(q,j)\}$ for $1 \le q \le n$ and $i = q-1, q+1, \ldots, n$,
- $\delta'((q,j), c_q) = \{(q,j), (q+1,j)\}$ for $1 \le q \le n-1$.

All transitions not listed above are undefined. Note that since in the distance d_r the cost of inserting/deleting a symbol is $r+1$ and $B_{n,r}$ recognizes a neighbourhood of radius r there are no error transitions corresponding to insertion/deletion. For the same reason the only error transitions for substitution correspond to substituting b_i with b_j, $i \ne j$. The distance between a_i and c_i is zero (no error), and all other substitutions have cost $r+1$.

For $0 \le k_i \le r+1$, $1 \le i \le n$, we define the string

$$w(k_1, \ldots, k_n) = a_1 b_1^{k_1} a_2 b_2^{k_2} \cdots a_{n-1} b_{n-1}^{k_{n-1}} b_n^{k_n}. \tag{3}$$

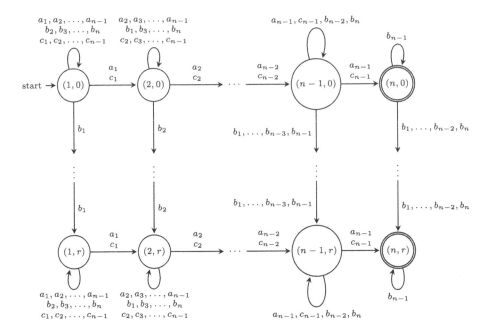

Fig. 2. The NFA $B_{n,r}$

The next lemma establishes a technical property of the computations of the NFA $B_{n,r}$ on the strings $w(k_1, \ldots, k_n)$. The property is then used to establish that the strings are pairwise inequivalent with respect to the language recognized by $B_{n,r}$.

Lemma 3.2. *If $k_i \leq r$, then there exists a computation C_i of the NFA $B_{n,r}$ which reaches the state (i, k_i) at the end of the input $w(k_1, \ldots, k_n)$, $1 \leq i \leq n$. There is no computation of $B_{n,r}$ on $w(k_1, \ldots, k_n)$ that reaches a state (i, k'_i) with $k'_i < k_i$. Furthermore, if $k_i = r + 1$, no computation of $B_{n,r}$ reaches at the end of $w(k_1, \ldots, k_n)$ a state where the first component is i.*

Proof. We verify that a computation C_i can reach state (i, k_i), $k_i \leq r$. First consider the case $i < n$. For $j = 1, \ldots, i - 1$, a_j takes state $(j, 0)$ to $(j+1, 0)$ and the next k_j symbols b_j are read using the self-loop in state $(j+1, 0)$. In this way the computation reaches state $(i, 0)$ where we read a_i using the self-loop and then reading the k_i symbols b_i the computation reaches (i, k_i). In state (i, k_i) the remaining suffix $a_{i+1} b_{i+1}^{k_{i+1}} \cdots a_{n-1} b_{n-1}^{k_{n-1}} b_n^{k_n}$ is consumed using the self-loops. Second, in the case $i = n$ similarly as above the computation after symbol a_{n-1} reaches state $(n, 0)$, the symbols b_{n-1} are read using self-loops and reading the k_n symbols b_n takes us to state (n, k_n).

To verify the second part of the lemma we first observe the following. The only transitions of $B_{n,r}$ which move from a state (i, j) to a state of the form $(i+1, j')$, $0 \leq j \leq r$, are on symbols a_i and c_i. Note that since the distance d_r

associates cost $r + 1$ to insertions and deletions, as well as to replacing a_i or c_i by any other symbol, the NFA $B_{n,r}$ does not have error transitions that change the first component of a state. Since $d_r(a_i, c_i) = 0$, we can treat them as the same letter and for convenience, we refer only to a_i. Thus, the only way to reach a state (q, j) for any $j \leq r$, is by taking transitions $((i, j), a_i, (i + 1, j))$ on each occurrence of a_i in $w(k_1, \ldots, k_n)$ for $i < q$. Otherwise, the computation remains in some state (i', j) for $i' < q$.

Now we show that there is no computation of $w(k_1, \ldots, k_n)$ that can reach a state (j, k_j') with $k_j' < k_j$. As discussed above, the only way for the computation to end in a state (j, i), $0 \leq i \leq r$, is by reaching the state $(j, 0)$ when consuming the prefix $a_1 b_1^{k_1} \cdots a_{j-1} b_{j-1}^{k_{j-1}}$ and then reading a_j using a self-loop. There is no other way to reach a state (j, i) for any i, since exiting the states with second components zero (corresponding to the original DFA) requires reading some $a_{j'}$ with a self-loop, after which there is no transition which can be taken to move to a state $(j' + 1, i)$. If in the state $(j, 0)$ the symbol a_j is not read with a self-loop then the first component becomes $j + 1$ and we cannot reach a state (j, i) with the remaining suffix. Thus, from $(j, 0)$ the NFA is forced to read the following k_j symbols b_j with error transitions, ending in the k_j-th level in the state (j, k_j).

Exactly the same argument verifies that in the case $k_j = r + 1$, no computation can end in a state where the first component is j. As above it is seen that to do this we must in the state $(j, 0)$ read the symbol a_j with a self-loop and after attempting to read the following $r + 1$ symbols b_j with an error transition the computation becomes undefined. $\square$

With the previous lemma we can now establish a lower bound for the state complexity of the neighbourhood of $L(A_n)$.

Lemma 3.3. *Let A_n be the DFA as in Figure 1. The strings $w(k_1, \ldots, k_n)$, $0 \leq k_i \leq r + 1$, $1 \leq i \leq n$, are all pairwise inequivalent with respect to the Kleene congruence of $E(L(A_n), d_r, r)$.*

Proof. We consider two distinct strings $w(k_1, \ldots, k_n)$ and $w(k_1', \ldots, k_n')$ with $0 \leq k_i, k_i' \leq r + 1$ for $i = 1, \ldots, n$. There exists an index j such that $k_j \neq k_j'$ and without loss of generality, we have $k_j < k_j'$. To distinguish the strings $w(k_1, \ldots, k_n)$ and $w(k_1', \ldots, k_n')$ consider the word $z = b_j^{r-k_j} a_{j+1} \cdots a_{n-1}$. The string z is well-defined since $k_j < k_j' \leq r + 1$ and so $r - k_j \geq 0$.

Let $B_{n,r}$ be the NFA constructed for $E(L(A), d_r, r)$ as in (2). We claim that $w(k_1, \ldots, k_n) \cdot z \in L(B_{n,r})$ but $w(k_1', \ldots, k_n') \cdot z \notin L(B_{n,r})$. We note that by Lemma 3.2, $B_{n,r}$ has a computation on $w(k_1, \ldots, k_n)$ that ends in state (j, k_j). Note that $k_j \leq r$. When continuing the computation on the string z, by reading the $r - k_j$ symbols b_j's, the machine is taken to the state (j, r). Then, reading the suffix $a_{j+1} \cdots a_{n-1}$ takes the machine to the accepting state (n, r).

To show $w(k_1', \ldots, k_n') \cdot z \notin L(B_{n,r})$, we consider from which states of $B_{n,r}$ an accepting state, that is, a state with first component n is reachable on the string z. We recall that in $B_{n,r}$ the transitions on b_j cannot change the first component of the state. (According to the definition of $B_{n,r}$ the reason for this is that d_r associates cost $r + 1$ to insertion/deletion or to subsitute a symbol a_i, c_i by b_j.)

Thus, for $B_{n,r}$ to reach an accepting state (with first component n) on the string $w(k'_1, \ldots, k'_n) \cdot z$, a computation must reach a state of the form (j, ℓ_j) on the prefix $w(k'_1, \ldots, k'_n)$. By Lemma 3.2, this is possible only if $\ell_j \geq k'_j$. From state (j, ℓ_j), $\ell_j \geq k'_j$, reading the substring $b_j^{r-k_j}$ takes the machine to an undefined state, as it is not possible to make $r - k_j$ error transitions on b_j in a state where the second component is $\ell_j > k_j$. This means that $B_{n,r}$ cannot accept the string $w(k'_1, \ldots, k'_n) \cdot z$.

Thus, each string $w(k_1, \ldots, k_n)$, $0 \leq i \leq r+1$, $1 \leq i \leq n$, defines a distinct equivalence class of $\equiv_{E(L(A), d_r, r)}$. $\qquad \square$

As a corollary of the proof of the previous lemma we get also a lower bound for the state complexity of the neighbourhood of an NFA-language with respect to an additive distance.

Lemma 3.4. *For $n, r \in \mathbb{N}$ there exists an additive distance d'_r and an NFA A'_n over an alphabet Σ'_n of size $2n - 1$ such that*

$$\mathrm{sc}(E(L(A'_n), d'_r, r)) \geq (r+2)^n.$$

Proof. Choose $\Sigma'_n = \{a_1, \ldots, a_{n-1}, b_1, \ldots, b_n\}$ and d'_r is the restriction of d_r to the alphabet Σ'_n (where d_r is the quasi-distance of Lemma 3.1). The function d'_r does not assign distance zero to any pair of distinct elements.

The NFA A'_n is obtained from the DFA A_n in Figure 1 by replacing all c_i-transitions by a_i-transitions, $1 \leq i \leq n - 1$. Thus, A'_n is nondeterministic. An NFA $B'_{n,r}$ for the neighbourhood $E(L(A'_n), d'_r, r)$ is obtained from the NFA $B_{n,r}$ in (2) simply by omitting all transitions on c_i, $1 \leq i \leq n - 1$. Note that in $B_{n,r}$ the transitions on c_i exactly coincide with the transitions on a_i, $1 \leq i \leq n - 1$, reflecting the situation that $d_r(a_i, c_i) = 0$.

The strings $w(k_1, \ldots, k_n)$ (as in (3)) did not involve any symbols c_i, and the proof of Lemma 3.3 remains the same, word for word, just by replacing $B_{n,r}$ with $B'_{n,r}$. $\qquad \square$

Now putting together Lemma 3.3, Lemma 3.4 and Proposition 2.1, we have:

Theorem 3.1. *If d is an additive quasi-distance, A is an NFA with n states and $r \in \mathbb{N}$,*

$$\mathrm{sc}(E(L(A), d, r) \leq (r+2)^n).$$

There exists an additive quasi-distance d_r and a DFA A with n states over an alphabet of size $3n - 2$ such that $\mathrm{sc}(E(L(A), d_r, r)) = (r+2)^n$.

There exists an additive distance d'_r and an NFA A' with n states over an alphabet of size $2n - 1$ such that $\mathrm{sc}(E(L(A'), d'_r, r)) = (r+2)^n$.

The lower bound construction has the trade-off of either using a DFA and a quasi-distance or an NFA and a distance, respectively. It would be interesting to know whether or not the general upper bound can be improved in cases where we are using a distance and the language is specified by a DFA.

4   State Complexity of Pattern Matching

We consider an extension of the pattern matching problem with mismatches in the sense of El-Mabrouk [6]. For a given finite automaton A and an additive quasi-distance d we construct a DFA for the language $\Sigma^* E(L(A), d, r)\Sigma^*$, that is, the set of strings that contain a substring within distance r from a string of $L(A)$. The construction gives an upper bound for the pattern matching problem and using a modification of the constructions in the previous section we show that the upper bound is optimal.

Lemma 4.1. *Let $A = (Q, \Sigma, \delta, q_0, F_A)$ be an n-state NFA with $k \geq 1$ final states and d is an additive quasi-distance. Then the language*

$$L_1 = \Sigma^* E(L(A), d, r)\Sigma^*$$

can be recognized by a DFA B with $(r + 2)^{n-1-k} + 1$ states.

Proof. Let $Q = \{q_0, q_1, \ldots, q_{n-1}\}$. If $q_0 \in F_A$, then $L_1 = \Sigma^*$ and there is nothing to prove. Thus, in the following we can assume that $F = \{q_{n-k}, q_{n-k+1}, \ldots, q_{n-1}\}, 1 \leq k \leq n - 1$. Furthermore, without loss of generality we assume that

$$(\forall w \in \Sigma^*) \, \delta(q_0, w) \cap F_A \neq \emptyset \text{ implies } d(\varepsilon, w) > r. \tag{4}$$

If the above condition does not hold, $\varepsilon \in E(L(A), d, r)$ and there is nothing to prove.

The DFA B recognizing L_1 operates as follows. Roughly speaking, B is looking for a substring of the input that belongs to $E(L(A), d, r)$. For this purpose, for all non-final states q_z of A, the deterministic computation of B keeps track of the smallest cumulative error between a string that takes q_0 to q_z and any suffix of the input processed thus far. Note that for the initial state q_0 this value is always zero and, hence, the states of P store the cumulative error only for the nonfinal states $q_1, \ldots, q_{n-k-1}$. When B has found a substring belonging to $E(L(A), d, r)$ the computation goes to the final state p_f and after that accepts an arbitrary suffix. Next we give the definition of B and after that include a brief correctness argument.

Define $B = (P, \Sigma, \gamma, p_0, F_B)$ with set of states

$$P = \{(i_1, \ldots, i_{n-k-1}) \mid 0 \leq i_j \leq r + 1, \, j = 1, \ldots, n - k - 1\} \cup \{p_f\},$$

the initial state is $p_0 = (h_1, \ldots, h_{n-k+1})$ where

$$h_z = \inf\{d(\varepsilon, w) \mid q_z \in \delta(q_0, w)\}, \quad 1 \leq z \leq n - k - 1,$$

and the set of final states is defined as $F_B = \{p_f\}$. Note that by (4) we know that $\varepsilon \notin L_1$. Next we define the transitions of B. First, $\gamma(p_f, b) = p_f$ for all $b \in \Sigma$. For $\mathbf{p} = (i_1, \ldots, i_{n-k-1}) \in P, 0 \leq i_z \leq r + 1, z = 1, \ldots, n - k - 1$, and $b \in \Sigma$ we define

(i) $\gamma(\mathbf{p}, b) = p_f$ if $(\exists 1 \leq z \leq n - k - 1)(\exists w \in \Sigma^*) \; \delta(q_z, w) \cap F_A \neq \emptyset$ and $i_z + d(b, w) \leq r$;
(ii) and if the conditions in (i) do not hold, then $\gamma(\mathbf{p}, b) = (j_1, \ldots, j_{n-k-1})$, where, for $x = 1, \ldots, n - k - 1$,

$$
\begin{aligned}
j_x \; = \; & \inf[\; \{i_z + d(b, w) \mid q_x \in \delta(q_z, w), 1 \leq z \leq n - k - 1\} \\
& \cup \; \{d(b, w) \mid q_x \in \delta(q_0, w)\} \;].
\end{aligned}
$$

In a state of the form $(i_1, \ldots, i_{n-k-1}) \in P$, the component $i_z, 1 \leq z \leq n-k-1$, keeps track of the smallest distance $d(u_{\mathrm{suf}}, w)$ where u_{suf} is a suffix of the input processed up to that point and w is a string that in A takes the initial state q_0 to state q_z. The smallest error between the suffix ε and a string that in A reaches q_0 is always zero and this value is not stored in the state of B. If the computation has found a substring in $E(L(A), d, r)$, the state of B will be p_f. $\square$

By modifying the construction used in the proof of Lemma 3.4 (and Lemma 3.3) we give a lower bound that matches the upper bound from Lemma 4.1.

Lemma 4.2. *For $n, r \in \mathbb{N}$, there exist an additive distance d and an NFA A with n states defined over an alphabet Σ of size $2n - 1$ such that the minimal DFA for $\Sigma^* E(L(A), d, r)\Sigma^*$ must have at least $(r + 2)^{n-2} + 1$ states.*

Proof. Choose $\Sigma_n = \{a_1, \ldots, a_{n-1}, b_1, \ldots, b_n\}$ and let A'_n and d'_r be as in the proof of Lemma 3.4. Let $B'_{n,r}$ be the NFA constructed for $E(L(A'_n), d'_r, r)$ in the proof of Lemma 3.4.[1] For $0 \leq k_i \leq r + 1$, $i = 1, 2, \ldots, n - 2$, define

$$
u(k_1, k_2, \ldots, k_{n-2}) = a_1 b_1^{k_1} a_2 b_2^{k_2} \cdots a_{n-2} b_{n-2}^{k_{n-2}}.
$$

Using the notations of (3) we have $u(k_1, \ldots, k_{n-2}) \cdot a_{n-1} = w(k_1, k_2, \ldots, k_{n-2}, 0, 0)$.

We claim that the strings $u(k_1, \ldots, k_{n-2})$ are all pairwise inequivalent with respect to the Kleene congruence of $\Sigma_n^* E(L(A'_n), d'_r, r)\Sigma_n^*$. Consider two strings $u(k_1, \ldots, k_{n-2})$ and $u(k'_1, \ldots, k'_{n-2})$ where for some $1 \leq j \leq n - 2$, $k_j < k'_j$.

Choose $z = b_j^{r-k_j} a_{j+1} \cdots a_{n-1}$. As in the proof of Lemma 3.2 it is observed that $B'_{n,r}$ has a computation on $u(k_1, \ldots, k_{n-2})$ that reaches state (j, k_j), and a computation started from state (j, k_j) on input z can reach the accepting state (n, r). Thus, $u(k_1, \ldots, k_{n-2}) \cdot z \in L(B'_{n,r}) = E(L(A'_n), d'_r, r)$. We claim that

$$
u(k'_1, \ldots, k'_{n-2}) \cdot z \notin \Sigma_n^* E(L(A'_n), d'_r, r)\Sigma_n^*. \tag{5}
$$

Note that the string $u(k'_1, \ldots, k'_{n-2}) \cdot z$ contains exactly one occurrence of both a_1 and a_{n-1} and these are, respectively, the first and the last symbol of the string. Since the distance d'_r associates cost $r + 1$ to any operation that substitutes, deletes or inserts a symbol a_i, if the negation of (5) holds then the only possibility is that $u(k'_1, \ldots, k'_{n-2}) \cdot z$ must be in $E(L(A'_n), d'_r, r)$. This, in turn, is possible

---

[1] $B'_{n,r}$ is obtained from the NFA of Fig. 2 by omitting all the transitions on c_i's.

only if the computation of $B'_{n,r}$ on the prefix $u(k'_1, \ldots, k'_{n-2})$ ends in a state of the form (j, x), $0 \leq x \leq r$. Now Lemma 3.2 implies that the second component x must be at least k'_j and it follows that the computation on the suffix z cannot end in an accepting state. (Lemma 3.2 uses $B_{n,r}$ but the same argument applies here because $B_{n,r}$ equals $B'_{n,r}$ when we omit the c_i-transitions.)

Finally we note that none of the strings $u(k_1, \ldots, k_{n-2})$, $0 \leq k_i \leq r + 1$, is in $\Sigma_n^* E(L(A'_n), d'_r, r) \Sigma_n^*$ and hence they are not equivalent with $a_1 a_2 \cdots a_{n-1}$ which then gives the one additional equivalence class.   □

Combining the previous lemmas we can state the main result of this section.

Theorem 4.1. *Let d be an additive quasi-distance on Σ^*. For any n-state NFA A and $r \in \mathbb{N}$ we have*

$$\mathrm{sc}(\Sigma^* \cdot E(L(A), d, r) \cdot \Sigma^*) \leq (r + 2)^{n-2} + 1).$$

For given $n, r \in \mathbb{N}$ there exists an additive distance d_r and an n-state NFA A defined over an alphabet of size $2n - 1$ such that $\mathrm{sc}(\Sigma^* E(L(A), d_r, r) \Sigma^*) = (r + 2)^{n-2} + 1$.

Proof. The upper bound of Lemma 4.1 is maximized by an NFA with one final state as $(r + 2)^{n-2} + 1$. The lower bound follows by Lemma 4.2.   □

Recall that Brzozowski et al. [3] have shown that, for an n-state DFA language L, the worst case state complexity of the two-sided ideal $\Sigma^* L \Sigma^*$ is $2^{n-2} + 1$. This corresponds to the case of having error radius zero ($r = 0$) in Theorem 4.1. Lemma 4.2 requires a linear size alphabet whereas the lower bound for the error free case is obtained with a three letter alphabet [3]. As in Theorem 3.1 in the lower bound result of Lemma 4.2 we can select A to be a DFA if we allow d to be a quasi-distance.

5   Conclusion

We have given a tight lower bound construction for the state complexity of a neighbourhood of a regular language. The construction uses a variable alphabet of size linear in the number of states of the NFA. The main open problem for further work is to develop lower bounds for neighbourhoods of languages over a fixed alphabet. For radius one Hamming neighbourhoods an improved upper bound and a good lower bound using a binary alphabet were given by Povarov [17].

Our lower bound for the approximate pattern matching problem was obtained by modifying the lower bound construction for neighbourhoods of a regular language. This was, roughly speaking, made possible by the choice of the distance function and the language definition where the strings must contain the symbols $a_1, \ldots, a_{n-1}$ in this particular order. Similar constructions will be more challenging if restricted to a fixed alphabet.

References

1. Bordihn, H., Holzer, M., Kutrib, M.: Determination of finite automata accepting subregular languages. Theoretical Computer Science **410**, 3209–3249 (2009)
2. Boyer, R.S., Moore, J.S.: A fast string searching algorithm. Communications of ACM **20**, 762–772 (1977)
3. Brzozowski, J., Jirásková, G., Li, B.: Quotient complexity of ideal languages. In: López-Ortiz, A. (ed.) LATIN 2010. LNCS, vol. 6034, pp. 208–221. Springer, Heidelberg (2010)
4. Calude, C.S., Salomaa, K., Yu, S.: Additive Distances and Quasi-Distances Between Words. Journal of Universal Computer Science **8**(2), 141–152 (2002)
5. Deza, M.M., Deza, E.: Encyclopedia of Distances. Springer-Verlag, Heidelberg (2009)
6. El-Mabrouk, N.: On the size of minimal automata for approximate string matching. Technical report, Institut Gaspard Monge, Université de Marne la Vallée, Paris (1997)
7. Han, Y.-S., Ko, S.-K., Salomaa, K.: The edit distance between a regular language and a context-free language. International Journal of Foundations of Computer Science **24**, 1067–1082 (2013)
8. Kari, L., Konstantinidis, S.: Descriptional complexity of error/edit systems. Journal of Automata, Languages, and Combinatorics **9**, 293–309 (2004)
9. Kari, L., Konstantinidis, S., Kopecki, S., Yang, M.: An efficient algorithm for computing the edit distance of a regular language via input-altering transducers. CoRR abs/1406.1041 (2014)
10. Konstantinidis, S.: Transducers and the properties of error detection, error-correction, and finite-delay decodability. Journal of Universal Computer Science **8**, 278–291 (2002)
11. Konstantinidis, S.: Computing the edit distance of a regular language. Information and Computation **205**, 1307–1316 (2007)
12. Konstantinidis, S., Silva, P.: Maximal error-detecting capabilities of formal languages. J. Automata, Languages and Combinatorics **13**, 55–71 (2008)
13. Konstantinidis, S., Silva, P.: Computing maximal error-detecting capabilities and distances of regular languages. Fundamenta Informaticae **101**, 257–270 (2010)
14. Levenshtein, V.I.: Binary codes capable of correcting deletions, insertions, and reversals. Soviet Physics Doklady **10**(8), 707–710 (1966)
15. Ng, T., Rappaport, D., Salomaa, K.: Quasi-distances and weighted finite automata. In: Shallit, J., Okhotin, A. (eds.) DCFS 2015. LNCS, vol. 9118, pp. 209–219. Springer, Heidelberg (2015)
16. Pighizzini, G.: How hard is computing the edit distance? Information and Computation **165**, 1–13 (2001)
17. Povarov, G.: Descriptive complexity of the hamming neighborhood of a regular language. In: Language and Automata Theory and Applications, pp. 509–520 (2007)
18. Salomaa, K., Schofield, P.: State Complexity of Additive Weighted Finite Automata. International Journal of Foundations of Computer Science **18**(06), 1407–1416 (2007)
19. Shallit, J.: A Second Course in Formal Languages and Automata Theory. Cambridge University Press (2009)
20. Yu, S.: Regular languages. In: Rozenberg, G., Salomaa, A. (Eds.) Handbook of Formal Languages, vol. I, pp. 41–110. Springer (1997)

Deterministic Ordered Restarting Automata that Compute Functions

Friedrich Otto[⊠] and Kent Kwee

Fachbereich Elektrotechnik/Informatik, Universität Kassel, 34109 Kassel, Germany
{otto,kwee}@theory.informatik.uni-kassel.de

Abstract. We present three methods for using deterministic ordered restarting automata to compute relations and functions. In the most general setting we obtain succinct representations for all rational relations, and in the most restricted setting we derive a succinct description for all rational functions that map the empty word to itself. In addition, we study the deterministic ordered restarting transducer that characterizes a proper superclass of the rational functions.

Keywords: Ordered restarting automaton · Rational function · Descriptional complexity

1 Introduction

The *deterministic ordered restarting automaton* (or *det-ORWW-automaton*) was introduced in [7] in the setting of picture languages. Such an automaton has a finite-state control, a tape with end markers that initially contains the input, and a window of size three. Based on its state and the content of its window, the automaton can either perform a *move-right step*, which shifts the window one position to the right and changes the state, or a *rewrite/restart step*, which replaces the symbol in the middle of the window by a symbol that is strictly smaller with respect to a predefined ordering on the working alphabet, moves the window back to the left end of the tape, and resets the state to the initial state, or an *accept step*, which causes the automaton to halt and accept. While the nondeterministic variant of this type of automaton even accepts some languages that are not context-free, the deterministic variant characterizes the regular languages.

In [8] an investigation of the descriptional complexity of the det-ORWW-automaton was initiated. Each det-ORWW-automaton can be simulated by an automaton of the same type that has only a single state, which means that for these automata, states are actually not needed. Accordingly, such an automaton is called a *stateless* det-ORWW-automaton (stl-det-ORWW-automaton). For these automata, the size of their working alphabets can be taken as a measure for their descriptional complexity, and it has been shown that these automata are polynomially related in size to the weight-reducing Hennie machines studied by Průša in [10]. For $n \geq 1$, there exists a regular language that is accepted

© Springer International Publishing Switzerland 2015
I. Potapov (Ed.): DLT 2015, LNCS 9168, pp. 401–412, 2015.
DOI: 10.1007/978-3-319-21500-6_32

by a stl-det-ORWW-automaton of size $O(n)$ such that each deterministic finite-state acceptor (DFA) for this language has at least 2^{2^n} many states. On the other hand, each stl-det-ORWW-automaton of size n can be simulated by an unambiguous nondeterministic finite-state acceptor (NFA) with $2^{O(n)}$ states [6], and therewith by a DFA with $2^{2^{O(n)}}$ states. Thus, the stl-det-ORWW-automaton is a deterministic device that is exponentially more succinct than NFAs.

Here we study det-ORWW-automata in the setting of relations and functions. First, we associate with such an automaton an *input-output relation*. It turns out that in this way we obtain succinct representations of all rational relations. In his PhD dissertation [4], Hundeshagen extended various types of restarting automata into *restarting transducers* by associating an output operation with each restart and each accept transition (see also [5]). Here we apply this concept to det-ORWW-automata showing that in this way we obtain a device that computes a proper superclass of the rational functions. Finally, we propose a way of associating transductions to det-ORWW-automata by using the result that is obtained by an accepting computation to determine the generated output directly. In this way we obtain a succint representation of all rational functions τ that satisfy the equality $\tau(\lambda) = \lambda$ (here λ denotes the empty word).

The paper is structured as follows. In Section 2 we recall the definition of the det-ORWW-automaton in short and summarize the main results on its descriptional complexity. In the next three sections we consider the three ways of associating relations and/or functions to det-ORWW-automata in turn. The paper closes with a short summary and some open problems.

2   Deterministic ORWW-Automata

A *det-ORWW-automaton* is given by an 8-tuple $M = (Q, \Sigma, \Gamma, \rhd, \lhd, q_0, \delta, >)$, where Q is a finite set of states, Σ is a finite input alphabet, Γ is a finite tape alphabet containing Σ, the symbols $\rhd, \lhd \notin \Gamma$ serve as markers for the left and right border of the work space, respectively, $q_0 \in Q$ is the initial state,

$$\delta : Q \times (((\Gamma \cup \{\rhd\}) \cdot \Gamma \cdot (\Gamma \cup \{\lhd\})) \cup \{\rhd\lhd\}) \dashrightarrow (Q \times \{\mathsf{MVR}\}) \cup \Gamma \cup \{\mathsf{Accept}\}$$

is the (partial) *transition function*, and $>$ is a *partial ordering* on Γ. The transition function describes three different types of transition steps:

(1) A *move-right step* has the form $\delta(q, a_1 a_2 a_3) = (q', \mathsf{MVR})$, where $q, q' \in Q$, $a_1 \in \Gamma \cup \{\rhd\}$ and $a_2, a_3 \in \Gamma$. It causes M to shift the window one position to the right and to enter state q'. Observe that no move-right step is possible, if the window contains the right sentinel $\lhd$.

(2) A *rewrite/restart step* has the form $\delta(q, a_1 a_2 a_3) = b$, where $q \in Q$, $a_1 \in \Gamma \cup \{\rhd\}$, $a_2, b \in \Gamma$, and $a_3 \in \Gamma \cup \{\lhd\}$ such that $a_2 > b$ holds. It causes M to replace the symbol a_2 in the middle of its window by the symbol b and to restart.

(3) An *accept step* $\delta(q, a_1 a_2 a_3) = \mathsf{Accept}$, where $q \in Q$, $a_1 \in \Gamma \cup \{\rhd\}$, $a_2 \in \Gamma$, and $a_3 \in \Gamma \cup \{\lhd\}$, causes M to halt and accept. In addition, we allow an accept step of the form $\delta(q_0, \rhd\lhd) = \mathsf{Accept}$.

If $\delta(q, u)$ is undefined for some $q \in Q$ and $u \in ((\Gamma \cup \{\triangleright\}) \cdot \Gamma \cdot (\Gamma \cup \{\triangleleft\})) \cup \{\triangleright \triangleleft\}$, then M necessarily halts, when it is in state q with u in its window, and we say that M *rejects* in this situation. The letters in $\Gamma \setminus \Sigma$ are called *auxiliary symbols*.

The det-ORWW-automaton M is called *stateless* if $Q = \{q_0\}$, that is, the initial state q_0 is the only state of M. For a stateless det-ORWW-automaton, we drop the components Q and q_0 from its description to simplify the notation. By stl-det-ORWW we denote the stateless det-ORWW-automata.

A *configuration* of a det-ORWW-automaton M is a pair of strings $(\alpha, q\beta)$, where $q \in Q$ and $|\beta| \geq 3$, and either $\alpha = \lambda$ and $\beta \in \{\triangleright\} \cdot \Gamma^+ \cdot \{\triangleleft\}$ or $\alpha \in \{\triangleright\} \cdot \Gamma^*$ and $\beta \in \Gamma \cdot \Gamma^+ \cdot \{\triangleleft\}$; here $q \in Q$ represents the current state, $\alpha\beta$ is the current content of the tape, and it is understood that the window of M contains the first three symbols of β. In addition, we admit the configuration $(\lambda, q_0 \triangleright \triangleleft)$. A *restarting configuration* has the form $(\lambda, q_0 \triangleright w \triangleleft)$; if $w \in \Sigma^*$, then the configuration $(\lambda, q_0 \triangleright w \triangleleft)$ is an *initial configuration*. Further, we use Accept to denote the *accepting configurations*, which are those configurations that M reaches by executing an accept step.

We observe that any computation of a det-ORWW-automaton M consists of certain phases. A phase, called a *cycle*, starts in a restarting configuration, the head moves along the tape performing MVR operations until a rewrite/restart operation is performed and thus, a new restarting configuration is reached. If no further rewrite operation is performed, any computation necessarily finishes in a halting configuration – such a phase is called a *tail*. By $\vdash_M^c$ we denote the execution of a complete cycle, and $\vdash_M^{c*}$ is the reflexive transitive closure of this relation. It can be seen as the *rewrite relation* that is realized by M on the set of restarting configurations.

An input $w \in \Sigma^*$ is accepted by M, if the computation of M which starts with the initial configuration $(\lambda, q_0 \triangleright w \triangleleft)$ ends with an Accept instruction. The language consisting of all input words that are accepted by M is denoted by $L(M)$.

As each cycle ends with a rewrite operation, which replaces a symbol a by a symbol b that is strictly smaller than a with respect to the given ordering $>$, we see that each computation of M on an input of length n consists of at most $(|\Gamma| - 1) \cdot n$ many cycles and a tail. Thus, M can be simulated by a deterministic single-tape Turing machine in time $O(n^2)$. Concerning the expressive power of det-ORWW-automata, the following results have been obtained (see [6, 8]).

Proposition 1. *For each DFA $A = (Q, \Sigma, q_0, F, \varphi)$, there exists a stl-det-ORWW-automaton $M = (\Sigma, \Gamma, \triangleright, \triangleleft, \delta, >)$ such that $L(M) = L(A)$ and $|\Gamma| = |Q| + |\Sigma|$.*

Proposition 2. *For each det-ORWW-automaton $M = (Q, \Sigma, \Gamma, \triangleright, \triangleleft, q_0, \delta, >)$, there exists a stateless det-ORWW-automaton $M' = (\Sigma, \Delta, \triangleright, \triangleleft, \delta', >')$ such that $L(M') = L(M)$ and $|\Delta| = |Q| \cdot |\Gamma|^2 + 2 \cdot |\Gamma|$.*

Theorem 3. *Let $M = (\Sigma, \Gamma, \triangleright, \triangleleft, \delta_M, >)$ be a stl-det-ORWW-automaton. Then an unambiguous NFA $A = (Q, \Sigma, \Delta_A, q_0, F)$ can be constructed from M such that $L(A) = L(M)$ and $|Q| \in 2^{O(|\Gamma|)}$.*

Thus, we have the following characterization.

Corollary 4. [7,8] REG $= \mathcal{L}(\text{stl-det-ORWW}) = \mathcal{L}(\text{det-ORWW})$.

Concerning the descriptional complexity of stl-det-ORWW-automata, the following results have been obtained, where we use the number of letters in the tape alphabet as the complexity measure for a stl-det-ORWW-automaton and the number of states as the complexity measure for NFAs and DFAs.

Corollary 5. *For converting a stl-det-ORWW-automaton with n letters into an equivalent NFA, $2^{O(n)}$ states are sufficient, and there are cases in which these many states are also necessary.*

Corollary 6. *For converting a stl-det-ORWW-automaton with n letters into an equivalent DFA, $2^{2^{O(n)}}$ states are sufficient, and there are cases in which these many states are also necessary.*

3   Relations Associated to det-ORWW-Automata

Let Γ be a finite alphabet, let Σ_1 and Σ_2 be two disjoint subalphabets of Γ, and let Pr^{Σ_1} and Pr^{Σ_2} denote the projections from Γ^* onto Σ_1^* and Σ_2^*. With a det-ORWW-automaton $M = (Q, \Sigma_1 \cup \Sigma_2, \Gamma, \triangleright, \triangleleft, q_0, \delta, >)$ we associate the (*input-output*) *relation* $Rel_{\text{io}}(M) \subseteq \Sigma_1^* \times \Sigma_2^*$ that is defined as follows:

$$Rel_{\text{io}}(M) := \{ (u, v) \in \Sigma_1^* \times \Sigma_2^* \mid \exists w \in L(M) : \mathsf{Pr}^{\Sigma_1}(w) = u \text{ and } \mathsf{Pr}^{\Sigma_2}(w) = v \}.$$

Thus, a pair $(u, v) \in \Sigma_1^* \times \Sigma_2^*$ belongs to the relation $Rel_{\text{io}}(M)$ if and only if there exists a word w in the shuffle of u and v such that w belongs to $L(M)$. We say that M *recognizes* the relation $Rel_{\text{io}}(M)$. A relation $R \subseteq \Sigma_1^* \times \Sigma_2^*$ is a *det-ORWW-io-relation* if there exists a det-ORWW-automaton M such that $Rel_{\text{io}}(M) = R$ holds.

A *rational transducer* is defined as $T = (Q, \Sigma, \Delta, q_0, F, E)$, where Q is a finite set of states, Σ is a finite input alphabet, Δ is a finite output alphabet, $q_0 \in Q$ is the initial state, $F \subseteq Q$ is the set of final states, and $E \subset Q \times (\Sigma \cup \{\lambda\}) \times \Delta^* \times Q$ is a finite set of transitions. The relation $Rel(T)$ computed by T consists of all pairs $(u, v) \in \Sigma^* \times \Delta^*$ such that there exists a computation of T that, starting from the initial state q_0, reaches a final state $q \in F$ reading the word u and producing the output v. By RatRel we denote the set of all *rational relations*, which is the set of binary relations that are computed by rational transducers (see, e.g., [2]).

Proposition 7. *Let Σ_1 and Σ_2 be two finite alphabets that are disjoint. Then a relation $R \subseteq \Sigma_1^* \times \Sigma_2^*$ is a det-ORWW-io-relation if and only if it is a stl-det-ORWW-io-relation if and only if it is a rational relation.*

Proof. By a theorem of Nivat (see, e.g., [1]) a binary relation $R \subseteq \Sigma_1^* \times \Sigma_2^*$ is rational if and only if there exists a regular language $L \subseteq (\Sigma_1 \cup \Sigma_2)^*$ such that $R = \{ (\mathsf{Pr}^{\Sigma_1}(w), \mathsf{Pr}^{\Sigma_2}(w)) \mid w \in L \}$. By Corollary 4 the class of regular languages coincides with the class of languages that are accepted by det-ORWW- and by stl-det-ORWW-automata. Thus, if M is an automaton of one of these two types such that $L(M) = L$, then R coincides with the relation $Rel_{\text{io}}(M)$.   □

Actually, from the simulation result of [6] we can derive the following result.

Theorem 8. *From a given stl-det-ORWW-automaton M on an alphabet of size n one can effectively construct a rational transducer with $2^{O(n)}$ states that computes the relation $Rel_{io}(M)$.*

Proof. From M we can construct an NFA A of size $2^{O(n)}$ for the language $L(M)$ by Theorem 3. As the subalphabets Σ_1 and Σ_2 are disjoint, we can transform A into a transducer T of the same size by replacing each transition that reads a letter $a \in \Sigma_1$ by a transition that reads a and outputs λ, and by replacing each transition that reads a letter $b \in \Sigma_2$ by a transition that reads λ and outputs b. Obviously, T computes the relation $Rel_{io}(M)$.   □

The language $U_n = \{a^{2^n}\}$ is accepted by a stl-det-ORWW-automaton M_n with $O(n)$ letters [6], and hence, the transduction $\hat{U}_n = \{(a^{2^n}, \lambda)\}$ coincides with the relation $Rel_{io}(M_n)$. On the other hand, each NFA for U_n, and therewith each rational transducer for $\hat{U}_n$, has at least $2^n + 1$ states. This means that the upper bound in Theorem 8 is sharp with respect to its order of magnitude.

4   Transductions Computed by det-ORWW-Transducers

Next we use the notion of *restarting transducer* as defined in [4].

Definition 9. *A det-ORWW-transducer $T = (Q, \Sigma, \Gamma, \Delta, \rhd, \lhd, q_0, \delta, >)$ is obtained from a det-ORWW-automaton $M = (Q, \Sigma, \Gamma, \rhd, \lhd, q_0, \delta_M, >)$ by introducing an output alphabet Δ and by extending the transition function into a (partial) function*

$$\delta : Q \times (((\Gamma \cup \{\rhd\}) \cdot \Gamma \cdot (\Gamma \cup \{\lhd\})) \cup \{\rhd\lhd\})$$
$$\dashrightarrow (Q \times \{\mathsf{MVR}\}) \cup ((\Gamma \cup \{\mathsf{Accept}\}) \times \Delta^*),$$

that is, each rewrite and accept step is extended by some output word from Δ^*.
The det-ORWW-transducer T is called proper *if, for each accept step, the associated output word is the empty word λ, that is, nonempty output can only be generated during rewrite/restart steps.*
A configuration of T is described by a triple $(\alpha, q\beta, z)$, where $(\alpha, q\beta)$ is a configuration of the underlying det-ORWW-automaton M and $z \in \Delta^*$ is an output word. If $(\alpha, q\beta) = (\rhd x, qa_1a_2a_3y\lhd)$ such that $\delta(q, a_1a_2a_3) = (b, v)$, then $(\alpha, q\beta, z) \vdash_T (\lambda, q_0 \rhd xa_1ba_3y\lhd, zv)$, that is, T rewrites a_2 into b, restarts, and produces the output v. Further, if $\delta(q, a_1a_2a_3) = (\mathsf{Accept}, v')$, then $(\alpha, qa_1a_2a_3y\lhd, z) \vdash_T (\mathsf{Accept}, zv')$, that is, T halts and accepts producing the output v'. An accepting computation of T consists of a finite sequence of cycles that is followed by an accepting tail computation, that is, it can be described as

$$(\lambda, q_0 \rhd w\lhd, \lambda) \vdash_T^c (\lambda, q_0 \rhd w_1\lhd, v_1) \vdash_T^c \cdots \vdash_T^c (\lambda, q_0 \rhd w_m\lhd, v_1 \cdots v_m)$$
$$\vdash_T^* (\mathsf{Accept}, v_1 \cdots v_m v_{m+1}).$$

With T we associate the following (input-output) relation

$$Rel(T) = \{ (w, z) \in \Sigma^* \times \Delta^* \mid (\lambda, q_0 \rhd w \lhd, \lambda) \vdash_T^* (\mathsf{Accept}, z) \}.$$

As T is deterministic, $Rel(T)$ is obviously the graph of a (partial) function f_T with domain $L(M)$. This function f_T is called the function computed by T.

We continue by presenting two examples.

Example 10. Let $T = (\Sigma, \Gamma, \Delta, \rhd, \lhd, \delta, >)$ be the stl-det-ORWW-transducer that is defined by taking $\Sigma = \{a, b\}$, $\Gamma = \Sigma \cup \{a', b'\}$, and $\Delta = \{a, b\}$, the ordering $>$ is given through $a > a'$ and $b > b'$, and the transition function is defined as follows, where $c, d, e \in \Sigma$:

$$\delta(\rhd\lhd) = (\mathsf{Accept}, \lambda), \quad \delta(\rhd c\lhd) = (c', c), \quad \delta(\rhd c'\lhd) = (\mathsf{Accept}, \lambda),$$
$$\delta(\rhd cd) = \mathsf{MVR}, \qquad \delta(cde) = \mathsf{MVR}, \qquad \delta(cd\lhd) = (d', d),$$
$$\delta(cde') = (d', d), \qquad \delta(\rhd cd') = (c', c), \quad \delta(\rhd c'd') = (\mathsf{Accept}, \lambda).$$

Given a word $w = a_1 a_2 \dots a_n$ as input, where $a_1, \dots, a_n \in \Sigma$, T executes the following accepting computation:

$$
\begin{array}{llll}
(\lambda, \rhd a_1 a_2 \dots a_n \lhd, \lambda) & \vdash_T & (\rhd, a_1 a_2 \dots a_n \lhd, \lambda) & \vdash_T^{n-2} \\
(\rhd a_1 \dots a_{n-2}, a_{n-1} a_n \lhd, \lambda) & \vdash_T & (\lambda, \rhd a_1 \dots a_{n-1} a_n' \lhd, a_n) & \vdash_T^{n-2} \\
(\rhd a_1 \dots a_{n-3}, a_{n-2} a_{n-1} a_n' \lhd, a_n) & \vdash_T & (\lambda, \rhd a_1 \dots a_{n-2} a_{n-1}' a_n' \lhd, a_n a_{n-1}) & \vdash_T^* \\
(\lambda, \rhd a_1 a_2' \dots a_n' \lhd, a_n \dots a_2) & \vdash_T & (\lambda, \rhd a_1' a_2' \dots a_n' \lhd, a_n \dots a_2 a_1) & \vdash_T \\
(\mathsf{Accept}, a_n \dots a_2 a_1) & = & (\mathsf{Accept}, w^R),
\end{array}
$$

which shows that T computes the function $^R : \Sigma^* \to \Delta^*$ that sends a word w to its mirror image w^R.

Observe that the stl-det-ORWW-transducer in the example above is *proper*, and that the relation $\{ (w, w^R) \mid w \in \{a, b\}^* \}$ is not rational.

Example 11. Let $\tau_1 : \{a\}^* \to \{b, c\}^*$ be defined by $\tau_1(a^{2n}) = b^{2n}$ and $\tau_1(a^{2n+1}) = c^{2n+1}$ for all $n \geq 0$. In [4] it is shown that τ_1 is computed by a proper monotone deterministic nf-RR(1)-transducer, but that it cannot be computed by any det-RR(1)-, nor by any R(1)-, nor by any monotone deterministic nf-R(1)-transducer. However, τ_1 is computed by the following proper stl-det-ORWW-transducer $T_1 = (\{a\}, \Gamma, \{b, c\}, \rhd, \lhd, \delta, >)$, where $\Gamma = \{a, a_0, a_1, b_2, c_2\}$, the ordering $>$ is given through $a > a_0 > a_1 > b_2$ and $a_1 > c_2$, and the transition function δ is defined as follows for all $i \in \{0, 1\}$:

$$\delta(\rhd\lhd) = (\mathsf{Accept}, \lambda), \ \delta(\rhd a\lhd) = (c_2, c), \ \delta(\rhd c_2\lhd) = (\mathsf{Accept}, \lambda),$$

$$\delta(\triangleright aa) = \mathsf{MVR}, \qquad \delta(aaa) = \mathsf{MVR}, \qquad \delta(aa\triangleleft) = (a_1, \lambda),$$
$$\delta(aaa_1) = (a_0, \lambda), \qquad \delta(aaa_0) = (a_1, \lambda), \qquad \delta(\triangleright aa_1) = (a_0, \lambda),$$
$$\delta(\triangleright aa_0) = (a_1, \lambda), \qquad \delta(\triangleright a_0 a_1) = (b_2, b), \qquad \delta(\triangleright a_1 a_0) = (c_2, c),$$
$$\delta(\triangleright b_2 a_1) = \mathsf{MVR}, \qquad \delta(\triangleright c_2 a_0) = \mathsf{MVR}, \qquad \delta(b_2 a_i a_{1-i}) = (b_2, b),$$
$$\delta(\triangleright b_2 b_2) = \mathsf{MVR}, \qquad \delta(b_2 b_2 b_2) = \mathsf{MVR}, \qquad \delta(b_2 b_2 a_i) = \mathsf{MVR},$$
$$\delta(b_2 a_1 \triangleleft) = (b_2, b), \qquad \delta(c_2 a_i a_{1-i}) = (c_2, c), \qquad \delta(\triangleright c_2 c_2) = \mathsf{MVR},$$
$$\delta(c_2 c_2 c_2) = \mathsf{MVR}, \qquad \delta(c_2 c_2 a_i) = \mathsf{MVR}, \qquad \delta(c_2 a_1 \triangleleft) = (c_2, c),$$
$$\delta(b_2 b_2 \triangleleft) = (\mathsf{Accept}, \lambda), \quad \delta(c_2 c_2 \triangleleft) = (\mathsf{Accept}, \lambda).$$

Given $w = a^n$ as input, T_1 first rewrites a^n into the word $a_0 a_1 a_0 a_1 \ldots a_0 a_1$, if n is even, and it rewrites w into the word $a_1 a_0 a_1 \ldots a_0 a_1$, if n is odd. Then T_1 rewrites this word again, letter by letter, from left to right. If the first letter is a_0, then T_1 replaces each letter by b_2, producing an output letter b for each such rewrite, and if the first letter is a_1, then T_1 replaces each letter by c_2, producing an output letter c for each such rewrite. It follows that T_1 does indeed compute the function τ_1.

Actually, stateless det-ORWW-transducers are just as expressive as det-ORWW-transducers, as shown by the following result, which can be shown by an adaptation of the proof of Theorem 2 of [8].

Theorem 12. *From a given det-ORWW-transducer T one can effectively construct a stl-det-ORWW-transducer T' that computes the same function as T.*

Thus, even the stl-det-ORWW-transducers are quite expressive. In fact, we have the following result on their expressive power, where RatF denotes the class of *rational functions*. These functions form an important subclass of RatRel as they are the rational relations that are partial functions. By RatF_0 we denote the class of all rational functions τ that satisfy the condition that $\tau(\lambda) = \lambda$. To prove our next result we need a characterization of this class that is due to Elgot and Mezei [3]. For stating it we introduce the following notions.

A *(left) sequential transducer* is a rational transducer $T = (Q, \Sigma, \Delta, q_0, Q, E)$ for which $E \subset Q \times \Sigma \times \Delta^* \times Q$ is a partial function from $Q \times \Sigma$ into $\Delta^* \times Q$. Observe that all states of a sequential transducer are final, and that in each step it reads a single symbol. Then the relation $Rel(T)$ is obviously a partial function. It is called a *(left) sequential function*, and by SeqF we denote the class of all sequential functions. By definition $(\lambda, \lambda) \in Rel(T)$ holds for each sequential transducer T. It is well known that SeqF is a proper subclass of RatF_0. Finally, a *(right) sequential transducer* is defined like a left sequential transducer that, however, processes its input from right to left, and that, accordingly, also produces its output from right to left. The relation computed by a right sequential transducer is called a *right sequential function*. According to [3], a partial function $\tau : \Sigma^* \to \Delta^*$ with $\tau(\lambda) = \lambda$ is a rational function if and only if there are a left sequential function $\rho : \Sigma^* \to \Theta^*$ that is total and length-preserving and a right sequential function $\sigma : \Theta^* \to \Gamma^*$ such that $\tau = \sigma \circ \rho$ (see also [1]).

Theorem 13. *Each rational function is computed by a stl-det-ORWW-transducer.*

Proof. Let $\tau : \Sigma^* \to \Delta^*$ be a rational function such that $\tau(\lambda) = \lambda$. Then by the result of Elgot and Mezei mentioned above, there exists a (left) sequential transducer $T_l = (Q, \Sigma, \Theta, q_0, Q, E)$ that computes a total and length-preserving function $\rho : \Sigma^* \to \Theta^*$, and there exists a right sequential transducer $T_r = (P, \Theta, \Delta, p_0, P, F)$ that computes a partial function $\sigma : \Theta^* \to \Delta^*$ such that $\tau = \sigma \circ \rho$. As ρ is total and length-preserving, we see that, for each $q \in Q$ and each letter $a \in \Sigma$, E contains a unique 4-tuple of the form (q, a, b, q') for some letter $b \in \Theta$ and a state $q' \in Q$. The transducer T_l reads a given input $w = a_1 a_2 \ldots a_n$ from Σ^+ letter by letter, from left to right, and it outputs a word $z = b_1 b_2 \ldots b_n = \rho(w)$ from Θ^+ in this process. Then the transducer T_r is given $z = b_1 b_2 \ldots b_n$ as input, and it reads z letter by letter from right to left, generating an output word $v = v_n \ldots v_2 v_1 \in \Delta^*$ from right to left. If T_r succeeds, then $v = \sigma(z) = \tau(w)$, otherwise, that is, if T_r gets stuck while reading z from right to left, $\sigma(z)$ is undefined, and accordingly, so is $\tau(w)$.

We now define a stl-det-ORWW-transducer $M = (\Sigma, \Gamma, \Delta, \rhd, \lhd, \delta, >)$ as follows, where $V = \{ v \in \Delta^* \mid \exists p, p' \in P, b \in \Theta : (p, b, v, p') \in F \}$:

- $\Gamma = \Sigma \cup \{ [q, b] \mid q \in Q, b \in \Theta \} \cup \{ [p, v], [p, v]' \mid p \in P, v \in V \}$, where we assume (w.l.o.g.) that P and Q are disjoint,
- the ordering $>$ is given through $a > [q, b] > [p, v] > [p, v]'$ for all $a \in \Sigma$, $q \in Q, b \in \Theta, p \in P$, and $v \in V$, and
- the transition function δ is defined as follows, where $a, a_1, a_2 \in \Sigma$, $q, q_1, q_2 \in Q, b, b_1, b_2 \in \Theta, p, p_1 \in P$, and $v, v_1, v_2 \in V$:

$$\delta(\rhd \lhd) = (\mathsf{Accept}, \lambda),$$
$$\delta(\rhd a \lhd) = ([q, b], \lambda), \quad \text{if } (q_0, a, b, q) \in E,$$
$$\delta(\rhd a a_1) = ([q, b], \lambda), \quad \text{if } (q_0, a, b, q) \in E,$$
$$\delta(\rhd [q, b] a) = \mathsf{MVR},$$
$$\delta([q, b] a_1 a_2) = ([q_1, b_1], \lambda), \quad \text{if } (q, a_1, b_1, q_1) \in E,$$
$$\delta(\rhd [q_1, b_1][q_2, b_2]) = \mathsf{MVR},$$
$$\delta([q, b][q_1, b_1][q_2, b_2]) = \mathsf{MVR},$$
$$\delta([q_1, b_1][q_2, b_2] a) = \mathsf{MVR},$$
$$\delta([q, b] a \lhd) = ([q_1, b_1], \lambda), \quad \text{if } (q, a, b_1, q_1) \in E,$$
$$\delta(\rhd [q, b] \lhd) = ([p, v], \lambda), \quad \text{if } (p_0, b, v, p) \in F,$$
$$\delta(\rhd [p, v] \lhd) = ([p, v]', v),$$
$$\delta(\rhd [p, v]' \lhd) = (\mathsf{Accept}, \lambda),$$
$$\delta([q_1, b_1][q, b] \lhd) = ([p, v], \lambda), \quad \text{if } (p_0, b, v, p) \in F,$$
$$\delta([q_1, b_1][q, b][p, v]) = ([p_1, v_1], \lambda), \quad \text{if } (p, b, v_1, p_1) \in F,$$
$$\delta(\rhd [q, b][p, v]) = ([p_1, v_1], \lambda), \quad \text{if } (p, b, v_1, p_1) \in F,$$
$$\delta(\rhd [p, v][p_1, v_1]) = ([p, v]', v),$$
$$\delta(\rhd [p, v]'[p_1, v_1]) = \mathsf{MVR},$$
$$\delta(\rhd [p, v]'[p_1, v_1]') = \mathsf{MVR},$$
$$\delta([p, v]'[p_1, v_1][p_2, v_2]) = ([p_1, v_1]', v_1),$$
$$\delta([p, v]'[p_1, v_1]'[p_2, v_2]) = \mathsf{MVR},$$
$$\delta([p, v]'[p_1, v_1]'[p_2, v_2]') = \mathsf{MVR},$$

$$\delta([p, v]'[p_1, v_1] \triangleleft) = ([p_1, v_1]', v_1),$$
$$\delta([p, v]'[p_1, v_1]' \triangleleft) = (\mathsf{Accept}, \lambda).$$

Given a word $w = a_1 a_2 \ldots a_n \in \Sigma^n$, where $n \geq 1$, M first rewrites w from left to right into the word $[q_1, b_1][q_2, b_2] \ldots [q_n, b_n]$, where $(q_{i-1}, a_i, b_i, q_i) \in E$ for all $i = 1, \ldots, n$, that is, M simulates the left sequential transducer T_l on input w, replacing each input letter a_i by the pair $[q_i, b_i]$, where q_i is the state that T_l reaches after reading the prefix $a_1 \ldots a_i$ of w, and b_i is the letter that T_l outputs during this computation when it reads the letter a_i. Thus, $b_1 b_2 \ldots b_n = \rho(w)$.

Then M rewrites the word $[q_1, b_1][q_2, b_2] \ldots [q_n, b_n]$ from right to left into the word $[p_n, v_n] \ldots [p_2, v_2][p_1, v_1]$, where $(p_{i-1}, b_{n+1-i}, v_i, p_i) \in F$ for all $i = 1, \ldots, n$, that is, M simulates the right sequential transducer T_r on input $\rho(w)$, and M succeeds if and only if T_r can read $\rho(w)$ completely. It follows that $v_n \ldots v_2 v_1 = \sigma(b_1 b_2 \ldots b_n) = \sigma(\rho(w)) = \tau(w)$. Finally, M rewrites the word $[p_n, v_n] \ldots [p_2, v_2][p_1, v_1]$ letter by letter, from left to right, into the word $[p_n, v_n]' \ldots [p_2, v_2]'[p_1, v_1]'$ thereby producing the output $v_n \ldots v_2 v_1 = \tau(w)$. Thus, M is a proper stl-det-ORWW-transducer that computes the function τ.

Finally, if $\tau(\lambda) = w \neq \lambda$, then we can simply change M by using the transition $\delta(\triangleright \triangleleft) = (\mathsf{Accept}, w)$. In this case, M is obviously not proper. □

5   Transductions Computed by det-ORWW-Automata

Finally, we use the rewriting process of a det-ORWW-automaton itself to obtain a transduction.

Definition 14. *Let $M = (Q, \Sigma, \Gamma, \triangleright, \triangleleft, q_0, \delta, >)$ be a det-ORWW-automaton, let Δ be a finite output alphabet, and let $\varphi : \Gamma^* \to \Delta^*$ be a morphism. For $w \in L(M)$, let $\hat{w}$ denote the tape inscription that M produces during its accepting computation on input w, that is, this accepting computation has the following form:*

$$(\lambda, q_0 \triangleright w \triangleleft) \vdash_M^{c^*} (\lambda, q_0 \triangleright \hat{w} \triangleleft) \vdash_M^* (\triangleright \hat{u}, q \hat{v} \triangleleft) \vdash_M \mathsf{Accept},$$

where $\hat{w} = \hat{u}\hat{v}$ holds. Then we associate with w the output word $\varphi(\hat{w})$, in this way defining a transduction $\varphi_M : L(M) \to \Delta^*$. We say that φ_M is the transduction that is defined by the pair (M, φ).

By $\mathcal{F}(\text{det-ORWW})$ we denote the class of all partial functions that can be realized by det-ORWW-automata (and morphisms) in the way described above, and by $\mathcal{F}(\text{stl-det-ORWW})$ we denote the class of all partial functions that can be realized by stl-det-ORWW-automata (and morphisms).

Example 15. By removing the output part from the transition function of the stl-det-ORWW-transducer T_1 in Example 11, we obtain a stl-det-ORWW-automaton M_1 such that $L(M_1) = a^*$. In fact, for each $n \geq 1$, M_2 rewrites a^n into b_2^n (if n is even) or into c_2^n (if n is odd).

Now let $\varphi_1 : \{b_2, c_2\}^* \to \{b, c\}^*$ be the morphism that is given though $b_2 \mapsto b$ and $c_2 \mapsto c$. Then $\varphi_{M_1}(a^n) = \tau_1(a^n)$ for all $n \geq 0$.

Concerning the functions that are computed by det-ORWW-automata with the aid of morphisms we have the following characterization.

Theorem 16. $\mathsf{RatF}_0 = \mathcal{F}(\text{stl-det-ORWW}) = \mathcal{F}(\text{det-ORWW})$.

The proof will be split into three lemmas.

Lemma 17. $\mathsf{RatF}_0 \subseteq \mathcal{F}(\text{stl-det-ORWW})$.

Proof. If $\tau : \Sigma^* \to \Delta^*$ is a rational function such that $\tau(\lambda) = \lambda$, then we can construct a stl-det-ORWW-automaton M that rewrites a given input word $w = a_1 a_2 \ldots a_n$ into the word $[p_n, v_n]' \ldots [p_2, v_2]'[p_1, v_1]'$, where $\tau(w) = v_n \ldots v_2 v_1$ (see the proof of Theorem 13). Now we define a morphism φ as follows:

$$\varphi([p, v]') = v \text{ for all } p \in P \text{ and } v \in V,$$
$$\varphi(A) = \lambda \text{ for all other letters.}$$

Then it follows immediately that the pair (M, φ) defines the function τ.   □

Lemma 18. $\mathcal{F}(\text{stl-det-ORWW}) \subseteq \mathsf{RatF}_0$.

Proof. Let $M = (\Sigma, \Gamma, \rhd, \lhd, \delta, >)$ be a stl-det-ORWW-automaton, let $\varphi : \Gamma^* \to \Delta^*$ be a morphism, and let $\tau : \Sigma^* \to \Delta^*$ be the transduction that is defined by the pair (M, φ). Then τ is a partial function, and by definition, $\tau(\lambda) = \lambda$ holds.

In [6] an algorithm is described that turns the stl-det-ORWW-automaton M into an equivalent unambiguous NFA $B = (Q, \Sigma, \Delta_B, q_0, F)$, where the set of states Q contains the initial state q_0, a designated final state q_F, and pairs of triples of the form $((L_1, W_1, R_1), (L_2, W_2, R_2))$, where, for $i = 1, 2$,

- W_i is a sequence of letters $W_i = (w_{i,1}, \ldots, w_{i,k_i})$ from Γ of length $1 \leq k_i \leq n$ such that $w_{i,1} > w_{i,2} > \cdots > w_{i,k_i}$, or $W_i = (\rhd)$ and $k_i = 1$,
- L_i is a sequence of positive integers $L_i = (l_{i,1}, \ldots, l_{i,k_i-1})$ of length $k_i - 1$ such that $l_{i,1} \leq l_{i,2} \leq \cdots \leq l_{i,k_i-1} \leq n$,
- R_i is a sequence of positive integers $R_i = (r_{i,1}, \ldots, r_{i,k_i-1})$ of length $k_i - 1$ such that $r_{i,1} \leq r_{i,2} \leq \cdots \leq r_{i,k_i-1} \leq n$,
- the sequences R_1 and L_2 are consistent, that is, $\text{order}(R_1, L_2) = \{1, 2, \ldots, k_1 + k_2 - 2\}$.

Given a word $w \in \Sigma^*$ as input, B guesses the corresponding states in such a way that the triples describe an accepting computation of M on input w. B is nondeterministic, but it is unambiguous as, for each word $w \in L(B) = L(M)$, it has only a single accepting computation. If in this computation, B enters the state $((\Lambda, (\rhd), \Lambda), (L_1, W_1, R_1))$ from its initial state q_0 on reading the first input letter a_1, then $W_1 = (a_1, A_2, \ldots, A_k)$ encodes the information that during the accepting computation of M on input w, the first letter is first rewritten into A_2, then later into A_3, and so on, and that A_k is the final letter in this position when M accepts. Accordingly, we obtain a transducer from B by simply adding the output component $\varphi(A_n)$ to the above transition of B. Analogously, if $\Delta_B\left(((L_1, W_1, R_1), (L_2, W_2, R_2)), x\right) \ni ((L_2, W_2, R_2), (L_3, W_3, R_3))$, where $x \in \Sigma$ and $W_3 = (a, C_2, \ldots, C_r)$, then we add the output component $\varphi(C_r)$ to this particular transition. Finally, to the final transitions of the form $q_F \in \Delta_B(((L_1, W_1, R_1), (L_2, W_2, R_2)), \lambda)$, we add the output component λ. Then it is easily seen that, on input $w \in \Sigma^*$, the transducer obtained from B accepts if and only if $w \in L(M)$, and in this case it produces the output $\tau(w)$. Thus, τ is a rational relation, and so, it is a rational function from RatF_0.   □

Lemmas 17 and 18 together show that the stl-det-ORWW-automata realize exactly the rational functions from RatF_0. In order to complete the proof of Theorem 16 it suffices to derive the following result (proof omitted).

Lemma 19. *For each det-ORWW-automaton M and each morphism φ, there exist a stateless det-ORWW-automaton M' and a morphism φ' such that the pair (M', φ') defines the same transduction as (M, φ).*

Let $\Sigma = \{0, 1, \#, \$\}$. For $n \geq 3$, let B_n be the following regular language:

$$B_n = \{\, v_1 \# v_2 \# \ldots \# v_m \$ u \mid m \geq 1,\ v_1, \ldots, v_m, u \in \{0,1\}^n,\ \exists i : v_i = u \,\}.$$

It can easily be shown that every NFA for B_n has at least 2^n states, and every DFA for B_n has at least 2^{2^n} states. On the other hand, B_n is accepted by a stl-det-ORWW-automaton with $O(n)$ letters [8]. Now let $\hat{B}_n$ be the function

$$\hat{B}_n = \{\, (v_1 \# \ldots \# v_m \$ u, v_{i_1} \ldots v_{i_r}) \mid v_1 \# \ldots \# v_m \$ u \in B_n \text{ and}$$
$$v_{i_j} \neq u, 1 \leq i_1 < \cdots < i_r \leq m \,\},$$

that is, the function $\hat{B}_n$ outputs the sequence of factors from v_1 to v_m that differ from the last factor u. In [9] a stl-det-ORWW-automaton M_n for the language B_n is presented that is of size $O(n)$ and that is reversible, which means that it also has a reverse transition function that can be used to undo cycles. In particular, this means that M_n keeps the information on the input letters encoded in the new letters whenever it performs a rewrite. At the end of an accepting computation, all syllables v_i for which $v_i = u$ holds are marked in one way (by a $+$), and the syllables v_j for which $v_j \neq u$ are marked in a different way (by a $-$). Now we can define a morphism that maps each letter containing the sign $-$ to its corresponding input letter (which is encoded in this auxiliary letter) and by mapping all other letters to λ. As each rational transducer for computing this relation entails an NFA for the language B_n, it is of size at least 2^n.

Proposition 20. *The function $\hat{B}_n$ is computed by a pair (M_n, φ), where M_n is a stl-det-ORWW-automaton that has a working alphabet of size $O(n)$, while each rational transducer for computing this function has at least 2^n states.*

On the other hand, it can be shown that from a pair (M, φ) computing a transduction τ one can effectively construct a rational transducer for τ that is of size $2^{O(n)}$ (see the proof of Lemma 18). Thus, we have an exponential trade-off for turning a stl-det-ORWW-automaton into a rational transducer, and this bound is sharp as witnessed by the functions $\hat{B}_n$.

6   Conclusion

We have presented three different ways of computing relations (or functions) by det-ORWW-automata. By using the input-output relations of these automata, we obtain succinct representations of all rational relations, but combining a det-ORWW-automaton with a morphism, we obtain succinct representations of all rational functions that map the empty word to itself, and by extending the det-ORWW-automaton into a transducer, we obtain a device that computes all rational functions and even some non-rational functions. For future work the following open problems remain:

– Which functions can be computed by det-ORWW-transducers?
– What is the descriptional complexity of operations on rational relations when expressed in terms of input-output relations of (stl-) det-ORWW-automata?
– What is the descriptional complexity of various operations on rational functions in terms of (stl-) det-ORWW-automata that compute them?

References

1. Berstel, J.: Transductions and Context-Free Languages. Teubner, Stuttgart (1979)
2. Choffrut, C., Culik II, K.: Properties of finite and pushdown transducers. SIAM J. Comput. **12**, 300–315 (1983)
3. Elgot, C.C., Mezei, G.: On relations defined by generalized finite automata. IBM Journal of Research and Development **9**, 47–65 (1965)
4. Hundeshagen, N.: Relations and Transductions Realized by RestartingAutomata. PhD thesis, Fachbereich Elektrotechnik/Informatik, Universität Kassel (2013)
5. Hundeshagen, N., Otto, F.: Characterizing the rational functions by restarting transducers. In: Dediu, A.-H., Martín-Vide, C. (eds.) LATA 2012. LNCS, vol. 7183, pp. 325–336. Springer, Heidelberg (2012)
6. Kwee, K., Otto, F.: On some decision problems for stateless deterministic ordered restarting automata. In: Shallit, J., Okhotin, A. (eds.) DCFS 2015. LNCS, vol. 9118, pp. 165–176. Springer, Heidelberg (2015)
7. Mráz, F., Otto, F.: Ordered restarting automata for picture languages. In: Geffert, V., Preneel, B., Rovan, B., Štuller, J., Tjoa, A.M. (eds.) SOFSEM 2014. LNCS, vol. 8327, pp. 431–442. Springer, Heidelberg (2014)
8. Otto, F.: On the descriptional complexity of deterministic ordered restarting automata. In: Jürgensen, H., Karhumäki, J., Okhotin, A. (eds.) DCFS 2014. LNCS, vol. 8614, pp. 318–329. Springer, Heidelberg (2014)
9. Otto, F., Wendlandt, M., Kwee, K.: Reversible ordered restarting automata. In: Krevine, J., Stefani, J.-B. (eds.) RC 2015. LNCS, vol. 9138, pp. 60–75. Springer, Heidelberg (2015)
10. Průša, D.: Weight-reducing Hennie machines and their descriptional complexity. In: Dediu, A.-H., Martín-Vide, C., Sierra-Rodríguez, J.-L., Truthe, B. (eds.) LATA 2014. LNCS, vol. 8370, pp. 553–564. Springer, Heidelberg (2014)

Weight Assignment Logic

Vitaly Perevoshchikov[(✉)]

Institut für Informatik, Universität Leipzig, 04109 Leipzig, Germany
perev@informatik.uni-leipzig.de

Abstract. We introduce a weight assignment logic for reasoning about quantitative languages of infinite words. This logic is an extension of the classical MSO logic and permits to describe quantitative properties of systems with multiple weight parameters, e.g., the ratio between rewards and costs. We show that this logic is expressively equivalent to unambiguous weighted Büchi automata. We also consider an extension of weight assignment logic which is expressively equivalent to nondeterministic weighted Büchi automata.

Keywords: Quantitative omega-languages · Quantitative logic · Multi-weighted automata · Büchi automata · Unambiguous automata

1 Introduction

Since the seminal Büchi theorem [5] about the expressive equivalence of finite automata and monadic second-order logic, a significant field of research investigates logical characterizations of language classes appearing from practically relevant automata models. In this paper we introduce a new approach to the logical characterization of quantitative languages of infinite words where every infinite word carries a value, e.g., a real number.

Quantitative languages of infinite words and various weighted automata for them were investigated by Chatterjee, Doyen and Henzinger in [7] as models for verification of quantitative properties of systems. Their weighted automata are automata with a single weight parameter where a computation is evaluated using measures like the limit average or discounted sum. Recently, the problem of analysis and verification of systems with multiple weight parameters, e.g. time, costs and energy consumption, has received much attention in the literature [3,5,16,17,20]. For instance, the setting where a computation is evaluated as the ratio between accumulated rewards and costs was considered in [3,5,17]. Another example is a model of energy automata with several energy storages [16].

Related Work. Droste and Gastin [9] introduced weighted MSO logic on finite words with constants from a semiring. In the semantics of their logic (which is a quantitative language of finite words) disjunction is extended by the sum operation of the semiring and conjunction is extended by the product. They show

---

Supported by DFG Research Training Group 1763 (QuantLA).

I. Potapov (Ed.): DLT 2015, LNCS 9168, pp. 413–425, 2015.
DOI: 10.1007/978-3-319-21500-6_33

that weighted MSO logic is more expressive than weighted automata [10] (the unrestricted use of weighted conjunction and weighted universal quantifiers leads to unrecognizability) and provide a syntactically restricted fragment which is expressively equivalent to weighted automata. This result was extended in [15] to the setting of infinite words. A logical characterization of the quantitative languages of Chatterjee, Doyen and Henzinger was given in [12] (again by a restricted fragment of weighted MSO logic). In [13], a multi-weighted extension of weighted MSO logic of [12] with the multiset-based semantics was considered.

Our Contribution. In this paper, we introduce a new approach to logic for quantitative languages, different from [9,12,13,15]. We develop a so-called *weight assignment logic* (WAL) on infinite words, an extension of the classical MSO logic to the quantitative setting. This logic allows us to assign weights (or multi-weights) to positions of an ω-word. Using WAL, we can, for instance, express that whenever a position of an input word is labelled by letter a, then the weight of this position is 2. As a weighted extension of the logical conjunction, we use the *merging* of partially defined ω-words. In order to evaluate a partially defined ω-word, we introduce a *default weight*, assign it to all positions with undefined weight, and evaluate the obtained totally defined ω-word, e.g., as the reward-cost ratio or discounted sum.

As opposed to the weighted MSO logic of [9], the weighted conjunction-like operators of WAL capture recognizability by weighted Büchi automata. We show that WAL is expressively equivalent to *unambiguous* weighted Büchi automata where, for every input ω-word, there exists at most one accepting computation. Unambiguous automata are of considerable interest for automata theory as they can have better decidability properties. For instance, in the setting of finite words, the equivalence problem for unambiguous max-plus automata is decidable [18] whereas, for nondeterministic max-plus automata, this problem is undecidable [19].

We also consider an extended version of WAL which captures nondeterministic weighted Büchi automata. In extended WAL we allow existential quantification over first-order and second-order variables in the prefix of a formula. The structure of extended WAL is similar to the structure of unweighted logics for, e.g., timed automata [22] and data automata [4].

For the proof of our expressiveness equivalence result, we establish a Nivat decomposition theorem for nondeterministic and unambiguous weighted Büchi automata. Recall that Nivat's theorem [21] is one of the fundamental characterizations of rational transductions and shows a connection between rational transductions and rational language. Recently, Nivat's theorem was proved for semiring-weighted automata on finite words [11], weighted multioperator tree automata [22] and weighted timed automata [14]. We obtain similar decompositions for WAL and extended WAL and deduce our results from the classical Büchi theorem [6]. Our proof is constructive and hence decidability properties for WAL and extended WAL can be transferred into decidability properties of weighted Büchi automata. As a side application of our Nivat theorem, we can easily show that weighted Büchi automata and weighted Muller automata are expressively equivalent.

Outline. In Sect. 2 we introduce a general framework for weighted Büchi automata and consider several examples. In Sect. 3 we prove a Nivat decomposition theorem for weighted Büchi automata. In Sect. 4 we define weight assignment logic and its extension. In Sect. 5 we state our main result and give a sketch of its proof for the unambiguous and nondeterministic cases.

2   Weighted Büchi Automata

Let $\mathbb{N} = \{0, 1, ...\}$ denote the set of all natural numbers. For an arbitrary set X, an ω-*word* over X is an infinite sequence $(x_i)_{i \in \mathbb{N}}$ where $x_i \in X$ for all $i \in \mathbb{N}$. Let X^ω denote the set of all ω-words over X. Any set $\mathcal{L} \subseteq X^\omega$ is called an ω-*language* over X.

A *Büchi automaton* over an alphabet Σ is a tuple $\mathcal{A} = (Q, I, T, F)$ where Q is a finite set of states, Σ is an alphabet (i.e. a finite non-empty set), $I, F \subseteq Q$ are sets of initial resp. accepting states, and $T \subseteq Q \times \Sigma \times Q$ is a transition relation. An (accepting) *run* $\rho = (t_i)_{i \in \mathbb{N}} \in T^\omega$ of $\mathcal{A}$ is defined as an infinite sequence of matching transitions which starts in an initial state and visits some accepting state infinitely often, i.e., $t_i = (q_i, a_i, q_{i+1})$ for each $i \in \mathbb{N}$, such that $q_0 \in I$ and $\{q \in Q \mid q = q_i$ for infinitely many $i \in \mathbb{N}\} \cap F \neq \emptyset$. Let $\mathrm{label}(\rho) := (a_i)_{i \in \mathbb{N}} \in \Sigma^\omega$, the *label* of ρ. We denote by $\mathrm{Run}_\mathcal{A}$ the set of all runs of $\mathcal{A}$ and, for each $w \in \Sigma^\omega$, we denote by $\mathrm{Run}_\mathcal{A}(w)$ the set of all runs ρ of $\mathcal{A}$ with $\mathrm{label}(\rho) = w$. Let $\mathcal{L}(\mathcal{A}) = \{w \in \Sigma^\omega \mid \mathrm{Run}_\mathcal{A}(w) \neq \emptyset\}$, the ω-language *accepted* by $\mathcal{A}$. We call an ω-language $\mathcal{L} \subseteq \Sigma^\omega$ *recognizable* if there exists a Büchi automaton $\mathcal{A}$ over Σ such that $\mathcal{L}(\mathcal{A}) = \mathcal{L}$.

We say that a monoid $\mathbb{K} = (K, +, \mathbb{0})$ is *complete* (cf., e.g., [15]) if it is equipped with infinitary sum operations $\sum_I : K^I \to K$ for any index set I, such that, for all I and all families $(k_i)_{i \in I}$ of elements of K, the following hold:

- $\sum_{i \in \emptyset} k_i = \mathbb{0}$, $\sum_{i \in \{j\}} k_i = k_j$, $\sum_{i \in \{p,q\}} k_i = k_p + k_q$ for $p \neq q$;
- $\sum_{j \in J}(\sum_{i \in I_j} k_i) = \sum_{i \in I} k_i$, if $\bigcup_{j \in J} I_j = I$ and $I_j \cap I_{j'} = \emptyset$ for $j \neq j'$.

Let $\overline{\mathbb{R}} = \mathbb{R} \cup \{-\infty, \infty\}$. Then, $\overline{\mathbb{R}}$ equipped with infinitary operations like *infimum* or *supremum* forms a complete monoid. Now we introduce an algebraic structure for weighted Büchi automata which is an extension of totally complete semirings [15] and valuation monoids [12] and covers various multi-weighted measures.

Definition 2.1. *A* valuation structure $\mathbb{V} = (M, \mathbb{K}, \mathrm{val})$ *consists of a non-empty set M, a complete monoid $\mathbb{K} = (K, +, \mathbb{0})$ and a mapping* $\mathrm{val} : M^\omega \to K$ *called henceforth a* valuation function.

In the definition of a valuation structure we have two weight domains M and K. Here M is the set of transition weights which in the multi-weighted examples can be tuples of weights (e.g., a reward-cost pair) and K is the set of weights of computations which can be single values (e.g., the ratio between rewards and costs).

Definition 2.2. *Let Σ be an alphabet and $\mathbb{V} = (M, (K, +, \mathbb{0}), \mathrm{val})$ a valuation structure. A* weighted Büchi automaton *(WBA) over $\mathbb{V}$ is a tuple $\mathcal{A} = (Q, I, T, F, \mathrm{wt})$ where (Q, I, T, F) is a Büchi automaton over Σ and $\mathrm{wt} : T \to M$ is a* transition weight function.

The behavior of WBA is defined as follows. Given a run ρ of this automaton, we evaluate the ω-sequence of transition weights of ρ (which is in M^ω) using the valuation function val and then resolve the nondeterminism on the weights of runs using the complete monoid $\mathbb{K}$. Formally, let $\rho = (t_i)_{i \in \mathbb{N}} \in T^\omega$ be a run of $\mathcal{A}$. Then, the *weight* of ρ is defined as $\mathrm{wt}_\mathcal{A}(\rho) = \mathrm{val}((\mathrm{wt}(t_i))_{i \in \mathbb{N}}) \in K$. The *behavior* of $\mathcal{A}$ is a mapping $[\![\mathcal{A}]\!] : \Sigma^\omega \to K$ defined for all $w \in \Sigma^\omega$ by $[\![\mathcal{A}]\!](w) = \sum(\mathrm{wt}_\mathcal{A}(\rho) \mid \rho \in \mathrm{Run}_\mathcal{A}(w))$. Note that the sum in the equation above can be infinite. Therefore we consider a complete monoid $(K, +, \mathbb{0})$. A mapping $\mathbb{L} : \Sigma^\omega \to K$ is called a *quantitative ω-language*. We say that $\mathbb{L}$ is (nondeterministically) *recognizable* over $\mathbb{V}$ if there exists a WBA $\mathcal{A}$ over Σ and $\mathbb{V}$ such that $[\![\mathcal{A}]\!] = \mathbb{L}$.

We say that a WBA $\mathcal{A}$ over Σ and $\mathbb{V}$ is *unambiguous* if $|\mathrm{Run}_\mathcal{A}(w)| \leq 1$ for every $w \in \Sigma^\omega$. We call a quantitative ω-language $\mathbb{L} : \Sigma^\omega \to K$ *unambiguously recognizable* over $\mathbb{V}$ if there exists an unambiguous WBA $\mathcal{A}$ over Σ and $\mathbb{V}$ such that $[\![\mathcal{A}]\!] = \mathbb{L}$.

Example 2.3. (a) The ratio measure was introduced in [5], e.g., for the modeling of the average costs in timed systems. In the setting of ω-words, we consider the model with two weight parameters: the *cost* and the *reward*. The rewards and costs of transitions are accumulated along every finite prefix of a run and their ratio is taken. Then, the weight of an infinite run is defined as the limit superior (or limit inferior) of the sequence of the computed ratios for all finite prefixes. To describe the behavior of these double-priced ratio Büchi automata, we consider the valuation structure $\mathbb{V}^{\mathrm{RATIO}} = (M, \mathbb{K}, \mathrm{val})$ where $M = \mathbb{Q} \times \mathbb{Q}_{\geq 0}$ models the reward-cost pairs, $\mathbb{K} = (\overline{\mathbb{R}}, \sup, -\infty)$ and $\mathrm{val} : M^\omega \to \overline{\mathbb{R}}$ is defined for every sequence $u = ((r_i, c_i))_{i \in \mathbb{N}} \in M^\omega$ by $\mathrm{val}(u) = \limsup_{n \to \infty} \frac{r_1 + \ldots + r_n}{c_1 + \ldots + c_n}$. Here, we assume that $\frac{r}{0} = -\infty$.

(b) *Discounting* [1,7] is a well-known principle which is used in, e.g., economics and psychology. In this example, we consider WBA with transition-dependent discounting, i.e., are two weight parameters: the cost and the discounting factor (which is not fixed and depends on a transition). In order to define WBA with discounting formally, we consider the valuation structure $\mathbb{V}^{\mathrm{DISC}} = (M, \mathbb{K}, \mathrm{val})$ where $M = \mathbb{Q}_{\geq 0} \times ((0, 1] \cap \mathbb{Q})$ models the pairs of a cost and a discounting factor, $\mathbb{K} = (\mathbb{R}_{\geq 0} \cup \{\infty\}, \inf, \infty)$, and val is defined for all $u = ((c_i, d_i))_{i \in \mathbb{N}} \in M^\omega$ as $\mathrm{val}(u) = c_0 + \sum_{i=1}^\infty c_i \cdot \prod_{j=0}^{i-1} d_j$.

(c) Since a valuation monoid $(K, (K, +, \mathbb{0}), \mathrm{val})$ of Droste and Meinecke [12] is a special case of valuation structures, all examples considered there also fit into our framework. □

3   Decomposition of WBA

In this section, we establish a Nivat decomposition theorem for WBA. We will need it for the proof of our main result. However, it also could be of independent interest.

Let Σ be an alphabet and $\mathbb{V} = (M, (K, +, \mathbb{0}), \mathrm{val})$ a valuation structure. For a (possibly different from Σ) alphabet Γ, we introduce the following operations. Let Δ be an arbitrary non-empty set and $h : \Gamma \to \Delta$ a mapping called henceforth a *renaming*. For any ω-word $u = (\gamma_i)_{i \in \mathbb{N}} \in \Gamma^{\omega}$, we let $h(u) = (h(\gamma_i))_{i \in \mathbb{N}} \in \Delta^{\omega}$. Now let $h : \Gamma \to \Sigma$ be a renaming and $\mathbb{L} : \Gamma^{\omega} \to K$ a quantitative ω-language. We define the *renaming* $h(\mathbb{L}) : \Sigma^{\omega} \to K$ for all $w \in \Sigma^{\omega}$ by $h(\mathbb{L})(w) = \sum (\mathbb{L}(u) \mid u \in \Gamma^{\omega} \text{ and } h(u) = w)$. For a renaming $g : \Gamma \to M$, the *composition* $\mathrm{val} \circ g : \Gamma^{\omega} \to K$ is defined for all $u \in \Gamma^{\omega}$ by $(\mathrm{val} \circ g)(u) = \mathrm{val}(g(u))$. Given a quantitative ω-language $\mathbb{L} : \Gamma^{\omega} \to K$ and an ω-language $\mathcal{L} \subseteq \Gamma^{\omega}$, the intersection $\mathbb{L} \cap \mathcal{L} : \Gamma^{\omega} \to K$ is defined for all $u \in \mathcal{L}$ as $(\mathbb{L} \cap \mathcal{L})(u) = \mathbb{L}(u)$ and for all $u \in \Gamma^{\omega} \setminus \mathcal{L}$ as $(\mathbb{L} \cap \mathcal{L})(u) = \mathbb{0}$. Given a renaming $h : \Gamma \to \Sigma$, we say that an ω-language $\mathcal{L} \subseteq \Gamma^{\omega}$ is *h-unambiguous* if for all $w \in \Sigma^{\omega}$ there exists at most one $u \in \mathcal{L}$ such that $h(u) = w$.

Our Nivat decomposition theorem for WBA is the following.

Theorem 3.1. *Let Σ be an alphabet, $\mathbb{V} = (M, (K, +, \mathbb{0}), \mathrm{val})$ a valuation structure, and $\mathbb{L} : \Sigma^{\omega} \to K$ a quantitative ω-language. Then*

(a) $\mathbb{L}$ *is unambiguously recognizable over $\mathbb{V}$ iff there exist an alphabet Γ, renamings $h : \Gamma \to \Sigma$ and $g : \Gamma \to M$, and a recognizable and h-unambiguous ω-language $\mathcal{L} \subseteq \Gamma^{\omega}$ such that $\mathbb{L} = h((\mathrm{val} \circ g) \cap \mathcal{L})$.*

(b) $\mathbb{L}$ *is nondeterministically recognizable over $\mathbb{V}$ iff there exist an alphabet Γ, renamings $h : \Gamma \to \Sigma$ and $g : \Gamma \to M$, and a recognizable ω-language $\mathcal{L} \subseteq \Gamma^{\omega}$ such that $\mathbb{L} = h((\mathrm{val} \circ g) \cap \mathcal{L})$.*

The proof idea is the following. To prove the recognizability of $h((\mathrm{val} \circ g) \cap \mathcal{L})$, one can show that recognizable quantitative ω-languages are closed under renaming, composition and intersection. For the converse direction, i.e., a decomposition of the behavior $[\![\mathcal{A}]\!]$ of a WBA $\mathcal{A}$, one can use a similar idea as in [11]. We let Γ be the set of all transitions of $\mathcal{A}$, h and g mappings assigning labels and weights, resp., to each transition and let $\mathcal{L}$ be the regular ω-language of words over Γ describing runs of $\mathcal{A}$.

Since Büchi automata are not determinizable, the most challenging part of the proof of Theorem 3.1 is to show that recognizable ω-languages are stable under intersection with ω-languages. To show this, we apply the result of [8] which states that every ω-recognizable language is accepted by an unambiguous Büchi automaton.

As a first application of Theorem 3.1 we show that WBA are equivalent to *weighted Muller automata* which are defined as WBA with the difference that a set of accepting states $F \subseteq Q$ is replaced by a set $\mathcal{F} \subseteq 2^{Q}$ of sets of accepting states. Then, for an accepting run ρ, the set of all states, which are visited in ρ infinitely often, must be in $\mathcal{F}$. Our expressiveness equivalence result extends the result of [15] for totally complete semirings. Whereas the proof of [15] was given by direct non-trivial automata transformation, our proof is based on the fact that weighted Muller automata permit the same decomposition as stated in Theorem 3.1 for WBA.

4   Weight Assignment Logic

4.1   Partial ω-words

Before we give a definition of the syntax and semantics of our new logic, we introduce some auxiliary notions about partial ω-words. Let X be an arbitrary non-empty set. A *partial ω-word* over X is a partial mapping $u : \mathbb{N} \dashrightarrow X$, i.e., $u : U \to X$ for some $U \subseteq \mathbb{N}$. Let $\text{dom}(u) = U$, the *domain* of u. We denote by $X^{\uparrow}$ the set of all partial ω-words over X. Clearly, $X^{\omega} \subseteq X^{\uparrow}$. A *trivial ω-word* $\top \in X^{\uparrow}$ is the partial ω-word with $\text{dom}(\top) = \emptyset$. For $u \in X^{\uparrow}$, $i \in \mathbb{N}$ and $x \in X$, the *update* $u[i/x] \in X^{\uparrow}$ is defined as $\text{dom}(u[i/x]) = \text{dom}(u) \cup \{i\}$, $u[i/x](i) = x$ and $u[i/x](i') = u(i')$ for all $i' \in \text{dom}(u) \setminus \{i\}$. Let $\theta = (u_j)_{j \in J}$ be an arbitrary family of partial ω-words $u_j \in X^{\uparrow}$ where J is an arbitrary index set. We say that θ is *compatible* if, for all $j, j' \in J$ and $i \in \text{dom}(u_j) \cap \text{dom}(u_{j'})$, we have $u_j(i) = u_{j'}(i)$. If θ is compatible, then we define the *merging* $u := (\bigsqcap_{j \in J} u_j) \in X^{\uparrow}$ as $\text{dom}(u) = \bigcup_{j \in J} \text{dom}(u_j)$ and, for all $i \in \text{dom}(u)$, $u(i) = u_j(i)$ whenever $i \in \text{dom}(u_j)$ for some $j \in J$. Let $\theta = \{u_j\}_{j \in \{1,2\}}$ be compatible. Then, we write $u_1 \uparrow u_2$. Clearly, the relation $\uparrow$ is reflexive and symmetric. In the case $u_1 \uparrow u_2$, for $\bigsqcap_{j \in \{1,2\}} u_j$ we will also use notation $u_1 \sqcap u_2$.

Example 4.1. Let $X = \{a, b\}$ with $a \neq b$ and $u_1 = a^{\omega} \in X^{\uparrow}$. Let $u_2 \in X^{\uparrow}$ be the partial ω-word whose domain $\text{dom}(u_2)$ is the set of all odd natural numbers and $u_2(i) = a$ for all $i \in \text{dom}(u_2)$. Let $u_3 \in X^{\uparrow}$ be the partial ω-word such that $\text{dom}(u_3)$ is the set of all even natural numbers and $u_3(i) = b$ for all $i \in \text{dom}(u_3)$. Then $u_1 \uparrow u_2$ and $u_2 \uparrow u_3$, but $\neg(u_1 \uparrow u_3)$. This shows in particular that the relation $\uparrow$ is not transitive if X is not a singleton set. Then, $u_1 \sqcap u_2 = a^{\omega}$ and $u_2 \sqcap u_3 = (ba)^{\omega}$.

4.2   WAL: Syntax and Semantics

Let V_1 be a countable set of *first-order variables* and V_2 a countable set of *second-order variables* such that $V_1 \cap V_2 = \emptyset$. Let $V = V_1 \cup V_2$. Let Σ be an alphabet and $\mathbb{V} = (M, (K, +, 0), \text{val})$ a valuation structure. We also consider a designated element $\mathbb{1} \in M$ which we call the *default weight*. We denote the pair $(\mathbb{V}, \mathbb{1})$ by $\mathbb{V}_{\mathbb{1}}$. The set $\mathbf{WAL}(\Sigma, \mathbb{V}_{\mathbb{1}})$ of formulas of *weight assignment logic* over Σ and $\mathbb{V}_{\mathbb{1}}$ is given by the grammar

$$\varphi ::= P_a(x) \mid x = y \mid x < y \mid X(x) \mid x \mapsto m \mid \varphi \Rightarrow \varphi \mid \varphi \sqcap \varphi \mid \sqcap x.\varphi \mid \sqcap X.\varphi$$

where $a \in \Sigma$, $x, y \in V_1$, $X \in V_2$ and $m \in M$. Such a formula φ is called a *weight assignment formula*.

Table 1. The auxiliary semantics of **WAL**-formulas

$$\langle\!\langle P_a(x)\rangle\!\rangle(w_\sigma) = \begin{cases} \top, & a_{\sigma(x)} = a \\ \bot, & \text{otherwise} \end{cases}$$

$$\langle\!\langle x \mapsto m\rangle\!\rangle(w_\sigma) = \top[\sigma(x)/m]$$

$$\langle\!\langle x = y\rangle\!\rangle(w_\sigma) = \begin{cases} \top, & \sigma(x) = \sigma(y) \\ \bot, & \text{otherwise} \end{cases}$$

$$\langle\!\langle \varphi_1 \Rightarrow \varphi_2\rangle\!\rangle(w_\sigma) = \begin{cases} \langle\!\langle\varphi_2\rangle\!\rangle(w_\sigma), & \langle\!\langle\varphi_1\rangle\!\rangle(w_\sigma) = \top \\ \top, & \text{otherwise} \end{cases}$$

$$\langle\!\langle x < y\rangle\!\rangle(w_\sigma) = \begin{cases} \top, & \sigma(x) < \sigma(y) \\ \bot, & \text{otherwise} \end{cases}$$

$$\langle\!\langle \varphi_1 \sqcap \varphi_2\rangle\!\rangle(w_\sigma) = \langle\!\langle\varphi_1\rangle\!\rangle(w_\sigma) \sqcap \langle\!\langle\varphi_2\rangle\!\rangle(w_\sigma)$$

$$\langle\!\langle \sqcap x.\varphi\rangle\!\rangle(w_\sigma) = \textstyle\bigsqcap_{i\in\mathrm{dom}(w)} \langle\!\langle\varphi\rangle\!\rangle(w_{\sigma[x/i]})$$

$$\langle\!\langle X(x)\rangle\!\rangle(w_\sigma) = \begin{cases} \top, & \sigma(x) \in \sigma(X) \\ \bot, & \text{otherwise} \end{cases}$$

$$\langle\!\langle \sqcap X.\varphi\rangle\!\rangle(w_\sigma) = \textstyle\bigsqcap_{I\subseteq\mathrm{dom}(w)} \langle\!\langle\varphi\rangle\!\rangle(w_{\sigma[X/I]})$$

Let $\varphi \in \mathbf{WAL}(\Sigma, \mathbb{V}_1)$. We denote by $\mathrm{CONST}(\varphi) \subseteq M$ the set of all weights $m \in M$ occurring in φ. The set $\mathrm{FREE}(\varphi) \subseteq V$ of *free variables* of φ is defined to be the set of all variables $\mathcal{X} \in V$ which appear in φ and are not bound by any quantifier $\sqcap\mathcal{X}$. We say that φ is a *sentence* if $\mathrm{FREE}(\varphi) = \emptyset$.

Note that the merging as defined before is a partially defined operation, i.e., it is defined only for compatible families of partial ω-words. In order to extend it to a totally defined operation, we fix an element $\bot \notin M^\uparrow$ which will mean the undefined value. Let $M_\bot^\uparrow = M^\uparrow \cup \{\bot\}$. Then, for any family $\theta = (u_j)_{j\in J}$ with $u_j \in M_\bot^\uparrow$, such that either $\theta \in (M^\uparrow)^J$ is not compatible or $\theta \in (M_\bot^\uparrow)^J \setminus (M^\uparrow)^J$, we let $\bigsqcap_{j\in J} u_j = \bot$.

For any ω-word $w \in \Sigma^\omega$, a *w-assignment* is a mapping $\sigma : V \to \mathrm{dom}(w) \cup 2^{\mathrm{dom}(w)}$ mapping first-order variables to elements in $\mathrm{dom}(w)$ and second-order variables to subsets of $\mathrm{dom}(w)$. For a first-order variable x and a position $i \in \mathbb{N}$, the w-assignment $\sigma[x/i]$ is defined on $V \setminus \{x\}$ as σ, and we let $\sigma[x/i](x) = i$. For a second-order variable X and a subset $I \subseteq \mathbb{N}$, the w-assignment $\sigma[X/I]$ is defined similarly. Let Σ_V^ω denote the set of all pairs (w, σ) where $w \in \Sigma^\omega$ and σ is a w-assignment. We will denote such pairs (w, σ) by w_σ.

The semantics of **WAL**-formulas is defined in two steps: by means of the auxiliary and proper semantics. Let $\varphi \in \mathbf{WAL}(\Sigma, \mathbb{V}_1)$. The *auxiliary semantics* of φ is the mapping $\langle\!\langle\varphi\rangle\!\rangle : \Sigma_V^\omega \to M_\bot^\uparrow$ defined for all $w_\sigma \in \Sigma_V^\omega$ with $w = (a_i)_{i\in\mathbb{N}}$ as shown in Table 1. Note that the definition of $\langle\!\langle..\rangle\!\rangle$ does not employ $+$ and val. The *proper semantics* $[\![\varphi]\!] : \Sigma_V^\omega \to K$ operates on the auxiliary semantics $\langle\!\langle\varphi\rangle\!\rangle$ as follows. Let $w_\sigma \in \Sigma_V^\omega$. If $\langle\!\langle\varphi\rangle\!\rangle(w_\sigma) \in M^\uparrow$, then we assign the default weight to all undefined positions in $\mathrm{dom}(\langle\!\langle\varphi\rangle\!\rangle(w_\sigma))$ and evaluate the obtained sequence using val. Otherwise, if $\langle\!\langle\varphi\rangle\!\rangle(w_\sigma) = \bot$, we put $[\![\varphi]\!](w_\sigma) = 0$. Note that if $\varphi \in \mathbf{WAL}(\Sigma, \mathbb{V}_1)$ is a sentence, then the values $\langle\!\langle\varphi\rangle\!\rangle(w_\sigma)$ and $[\![\varphi]\!](w_\sigma)$ do not depend on σ and we consider the auxiliary semantics of φ as the mapping $\langle\!\langle\varphi\rangle\!\rangle : \Sigma^\omega \to M_\bot^\uparrow$ and the proper semantics of φ as the quantitative ω-language $[\![\varphi]\!] : \Sigma^\omega \to K$. Note that $+$ was not needed for the semantics of **WAL**-formulas. This operation will be needed in the next section for the extension of **WAL**. We say that a quantitative ω-language $\mathbb{L} : \Sigma^\omega \to K$ is **WAL**-*definable* over $\mathbb{V}$ if there exist a default weight $\mathbb{1} \in M$ and a sentence $\varphi \in \mathbf{WAL}(\Sigma, \mathbb{V}_1)$ such that $[\![\varphi]\!] = \mathbb{L}$.

Example 4.2. Consider a valuation structure $\mathbb{V} = (M, (K, +, \mathbb{0}), \mathrm{val})$ and a default weight $\mathbb{1} \in M$. Consider an alphabet $\Sigma = \{a, b, ...\}$ of actions. We assume that the cost of a is $c(a) \in M$, the cost of b is $c(b) \in M$, and the costs of all other actions x in Σ are equal to $c(x) = \mathbb{1}$ (which can mean, e.g., that these actions do not invoke any costs). Then every ω-word w induces the ω-word of costs. We want to construct a sentence of our WAL which for every such an ω-word will evaluate its sequence of costs using val. The desired sentence $\varphi \in \mathbf{WAL}(\Sigma, \mathbb{V}_{\mathbb{1}})$ is $\varphi = \sqcap x.([P_a(x) \Rightarrow (x \mapsto c(a))] \sqcap [P_b(x) \Rightarrow (x \mapsto c(b))])$. Then, for every $w = (a_i)_{i \in \mathbb{N}} \in \Sigma^\omega$, the auxiliary semantics $\langle\!\langle \varphi \rangle\!\rangle(w)$ is the partial ω-word over M where all positions $i \in \mathbb{N}$ with $a_i = a$ are labelled by $c(a)$, all positions with $a_i = b$ are labelled by $c(b)$, and the labels of all other positions are undefined. Then, the proper semantics $[\![\varphi]\!](w)$ assigns $\mathbb{1}$ to all positions with undefined labels and evaluates it by means of val.

4.3 WAL: Relation to MSO Logic

Let Σ be an alphabet. We consider monadic second-order logic $\mathbf{MSO}(\Sigma)$ over ω-words to be the set of formulas

$$\varphi ::= P_a(x) \mid x = y \mid x < y \mid X(x) \mid \varphi \wedge \varphi \mid \neg\varphi \mid \forall x.\varphi \mid \forall X.\varphi$$

where $a \in \Sigma$, $x, y \in V_1$ and $X \in V_2$. For $w_\sigma \in \Sigma_V^\omega$, the satisfaction relation $w_\sigma \models \varphi$ is defined as usual. The usual formulas of the form $\varphi_1 \vee \varphi_2$, $\exists \mathcal{X}.\varphi$ with $\mathcal{X} \in V$, $\varphi_1 \Rightarrow \varphi_2$ and $\varphi_1 \Leftrightarrow \varphi_2$ can be expressed using $\mathbf{MSO}$-formulas.

For any formula $\varphi \in \mathbf{MSO}(\Sigma)$, let $W(\varphi)$ denote the $\mathbf{WAL}$-formula obtained from φ by replacing $\wedge$ by $\sqcap$, $\forall \mathcal{X}$ (with $\mathcal{X} \in V$) by $\sqcap\mathcal{X}$, and every subformula $\neg\psi$ by $\psi \Rightarrow \mathbf{false}$. Here $\mathbf{false}$ can be considered as abbreviation of the sentence $\sqcap x.(x < x)$. Note that $W(\varphi)$ does not contain any assignment formulas $x \mapsto m$ and $\langle\!\langle W(\varphi) \rangle\!\rangle(w_\sigma) \in \{\top, \bot\}$ for every $w_\sigma \in \Sigma_V^\omega$. Moreover, it can be easily shown by induction on the structure of φ that, for all $w_\sigma \in \Sigma_V^\omega$: $w_\sigma \models \varphi$ iff $\langle\!\langle W(\varphi) \rangle\!\rangle(w_\sigma) = \top$. This shows that MSO logic on infinite words is subsumed by $\mathbf{WAL}$. For the formulas which do not contain any assignments of the form $x \mapsto m$, the merging $\sqcap$ can be considered as the usual conjunction and the merging quantifiers $\sqcap\mathcal{X}$ as the usual universal quantifiers $\forall\mathcal{X}$. Moreover, $\top$ corresponds to the boolean true value and $\bot$ to the boolean false value. For a $\mathbf{WAL}$-formula φ, we will consider $\neg\varphi$ as abbreviation for $\varphi \Rightarrow \mathbf{false}$.

4.4 Extended WAL

Here we extend $\mathbf{WAL}$ with weighted existential quantification over free variables in $\mathbf{WAL}$-formulas. Let Σ be an alphabet, $\mathbb{V} = (M, (K, +, \mathbb{0}), \mathrm{val})$ a valuation structure and $\mathbb{1} \in M$ a default weight. The set $\mathbf{eWAL}(\Sigma, \mathbb{V}_{\mathbb{1}})$ of formulas of *extended weight assignment logic* over Σ and $\mathbb{V}_{\mathbb{1}}$ consists of all formulas of the form $\sqcup\mathcal{X}_1. ... \sqcup\mathcal{X}_k.\varphi$ where $k \geq 0$, $\mathcal{X}_1, ..., \mathcal{X}_k \in V$ and $\varphi \in \mathbf{WAL}(\Sigma, \mathbb{V}_{\mathbb{1}})$. Given a formula $\varphi \in \mathbf{eWAL}(\Sigma, \mathbb{V}_{\mathbb{1}})$, the *semantics* of φ is the mapping $[\![\varphi]\!] : \Sigma_V^\omega \to K$ defined inductively as follows. If $\varphi \in \mathbf{WAL}(\Sigma, \mathbb{V}_{\mathbb{1}})$, then $[\![\varphi]\!]$ is defined as the proper semantics for **WAL**. If φ contains a prefix $\sqcup x$ with $x \in V_1$ or $\sqcup X$ with $X \in V_2$, then, for all $w_\sigma \in \Sigma_V^\omega$, $\llbracket \varphi \rrbracket (w_\sigma)$ is defined inductively as follows:

$$\llbracket \sqcup x.\varphi \rrbracket (w_\sigma) = \sum \left(\llbracket \varphi \rrbracket (w_{\sigma[x/i]}) \mid i \in \mathrm{dom}(w) \right)$$
$$\llbracket \sqcup X.\varphi \rrbracket (w_\sigma) = \sum \left(\llbracket \varphi \rrbracket (w_{\sigma[X/I]}) \mid I \subseteq \mathrm{dom}(w) \right)$$

Again, if φ is a sentence, then we can consider its semantics as the quantitative ω-language $\llbracket \varphi \rrbracket : \Sigma^\omega \to K$. We say that a quantitative ω-language $\mathbb{L} : \Sigma^\omega \to K$ is **eWAL**-*recognizable* over V if there exist a default weight $\mathbb{1} \in M$ and a sentence $\varphi \in \mathbf{eWAL}(\Sigma, V_1)$ such that $\llbracket \varphi \rrbracket = \mathbb{L}$.

Example 4.3. Let $\Sigma = \{a\}$ be a singleton alphabet, $V = V^{\mathrm{Disc}}$ as defined in Example 2.3(b). Assume that, for every position of an ω-word, we can either assign to this position the cost 5 and the discounting factor 0.5 or we assign the cost the smaller cost 2 and the bigger discounting factor 0.75. After that we compute the discounted sum using the valuation function of V^{Disc}. We are interested in the infimal value of this discounted sum. We can express it by means of the **eWAL**-formula $\varphi = \sqcup X.\sqcap x.([X(x) \Rightarrow (x \mapsto (5, 0.5))] \sqcap [(\neg X(x)) \Rightarrow (x \mapsto (2, 0.75))])$ i.e. $\llbracket \varphi \rrbracket$ (a^ω) is the desired infimal value.

5 Expressiveness Equivalence Result

In this section we state and prove the main result of this paper.

Theorem 5.1. *Let Σ be an alphabet, $V = (M, (K, +, 0), \mathrm{val})$ a valuation structure and $\mathbb{L} : \Sigma^\omega \to K$ a quantitative ω-language. Then*

(a) $\mathbb{L}$ is **WAL**-*definable over V iff $\mathbb{L}$ is unambiguously recognizable over V.*
(b) $\mathbb{L}$ is **eWAL**-*definable over V iff $\mathbb{L}$ is recognizable over V.*

5.1 Unambiguous Case

In this subsection, we sketch the proof of Theorem 5.1 (a). First we show **WAL**-definability implies unambiguous recognizability. We establish a decomposition of **WAL**-formulas in a similar manner as it was done for unambiguous WBA in Theorem 3.1 (a). Assume that $\mathbb{L} = \llbracket \varphi \rrbracket$ where $\varphi \in \mathbf{WAL}(\Sigma, V_1)$. We show that there exist an alphabet Γ, renamings $h : \Gamma \to \Sigma$ and $g : \Gamma \to M$, and a sentence $\beta \in \mathbf{MSO}(\Gamma)$ such that $\llbracket \varphi \rrbracket = h((\mathrm{val} \circ g) \cap \mathcal{L}(\beta))$ where $\mathcal{L}(\beta) \subseteq \Gamma^\omega$ is the h-unambiguous ω-language defined by β. Then, applying the classical Büchi theorem (which states that $\mathcal{L}(\beta)$ is recognizable) and our Nivat Theorem 3.1(a), we obtain that $\mathbb{L}$ is recognizable over V. Let $\# \notin M$ be a symbol which we will use to mark all positions whose labels are undefined in the auxiliary semantics of **WAL**-formulas. Let $\Delta_\varphi = \mathrm{Const}(\varphi) \cup \{\#\}$. Then our extended alphabet will be $\Gamma = \Sigma \times \Delta_\varphi$. We define the renamings h, g as follows. For all $u = (a, b) \in \Gamma$, we let $h(u) = a$, $g(u) = b$ if $b \in M$, and $g(u) = \mathbb{1}$ if $m = \#$. The main difficulty is to construct the sentence β. For any ω-word $w = (a_i)_{i \in \mathbb{N}} \in \Sigma^\omega$ and any partial ω-word $\eta \in (\text{CONST}(\varphi))^\dagger$, we encode the pair (w, η) as the ω-word $\text{code}(w, \eta) = ((a_i, b_i))_{i \in \mathbb{N}} \in \Gamma^\omega$ where, for all $i \in \text{dom}(\eta)$, $b_i = \eta(i)$ and, for all $i \in \mathbb{N} \setminus \text{dom}(\eta)$, $b_i = \#$. In other words, we will consider ω-words of Γ as convolutions of ω-words over Σ with the encoding of the auxiliary semantics of φ.

Lemma 5.2. *For every subformula ζ of φ, there exists a formula $\Phi(\zeta) \in \mathbf{MSO}(\Sigma \times \Delta_\varphi)$ such that $\text{FREE}(\Phi(\zeta)) = \text{FREE}(\zeta)$ and, for all $w_\sigma \in \Sigma_V^\omega$ and $\eta \in (\text{CONST}(\varphi))^\dagger$, we have: $\langle\!\langle \zeta \rangle\!\rangle(w_\sigma) = \eta$ iff $(\text{code}(w, \eta))_\sigma \models \Phi(\zeta)$.*

Proof (Sketch). Let $Y \in V_2$ be a fresh variable which does not occur in φ. First, we define inductively the formula $\Phi_Y(\zeta) \in \mathbf{MSO}(\Gamma)$ with $\text{FREE}(\Phi_Y(\zeta)) = \text{FREE}(\zeta) \cup \{Y\}$ which describes the connection between the input ω-word w and the output partial ω-word η; here the variable Y keeps track of the domain of η.

- For $\zeta = P_a(x)$, we let $\Phi_Y(\zeta) = \bigvee_{b \in \Delta_\varphi} P_{(a,b)}(x) \wedge Y(\emptyset)$ where $Y(\emptyset)$ is abbreviation for $\forall y.\neg Y(y)$. Here we demand that the first component of the letter at position x is a and the second component is an arbitrary letter from Δ_φ and that the auxiliary semantics of ζ is the trivial partial ω-word $\top$.
- Let ζ be one of the formulas of the form $x = y$, $x < y$ or $X(x)$. Then, we let $\Phi_Y(\zeta) = \zeta \wedge Y(\emptyset)$.
- For $\zeta = (x \mapsto m)$, we let $\Phi_Y(\zeta) = \bigvee_{a \in \Sigma} P_{(a,m)}(x) \wedge \forall y.(Y(y) \Leftrightarrow x = y)$. This formula describes that position x of η must be labelled by m and all other positions are unlabelled.
- Let $\zeta = (\zeta_1 \Rightarrow \zeta_2)$. Let $Z \in V_2$ be a fresh variable. Consider the formula $\kappa = \exists Z.[\Phi_Z(\zeta_1) \wedge Z(\emptyset)]$ which checks whether the value of the auxiliary semantics of ζ_1 is $\top$. Then, we let $\Phi_Y(\zeta) = (\kappa \wedge \Phi_Y(\zeta_2)) \vee (\neg \kappa \wedge Y(\emptyset))$.
- Let $\zeta = \zeta_1 \sqcap \zeta_2$. Let $Y_1, Y_2 \in V_2$ be two fresh distinct variables. Then, we let $\Phi_Y(\zeta) = \exists Y_1.\exists Y_2.(\Phi_{Y_1}(\zeta_1) \wedge \Phi_{Y_2}(\zeta_2) \wedge [Y = Y_1 \cup Y_2])$. Note that the property $Y = Y_1 \cup Y_2$ is MSO-definable.
- The most interesting case is a formula of the form $\zeta = \sqcap \mathcal{X}.\zeta'$ with $\mathcal{X} \in V$. Here, every value of $\mathcal{X}$ induces its own value of $Y(\mathcal{X})$ and we have to merge infinitely many partial ω-words, i.e., to express that Y is the infinite union of $Y(\mathcal{X})$ over all sets $\mathcal{X}$. We can show that Y must be the minimal set which satisfies the formula $\xi(Y) = \forall \mathcal{X}.\exists Y'.(\Phi_{Y'}(\zeta') \wedge (Y' \subseteq Y))$ where $Y' \in V_2$ is a fresh variable. Then, we let $\Phi_Y(\zeta) = \xi(Y) \wedge \forall Z.(\xi(Z) \Rightarrow (Y \subseteq Z))$.

Finally, we construct $\Phi(\zeta)$ from $\Phi_Y(\zeta)$ by labelling all positions not in Y by $\#$:
$$\Phi(\zeta) = \exists Y.(\Phi_Y(\zeta) \wedge \forall x.(Y(x) \vee \bigvee_{a \in \Sigma} P_{(a,\#)}(x))). \qquad \square$$

Now we apply Lemma 5.2 to the case $\zeta = \varphi$. Then, $\Phi(\varphi)$ is a sentence and $\mathcal{L}(\Phi(\varphi)) = \{\text{code}(w, \eta) \mid \langle\!\langle \varphi \rangle\!\rangle(w) = \eta \neq \bot\}$. Note that $\mathcal{L}(\Phi(\varphi))$ is h-unambiguous, since for every $w \in \Sigma^\omega$ there exists at most one $u \in \mathcal{L}(\Phi(\varphi))$ with $h(u) = w$. If we let $\beta = \Phi(\varphi)$, then we obtain the desired decomposition $[\![\varphi]\!] = h((\text{val} \circ g) \cap \mathcal{L}(\beta))$. Hence **WAL**-definability implies unambiguous recognizability.

Now we show the converse part of Theorem 5.1 (a), i.e., we show that unambiguous recognizability implies **WAL**-definability. Let $\mathcal{A} = (Q, I, T, F, \text{wt})$ be an unambiguous WBA over Σ and $\mathbb{V}$. First, using the standard approach, we describe runs of $\mathcal{A}$ by means of MSO-formulas. For this, we fix an enumeration $(t_i)_{1 \le i \le m}$ of T and associate with every transition t_i a second-order variable X_i which keeps track of positions where t is taken. Then, a run of $\mathcal{A}$ can be described using a formula $\beta \in \mathbf{MSO}(\Sigma)$ with $\mathrm{FREE}(\beta) = \{X_1, ..., X_m\}$ which demands that values of the variables $X_1, ..., X_m$ form a partition of the domain of an input word, the transitions of a run are matching, the labels of transitions of a run are compatible with an input word, a run starts in I and visits some state in F infinitely often. Let $\mathbb{1} \in M$ be an arbitrary default weight. Consider the $\mathbf{WAL}(\Sigma, \mathbb{V}_{\mathbb{1}})$-sentence

$$\varphi = W(\exists X_1 ... \exists X_m.\beta) \sqcap (\sqcap X_1 ... \sqcap X_m.[W(\beta) \Rightarrow \sqcap x.\textstyle\bigsqcap_{i=1}^m X_i(x) \Rightarrow (x \mapsto \mathrm{wt}(t_i))]).$$

It can be shown that $[\![\varphi]\!] = [\![\mathcal{A}]\!]$. Hence unambiguous recognizability implies $\mathbf{WAL}$-definability.

5.2 Nondeterministic Case

Now we sketch of the proof of Theorem 5.1 (b). First we show that $\mathbf{eWAL}$-definability implies nondeterministic recognizability. The idea of our proof is similar to the unambiguous case, i.e., via a decomposition of a $\mathbf{eWAL}$-sentence. Let $\mathbb{1} \in M$ be a default weight and $\psi \in \mathbf{eWAL}(\Sigma, \mathbb{V}_{\mathbb{1}})$ a sentence. We may assume that $\psi = \sqcup x_1 ... \sqcup x_k.\sqcup X_1 ... \sqcup X_l.\varphi$ where $\varphi \in \mathbf{WAL}(\Sigma, \mathbb{V}_{\mathbb{1}})$ and $x_1, ..., x_k$, $X_1, ..., X_l$ are pairwise distinct variables. Again, we will establish a decomposition $[\![\varphi]\!] = h((\mathrm{val} \circ g) \cap \mathcal{L}(\beta))$ for some alphabet Γ, renamings $h : \Gamma \to \Sigma$ and $g : \Gamma \to M$, and an MSO-sentence β over Γ. Note that, as opposed to the unambiguous case, the ω-language $\mathcal{L}(\beta)$ is not necessarily h-unambiguous. Then, the quantitative ω-language $\mathbb{L}$ is recognizable over $\mathbb{V}$ by Theorem 3.1 (b) and the classical Büchi theorem (which states that $\mathcal{L}(\beta)$ is a recognizable ω-language). As opposed to the unambiguous case, the extended alphabet Γ must also keep track of the values of the variables $x_1, ..., x_k, X_1, ..., X_l$. Let $\mathcal{V} = \{x_1, ..., x_k, X_1, ..., X_l\}$ and Δ_φ be defined as in the unambiguous case. Then we let $\Gamma = \Sigma \times \Delta_\varphi \times 2^{\mathcal{V}}$ and define h, g as in the unambiguous case ignoring the new component $2^{\mathcal{V}}$. Finally we construct the MSO-sentence β over Γ. The construction of β will be based on Lemma 5.2. Let $\Phi(\varphi) \in \mathbf{MSO}(\Sigma \times \Delta_\varphi)$ be the formula constructed in Lemma 5.2 for $\zeta = \varphi$. By simple manipulations with the predicates $P_{(a,b)}(x)$ of $\Phi(\varphi)$ (describing that the $2^{\mathcal{V}}$-component is arbitrary), we transform $\Phi(\varphi)$ to the formula $\overline{\Phi(\varphi)} \in \mathbf{MSO}(\Gamma)$. Using the standard Büchi encoding technique we construct a formula $\phi \in \mathbf{MSO}(\Gamma)$ which encodes the values of $\mathcal{V}$-variables in the $2^{\mathcal{V}}$-component of an ω-word over Γ. Then we let $\beta = \exists x_1 ... \exists x_k.\exists X_1 ... \exists X_l.(\phi \wedge \overline{\Phi(\varphi)})$. It can be shown that $[\![\varphi]\!] = h((\mathrm{val} \circ g) \cap \mathcal{L}(\beta))$. Hence $\mathbf{eWAL}$-definability implies recognizability.

Now we show that recognizability implies $\mathbf{eWAL}$-definability. Our proof is a slight modification of our proof for the unambiguous case. Let $\mathcal{A} = (Q, I, T, F, \mathrm{wt})$ be a nondeterministic WBA. Adopting the notations from the corresponding proof of Subsect. 5.1, we construct the $\mathbf{eWAL}(\Sigma, \mathbb{V}_{\mathbb{1}})$-sentence

$$\varphi = \sqcup X_1...\sqcup X_m.\big(W(\beta) \Rightarrow \sqcap x.\textstyle\prod_{i=1}^{m} X_i(x) \Rightarrow (x \mapsto \mathrm{wt}(t_i))\big).$$

It can be shown that $[\![\varphi]\!] = [\![\mathcal{A}]\!]$. Hence recognizabilty implies **eWAL**-definability.

6 Discussion

In this paper we introduced a weight assignment logic which is a simple and intuitive logical formalism for reasoning about quantitative ω-languages. Moreover, it works with arbitrary valuation functions whereas in weighted logics of [12], [14] some additional restrictions on valuation functions were added. We showed that WAL is expressively equivalent to unambiguous weighted Büchi automata. We also considered an extension of WAL which is equivalent to non-deterministic Büchi automata. Our expressiveness equivalence results can be helpful to obtain decidability properties for our new logics. The future research should investigate decidability properties of nondeterministic and unambiguous weighted Büchi automata with the practically relevant valuation functions. Although the weighted ω-automata models [7] do not have a Büchi acceptance condition, it seems likely that their decidability results about the threshold problems hold for Büchi acceptance condition as well. It could be also interesting to study our weight assignment technique in the context of temporal logic like LTL. Our results obtained for ω-words can be easily adopted to the structures like finite words and trees.

References

1. Andersson, D.: Improved combinatorial algorithms for discounted payoff games. Master's thesis, Uppsala University, Department of Information Technology (2006)
2. Bloem, R., Greimel, K., Henzinger, T.A., Jobstmann, B.: Synthesizing robust systems. In: FMCAD 2009, pp. 85–92. IEEE (2009)
3. Bouyer, P.: A logical characterization of data languages. Inf. Process. Lett. **84**(2), 75–85 (2002)
4. Bouyer, P., Brinksma, E., Larsen, K.G.: Optimal infinite scheduling for multi-priced timed automata. Formal Methods in System Design **32**, 3–23 (2008)
5. Büchi, J.R.: Weak second-order arithmetic and finite automata. Z. Math. Logik und Grundl. Math. **6**, 66–92 (1960)
6. Chatterjee, K., Doyen, L., Henzinger, T.A.: Quantitative languages. In: Kaminski, M., Martini, S. (eds.) CSL 2008. LNCS, vol. 5213, pp. 385–400. Springer, Heidelberg (2008)
7. Carton, O., Michel, M.: Unambiguous Büchi automata. In: Gonnet, G.H., Viola, A. (eds.) LATIN 2000. LNCS, vol. 1776, pp. 407–416. Springer, Heidelberg (2000)
8. Droste, M., Gastin, P.: Weighted automata and weighted logics. Theoret. Comp. Sci. **380**(1–2), 69–86 (2007)
9. Droste, M., Kuich, W., Vogler, H. (eds.): Handbook of Weighted Automata. EATCS Monographs on Theoretical Computer Science. Springer (2009)

10. Droste, M., Kuske, D.: Weighted automata. In: Pin, J.-E. (ed.) Handbook: "Automata: from Mathematics to Applications". European Mathematical Society (to appear)
11. Droste, M., Meinecke, I.: Weighted automata and weighted MSO logics for average and long-time behaviors. Inf. Comput. **220–221**, 44–59 (2012)
12. Droste, M., Perevoshchikov, V.: Multi-weighted automata and MSO logic. In: Bulatov, A.A., Shur, A.M. (eds.) CSR 2013. LNCS, vol. 7913, pp. 418–430. Springer, Heidelberg (2013)
13. Droste, M., Perevoshchikov, V.: A Nivat theorem for weighted timed automata and weighted relative distance logic. In: Esparza, J., Fraigniaud, P., Husfeldt, T., Koutsoupias, E. (eds.) ICALP 2014, Part II. LNCS, vol. 8573, pp. 171–182. Springer, Heidelberg (2014)
14. Droste, M., Rahonis, G.: Weighted automata and weighted logics on infinite words. In: Ibarra, O.H., Dang, Z. (eds.) DLT 2006. LNCS, vol. 4036, pp. 49–58. Springer, Heidelberg (2006)
15. Fahrenberg, U., Juhl, L., Larsen, K.G., Srba, J.: Energy games in multiweighted automata. In: Cerone, A., Pihlajasaari, P. (eds.) ICTAC 2011. LNCS, vol. 6916, pp. 95–115. Springer, Heidelberg (2011)
16. Filiot, E., Gentilini, R., Raskin, J.-F.: Quantitative languages defined by functional automata. In: Koutny, M., Ulidowski, I. (eds.) CONCUR 2012. LNCS, vol. 7454, pp. 132–146. Springer, Heidelberg (2012)
17. Hashiguchi, K., Ishiguro, K., Jimbo, S.: Decidability of the equivalence problem for finitely ambiguous finance automata. Int. Journal of Algebra and Computation **12**(3), 445–461 (2002)
18. Krob, D.: The equality problem for rational series with multiplicities in the tropical semiring is undecidable. International Journal of Algebra and Computation **4**(3), 405–425 (1994)
19. Larsen, K.G., Rasmussen, J.I.: Optimal conditional reachability for multi-priced timed automata. In: Sassone, V. (ed.) FOSSACS 2005. LNCS, vol. 3441, pp. 234–249. Springer, Heidelberg (2005)
20. Nivat, M.: Transductions des langages de Chomsky. Ann. de l'Inst. Fourier **18**, 339–456 (1968)
21. Stüber, T., Vogler, H., Fülöp, Z.: Decomposition of weighted multioperator tree automata. Int. J. Foundations of Computer Sci. **20**(2), 221–245 (2009)
22. Wilke, T.: Specifying timed state sequences in powerful decidable logics and timed automata. In: Langmaack, H., de Roever, W.-P., Vytopil, J. (eds.) FTRTFT 1994 and ProCoS 1994. LNCS, vol. 863, pp. 694–715. Springer, Heidelberg (1994)

Complexity Bounds of Constant-Space Quantum Computation
(Extended Abstract)

Tomoyuki Yamakami[(✉)]

Department of Information Science, University of Fukui,
3-9-1 Bunkyo, Fukui 910-8507, Japan
tomoyukiyamakami@gmail.com

Abstract. We model constant-space quantum computation as measure-many two-way quantum finite automata and evaluate their language recognition power by analyzing their behaviors and explore their properties. In particular, when the automata halt "in finite steps," they must terminate in worst-case liner time. Even if all computation paths of bounded-error automata do not terminate, it suffices to focus only on computation paths that terminate after exponentially many steps. We present a classical simulation of those automata on multi-head probabilistic finite automata with cut points. Moreover, we discuss how the power of the automata varies as the automata's acceptance criteria change to error free, one-sided error, bounded error, and unbounded error.

1 Quick Overview

Computer scientists have primarily concerned with "resources" used up to execute desired computations on a given computing device. Particularly, we are keen to *memory space* that stores information or data necessary to carry out a carefully designed single protocol on the device. In quantum computing, when a protocol requires only a constant amount of memory space (independent of input size), we are used to view such device as *quantum finite automata* (qfa), mainly because they are still capable of storing useful information by way of manipulating a few number of "inner states" without equipping an additional memory tape. A qfa proceeds its computation by applying a finite-dimensional unitary transition matrix and projective measurements. Such simplicity of the fundamental structure of qfa's is ideal for us to conduct a deeper analysis on their algorithmic procedures. In this extended abstract, we are focused mostly on *measure-many two-way quantum finite automata* (or *2qfa's*, for brevity) of Kondacs and Watrous [4] because of the simplicity of their definition.

A computation of the 2qfa evolves linearly by applying a transition matrix to a superposition of *configurations* in a finite-dimensional Hilbert space (called a *configuration space*). Unlike a model in [8], Kondacs and Watrous' qfa model further uses a classical operation of observing halting inner states at every step. Allowing its tape head to move in all directions, in fact, enables the 2qfa's to

© Springer International Publishing Switzerland 2015
I. Potapov (Ed.): DLT 2015, LNCS 9168, pp. 426–438, 2015.
DOI: 10.1007/978-3-319-21500-6_34

attain a significant increase of computational power over 2-way deterministic finite automata [4]. Despite our efforts over the past 20 years, the behaviors of 2qfa's have been largely enigmatic and the 2qfa's are still awaiting for full investigation.

There are three important issues that we wish to address in depth.

(1) *Acceptance criteria issue.* The first issue is that, in traditional automata theory, the language recognition has been concerned with a threshold of the acceptance probability of underlying automata under the term of "(isolated) cut point." In quantum automata theory, on the contrary, qfa's were originally defined in terms of "bounded-error probability." The family of languages recognized by such bounded-error 2qfa's is shorthandedly denoted by 2BQFA. When we further modify 2qfa's error probability, we obtain 2EQFA, 2PQFA, and 2C$_=$QFA for the language families[1] induced by 2qfa's with error-free (or exact), unbounded-error, and equal probabilities, respectively. Here, we present in Section 3 various inclusion and collapsing relationships among those families.

(2) *Termination issue.* In the literature, space-bounded quantum computation has been discussed mostly in the case of *absolutely halting* (i.e., eventual termination of all computation paths) [12]. Bounded-error 2qfa's that halt absolutely define a language family 2BQFA(*abs-halt*); in contrast, 2qfa's whose computation paths terminate with probability 1 (i.e., the probability of non-halting computation is 0) are said to *halt completely* and induce 2BQFA(*comp-halt*). Here, we demonstrate in Section 4 that, when a 2qfa makes bounded errors, most computation paths of the 2qfa can terminate after exponentially many steps. A key to the proof of this result is the *Dimension Lemma* of Yao [16], presented in Section 3. This is a direct consequence of an analysis of 2qfa's transition matrices. In particular, when 2qfa's halt absolutely, we can upper-bound the "worst case" running time of the 2qfa's by $O(n)$, where n is the input length.

(3) *Classical simulation issue.* Watrous [12] gave a general procedure of simulating space-bounded unbounded-error quantum Turing machines. As noted in [9], this simulation leads to the containment 2BQFA$_A$ $\subseteq$ PL, where PL is the family of all languages recognized by unbounded-error probabilistic Turing machines with $\{0, 1/2, 1\}$-transition probabilities using $O(\log n)$ space. In Section 5, we give a better complexity upper bound to 2PQFA (and therefore 2BQFA) using multi-head 2-way probabilistic finite automata with cut points. For this purpose, we make an appropriate implementation of a GapL-algorithm of [7, Theorem 4] that computes integer determinants. For our implementation, we need to make various changes to the original algorithm. Such changes are necessary because a target matrix is given as "input" in [7]; however, in our case, we must generate "probabilities" to express the desired determinants.

---

[1] These notations are analogous to EQP, PP, and C$_=$P in complexity theory.

2 Basic Notions and Notation

2.1 General Definitions

Let $\mathbb{N}$ be the set of all *natural numbers* (that is, nonnegative integers) and set $\mathbb{N}^+ = \mathbb{N} - \{0\}$. Moreover, let $\mathbb{Z}, \mathbb{Q}, \mathbb{R}$, and $\mathbb{C}$ be the sets of all *integers*, of all *rational numbers*, of all *real numbers*, and of all *complex numbers*, respectively. The notation $\mathbb{A}$ stands for the set of all *algebraic complex numbers*. We write $\imath$ for $\sqrt{-1}$. Given two numbers $m, n \in \mathbb{Z}$ with $m \leq n$, $[m, n]_{\mathbb{Z}}$ expresses an *integer interval* between m and n; i.e., the set $\{m, m+1, m+2, \ldots, n\}$. For any finite set Q, $|Q|$ denotes the *cardinality* of Q. All vectors in $\mathbb{C}^n$ are expressed as *column vectors*. Given a number $n \in \mathbb{N}^+$, $M_n(\mathbb{C})$ stands for the set of all $n \times n$ complex matrices. For any complex matrix A, the notation A^T and $A^\dagger$ respectively denote the *transpose* and the *Hermitian adjoint* of A.

2.2 Classical Finite Automata and Cut Point Formulation

We assume the reader's familiarity with *2-way probabilistic finite automata* (or 2pfa's) with real transition probabilities. Here, we formulate such 2pfa's as $M = (Q, \Sigma, \delta, q_0, Q_{acc}, Q_{rej})$ by including Q_{rej}, which was not in the original definition of Rabin [10]. When all transition probabilities of M are drawn from a designated set K ($K \subseteq \mathbb{R}$), we say that a 2pfa takes *K-transition probabilities*. Implicitly, we always assume that $\{0, 1/2, 1\} \subseteq K$. Given a 2pfa M, $p_{M,acc}(x)$ and $p_{M,rej}(x)$ respectively denote the acceptance probability and the rejection probability of M on input x.

As a variant of 2pfa's, we also define a *k-head two-way probabilistic finite automaton* (or a $k$head-2pfa, in short) by allowing a 2pfa to use k tape heads that move separately along a single input tape [6]. The notation $2\text{PPFA}_K(k\text{-}head)$ denotes the family of all languages recognized with "cut points" in $K \cap [0, 1]$ by $k$head-2pfa's. In a similar way, $2\text{C}_=\text{PFA}_K(k\text{-}head)$ is defined using "exact cut points"[2] in place of "cut points."

The notation $\#2\text{PFA}_K$ denotes the collection of *stochastic functions*, which are of the form $p_{M,acc}$ for certain 2pfa's M with K-transition probabilities. (See [5] for the case of 1pfa's.)

2.3 Quantum Finite Automata and Bounded Error Formulation

We briefly give the formal definition of *2-way quantum finite automata* (or 2qfa's, in short). Formally, a 2qfa M is described as a sextuple $(Q, \Sigma, \delta, q_0, Q_{acc}, Q_{rej})$, where Q is a finite set of inner states with $Q_{acc} \cup Q_{rej} \subseteq Q$ and $Q_{acc} \cap Q_{rej} = \emptyset$, Σ is a finite alphabet, q_0 is the initial inner state, and δ is a transition function mapping from $Q \times \check{\Sigma} \times Q \times D$ to $\mathbb{C}$, where $\check{\Sigma} = \Sigma \cup \{\mathcal{c}, \$\}$ and $D = \{-1, 0, +1\}$. The transition function δ specifies *transitions* whose output values $\delta(p, \sigma, q, d)$

---

[2] A number $\eta \in [0, 1]$ is an *exact cut point* for L if the following condition holds: for all $x \in \Sigma^*$, $x \in L$ if and only if $p_{M,acc}(x) = \eta$.

are called *amplitudes*). Each transition $\delta(p, \sigma, q, d) = \gamma$ indicates that, assuming that the 2qfa M is in inner state p scanning a symbol σ, M at the next step changes its inner state to q and moves its head in direction d with amplitude γ. The set Q is partitioned into three sets: Q_{acc}, Q_{rej}, and Q_{non}. Inner states in Q_{acc} (resp., in Q_{rej}) are called *accepting states* (resp., *rejecting states*). A *halting state* refers to an inner state in $Q_{acc} \cup Q_{rej}$. The rest of inner states, denoted by Q_{non}, are *non-halting states*. We say that M has K-*amplitudes* if all amplitudes of M are in set K ($\subseteq \mathbb{C}$), provided that $\{0, 1/2, 1\} \subseteq K$.

An input tape has two endmarkers ¢ and $ and is indexed by integers between 0 and $n + 1$, including the endmarkers, when an input of length n is given. For technical convenience, we assume that the input tape is *circular* (as originally defined in [4]). A computation of M is a series of superpositions of (classical) configurations, each of which evolves by an application of δ to its predecessor (if not the initial configuration) and by an application of a projective (or von Neumann) measurement. From δ and input $x \in \Sigma^n$, we define a *time-evolution operator* $U_\delta^{(x)}$ as a linear operator acting on $span\{|q, \ell\rangle \mid q \in Q, \ell \in [0, n+1]_{\mathbb{Z}}\}$ (called a *configuration space*) in the following way: for each $(p, i) \in Q \times [0, n+1]_{\mathbb{Z}}$, $U_\delta^{(x)}|p, i\rangle$ equals $\sum_{(q,d) \in Q \times \{0, \pm 1\}} \delta(p, x_i, q, d)|q, i + d \pmod{n + 2}\rangle$, where $x_0 = $ ¢, $x_{n+1} = $ \$, and x_i is the ith symbol of x for each index $i \in [1, n]_{\mathbb{Z}}$. Throughout this extended abstract, we always assume $U_\delta^{(x)}$ to be *unitary* for any x. Three projections Π_{acc}, Π_{rej}, and Π_{non} are linear maps that project onto $W_{acc} = span\{|q\rangle \mid q \in Q_{acc}\}$, $W_{rej} = span\{|q\rangle \mid q \in Q_{rej}\}$, and $W_{non} = span\{|q\rangle \mid q \in Q_{non}\}$, respectively. A *computation* of M on input x proceeds as follows. The 2qfa M starts with its initial configuration $|\phi_0\rangle = |q_0\rangle|0\rangle$ (where 0 means the head is scanning ¢). At Step i, M applies $U_\delta^{(x)}$ to $|\phi_{i-1}\rangle$ and then applies $\Pi_{acc} \oplus \Pi_{rej} \oplus \Pi_{non}$. We say that M *accepts* (resp., *rejects*) at Step i with probability $p_{M,acc,i}(x) = \|\Pi_{acc}U_\delta^{(x)}|\phi_{i-1}\rangle\|^2$ (resp., $p_{M,rej,i}(x) = \|\Pi_{rej}U_\delta^{(x)}|\phi_{i-1}\rangle\|^2$). The ith unnormalized quantum state $|\phi_i\rangle$ becomes $\Pi_{non}U_\delta^{(x)}|\phi_{i-1}\rangle$. The *acceptance probability* of M on x is $p_{M,acc}(x) = \sum_{i=1}^{\infty} p_{M,acc,i}(x)$. The *rejection probability* is defined similarly and is denoted by $p_{M,rej}(x)$.

Let ε be any constant in $[0, 1/2)$ and let L be any language over alphabet Σ. We say that a 2qfa M *recognizes* L *with error probability at most* ε if (i) for every $x \in L$, $p_{M,acc}(x) \geq 1 - \varepsilon$ and (ii) for every $x \in \overline{L}$ ($= \Sigma^* - L$), $p_{M,rej}(x) \geq 1 - \varepsilon$. When such an ε exists, we also say that M recognizes L with *bounded-error probability*. We define 2BQFA$_K$ as the family of all languages that can be recognized by bounded-error 2qfa's with K-amplitudes.

We say that a 2qfa *halts completely* if its halting probability equals 1, whereas a 2qfa *halts absolutely* if all the computation paths of the 2qfa eventually terminate in halting inner states. If a 2qfa halts absolutely, then it must halt completely, but the converse is not always true since a 2qfa that halts completely might possibly have a few computation paths that do not terminate.

When we place various restrictions specified as ⟨*restrictions*⟩ on 2qfa's, we intend to use a conventional notation of the form 2BQFA$_K$(*restrictions*). Let

⟨*comp-halt*⟩ and ⟨*abs-halt*⟩ respectively mean that a target 2qfa halts completely and absolutely. We often drop the subscript K whenever $K = \mathbb{C}$.

We discuss four more language families. The error-free language family $2\mathrm{EQFA}_K$ is obtained from $2\mathrm{BQFA}_K$ by always setting $\varepsilon = 0$ (i.e., either $p_{M,acc}(x) = 1$ or $p_{M,rej}(x) = 1$ for all $x \in \Sigma^*$). We also obtain the unbounded-error language family $2\mathrm{PQFA}_K$ by requiring that $p_{M,acc}(x) > 1/2$ for all $x \in L$ and $p_{M,rej}(x) \geq 1/2$ for all $x \in \overline{L}$. Similarly, the equality language family $2\mathrm{C}_=\mathrm{QFA}_K$ is derived from $2\mathrm{BQFA}_K$ by demanding that, for all x, $x \in L$ iff $p_{M,acc}(x) = 1/2$. In contrast, the one-sided error language family $2\mathrm{RQFA}_K$ requires the existence of a constant $\varepsilon \in [0, 1/2]$ satisfying that $p_{M,acc}(x) = 1$ for all $x \in L$ and $p_{M,rej}(x) \geq 1 - \varepsilon$ for all $x \in \overline{L}$.

By the definitions of the aforementioned language families, the following properties hold. Recall that $\{0, 1/2, 1\} \subseteq K$ is implicitly assumed.

Lemma 1. *Let K be any subset of $\mathbb{C}$.*

1. $2\mathrm{EQFA}_K \subseteq 2\mathrm{RQFA}_K \subseteq 2\mathrm{BQFA}_K \subseteq 2\mathrm{PQFA}_K$.
2. $2\mathrm{EQFA}_K = co\text{-}2\mathrm{EQFA}_K$ *and* $2\mathrm{BQFA}_K = co\text{-}2\mathrm{BQFA}_K$.
3. $2\mathrm{EQFA}_K = 2\mathrm{EQFA}_K(comp\text{-}halt) \subseteq 2\mathrm{C}_=\mathrm{QFA}_K(comp\text{-}halt) \cap co\text{-}2\mathrm{C}_=\mathrm{QFA}_K(comp\text{-}halt)$.
4. $2\mathrm{RQFA}_K \cup co\text{-}2\mathrm{RQFA}_K \subseteq 2\mathrm{BQFA}_K$ *if* $K \supseteq \mathbb{Q} \cap [0, 1]$.
5. $2\mathrm{C}_=\mathrm{QFA}_K \subseteq co\text{-}2\mathrm{PQFA}_K$.

The notion of *quantum functions* given in [14] generated by quantum Turing machines is quite useful in describing various language families. Similarly to the notation #BQP used in [14], we define $\#2\mathrm{QFA}_K$ to be the set of quantum functions $p_{M,acc} : \Sigma^* \to [0, 1]$ for any K-amplitude 2qfa M. This can be seen as an extension of $\#2\mathrm{PFA}_K$ of stochastic functions in Section 2.2. Note that, by exchanging Q_{acc} and Q_{rej} of M, the function $p_{M,rej}$ also belongs to $\#2\mathrm{QFA}_K$.

3   Behaviors of Absolutely Halting QFAs

We will discuss an issue on termination criteria of 2qfa's. We begin with the case where 2qfa's *halt absolutely* (that is, all computation paths of M halt on all inputs within a finite number of steps). Since those 2qfa's are relatively easy to handle, we obtain several intriguing properties of them. In what follows, the notation REG represents the collection of all *regular languages*. We write $\mathrm{AM}(2pfa, poly\text{-}time)$ for the family of all languages recognized by Dwork-Stockmeyer *interactive proof systems* using 2pfa verifiers with $\mathbb{Q}$-transition probabilities running in expected polynomial time [3].

Proposition 2. REG $\subseteq$ $2\mathrm{EQFA}_\mathbb{Q}(abs\text{-}halt)$ $\subseteq$ $2\mathrm{RQFA}_\mathbb{Q}(abs\text{-}halt)$ $\nsubseteq$ $\mathrm{AM}(2pfa, poly\text{-}time)$.

Next, we will give a precise bound on the running time of the 2qfa's when they halt absolutely. For convenience, we say that a 2qfa halts in *worst-case linear time* if every computation path terminates within time linear in input size. In this case, we use another notation $2\text{BQFA}_K[\textit{lin-time}]$ to differentiate it from the case of expected liner-time computation. As before, we omit the subscript K whenever $K = \mathbb{C}$. We introduce similar notations for 2EQFA, 2RQFA, $2\text{C}_=\text{QFA}$, and 2PQFA. In the following theorem, we prove that every absolutely-halting 2qfa terminates in worst-case linear time.

Theorem 3. *For any set $K \subseteq \mathbb{C}$, $2\text{BQFA}_K(\textit{abs-halt}) = 2\text{BQFA}_K[\textit{lin-time}]$. The same is true for 2EQFA, 2RQFA, $2\text{C}_=\text{QFA}$, and 2PQFA.*

For the proof of Theorem 3, we need to observe the behaviors of 2qfa's that halt absolutely. Back in 1998, Yao [16] made the following observation.

Lemma 4. *Any $\mathbb{C}$-amplitude 2qfa with a set Q of inner states should halt within worst-case $|Q|(n + 2) + 1$ steps if all (non-zero amplitude) computation paths of the 2qfa eventually terminate, where n is input length.*

Proof of Theorem 3. This theorem directly follows from Lemma 4 as follows. Let L be any language in $2\text{BQFA}_K(\textit{abs-halt})$ recognized by a certain K-amplitude 2qfa, say, M with bounded-error probability. Assume that M halts absolutely. By Lemma 4, we conclude that M halts within worst-case $O(n)$ steps. This indicates that L belongs to $2\text{BQFA}_K[\textit{lin-time}]$. Thus, we immediately obtain $2\text{BQFA}_K(\textit{abs-halt}) \subseteq 2\text{BQFA}_K[\textit{lin-time}]$. Since the converse containment is trivial, it follows that $2\text{BQFA}_K(\textit{abs-halt}) = 2\text{BQFA}_K[\textit{lin-time}]$. □

Lemma 4 also leads to various consequences. We denote by REC the family of all *recursive languages*.

Corollary 5. *1. $2\text{PQFA}_\mathbb{A}(\textit{abs-halt}) = \text{co-}2\text{PQFA}_\mathbb{A}(\textit{abs-halt})$.*
2. $2\text{C}_=\text{QFA}_\mathbb{A}(\textit{abs-halt}) \cup \text{co-}2\text{C}_=\text{QFA}_\mathbb{A}(\textit{abs-halt}) \subseteq 2\text{PQFA}_\mathbb{A}(\textit{abs-halt})$.
3. $2\text{EQFA}_\mathbb{C}(\textit{abs-halt}) = 2\text{EQFA}_{\mathbb{A} \cap \mathbb{R}}(\textit{abs-halt}) \subseteq \text{REC}$.

In Corollary 5(1–2), we do not know whether the amplitude set $\mathbb{A}$ can be replaced by $\mathbb{C}$. Here, we want to prove this corollary.

Proof Sketch of Corollary 5. (1) Let L be any language in $2\text{PQFA}_\mathbb{A}(\textit{abs-halt})$ witnessed by a certain 2qfa M of the form $(Q, \Sigma, \delta, q_0, Q_{acc}, Q_{rej})$. Lemma 4 shows that all computation paths of M on inputs of length n terminate within $|Q|(n + 2) + 1$ steps. Let F_M be a set of all amplitudes used by M. Since $F_M \subseteq \mathbb{A}$, choose $\alpha_1, \alpha_2, \ldots, \alpha_e \in \mathbb{A}$ such that $F_M \subseteq \mathbb{Q}(\alpha_1, \ldots, \alpha_e)/\mathbb{Q}$, where $e \leq |Q||\check{\Sigma}||D|$. Let $\alpha_x = p_{M,acc}(x) - 1/2$ if $x \in L$; $p_{M,rej}(x) - 1/2$ otherwise. It is possible to express α_x as a certain polynomial in $(\alpha_1, \ldots, \alpha_e)$ whose degree is at most $2|Q|(n + 2) + 2$.

Next, we use the following known result taken from Stolarsky's textbook [11].

Lemma 6. *Let $\alpha_1, \ldots, \alpha_e \in \mathbb{A}$. Let h be the degree of $\mathbb{Q}(\alpha_1, \ldots, \alpha_e)/\mathbb{Q}$. There exists a constant $c > 0$ that satisfies the following statement: for any complex number α of the form $\sum_k a_k \left(\prod_{i=1}^{e} \alpha_i^{k_i} \right)$, where $k = (k_1, \ldots, k_e)$ ranges over*

$\mathbb{Z}_{[N_1]} \times \cdots \times \mathbb{Z}_{[N_e]}$, $(N_1, \ldots, N_e) \in \mathbb{N}^e$, and $a_k \in \mathbb{Z}$, if $\alpha \neq 0$, then $|\alpha| \geq (\sum_k |a_k|)^{1-h} \prod_{i=1}^e c^{-hN_i}$.

Apply Lemma 6 and we then obtain a constant c with $0 < c < 1$ satisfying that $\alpha_x \geq c^{|x|+1}$ for all x with $\alpha_x \neq 0$. Using a standard technique used for the complexity class PP, we can build a new 2qfa N that forces L to fall into co-2PQFA$_\mathbb{A}$(abs-halt).

(2) It follows from (the proof of) Lemma 1(5) that (*) $2C_=QFA_\mathbb{A}$(abs-halt) $\subseteq$ co-2PQFA$_\mathbb{A}$(abs-halt). Since 2PQFA$_\mathbb{A}$(abs-halt) is closed under complementation by (1), we conclude from (*) that $2C_=QFA_\mathbb{A}$(abs-halt) $\subseteq$ 2PQFA$_\mathbb{A}$(abs-halt). Moreover, (*) implies that co-$2C_=QFA_\mathbb{A}$(abs-halt) $\subseteq$ 2PQFA$_\mathbb{A}$(abs-halt). Thus, the desired result follows.

(3) This claim can be proven in a similar argument as in [1].   □

Hereafter, we will discuss how to prove Lemma 4. The core of the proof of this lemma is the *Dimension Lemma* (Lemma 7), which relates to the eventual behavior of each 2qfa, which performs a series of unitary operations and projective measurements. This lemma is an important ingredient in proving Lemma 9 in Section 4 and we thus need to zero in to the lemma. To state this lemma, nonetheless, we need to introduce a few notions. Let $V = \mathbb{C}^N$ be a Hilbert space and let U be any $N \times N$ unitary matrix over V. Let W be any fixed nonempty subspace of V and let $W^\perp$ be its dual space; namely, $V = W \oplus W^\perp$. Let $P_{W^\perp}$ be the projection operator onto $W^\perp$. Obviously, $P_{W^\perp}(W) = \{0\}$ because of $W \perp W^\perp$. Let us consider the operation $U_W =_{def} UP_{W^\perp}$. Moreover, let $U_W^0(w) = w$ and let $U_W^{i+1}(w) = U_W(U_W^i(w))$ for any $i \in \mathbb{N}$ and any $w \in V$. Given each $i \in \mathbb{N}$, define $W_i = \{w \in V \mid U_W^{i+1}(w) = 0\}$ and set $W_{max} = \bigcup_{i \in \mathbb{N}} W_i$; in other words, $W_{max} = \{w \in V \mid \exists i \in \mathbb{N} \, [U_W^{i+1}(w) = 0]\}$.

Lemma 7. [Dimension Lemma] *There exists a number* $d \in [0, N]_\mathbb{Z}$ *for which* $W_{max} = W_d$.

Proof of Lemma 4. Let $M = (Q, \Sigma, \delta, q_0, Q_{acc}, Q_{rej})$ be any $\mathbb{C}$-amplitude 2qfa that halts absolutely. Let $x \in \Sigma^n$ and define $CONF_n = Q \times [0, n+1]_\mathbb{Z}$.

Consider a time-evolution matrix $U_\delta^{(x)}$ induced from δ and a halting configuration space $W = span\{|q\rangle|h\rangle \mid q \in Q_{acc} \cup Q_{rej}, h \in [0, n+1]_\mathbb{Z}\}$. Let $V = \mathbb{C}^N$, where $N = |CONF_n|$. Since $P_{W^\perp} = \Pi_{non}$, it follows that $U_W = U_\delta^{(x)} \Pi_{non}$. Hence, $W_i = \{w \in V \mid (U_\delta^{(x)} \Pi_{non})^{i+1}(w) = 0\}$. set $w_0 = |q_0\rangle|0\rangle$. Since all computation paths of M on x terminate eventually, we obtain $w_0 \in W_{max}$. Lemma 7 implies that $W_{max} = W_d$ for a certain index $d \in [0, N]_\mathbb{Z}$. Thus, $w_0 \in W_d \subseteq W_N$. This means that all the computation paths terminate within $N + 1$ steps.   □

4   Runtime Bounds of 2qfa's

We have shown in Section 3 a runtime upper bound of absolutely-halting 2qfa's. Here, we want to show that, although certain computation paths of general bounded-error 2qfa's may not even terminate, it suffices to consider only computation paths that actually terminate in exponential time.

To state our result formally, we need to define a restricted form of 2qfa's. A $t(n)$ *time-bounded* 2qfa M is a variant of 2qfa that satisfies the following condition: we force M to "halt" after exactly $t(n)$ steps (unless it halts earlier) and ignore any computation step after this point, where any computation path that does not enter a halting state within $t(n)$ steps is considered "unhalting" and excluded from any calculation of acceptance/rejection probability.

Theorem 8. *Any language in* $2BQFA_A$ *can be recognized by a certain* $2^{O(n)}$ *time-bounded 2qfa with bounded-error probability.*

We show how to estimate the runtime of a given 2qfa by evaluating *eigenvalues*, which are associated with its time-evolution matrix.

Lemma 9. *Let M be any A-amplitude 2qfa with a set Q of inner states with error probability at most ε, where $\varepsilon \in [0, 1/2]$. Let $\varepsilon' = (1 - 2\varepsilon)/4$. There exist a constant $c > 0$ and a $c^{|Q|(n+2)}$ time-bounded 2qfa N that satisfy the following: for any input x, (i) M accepts (resp., rejects) x with probability at least $1 - \varepsilon$ if and only if N accepts (resp., rejects) x with probability at least $1 - \varepsilon'$.*

Proof of Theorem 8. Consider a language L in $2BQFA_A$ and take an A-amplitude 2qfa M that recognizes L with error probability at most $\varepsilon \in [0, 1/2)$. By Lemma 9, there is another 2qfa, say, N that is $c^{|Q|(n+2)}$ time-bounded and $p_{M,e}(x) \geq 1 - \varepsilon$ iff $p_{N,e}(x) \geq 1 - \varepsilon'$ for each type $e \in \{acc, rej\}$, where $\varepsilon' = (1 - 2\varepsilon)/4$. Since $\varepsilon' \in [0, 1/2)$, L can be recognized by N with bounded-error probability. □

To complete the proof of Theorem 8, hereafter, we intend to prove Lemma 9. In the following proof of the lemma, we will use the same terminology introduced in Section 3.

Proof Sketch of Lemma 9. First, let $M = (Q, \Sigma, \delta, q_0, Q_{acc}, Q_{rej})$ denote any 2qfa with A-amplitudes with error probability at most $\varepsilon \in [0, 1/2]$. In what follows, fix $n \in \mathbb{N}$ and let N be the total number of configurations of M on inputs of length n; that is, $N = |Q|(n + 2)$. For simplicity, let $V = \mathbb{C}^N$ be the configuration space of M on inputs of length n. Hereafter, we fix x in Σ^n and abbreviate $U_\delta^{(x)}$ as U. Recall the notations W_{acc}, W_{rej}, and W_{non} from Section 2.3. By setting $W = W_{acc} \oplus W_{rej}$ and $W^\perp = W_{non}$, we obtain U_W, W_i and W_{max}, stated in Section 3. By Lemma 7, there exists a number $d' \in [0, N]_{\mathbb{Z}}$ satisfying that $W_{max} = W_{d'}$. In other words, any element $v \in W_{max}$ is mapped to W within $d' + 1$ steps. In what follows, we set $\tilde{U}_W = U_W^{d'+1}$.

Without loss of generality, we assume that $\tilde{U}_W = \begin{pmatrix} A & O \\ B & O \end{pmatrix}$, using an $m \times m$ matrix A and an $(N - m) \times m$ matrix B. For any $v = (w, 0, \ldots, 0)^T \in W_{max}^\perp$ with $w \in \mathbb{C}^m$, we obtain $\tilde{U}_W(v) = (Aw, Bw)^T$. More generally, for each $k \in \mathbb{N}^+$, it follows that $\tilde{U}_W^k(v) = (A^k w, BA^{k-1}w)^T$. Notice that $(A^k w, 0, \ldots, 0)^T \in W_{max}^\perp$

and $(0, \ldots, 0, BA^{k-1}w)^T \in W_{max}$. Since $\tilde{U}_W$ maps $(0, \cdots, 0, BA^{k-1}w)^T$ to W, the vector $(0, \ldots, 0, BA^{k-1}w)^T$ must be mapped by M to W within $N+1$ steps.

Since A is diagonalizable in $\mathbb{C}$, we denote by $\{\lambda_1, \ldots, \lambda_m\}$ the set of all eigenvalues of A and let $\{v_1, \ldots, v_m\}$ be the set of their corresponding *unit-length* eigenvectors. For convenience, we first sort these eigenvalues in increasing order according to their absolute values. Let i_0 be the maximal index such that $|\lambda_i| < 1$ for all $i \leq i_0$ and $|\lambda_i| = 1$ for all $i > i_0$. Take an appropriate unitary matrix P forcing $A = P^\dagger CP$ with a diagonal matrix C composed of the eigenvalues $\lambda_1, \ldots, \lambda_m$. Consider the *undetermined space* $D_{und} = \text{span}\{v_1, v_2, \ldots, v_{i_0}\}$ and the *stationary space* $D_{sta} = \text{span}\{v_{i_0+1}, v_{i_0+2}, \ldots, v_m\}$. Obviously, $\mathbb{C}^m = D_{und} \oplus D_{sta}$. Note also that if $w \in D_{sta}$ then $\|Aw\| = \|w\|$, implying $Bw = 0$. This means that, once w falls in D_{sta}, $\tilde{U}_W((w, 0, \ldots, 0)^T)$ is also in $D_{sta} \otimes \{0\}^{m-i_0}$. However, when $w \in D_{und}$, since w is of the form $\sum_{1 \leq j \leq i_0} \alpha_j v_j$ for certain coefficients $\alpha_1, \ldots, \alpha_{i_0}$, it follows that $Aw = \sum_j \alpha_j \lambda_j v_j$. Let λ_{max} be one such that $|\lambda_{max}| = \max_{1 \leq j \leq i_0}\{|\lambda_j|\}$. By the choice of $\{\lambda_j\}_j$, we obtain $|\lambda_{max}| < 1$. Thus, it follows that $\|Aw\|^2 \leq |\lambda_{max}|^2 \sum_j |\alpha_j|^2 = |\lambda_{max}|^2 \|w\|^2$ (since $\|v_j\| = 1$), implying $\|Aw\| \leq |\lambda_{max}|\|w\|$. From this fact, we conclude that $\|A^k w\| \leq |\lambda_{max}|^k \|w\|$ for any $k \geq 1$. This implies that $\lim_{k \to \infty} \|A^k w\| \leq \lim_{k \to \infty} |\lambda_{max}|^k \|w\| = 0$.

Let $\varepsilon' = \frac{1}{2}(\frac{1}{2} - \varepsilon) > 0$. Note that ε' is a constant because so is ε. Since our 2qfa M halts with probability at least $1 - \varepsilon$, we conclude that, for any sufficiently large k, $\|\tilde{A}^k v\|^2 = \|A^k w\|^2 \leq |\lambda_{max}|^{2k} \leq \varepsilon'$ for all $v = (w, 0, \ldots, 0)^T \in W_{max}^\perp$. Such a k must satisfy that $k \leq (\log \varepsilon')/(2 \log |\lambda_{max}|)$.

Next, we need to show the desired upper-bound on the value $|\lambda_{max}|$. By the definition of λ_{max}, the value $|\lambda_{max}|$ must be described by $O(N)$ applications of arithmetic operations. Letting $\alpha = 1 - |\lambda_{max}|$, since $\alpha \neq 0$, we conclude by Lemma 6 that $|\alpha| \geq c^{-N}$ holds for a suitable constant $c > 0$; in other words, $|\lambda_{max}| \leq 1 - c^{-N}$ holds. This implies that $\log |\lambda_{max}|^{-1} \geq \log(1 - c^{-N})^{-1} \geq c^{-N}$. Hence, we have $k \leq (\log \varepsilon')/(2 \log |\lambda_{max}|) = (\log (\varepsilon')^{-1})/(2 \log |\lambda_{max}|^{-1}) \leq c'c^N$ for another appropriate constant $c' > 0$. $\qquad\qquad\Box$

5   Classical Simulations of 2qfa's

We wish to present two classical-complexity upper-bounds of 2PQFA and 2C$_=$QFA for any amplitude set K. As noted in Section 2.2, $\{0, 1/2, 1\} \subseteq K$ is assumed. Given such a subset K of $\mathbb{R}$, the notation $\hat{K}$ denotes the minimal set that contains K and is closed under *multiplication* and *addition*. Recall that the notation "$[t(n)\text{-}time]$" refers to a worst-case time bound $t(n)$.

Theorem 10. *There exists an integer $k \geq 2$ such that, for any set $K \subseteq \mathbb{R}$, the following statements hold.*

1. *$2PQFA_K \subseteq 2PPFA_{\hat{K}}(k\text{-head})[poly\text{-}time]$.*
2. *$2C_=QFA_K \subseteq 2C_=PFA_{\hat{K}}(k\text{-head})[poly\text{-}time]$.*

An immediate corollary of Theorem 10 is given below, using Lemma 1.

Corollary 11. *There is a constant $k \geq 2$ such that, for any set $K \subseteq \mathbb{C}$, $2\text{EQFA}_K \subseteq 2\text{C}_=\text{PFA}_{\widehat{K}}(k\text{-head})[poly\text{-}time] \cap \text{co-}2\text{C}_=\text{PFA}_{\widehat{K}}(k\text{-head})[poly\text{-}time]$ and $2\text{BQFA}_K \subseteq 2\text{PPFA}_{\widehat{K}}(k\text{-head})[poly\text{-}time] \cap \text{co-}2\text{PPFA}_{\widehat{K}}(k\text{-head})[poly\text{-}time]$.*

From [6, Lemmas 1–2&Theorem 2], it follows that the language family $2\text{PPFA}_{\mathbb{Q}}(k\text{-head})[poly\text{-}time]$ is *properly* contained within PL. Since it is already known that $2\text{BQFA}_{\mathbb{A}} \subseteq \text{PL}$ [9,12], we obtain the following separation.

Corollary 12. $2\text{BQFA}_{\mathbb{Q}} \subsetneq \text{PL}$.

Theorem 10 is a direct consequence of the following technical lemma regarding a classical simulation of 2qfa's on multi-head 2pfa's.

Lemma 13. *Let K be any subset of $\mathbb{R}$. There exists an index $k \geq 2$ that satisfies the following. Given a K-amplitude 2qfa M, there exist two khead-2pfa's N_1 and N_2 such that (i) N_1 and N_2 have nonnegative $\widehat{K}$-transition probabilities, (ii) N_1 and N_2 halt in worst-case $n^{O(1)}$ time, and (iii) it holds that, for every x, $(p_{N_1,acc}(x) - p_{N_1,rej}(x))p_{M,acc}(x) = p_{N_2,acc}(x) - p_{N_2,rej}(x)$.*

Concerning each quantum function f in $\#2\text{QFA}_K$ generated by an appropriate 2qfa M, we apply Lemma 13 and then obtain two khead-2pfa's N_1 and N_2. By setting $g_1(x) = p_{N_1,acc}(x)$, $g_2(x) = p_{N_1,rej}(x)$, $h_1(x) = p_{N_2,acc}(x)$, and $h_2(x) = p_{N_2,rej}(x)$, it immediately follows that $(g_1(x) - g_2(x))f(x) = h_1(x) - h_2(x)$. Clearly, g_1, g_2, h_1, h_2 all belong to $\#2\text{PFA}_{\widehat{K}}(k\text{-head})[poly\text{-}time]$. This further yields the next corollary.

Corollary 14. *Let $K \subseteq \mathbb{C}$. There exists an index $k \geq 2$ such that, for any $f \in \#2\text{QFA}_K$, there are four functions $g_1, g_2, h_1, h_2 \in \#2\text{PFA}_{\widehat{K}}(k\text{-head})[poly\text{-}time]$ satisfying $(g_1(x) - g_2(x))f(x) = h_1(x) - h_2(x)$ for every x.*

Assuming that Lemma 13 is true, let us prove Theorem 10 firstly.

Proof Sketch of Theorem 10. Here, we plan to prove only the first containment $2\text{PQFA}_K \subseteq 2\text{PPFA}_{\widehat{K}}(k\text{-head})[poly\text{-}time]$. Take an arbitrary language L in 2PQFA_K, witnessed by a certain K-amplitude 2qfa, say, M. By Lemma 13, there are two appropriate khead-2pfa's N_1 and N_2 that satisfy Conditions (i)–(iii) of the lemma. Define a new khead-2pfa N as follows. On input x, from q_0, enter q_2 with probability $1/2$, and q_1 and q_3 with probability $1/4$ each. From q_1, run N_1 on x. From q_2, simulate N_2 but flips its outcome (i.e., either accepting or rejecting states). From q_3, enter q_{acc} and q_{rej} with equal probability $1/2$. By the definition, it follows that $p_{N,acc}(x) = \frac{1}{4}p_{N_1,acc}(x) + \frac{1}{2}p_{N_2,rej}(x) + \frac{1}{8}$ and $p_{N,rej}(x) = \frac{1}{4}p_{N_1,rej}(x) + \frac{1}{2}p_{N_2,acc}(x) + \frac{1}{8}$. From those equalities, it is not difficult to show that $x \in L$ iff $p_{N,acc}(x) > 1/2$. We therefore conclude that L belongs to $2\text{PPFA}_{\widehat{K}}(k\text{-head})[poly\text{-}time]$. $\square$

Finally, let us return to Lemma 13. Our proof of this lemma is based on a dextrous implementation of the Mahajan-Vinay algorithm [7] that efficiently computes an integer determinant on multi-head 2pfa's.

Proof Sketch of Lemma 13. Let L be any language and let n be any number in $\mathbb{N}$. Let M be any $\mathbb{R}$-amplitude 2qfa having acceptance probability $p_{M,acc}(x)$

and rejection probability $p_{M,rej}(x)$ on each input x. In what follows, let M be of the form $(Q, \Sigma, \delta, q_0, Q_{acc}, Q_{rej})$. Assuming $Q = \{q_0, q_1, \ldots, q_c\}$, define two index sets A and R for which $Q_{acc} = \{q_j \mid j \in A\}$ and $Q_{rej} = \{q_j \mid j \in R\}$. For later use, we set $\delta_{+1}(q, \sigma, p, h) = \delta(q, \sigma, p, h)$ if $\delta(q, \sigma, p, h) > 0$; 0 otherwise. Similarly, let $\delta_{-1}(q, \sigma, p, h) = -\delta(q, \sigma, p, h)$ if $\delta(q, \sigma, p, h) < 0$; 0 otherwise. Note that $\delta(q, \sigma, p, h) = \sum_{e \in \{\pm 1\}} \delta_e(q, \sigma, p, h)$.

We denote by x an arbitrary input of length n. Let $CONF_n = Q \times [0, n+1]_{\mathbb{Z}}$ and $N = |CONF_n|$. First, we review how to evaluate the acceptance probability $p_{M,acc}(x)$ of M on x. Recall that each $((q, \ell), (p, m))$-entry of $U_\delta^{(x)}$ is exactly $\delta(q, x_\ell, p, m - \ell)$, provided that $|m - \ell| \leq 1$. Let P_{non} denote the projection operator onto the space spanned by non-halting configurations. Let $D_x = U_\delta^{(x)} P_{non}$, which describes one step of M's move on this input x unless its inner state is a halting state.

It is easy to see that the acceptance probability of M on input x obtained at time k (i.e., $\sum_{q,\ell} |\langle q, \ell | D_x^k | q_0, 0 \rangle|^2$) equals

$$\sum_{q,\ell} \langle q, \ell | D_x^k | q_0, 0 \rangle \langle q, 0 | D_x^k | q_0, 0 \rangle = \sum_{q,\ell} \langle (q, \ell), (q, \ell) | (D_x^k \otimes D_x^k) | (q_0, 0), (q_0, 0) \rangle,$$

where $q \in Q_{acc}$ and $\ell \in [0, n+1]_{\mathbb{Z}}$. Take the vector $y_{ini} = |q_0, 0\rangle |q_0, 0\rangle$ associated with the initial configuration and, for each $j \in A$, let $y_{acc,j,\ell}$ denote the vector of the form $|q_j, \ell\rangle |q_j, \ell\rangle$. Since $(D_x \otimes D_x)^k = D_x^k \otimes D_x^k$ for any $k \in \mathbb{N}$, the total acceptance probability $p_{M,acc}(x)$ of M on x exactly matches

$$\sum_{k=0}^\infty \left(\sum_{j \in A} \sum_{\ell \in [0, n+1]_{\mathbb{Z}}} y_{acc,j,\ell}^T (D_x \otimes D_x)^k y_{ini} \right) = y_{acc}^T \left(\sum_{k=0}^\infty (D_x^k \otimes D_x^k) \right) y_{ini},$$

where $y_{acc} = \sum_{j \in A} \sum_{\ell \in [0, n+1]_{\mathbb{Z}}} y_{acc,j,\ell}$.

We are now focused on the term $\sum_{k=0}^\infty (D_x^k \otimes D_x^k)$. Let us express D_x as a difference of two nonnegative real matrices. We first define $D_x^+[i, j] = D_x[i, j]$ if $D_x[i, j] > 0$; $D_x^+[i, j] = 0$ otherwise. Similarly, let $D_x^-[i, j] = -D_x[i, j]$ if $D_x[i, j] < 0$; $D_x^-[i, j] = 0$ otherwise. In addition, we set $\tilde{D}_x = \begin{pmatrix} D_x^+ & D_x^- \\ D_x^- & D_x^+ \end{pmatrix}$. To specify each entry in $\tilde{D}_x$, we conveniently use an index set $Q \times [0, n+1]_{\mathbb{Z}} \times \{\pm 1\}$ so that, for any $(p, m, b), (q, \ell, a) \in Q \times [0, n+1]_{\mathbb{Z}} \times \{\pm 1\}$, $\tilde{D}_x[(p, m, b), (q, \ell, a)]$ equals $\delta_{ab}(q, x_\ell, p, m - \ell)$ if $|m - \ell| \leq 1$; 0 otherwise. Let $CONF_* = (CONF_n \times \{\pm 1\})^2$.

Since the infinite sum $\sum_{k=0}^\infty (\tilde{D}_x \otimes \tilde{D}_x)^k$ converges and $\|\tilde{D}_x \otimes \tilde{D}_x\| < 1$ for a suitable matric norm, it follows that $I - \tilde{D}_x \otimes \tilde{D}_x$ is invertible and, moreover, $(I - \tilde{D}_x \otimes \tilde{D}_x)^{-1} = \sum_{k=0}^\infty (\tilde{D}_x \otimes \tilde{D}_x)^k$. Let $i_{0,a} = ((q_0, 0, a), (q_0, 0, a))$ and $\hat{j}_{a,\ell} = ((q_j, \ell, a), (q_j, \ell, a))$ for $j \in A$, $\ell \in [0, n+1]_{\mathbb{Z}}$, and $a \in \{\pm 1\}$. For simplicity, we assume an appropriate ordering on $CONF_*$ and identify each element in $CONF_*$ with its index specified by this ordering. We therefore obtain that $p_{M,acc}(x) = \sum_{a \in \{\pm 1\}} \sum_{j \in A} (I - \tilde{D}_x \otimes \tilde{D}_x)^{-1} [\hat{j}_{a,\ell}, i_{0,a}]$. By Laplace's formula (i.e., C^{-1} equals the *adjugate* of C divided by $det(C)$), $p_{M,acc}(x)$ equals

$$\sum_{a\in\{\pm1\}}\sum_{j\in A}\sum_{\ell\in[0,n+1]_{\mathbb{Z}}}(-1)^{i_{0,a}+\hat{j}_{a,\ell}}det[(I-\tilde{D}_x\otimes\tilde{D}_x)_{i_{0,a},\hat{j}_{a,\ell}}]/det(I-\tilde{D}_x\otimes\tilde{D}_x),\quad(1)$$

where "$C_{i,j}$" is obtained from matrix C by deleting row i and column j.

Let us state our key lemma. In the lemma, f denotes a function defined by $f(x)=(\frac{1}{2|A|}2^{-2\lceil\log_2(n+2)\rceil})(\frac{1}{4|Q|^2}2^{-2\lceil\log_2(n+2)\rceil})8|Q|^2(n+2)^2-1$ for every $x\in\Sigma^*$.

Lemma 15. *There exist a $k\geq2$ and a khead-2pfa N_1 such that $det[I-\tilde{D}_x\otimes\tilde{D}_x]=f(x)[p_{N_1,acc}(x)-p_{N_1,rej}(x)]$ for all x. Similarly, a certain khead-2pfa N_2 satisfies that $\sum_{a\in\{\pm1\}}\sum_{j\in A}\sum_{\ell\in[0,n+1]_{\mathbb{Z}}}(-1)^{i_{0,a}+\hat{j}_{a,\ell}}det[(I-\tilde{D}_x\otimes\tilde{D}_x)_{i_{0,a},\hat{j}_{a,\ell}}]=f(x)[p_{N_2,acc}(x)-p_{N_2,rej}(x)]$ for all x.*

Using Lemma 15, we can complete the proof of Lemma 13 as follows. Take $k$head-2pfa's N_1 and N_2 given in Lemma 15. By Eq.(1), it follows that $p_{M,acc}(x)=(p_{N_2,acc}(x)-p_{N_2,rej}(x))/(p_{N_1,acc}(x)-p_{N_1,rej}(x))$, as requested.

To prove Lemma 15, we wish to use an elegant algorithm of Mahajan and Vinay [7], who demonstrated how to compute the determinant of an integer matrix using "closed walk (clow)." We intend to implement on $k$head-2pfa's a GapL-algorithm given in the proof of [7, Theorem 4], which produces a nondeterministic computation tree whose accepting/rejecting computation paths contribute to the calculation of the determinant of a given integer matrix. As noted in Section 1, our implementation, nevertheless, requires significant changes to the original algorithm, because the original algorithm takes a target matrix as "input," whereas we must deal probabilistically with elements of the matrix. □

(*) All omitted or abridged proofs in this extended abstract will appear in a forthcoming full paper.

References

1. Adleman, L.M., DeMarrais, J., Huang, M.A.: Quantum computability. SIAM J. Comput. **26**, 1524–1540 (1997)
2. Ambainis, A., Freivalds, R.: 1-way quantum finite automata: strengths, weaknesses and generalizations. In: FOCS 1998, pp. 332–341 (1998)
3. Dwork, C., Stockmeyer, L.: Finite state verifier I: the power of interaction. J. ACM **39**, 800–828 (1992)
4. Kondacs, A., Watrous, J.: On the power of quantum finite state automata. In: FOCS 1997, pp. 66–75 (1997)
5. Macarie, I.: Closure properties of stochastic languages. Technical Report No.441, Computer Science Department, University of Rochester (1993)
6. Macarie, I.I.: Multihead two-way probabilistic finite automata. Theory Comput. Syst. **30**, 91–109 (1997)
7. Mahajan, M., Vinay, V.: Determinant: combinatorics, algorithms, and complexity. Chicago J. Theoret. Comput. Sci. **1997**, Article no. 1997-5 (1997)

8. Moore, C., Crutchfield, J.: Quantum automata and quantum grammar. Theor. Comput. Sci. **237**, 275–306 (2000)
9. Nishimura, H., Yamakami, T.: An application of quantum finite automata to interactive proof systems. J. Comput. System Sci. **75**, 255–269 (2009)
10. Rabin, M.O.: Probabilistic automata. Inform. Control **6**, 230–244 (1963)
11. Stolarsky, K.B.: Algebraic Numbers and Diophantine Approximations. Marcel Dekker (1974)
12. Watrous, J.: On the complexity of simulating space-bounded quantum computations. Computational Complexity **12**, 48–84 (2003)
13. Yakaryilmaz, A., Say, A.C.C.: Unbounded-error quantum computation with small space bounds. Inf. Comput. **209**, 873–892 (2011)
14. Yamakami, T.: Analysis of quantum functions. Internat. J. Found. Comput. Sci. **14**, 815–852 (2003)
15. Yamakami, T., Yao, A.C.: $NQP_{\mathbb{C}} = \text{co-}C_{=}P$. Inf. Process. Lett. **71**, 63–69 (1999)
16. Yao, A.C.: Class Note. Unpublished. Princeton University (1998)

Author Index